Biodiversity, Biological Systems and Conservation

Biodiversity, Biological Systems and Conservation

Edited by **Neil Griffin**

New York

Published by Callisto Reference,
106 Park Avenue, Suite 200,
New York, NY 10016, USA
www.callistoreference.com

Biodiversity, Biological Systems and Conservation
Edited by Neil Griffin

International Standard Book Number: 978-1-63239-733-1 (Hardback)

Printed in the United States of America.

Contents

Preface

Biodiversity is the variety and variability of life forms present in different geographical regions. It is a result of different climatic patterns and geographic settings. Biological systems may range from micro to macro that is they can exist in a form as small as mitochondria and as big as rainforests. Conservation aims at protecting habitats, ecosystems and species to maintain ecological balance. This book strives to provide extensive information on abundance, diversity and distribution of genes, species and ecosystems, components of biodiversity and effect of disturbance on ecosystems, species and populations, etc. It is compiled in such a manner, that it will provide in-depth knowledge about the theory and practice of this field. Those in search of information to further their knowledge will be greatly assisted by this book. It will serve as an essential guide for environmentalists, ecologists, conservationists, biologists, researchers and students alike.

This book is a result of research of several months to collate the most relevant data in the field.

When I was approached with the idea of this book and the proposal to edit it, I was overwhelmed. It gave me an opportunity to reach out to all those who share a common interest with me in this field. I had 3 main parameters for editing this text:

1. Accuracy – The data and information provided in this book should be up-to-date and valuable to the readers.

2. Structure – The data must be presented in a structured format for easy understanding and better grasping of the readers.

3. Universal Approach – This book not only targets students but also experts and innovators in the field, thus my aim was to present topics which are of use to all.

Thus, it took me a couple of months to finish the editing of this book.

I would like to make a special mention of my publisher who considered me worthy of this opportunity and also supported me throughout the editing process. I would also like to thank the editing team at the back-end who extended their help whenever required.

Editor

Revision of the genus *Sphenostylis* (Fabaceae: Phaseoleae) in South Africa and Swaziland

A.N. MOTEETEE*† and B.-E. VAN WYK*

Keywords: Fabaceae, Leguminosae, *Nesphostylis* Verdc., Phaseoleae, Phaseolinae, southern Africa, *Sphenostylis* E.Mey., *Vigna* Savi

ABSTRACT

A taxonomic revision of the genus *Sphenostylis* E.Mey. (tribe Phaseoleae) in southern Africa is presented. The genus is distinguishable by its characteristic wedge-shaped, dorsiventrally flattened style tip. Of the seven known species in the genus, only two, *S. angustifolia* and *S. marginata*, occur in South Africa and Swaziland; and the former is endemic to these countries. The two species differ in the size, shape, and venation of the leaflets; as well as the length of the petiole, peduncle, and pod. A key to the two species is provided and the correct nomenclature, typification, and known geographical distributions are given.

INTRODUCTION

Sphenostylis E.Mey. belongs to the subtribe Phaseolinae of tribe Phaseoleae (Fabaceae). It is named for the wedge-shaped, dorsiventrally flattened style (from the Greek word *spheno* = wedge). It is a small genus comprising seven species, all occurring in the tropics and southern parts of the African continent (Gillett *et al.* 1971; Potter 1992; Potter & Doyle 1994; Schrire 2005). The Indian species *S. bracteata* (Baker) Gillett, transferred from *Dolichos* L. to *Sphenostylis* by Gillett (1966), was moved to *Nesphostylis* Verdc. by Potter & Doyle (1994) based on a cladistic analysis of morphological data. Three species of *Sphenostylis* are used as food sources in Africa. *Sphenostylis stenocarpa* (Hochst. ex A.Rich.) Harms is known as African yam bean or *girigiri* (Burkhill 1995), and cultivars of this tropical African species are grown for their seeds and tubers in tropical and West Africa. The leaves, flowers, pods, and seeds of *S. schweinfurthii* Harms and *S. erecta* Hutch. ex Baker f. are wild-harvested. The roots of *S. erecta* are also used as fish poison (Potter 1992).

Sphenostylis appears to have a close relationship with *Nesphostylis* (Potter & Doyle 1994) with which it shares the dorsiventrally flattened style tip; but it differs by features of the calyx (Lackey 1981), the standard petals, and the stamens (Potter 1992). In *Nesphostylis*, the inner surface of the calyx is pubescent, standard appendages are present, the base of the vexillary stamen is hooked, and the stamen apices are dilated. None of these characters are present in *Sphenostylis*. Based mainly on the narrow pods, several botanists including Harvey (1862), Bentham (1865) and Taubert (1894), relegated *Sphenostylis* into synonymy with the genus *Vigna* Savi (Gillett 1966). Harms (1899) reinstated the genus and expanded it to include related species previously placed in *Dolichos* and *Vigna*. *Sphenostylis* differs from these two genera by its distinctive style tip. Furthermore, *Vigna* generally has peltate stipules (stipules not peltate in *Sphenostylis*). In addition, a phylogeny based on molecular data shows that *Vigna* and *Sphenostylis* are not as closely related as previously thought, and that *Sphenostylis* is rather closer to *Dolichos* and *Macrotyloma* (Wight & Arn.) Verdc. (Wojciechowski *et al.* 2004). *Sphenostylis* can be distinguished from *Dolichos*, *Macrotyloma*, and *Vigna* by the absence of appendages on the standard petals (appendages present on the standard petals in the latter three genera).

Sphenostylis is represented in South Africa and Swaziland by two species, namely *S. angustifolia* Sond. and *S. marginata* E.Mey.; the former being endemic to the two countries.

MATERIALS AND METHODS

Plant material was studied mainly from herbarium specimens housed in JRAU, NH, NU, and PRE (acronyms after Holmgren *et al.* 1990); and also in the field. Habit affinities are described according to Mucina & Rutherford (2006).

TAXONOMY

Sphenostylis *E.Mey.*, Commentariorum de Plantis Africae Australioris: 148 (1836); Baker f.: 670 (1929); Burtt Davy: 418 (1932); E.Phillips: 427 (1951); Wilczek: 273 (1954); Verdc.: 389 (1970); J.B.Gillett et al.: 670 (1971); Compton: 286 (1975); R.A.Dyer: 275 (1975); Potter & Doyle: 389–406 (1994); Germish.: 296 (2000); Verdc. & Døygaard: 68 (2001). Type species: *S. marginata* E.Mey.

Prostrate, climbing (twining), or erect shrubs or perennial herbs, arising from a thick, woody rootstock. *Leaves* pinnately 3-foliolate (Figure 1D); stipules persistent, ovate-lanceolate or ovate-acuminate; leaflets ovate, elliptic or linear, with four linear stipels, one at the base of each lateral leaflet and two at the base of the terminal one (Figure 1E), appressed-pubescent when young but becoming glabrescent with age. *Inflorescence*: flowers in congested heads on long peduncles, peduncles much longer than leaves; bracts linear or oblong-lanceolate, small, falling off at a very early stage or absent; bracteoles 2, ovate-lanceolate or oblong-ovate, caducous or persistent. *Calyx* tube campanulate, bilabiate, the two

* Department of Botany and Plant Biotechnology, University of Johannesburg, P.O. Box 524, Auckland Park, 2006 Johannesburg.
†amoteetee@uj.ac.za.

FIGURE 1.—*Sphenostylis marginata* (A, B, D) and *S. angustifolia* (C, E–H): A, flowers showing the twisted standard petals; B, flower showing the cuneate style tip; C, keel tip and cuneate style tip; D, leaf in adaxial view; E, flowering branch (note the stipules and stipels); F, pod; G, flower in front view (note the twisted petals); H, seed. Scale bars: A, 8 mm; B, 12 mm; C, E–G: 5 mm; D, H: 2 mm. Photographs: A, B, D by David Styles; C, E–H by Ben-Erik van Wyk.

lobes of the upper lip partially or entirely connate, lobes very short and blunt. *Corolla* purple, purplish pink, violet, whitish pink, or yellow; standard suborbicular, symmetrical or twisted (Figure 1A & 1G) with 2 inflexed auricles, but without appendages, with well-developed, channelled claw; wings obliquely obovate, eared near base, with short linear claw; keel incurved, ± concave-convex beaked, with short linear claw. *Androecium* diadelphous (with 9 stamens fused into a tube, vexillary stamen free), 5 basifixed anthers alternating with 5 dorsifixed anthers. *Ovary* narrowly oblong, pubescent, 3–12-ovuled; style penicillate below the stigma; stigma dorsiventrally flattened, ciliate on the margins (Figures 1B & 1C). *Fruit* linear, compressed, twisting after dehiscence, glabrescent to densely silky, with persistent style at tip (Figure 1F), 2–several-seeded, dehiscent. *Seeds* oblong, uniformly black or brown to reddish brown speckled black, minutely papillose (Figure 1H).

Key to species of *Sphenostylis* in South Africa and Swaziland:

1a Leaflets 20–45 mm wide; petiole 50–70 mm long; peduncles
    220–300 mm long; fruit 95–120 mm long . . . . *S. marginata*
1b Leaflets 10–19 mm wide; petiole 8–16 mm long; peduncles
    75–115 mm long; fruit 55–85 mm long . . . . *S. angustifolia*

1. **Sphenostylis angustifolia** *Sond.* in Linnaea 23: 33 (1850); R.A.Dyer: t 1010 (1947); Burtt Davy: 418 (1932); Compton: 287 (1975). *Vigna angustifolia* (Sond.) Benth. ex Harv.: 240 (1862). Type: South Africa, Gauteng, 2528 (Pretoria): Magaliesberg, (–DC), *Zeyher 524* (S, specimen on the left, lecto.!, here designated; BM!, K!, isolecto.). Syntype: KwaZulu-Natal, 2931 (Stanger): Port Natal [now Durban], (–CC), *Gueinzius 624* (S!).

Note: the Zeyher specimen in the Sonder Herbarium in S is chosen as lectotype because this is probably the specimen that Sonder used in his description. The twig on the left is chosen because it bears a flower and some immature fruits. Curiously Potter (in sched.) chose the Zeyher specimen in K as lectotype but this choice was apparently never published.

Erect suffrutex, rarely climbing and twining, much-branched, up to 0.5 m tall, spreading. *Leaflets* oblong or linear-lanceolate, 30–60 × (7–)10–19 mm; petiole 8–16 mm long; stipules ovate-acuminate, 3.5–5.0 × 2–3 mm. *Inflorescences* subumbellate axillary racemes; peduncles (55–)75–115 mm long, with 2–4 flowers; flowers pink or purple, with a whitish or yellowish centre, 12–25 mm long; bracts linear, ± 1.5 mm long, caducous; bracteoles ovate-lanceolate, 2.0–2.5 × ± 1.5 mm, persistent. *Calyx* sparsely pubescent, with ± equal lips, upper lip 6–8 mm long, lower lip 6–8 mm long; lobes rounded, those of the upper lip joined for almost their entire length. *Corolla* pinkish red; standard suborbicular, 15–25 × 15–26 mm,

glabrous; wings obovate, 14–24 × 4–7 mm, without surface sculpturing, apex rounded; keel falcate, 14–24 × 6–10 mm, apex rounded, pocket absent. *Ovary* 9–13 mm long, linear-oblong. *Fruit* linear, 55–85 × 5–6 mm, glabrescent, many-seeded. *Seeds* reniform, ± 7 × ± 4 mm, brown, reddish brown speckled black, or uniformly black (Figure 2). *Flowering time*: Sept.–Feb.

Diagnostic characters: Based on Potter & Doyle's cladistic analysis (1994), *Sphenostylis angustifolia* appears to have close affinities with *S. zimbabweensis* R.Mithen, which is restricted to the Highlands of Zimbabwe. The two species share the deciduous bracts and the persistent bracteoles. *Sphenostylis zimbabweensis* is, however, differentiated by the more prostrate growth form (as opposed to the erect growth form of *S. angustifolia*), and the broader, ovate to elliptic leaflets (leaflets narrower and oblong to linear-lanceolate in *S. angustifolia*). *Sphenostylis angustifolia* can be distinguished from *S. marginata* by the narrower leaflets and the much shorter petioles, peduncles, and pods (dimensions are given in the key).

Distribution and habitat: *Sphenostylis angustifolia* occurs in South Africa (Limpopo, North-West, Gauteng, Mpumalanga and KwaZulu-Natal Provinces) and Swaziland (Figure 3). It grows scattered in Zeerust Thornveld (SVcb 3), Central Sandy Bushveld (SVcb 12), Soutpansberg Mountain Bushveld (SVcb 21), Polokwane Plateau Bushveld (SVcb 23), Granite Lowveld (SVl 3), Tza-

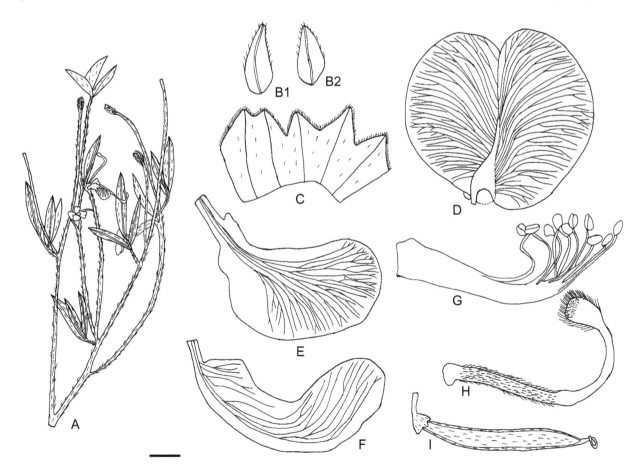

FIGURE 2.—Vegetative and reproductive morphology of *Sphenostylis angustifolia*: A, flowering branch; B1 & B2, abaxial view of bracteoles; C, calyx opened out with upper lobes to left; D, standard petal; E, wing petal; F, keel petal; G, stamens; H, pistil; I, lateral view of pod. Vouchers: A from *A.O.D. Mogg 35461* (JRAU); B–I from *B-E. van Wyk 1438* (JRAU). Scale bar: A, 30 mm; B, 1 mm; C–F, H, 3 mm; G, 2 mm; I, 10 mm.

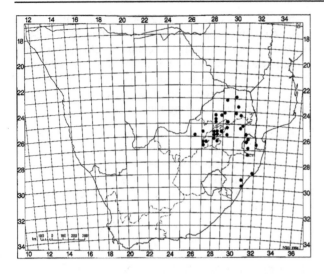

FIGURE 3.—Distribution of *Sphenostylis angustifolia*.

neen Sour Bushveld (SVl 8), Legogote Sour Bushveld (SVl 9), Swaziland Sour Bushveld (SVl 14), Southern Lebombo Bushveld (SVl 16), Carletonville Dolomite Grassland (Gh 15), Soweto Highveld Grassland (Gm 8), Rand Highveld Grassland (Gm 11), KaNgwane Montane Grassland (Gm 16), Maputaland Coasal Belt (CB 1), KwaZulu-Natal Coastal Belt (CB 3).

Additional specimens examined

LIMPOPO.—2230 (Messina): Soutpansberg, Entabeni Forestry Station at Muchinudi Fall, (–CC), 27 Jan. 1954, *L.E. Codd 8392* (PRE). 2329 (Polokwane): Polokwane, Farm Eersteling, on slope of mtn, (–AB), 1 Jan. 1992, *G.J. du Toit 2238* (NH). 2330 (Tzaneen): Modjadjis Reserve near Duiwelskloof, (–CB), 24 May 1938, *J.D. Krige 164* (PRE). 2428 (Nylstroom): Waterberg, (–AD), 6 Nov. 1978, *G. Germishuizen 909* (PRE); Naboomfontein, (–BD), 13 Dec. 1934, *E.E. Galpin 133130* (PRE); between Warmbaths and Pietersburg [Polokwane], (–CB), 4 Nov. 1985, *B.J. Pienaar 636* (PRE); 18 km from Nylstroom on road to Warmbaths near Groot Nyl turn-off, (–CD), 4 Nov. 1985, *G. Germishuizen 3347* (PRE). 2429 (Zebediela): near Daggakraal 50 m NE of Potgietersrus, (–AA), 3 Jan. 1954, *B. Maguire 2569* (PRE); Arabie, Camp 1, (–CD), 7 Jan. 2007, *W. Ellery 357* (PRE). 2430 (Pilgrim's Rest): Lekgalameetse Nature Reserve, The Downs, SE of Makwens, (–AA), 16 Oct. 1985, *M. Stalmans 718* (PRE).

NORTH-WEST.—2526 (Zeerust): Marico Distr., ± 10 km NE of Wondergat, (–CD), 6 Feb. 1983, *C. Reid 676* (PRE). 2527 (Rustenburg): Rustenburg Nature Reserve, (–CA), 27 Feb. 1970, *N. Jacobsen 845* (PRE). 2627 (Potchefstroom): on road from Frederikstad to Rysmierbult, (–AC), 31 Oct. 1978, *B. Ubbink 733* (PRE); Ventersdorp, Goedgedacht, (–CA), 29 Dec. 1930, *J.D. Sutton 511* (PRE).

GAUTENG.—2528 (Pretoria): La Montagne Rand, N of Chambord W/S, (–CA), 30 Sept. 1978, *A.E. van Wyk 2405* (PRE); 28 miles [45 km] from the National Herbarium on road to Boekenshoutskloof, (–CB), 15 Apr. 1971, *L.A. Coetzer 73* (PRE); Fountains Valley, (–CC), 3 Oct. 1948, *J.M. Watt 4565* (PRE); Doornkloof, Smutskoppie, (–CD); 24 Nov. 1985, *B-E. van Wyk 1438* (JRAU); Renosterkop, NE of Bronkhorstspruit, (–DB); 6 Dec. 1987, *B-E. van Wyk 2730* (JRAU). 2628 (Johannesburg): Melville Koppies Nature Reserve, (–AA), 4 Feb. 1987, *B-E.van Wyk 2600–2604* (JRAU); Heidelberg, (–AD), Nov. 1927, *A. Thode A1311* (PRE).

MPUMALANGA.—2430 (Pilgrim's Rest): Drie Rondavels lookout, (–BC), 2 Feb. 1982, *A.E. van Wyk, R. Dahlgren & P.D.F. Kok 5487* (PRE). 2529 (Witbank): Loskopdam Game Reserve, (–AD), 13 Dec. 1966, *G.K. Theron 728* (JRAU, PRE); between Witbank and Middelburg along N4 highway, (–CD), 20 Feb. 1991, *P. Herman 1338* (PRE). 2530 (Lydenburg): 7 km S of Sabie, (–BB), 29 Sept. 2005, *J.J. Meyer 4530* (PRE); Wonderkloof Nature Reserve, (–BC), 17 Nov. 1978, *J.P. Kluge 1408* (PRE). 2531 (Komatipoort): KaNgwane, Songimvelo Game Reserve, (–CC), 8 Dec. 1992, *N.L. Meyer 29* (PRE).

SWAZILAND.—2631 (Mbabane): Hhohho Dist., Nyokane, 15 km from Piggs Peak-Mbabane turnoff on road to Maphalaleni, (–AB), 23 Oct. 1963, *R.H. Compton 31707* (PRE); Dabriach, (–AC), 18 Oct. 1958, *R.H. Compton 28097* (PRE); Hlatikulu, (–CD), Oct. 1910, *M. Steward 10081* (PRE). 2632 (Bela Vista): Lebombo Mountains, near fence separating Swaziland and Mozambique, (–CA), 22 Nov. 2002, *M.K. Maserumule 74* (PRE).

KWAZULU-NATAL.—2930 (Pietermaritzburg): Beacon Hill, off Panorama Terrace, Wyebank, (–DD), 10 Jan. 2005, *D. Styles 2180* (NH).

2. **Sphenostylis marginata** *E.Mey.*, Commentariorum de Plantis Africae Australioris: 148 (Feb. 1836); Baker f.: 148 (1929); Wilczek 6: 274 (1954); Verdoorn: t. 1521 (1968); J.B.Gillett et al.: 671 (1971); Compton: 287 (1975); Lock: 438 (1989); Potter & Doyle: 403 (1994); Verdc. & Døygaard.: 71 (2001). *Vigna marginata* (E.Mey.) Benth. ex Harv.: 240 (1862); De Wild.: 98 (1921). Type: KwaZulu-Natal, 3030 (Port Shepstone): 'in graminosis ad ostia fluvii Omsamculo' [mouth of Umzimkulu River], (–CB), *Drège s.n.* (K!, lecto., designated by Potter & Doyle (1994) [as 'holo'.], G!, MO!, P!, isolecto.)

Note: although the P specimen which bears an original *Drège* label with locality details corresponding exactly to those given in the prognosis as well as Meyer's handwriting ('*mihi*') would have been more appropriately selected as lectotype, Potter & Doyle's (1994) designation of the K specimen as the holotype constitutes effective lectotypification (Art. 9.8).

Prostrate suffrutex with twining stems up to 1.5 m long, arising from a woody rootstock. *Leaflets* ovate, elliptic or oblong, 45–110(–125) × 20–40(–45) mm; petiole (15–)50–70 mm long; stipules oblong-lanceolate or ovate, 3–7 × 2–4 mm. *Inflorescence* pseudo-umbellate, axillary raceme, long-stalked; peduncles 220–300 mm long, few-flowered; flowers mauve, 14–16 mm long; bracts oblong-lanceolate, ± 1.5 mm long, caducous, bracteoles ovate-lanceolate, 2–3 × 1.5–2.0 mm, persistent. *Calyx* pubescent, lobes rounded, with equal lips, upper and lower lips 5–6 mm long, lobes of upper lip joined for almost their entire length. *Corolla* pinkish red; standard broadly obovate, 13–17 × 14–23 mm, glabrous; wings obovate, deep purplish pink 10–17 × 4–6 mm, without surface sculpturing, apex rounded; keel ± equal to wings, 12–18 × 4–6 mm, paler or white, apex rounded, pocket absent. *Ovary* 10–15 mm long, linear-oblong. *Fruit* linear, 95–120 × 4–8 mm, glabrescent, 5–8-seeded. *Seeds* reniform, ± 5.0 × ± 3.5 mm, brown or reddish brown speckled black (Figure 4). *Flowering time*: Nov.–Feb.

Diagnostic characters: *Sphenostylis marginata* differs from *S. angustifolia* in having broader leaflets and longer petioles, peduncles, and pods (see key for dimensions).

Verdcourt (1970) divided what he called the *Sphenostylis marginata* complex, which included the two central and East African species *S. erecta* (Baker f.) Baker f. and *S. obtusifolia* Harms, into three subspecies, namely subsp. *marginata* (occurring only in South Africa and Swaziland), subsp. *obtusifolia* (Harms) Verdc., and subsp *erecta* (Baker f.) Verdc. However, based on morphological and DNA data, Potter & Doyle (1994) pro-

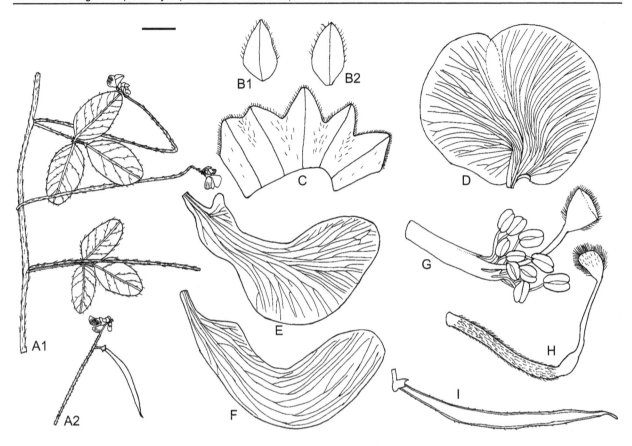

FIGURE 4.—Vegetative and reproductive morphology of *Sphenostylis marginata*: A1, flowering branch; A2, inflorescence; B1 & B2, abaxial view of bracteoles; C, calyx opened out with upper lobes to left; D, standard petal; E, wing petal; F, keel petal; G, stamens; H, pistil; I, lateral view of pod. Vouchers: A from *N. Grobbelaar 1648* (PRE); B–F from *Commins 855* (PRE); H from *Acocks 20928* (PRE); I from *N. Grobbelaar 1648* (PRE). Scale bar: A1, A2, 30 mm; B, 1 mm; C, 2 mm; D–F, H, 3 mm; G, 4 mm; I, 15 mm.

posed a narrow view of *S. marginata* by limiting the species concept to subsp. *marginata* (the other two subspecies were excluded). *Sphenostylis marginata* subsp. *erecta* was restored to species level and subspecies *obtusifolia* was transferred to *S. erecta* (as subspecies).

Distribution and habitat: Sphenostylis marginata occurs in South Africa (Limpopo, Mpumalanga, Kwa-Zulu-Natal and Eastern Cape Provinces), and Swaziland (Figure 5). It grows in Central Sandy Bushveld (SVcb 12), Ohrigstad Mountain Bushveld (SVcb 26), Gran-ite Lowveld (SVl 3), Legogote Sour Bushveld (SVl 9), Zululand Lowveld (SVl 23), PaulPietersburg Moist Grassland (Gm 15), KaNgwane Montane Grassland (Gm 16), Income Sandy Grassland (Gs 7), Ngongoni Veld (SVs 4), Eastern Valley Bushveld (SVs 6), Maputaland Coastal Belt (CB 1), and KwaZulu-Natal Coastal Belt (CB 3).

Additional specimens examined

LIMPOPO.—2428 (Nylstroom): 13 km NE of Warmbaths [Bela-Bela] on road to Nylstroom, (–CD), 18 Nov. 1981, *C. Reid 439* (PRE).

MPUMALANGA.—2430 (Pilgrim's Rest): foothills SE of Magalieskop, Mariepskop Dist., (–DB), 6 Dec. 1990, *H.P. van der Schijf 5878* (PRE). 2531 (Komatipoort): 25 km from White River to Hazyview, near White Waters Forest Station, (–AA), 3 Jan. 1984, *M. Jordaan 298* (PRE); Kruger National Park, Lower Sabie dam, (–BB), 26 Nov. 1990, *V.R. Bredenkamp 468* (PRE); Eerste Geluk no. 16, Uitkyk, (–CA), 26 Mar. 1975, *C.H. Stirton 1729* (PRE).

SWAZILAND.—2631 (Mbabane): Hhohho Dist., Masilela area, on Maphalaleni Rd., (–AB), 27 Jan. 1994, *G. Germishuizen 7152* (PRE); Little Usutu River, (–AC), 27 Oct. 1956, *R.H. Compton 26163* (NH, PRE); Stegi, (–BD), 22 Dec. 1960, *R.H. Compton 30388* (NH, PRE); S of Mankai-ana, (–CA), 6 Nov. 1949, *J.L. Sidey 1933* (PRE). 2731 (Louws-burg): at camp on hilltop before reaching Hluti, (–BA), 8 1931, *I.B. Pole-Evans 3364* (PRE).

KWAZULU-NATAL.—2730 (Vryheid): 0.9 km towards Vryheid from Natal Spa, Freddie Coetzee's farm, (–BD), 2 Dec. 1988, *P.D.F. Kok & B.J. Pienaar 1282* (PRE). 2731 (Louwsburg): Nongoma, (–DC), 20 Nov. 1960, *M.J. Wells 2060* (PRE). 2830 (Dundee): hill near Glencoe, (–AA), 22 Feb. 1993, *J. Medley-Wood 4823* (PRE); Dundee Dist., on banks of Buffalo River, 1 km S of P.O. Vantsdrift, (–AB), 22 Dec. 1946, *L.E. Codd 2377* (PRE); Elandslaagte, Blanerne Farm,

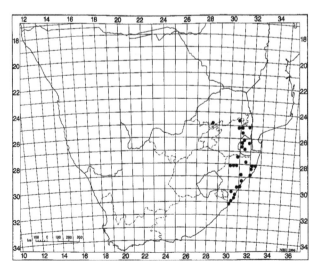

FIGURE 5.—Distribution of *Sphenostylis marginata*.

32 km from Ladysmith on road to Newcastle, (–BA), 15 Nov. 1994, *A.M. Ngwenya 1295* (PRE). 2831 (Nkandla): Babanango Dist., valley N of Izulu Hill, (–CC), 16 Jan. 1946, *J.P.H. Acocks 12317* (PRE). 2832 (Mtubatuba): Masundwini, Hluhluwe Game Reserve, (–AA), 7 Nov. 1971, *P.M. Hitchins 635* (NU, PRE); Palm Ridge Farm, (–AC), 3 Oct. 1967, *E.R. Harrison 128* (PRE). 2931 (Stanger): Twinstreams Farm, Mtunzini, (–DC), 12 Oct. 1984, *I. Garland s.n,. & G. Nichols 805* (PRE). 3030 (Port Shepstone): St Michaels-on-Sea, (–AB), 1 Feb. 1985, *B.J. Pienaar 587* (PRE); Port Shepstone (–CD), 24 Mar. 1967, *R.G. Strey 7420* (PRE); Hibberdene, (–DA), 09 Mar. 1970, *R.G. Strey 9701* (NU).

ACKNOWLEDGEMENTS

The curator of PRE is gratefully acknowledged for assistance with herbarium material and the curator of P is thanked for the image of the *Sphenostylis marginata* type specimen. The University of Johannesburg and the National Research Foundation provided financial support. We thank David Styles for showing us a population of *S. marginata* and for providing photographs of the species.

REFERENCES

BAKER, E.G. 1929. *The Leguminosae of tropical Africa* 1: 148. Erasmus Press, Gent.

BENTHAM, G. 1862. Leguminosae. In W.H. Harvey & O.W. Sonder (eds), *Flora capensis* 2: 240. Hodges, Smith, Dublin.

BENTHAM, G. 1865. LVII. Leguminosae. *Genera plantarum* 1, 2: 539. Lovell Reeve, London.

BURKHILL, H.M. 1995. *The useful plants of West tropical Africa*, edn 2: 447. Royal Botanic Gardens, Kew.

BURTT DAVY, J. 1932. Papilionaceae. *Manual of the Flowering Plants and Ferns of the Transvaal with Swaziland, South Africa*: 418. Longmans, Green & Co., London.

COMPTON, R.H. 1975. Flora of Swaziland. *South African Journal of Botany*, suppl. 11: 286.

DE WILDEMAN, E.A.J. 1921. *Contribution à l'etude de la flore du Katanga*: 98. D. Reynaert, Bruxelles [Brussels].

DYER, R.A. 1947. *Sphenostylis angustifolia. Flowering plants of Africa*: t. 1010.

DYER, R.A. 1975. *The genera of southern African flowering plants*: 275. Department of Agricultural Technical Services, Botanical Research Institute, Pretoria.

GERMISHUIZEN, G. 2000. Fabaceae. In O.A. Leistner (ed.), *Seed plants of southern Africa: families and genera. Strelitzia* 10: 296. South African National Biodiversity Institute, Pretoria.

GILLETT, J.B. 1966. Notes on Leguminoseae (Phaseoleae). 1. *Sphenostylis* E.Mey., a leguminous genus hitherto unrecognized in India. *Kew Bulletin* 20: 103–111.

GILLETT, J.B., POLHILL, R.M. & VERDCOURT, B. (eds). 1971. *Flora of tropical East Africa* 3: 671. Royal Botanic Gardens, Kew.

HARMS, H. 1899. Leguminosae africanae II. *Botanische Jahrbücher für Systematik, Pflanzengeschichte und Pflanzengeographie*, 26: 253–324.

HARVEY, W. 1862. Leguminosae. In W.H. Harvey & O.W. Sonder (eds), *Flora capensis* 2: 239. Hodges, Smith, Dublin.

LACKEY, J.A. 1981. Tribe 10. Phaseoleae. In R.M. Polhill & P.H. Raven (eds), *Advances in legume systematics*, 1: 301–327. Royal Botanic Gardens, Kew.

LOCK, J.M. 1989. *Legumes of Africa: a check-list*. Royal Botanic Gardens, Kew.

MEYER, E.H.F. 1836. *Commentariorum de plantes Africae australis* 1: 65–75. Leopoldum voss, Leipzig.

MUCINA, L. & RUTHERFORD, M.C. (eds). 2006. The vegetation of South Africa, Lesotho and Swaziland. *Strelitzia* 19. South African National Biodiversity Institute, Pretoria.

PHILLIPS, E.P. 1951. *The genera of South African flowering plants. Botanical Survey Memoir* no. 25: 427. Department of Agriculture, South Africa.

POTTER, D. 1992. Economic botany of *Sphenostylis* (Leguminosae). *Economic Botany* 46: 262–275.

POTTER, D. & DOYLE, J.J. 1994. Phylogeny and systematics of *Sphenostylis* and *Nesphostylis* (Leguminosae: Phaseoleae) based on morphological and chloroplast DNA data. *Systematic Botany* 19: 389–406.

SAVI, G. 1824. Phaseoli. Nuovo Giornale dei Letterati 8: 113. Sebastiano Nistri, Pisa.

SCHRIRE, B.D. 2005. Phaseoleae. In G. Lewis, B.D. Schrire, B. Mackinder, & M. Lock, *Legumes of the World*: 393–430. Royal Botanic Gardens, Kew.

SONDER, O.W. 1850. Beiträge zur Flora von Südafrica. *Linnaea* 23: 33. Berlin.

TAUBERT, P. 1894. Leguminosae. In A. Engler & K. Prantl, *Die natürlichen Pflanzenfamilien* 3: 70–399. Engelmann, Leipzig.

VERDCOURT, B. 1970. Studies in the Leguminosae–Papilionoideae for the Flora of tropical east Africa: III. *Kew Bulletin* 24: 380–442.

VERDCOURT, B. & DØYGAARD, S. 2001. *Sphenostylis*. In G.V. Pope & R.M. Polhill (eds), *Flora Zambesiaca* 3, 5: 68. Royal Botanic Gardens, Kew.

VERDOORN, I.C. 1968. *Sphenostylis marginata. Flowering plants of Africa*: t. 1521.

WILCZEK, R. 1954. Phaseolinae. *Flore du Congo Belge et du Ruanda-Urundi* VI: 274. Institut National pour l'Etude Agronomique du Congo Belge, Brussels.

WOJCIECHOWSKI, M.F., LAVIN, M. & SANDERSON, M.J. 2004. A phylogeny of legumes (Leguminosae) based on analysis of the plastid *matK* gene resolves many well-supported subclades within the family. *American Journal of Botany* 91: 1846–1862.

2

Recircumscription and distribution of elements of the 'Ceterach cordatum' complex (Asplenium: Aspleniaceae) in southern Africa

R.R. KLOPPER* and N.R. CROUCH**

Keywords: Aspleniaceae, *Asplenium*, *Ceterach*, ferns, pteridophytes, southern Africa, xerophytes

ABSTRACT

Ceterachoid aspleniums in southern Africa have long been treated as a single widespread and variable taxon, *Asplenium cordatum* (Thunb.) Sw. (= *Ceterach cordatum* Thunb.). In addition to *A. cordatum*, a further two ceterachoid taxa are now recognized as occurring in the *Flora of southern Africa* (FSA) region, namely *A. capense* (Kunze) Bir, Fraser-Jenk. & Lovis and *A. phillipsianum* (Kümmerle) Bir, Fraser-Jenk. & Lovis. We provide full descriptions and distributions of these three taxa.

INTRODUCTION

Pteridophyte treatments for the *Flora of southern Africa* (FSA) (Roux 1986; Schelpe & Anthony 1986; Burrows 1990) and *Flora zambesiaca* (FZ) (Schelpe 1970) regions, have considered a single xerophytic rock fern, *Ceterach cordatum* Thunb., to occur widely throughout the region and to show great morphological variability. *Ceterach* is currently treated as a subgenus of *Asplenium* L. (Crabbe *et al*. 1975; Bir *et al*. 1985; Roux 2001). Subgenus *Ceterach* is distinguished from subgenus *Asplenium* in our region by its lack of indusia and the presence of densely-set scales (paleae) on the abaxial lamina surface, versus indusiate sori and sparsely-set scales in subgenus *Asplenium* (Roux 2001). Moore (1857) early recognized that the strictly African ceterachoid species were anatomically and morphologically distinct from the Eurasian-Macronesian elements, and therefore excluded them from *Asplenium* subgen. *Ceterach*. More recent molecular analyses have demonstrated the polyphyly of subgenus *Ceterach*, implicating homoplasy in the dense scale cover and in pinnatisect laminae of these asplenioid ferns (Pinter *et al*. 2002; Van den Heede *et al*. 2003). The taxonomic implications are that southern African taxa referred earlier to subgenus *Ceterach* should either be accommodated in a new subgenus, or that distinction at a subgeneric level (*sensu* Roux 2001) should be abandoned altogether (Van den Heede *et al*. 2003).

The multiple origins in *Asplenium* of redundant indusia and dense abaxial scales have been attributed to independent adaptation to rocky xeroseres across its range (Van den Heede *et al*. 2003). We further postulate that the pinnatisect fronds common to both groups is a poikilohydric modification that allows for frond integrity to be maintained in the inrolled, dessicated state (Figure 1), and for even restoration following rehydration. Ceterachoid taxa in southern Africa are extremely desiccation-tolerant; a member of this group has been shown by Gaff (1977) to tolerate relative humidities in the 0–5% range for at least six months, with an initial water potential (ψ) of 18%, expressed in terms of the relative humidity at 28°C.

We concur with Roux (2009a) in his reinstatement of *Asplenium capense* (Kunze) Bir, Fraser-Jenk. & Lovis as distinct from *A. cordatum* (Thunb.) Sw., but we also identify *A. phillipsianum* (Kümmerle) Bir, Fraser-Jenk. & Lovis as occurring in the FSA region (Table 1). This taxon extends from the island of Socotra off Somalia through East and Central Africa to the northern provinces of South Africa, as far south as the northern regions of Mpumalanga, Gauteng, and North West Province. Pappe & Rawson (1858) recognized the widespread European and North African taxon *Asplenium ceterach* L. (= *Ceterach officinarum* Lam. & DC) as South African, based on a Krebs collection made on the Baviaans River. This species is not currently considered to occur south of the Sahara, and has been taken as a misidentification (Roux 1986, 2009a).

We provide the recircumscription and distribution of the three ceterachoid elements of *Asplenium* currently known from southern Africa.

* Biosystematics Research and Biodiversity Collections Division, South African National Biodiversity Institute, Private Bag X101, 0001 Pretoria / Department of Plant Science, University of Pretoria, 0002 Pretoria. E-mail: r.klopper@sanbi.org.za.
** Ethnobotany Unit, South African National Biodiversity Institute, P.O. Box 52099, Berea Road, 4007 Durban / School of Chemistry, University of KwaZulu-Natal, 4041 Durban. E-mail: n.crouch@sanbi.org.za.

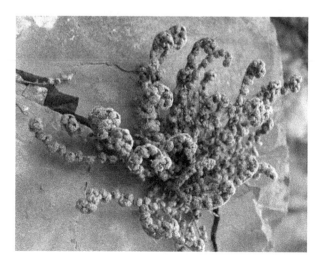

FIGURE 1.—Inrolled pinnatisect leaves of a desiccated plant of *Asplenium cordatum*, Nieu Bethesda, Eastern Cape Province. Photo: N. Crouch.

TABLE 1.—Characters distinguishing ceterachoid members of *Asplenium* occurring in the FSA region.

Character/taxon	*A. cordatum*	*A. capense*	*A. phillipsianum*
Frond division	shallowly to very deeply 2-pinnatifid, sometimes 2-pinnate	pinnatisect to pinnate, sometimes shallowly 2-pinnatifid	pinnatisect to very shallowly 2-pinnatifid
Frond apex division	to near apex	to some way below apex	to some way below apex
Pinnae width	narrower at base	broader at base	broader at base
Pinnae shape	narrowly oblong to oblong-lanceolate, apex bluntly acute; petiolulate throughout; base flared to somewhat auriculate-cordate	ovate-oblong, apex obtuse; apical pinnae adnate with base decurrent; basal 1–2 pinnae pairs becoming petiolulate with base somewhat flared to auriculate-cordate	narrowly ovate-oblong, apex obtuse; base adnate and decurrent throughout
Scale density on abaxial surface	very dense	very sparse	sparse
Scale shape	deltate, attenuate	deltate to lanceolate, attenuate	lanceolate
Scale margins	serrate	finely serrate	finely serrate
Scale composition	cells short roundish to oval	cells narrowly oblong	cells oblong
Scale colour on stipe	not darker towards apex	darker towards apex	darker towards apex
Rhachis robustness	strong, rigid	strong, fairly rigid	weaker, flexible
Sori size	small, up to 2 mm, not confluent	very large, 3–6 mm, later confluent	large, 2–3 mm, later confluent
Gemmae	absent	absent	sometimes present near apex
Stipe length	17–40 mm	15–35 mm	5–20 mm

MATERIALS AND METHODS

All ceterachoid *Asplenium* material held in the following herbaria (totalling almost 600 specimens) were studied: Geo-Potts Herbarium (BLFU), University of the Free State, Bloemfontein; Buffelskloof Nature Reserve Herbarium (BNRH), Lydenburg; Bolus Herbarium (BOL), University of Cape Town, Cape Town; Selmar Schönland Herbarium (GRA), Albany Museum, Grahamstown; the Herbarium at the Royal Botanic Gardens, Kew (K), London, United Kingdom; Mpumalanga Parks Board Herbarium (LYD), Lydenburg; Compton Herbarium (NBG), SANBI, Cape Town; KwaZulu-Natal Herbarium (NH), SANBI, Durban; National Museum Herbarium (NMB), Bloemfontein; Bews Herbarium (NU), University of KwaZulu-Natal, Pietermaritzburg; National Herbarium (PRE), SANBI, Pretoria; H.G.W.J. Schweickerdt Herbarium (PRU), University of Pretoria, Pretoria; A.P. Goossens Herbarium (PUC), North West University, Potchefstroom; and the South African Museum Herbarium (SAM), SANBI, Cape Town.

The JSTOR Plant Science website (http://plants.jstor.org) was consulted for type material held in other herbaria. Images of type specimens were directly obtained from The Museum of Evolution Herbarium (UPS), Uppsala University, Uppsala, Sweden. Types seen electronically are cited as e!

Herbarium acronyms follow Holmgren *et al.* (1990). Author citations used follow the standardized author abbreviations provided by the International Plant Names Index (http://www.ipni.org).

TAXONOMY

Three taxa are recognized in southern Africa, all chasmophytic saxicoles (Jacobsen 1983):

Key to the species

1a. Fronds always pinnate, with stronger, rigid rachis; pinnae free from rachis (petiolulate), with flared to somewhat auriculate/cordate base, abaxially very densely covered with overlapping, broadly deltate scales; sori small, discrete . . . . 1. *A. cordatum*
1b. Fronds pinnatisect to pinnate, with weaker, flexible rachis; pinnae adnate to rachis, with broader decurrent base (at least in distal half of frond), abaxially sparsely covered with ovate lanceolate scales; sori large and confluent:
2a. Pinnae adnate to rachis and decurrent apically, becoming free from rachis (petiolulate) with ± flared to auriculate/cordate base basiscopically; rachis winged apically . . . . . . 2. *A. capense*
2b. Pinnae adnate to rachis and decurrent throughout; rachis almost winged throughout . . . . 3. *A. phillipsianum*

1. **Asplenium cordatum** *(Thunb.) Sw.* in Journal für die Botanik 1800,2: 54 (1801). *Acrostichum cordatum* Thunb.: 171 (1800). *Grammitis cordata* (Thunb.) Sw.: 23, 217 (1806). *Cincinalis cordata* (Thunb.) Desv. (1811). *Notholaena cordata* (Thunb.) Desv.: 92 (1813). *Gymnogramma cordata* (Thunb.) Schltdl.: 16 (1825). *Ceterach cordatum* (Thunb.) Desv.: 223 (1827). Type: South Africa, 'e Cap bonae Spei', *Thunberg s.n.* (UPS-Thunb 24439, holo. e!).

Ceterach crenata Kaulf.: 85, 86 (1824), nom. illegit. superfl. Type: as for *Acrostichum cordatum* Thunb. [McNeill *et al.* (2006) Art. 7.5].

Gymnogramma namaquensis Pappe & Rawson: 42 (1858). *Gymnogramma cordata* var. *namaquensis* (Pappe & Rawson) Sim: 212 (1892). *Ceterach cordatum* var. *namaquensis* (Pappe & Rawson) Sim: 176 (1915). Type: South Africa, Namaqualand, between rocks near Modderfontein, 1856, *Whitehead s.n.* (K, holo.!).

Grammitis cordata [var. and] subvar. *nudiuscula* Hook.: t.7 (1860), nom. illegit. superfl. Type: as for *Gymnogramma namaquensis* Pappe & Rawson [see Roux (2009a)].

Grammitis cordata var. *pinnato-pinnatifida* Hook.: t.7 (1860), nom. inval. [McNeill *et al.* (2006) Art. 26.2].

Gymnogramma cordata var. *subbipinnata* Hook.: t.7 (1860). *Ceterach cordatum* var. *subbipinnata* (Hook.) Kümmerle: 289 (1909). Type: South Africa, 'elevated mountain of Macaliesberg', *Ecklon & Burke s.n.* [missing, see Schelpe & Anthony (1986: 206) and Roux (2009: 83)].

Gymnogramma cordata var. *bipinnata* Sim: 212 (1892). Type: South Africa, Namaqualand, without precise locality, *Holland s.n.* [NBG, lecto.!, designated by Schelpe & Anthony (1986: 206)].

Notholaena inaequalis Kunze γ *eckloniana* (Kunze) Kuntze var. *rawsonii* (Pappe) Kuntze forma *minor* Kuntze: 379 (1898). Type: South Africa, [Eastern Cape], 'Capland', Cradock, 940 m, 12 Feb. 1894, *Kuntze s.n.* [NY, lecto., designated by Roux (2009b: 228)].

Rhizome to 3–6 mm diam., erect or procumbent; scales sessile, clathrate, very narrowly lanceolate, 2–3 × 0.5–0.6 mm, acuminate, frequently with a hair point, irregularly serrate, bicolorous, chestnut-brown distally, brown proximally, with paler margin throughout, with narrowly oblong to ovate cells. *Fronds* tufted, suberect to erect; *stipe* (10–)17–40(–60) mm long, dark chestnut-brown, densely scaled when young, becoming subglabrous with age, scales sessile, lanceolate, 3.0–3.5 × 0.7–0.9 mm, acuminate, irregularly serrate, glossy, sometimes bicolorous, chestnut-brown, sometimes with narrow paler margin, with oblong to ovate cells; *lamina* subcoriaceous, involute and inrolled when dry, elliptic or narrowly elliptic to oblanceolate in outline, shallowly to very deeply 2-pinnatifid, sometimes 2-pinnate, (20–)50–120(–150) × (10–)15–40(–50) mm, basal pinnae gradually decrescent; *rachis* not winged between pinnae, scales as for stipe but lanceolate, 3.0 × 0.9 mm; *pinnae* (4–)6–19(–30) × (2–)3–7(–13) mm, free from rachis (petiolulate), with flared to ± auriculate-cordate base, narrowly oblong to oblong-lanceolate, bluntly acute, margin irregularly scalloped to incised, glabrous above at maturity, abaxially very densely scaled, scales sessile, deltate, ± 2.0 × 1.5 mm, attenuate, serrate, glossy, light reddish brown, with short roundish to oval cells; *sori* linear along (obscure) veins, up to 2 mm long, exindusiate, almost totally obscured by scales. Figure 2A–D.

Etymology: *cordatum* = heart-shaped, referring to the basally lobed pinnae.

Distribution and ecology: *Asplenium cordatum* is widespread in South Africa, Lesotho, Swaziland, Namibia, Botswana, and Zimbabwe (Figure 3); also in Angola, Tanzania, Kenya, Uganda, and Ethiopia.

It occurs in rocky crevasses in exposed, hot, and dry habitats, often at the base of boulders, well away from water. A dense scale cover on the abaxial surface of the fronds serves to protect the sori of this xerophytic species.

Asplenium cordatum is very variable in size, with the largest specimens originating from the dry areas of the north-western parts of the Western and Northern Cape provinces. This immense variability is reflected in the long list of synonyms for this taxon. Burrows (1990) noted that northwards of the Limpopo River it becomes progressively rarer, reaching the extreme of its range in Ethiopia.

Although purported to occur in Madagascar (Roux 2009a), where a ceterachoid element has been reported from a single locality on Mount Morahahiva (Tardieu-Blot 1957), this collection best matches *Asplenium phillipsianum* (Tardieu-Blot 1958: Fig. XVII).

Vouchers: *H.H. Burrows 3294* (GRA); *E. Esterhuysen 25624* (BOL); *D. Galpin 4782a* (BLFU); *H.H.W. Pearson 8557* (BOL, K, NBG); *L.E. Taylor 2913* (NBG).

2. **Asplenium capense** *(Kunze) Bir, Fraser-Jenk. & Lovis* in Fern Gazette 13,1: 61 (1985). *Ceterach capense* Kunze: 496 (1836). *Grammitis capensis* (Kunze) T.Moore: 1xiii (1857). Type: South Africa, 'Port Natal et Afrique meridionalea', *Drège s.n.* [G, lecto.!, isolecto., designated by Roux (1986: 352)]; 'Ceded territory, bergwaldungen an den Quellen des Katriver, Oberhalb Philipstown', *Ecklon & Zeyher s.n.* (UPS, syn. e!).

Gymnogramma capensis Spreng. ex Kaulf.: 183 (1831), nom. nud. *Ceterach cordatum* var. *capense* (Spreng. ex Kaulf.) Hieron. ex Kümmerle: 287 (1909). Type: South Africa, 'Cap. Bon spei: in einer Felsenritze am Löwenberg', *Zeyher s.n.* Fl. Cap. No. 273 (HAL, holo. e!; BOL, iso.!).

Grammitis cordata var. *pinnata* Hook.: t.7 (1860), Type: as for *Ceterach capense* Kunze [see Roux (2009a)].

Rhizome to 4 mm diam., erect or procumbent; scales sessile, clathrate, lanceolate, 3–5 × 1.0–1.5 mm, acuminate, frequently with a hair point, irregularly serrate, bicolorous, with dark brown central region and paler margins, with narrowly oblong cells. *Fronds* tufted, erect to suberect; *stipe* 15–35(–60) mm long, chestnut-brown to dark chestnut-brown, densely scaled, scales sessile, narrowly triangular, 3–5 × 0.5–1.0 mm, acuminate, irregularly finely serrate, glossy, sometimes bicolorous, with narrow rust coloured central region and broad straw coloured margin, with oblong to ovate cells; *lamina* herbaceous, involute and inrolled when dry, elliptic to narrowly obovate in outline, pinnatisect to pinnate, sometimes shallowly 2-pinnatifid, (80–)90–120(–190) × (20–)30–45(–52) mm, basal pinnae gradually decrescent; *rachis* somewhat winged apically, not winged between widely spaced pinnae basally, scales as for stipe but 2–4 × 0.7–1.0 mm; *pinnae* (10–)15–25(–28) × (4–)5–10(–11) mm, adnate to rachis with decurrent base apically, becoming free from rachis (petiolulate) with somewhat flared to auriculate-cordate base basiscopically, ovate-oblong, obtuse, margin sinuate to scalloped, glabrous above at maturity, abaxially very sparsely scaled, scales sessile, deltate to lanceolate, 1.5–2.5 × 0.4–0.7 mm, attenuate, finely serrate, glossy, bicolorous, with narrow rust coloured central region and broad straw coloured margin, with narrowly oblong cells; *sori* linear along (obscure) veins, 3–6 mm long, becoming confluent at maturity, exindusiate, not obscured by scales. Figure 2E–H.

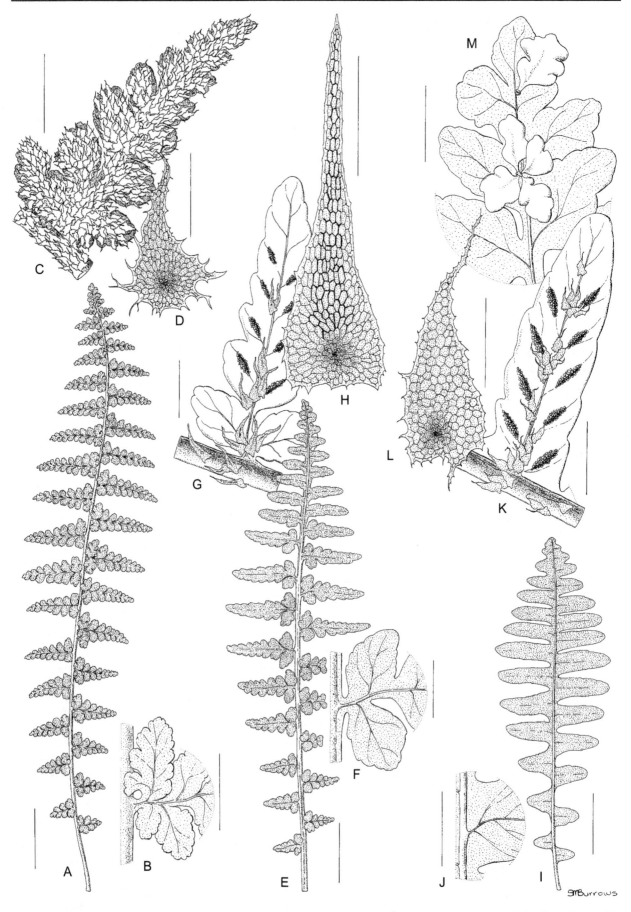

FIGURE 2.—A–D, *Asplenium cordatum*, *J.E. Burrows 1110* (BNRH): A, complete frond; B, cordate pinna base; C, abaxial surface of pinna with dense scales; D, lamina scale. E–H, *Asplenium capense*, *H.H. Burrows 2891* (BNRH): E, complete frond; F, pinna base; G, abaxial surface of pinna with sparse scales; H, lamina scale. I–M, *Asplenium phillipsianum*, *J.E. Burrows & S.M. Burrows s.n.* (BNRH): I, complete frond; J, adnate pinna base; K, abaxial surface of pinna with sparse scales; L, lamina scale; M, gemmae at frond apex. Scale bar: A, E, I, 20 mm; B, C, F, G, J, K, M, 5 mm; D, H, L, 1 mm. Artist: Sandra Burrows.

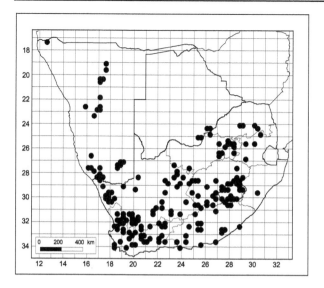

FIGURE 3.—Distribution of *Asplenium cordatum* in the FSA region.

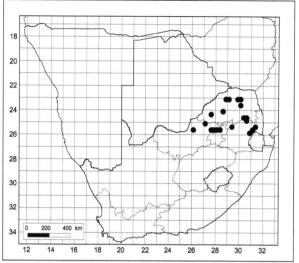

FIGURE 5.—Distribution of *Asplenium phillipsianum* in the FSA region.

Etymology: *capense* = pertaining to the Cape.

Distribution and ecology: *Asplenium capense* occurs from the Cape Peninsula through the Western and Eastern Cape, KwaZulu-Natal, the Free State and northern provinces of South Africa (Figure 4), extending sporadically to central and tropical East Africa.

This species typically grows in sandy soil on forest floors, as well as under coastal dune scrub, often fairly close to streams and under trees on steep damp earth banks. It is usually associated with riparian or open forest, most often in lightly rather than deeply shaded conditions.

Vouchers: *Th.C.E. Fries, T. Norlich & H. Weimark 30-8* (BOL, K); *D.B. Müller 890* (NMB; PRE); *R. Schlechter 2703* (GRA, K); *E.M. van Zinderen Bakker 1135* (BLFU, PRE); *C.J. Ward 12400* (NH, NU, PRE).

3. Asplenium phillipsianum *(Kümmerle) Bir, Fraser-Jenk. & Lovis* in Fern Gazette 13,1: 62 (1985).

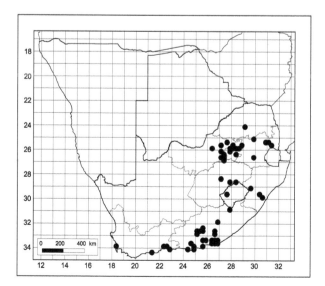

FIGURE 4.—Distribution of *Asplenium capense* in the FSA region.

Ceterach phillipsianum Kümmerle: 287 (1909). Type: Somalia, 'Ferns from deep shade Wagga Mountain, anno 1897', *Phillips s.n.* (BM, holo.; K, iso.!)

Rhizome to 4 mm diam., erect or procumbent; scales sessile, clathrate, lanceolate, 3–4 × 0.6–1.0 mm, acuminate, frequently with a hair point, irregularly serrate, sometimes bicolorous, with dark chestnut-brown central region, sometimes with darker apex, with paler margins throughout, with narrowly oblong to ovate cells. *Fronds* tufted, suberect to spreading; *stipe* 5–20(–30) mm long, dark chestnut-brown, densely scaled, scales sessile, lanceolate, 3–4 × 0.7–1.0 mm, acuminate, irregularly serrate, glossy, bicolorous, distally dark brown and proximally chestnut-brown with paler margins throughout, with narrowly oblong to ovate cells; *lamina* herbaceous, involute and inrolled when dry, elliptic to obovate in outline, pinnatisect to very shallowly 2-pinnatifid, (5–)70–100(–130) × (16–)24–33(–50) mm, lower pinnae gradually decrescent, occasionally produces 1–3 gemmae situated adaxially in the sinus of the distal pinnules; *rachis* almost winged throughout, scales as for stipe but 2.5–3.0 × 0.6–1.0 mm, concolorous reddish brown, sometimes chestnut-brown towards apex; *pinnae* (6–)10–20(–25) × (3–)4–6(–9) mm, adnate to rachis with decurrent base throughout, narrowly ovate-oblong, obtuse, margin entire to weakly sinuate, glabrous above at maturity, abaxially sparsely scaled, scales sessile, lanceolate, 1.7–2.4 × 0.6–0.8 mm, attenuate, finely serrate, glossy, brown, with oblong cells; *sori* linear along (obscure) veins, 2–3 mm long, becoming confluent at maturity, exindusiate, not totally obscured by scales. Figure 2I–M.

Etymology: *phillipsianum* = commemorates the English explorer Ethelbert Lort-Phillips (1857–1944) who in 1897 collected the type specimen in Somalia.

Distribution and ecology: *Asplenium phillipsianum* occurs from the northern provinces of South Africa (Limpopo, North West, Gauteng, and Mpumalanga) (Figure 5), through central and tropical Africa as far north as Socotra; also known from Réunion and Madagascar.

This species is typically found close to streams, shaded under trees on steep damp earth banks.

Of the three regional ceterachoid species, *A. phillipsianum* is the only member noted to be gemmiferous (J. Nel, pers. comm.) (Figure 2M).

Vouchers: *J.E. Burrows & S.M. Burrows s.n.* (BNRH); *M.F. Glen PRE62127* (PRE); *E. Retief & S.E. Strauss 2154* (PRE); *J.P. Roux 3195* (NBG); *A. Winterboer s.n.* (PRU).

Excluded name

Ceterach cordatum var. *pinnatifida* Sim: 177 (1915). The syntypes cited by Sim (1915) include representatives of both *A. capense* and *A. phillipsianum*.

ACKNOWLEDGEMENTS

Dr Hugh Glen, KwaZulu-Natal Herbarium, SANBI, Durban, is thanked for translating Latin texts and for providing nomenclatural advice; Prof. John McNeill, Royal Botanic Gardens, Edinburgh, for providing nomenclatural advice; Ms Hester Steyn, National Herbarium, SANBI, Pretoria, for producing updated distribution maps. The Curators of BLFU, BNRH, BOL, GRA, K, LYD, NBG, NH, NMB, NU, PRE, PRU, PUC and SAM kindly facilitated access to their collections.

REFERENCES

BIR, S.S., FRASER-JENKINS, C.R. & LOVIS, J.D. 1985. *Asplenium punjabense* sp. nov. and its significance for the status of *Ceterach* and *Ceterachopsis*. *Fern Gazette* 13,1: 53–63.

BURROWS, J.E. 1990. *Southern African ferns and fern allies*. Frandsen, Sandton.

CRABBE, J.A., JERMY, A.C. & MICKEL, J.M. 1975. A new generic sequence for the pteridophyte herbarium. *Fern Gazette* 11: 141–162.

DESVAUX, N.A. 1811. Observations sur quelques nouveaux genres de fougéres et sur plusieures espèces nouvelles de la même famille. *Magazin für die neuesten Entdeckungen in der gesammten Naturkunde, Gesellschaft Naturforschender Freunde zu Berlin* 5: 297–330, figs 4–7.

DESVAUX, N.A. 1813. Espèces de fougéres à ajouter au genre *Notholaena*. *Journal de Botanique, Appliquée à l'Agriculture, à la Pharmacie, à la Médicine et aux Arts* 1: 91–93.

DESVAUX, N.A. 1827. Prodrome de la familles des fougères. *Mémoires de la Société Linnéenne de Paris* 6(2): 171–337.

GAFF, D.F. 1977. Desiccation tolerant plants of southern Africa. *Oecologia* 31: 95–109.

HOLMGREN, P.K., HOLMGREN, N.H. & BARNETT, L.C. 1990. *Index herbariorum, part 1: The herbaria of the world*, edn 8. New York Botanical Garden. New York.

HOOKER, W.J. 1860. *A second century of ferns*: [i]-xii, t.1–100, Pamplin, London.

JACOBSEN, W.B.G. 1983. *The ferns and fern allies of southern Africa*. Butterworths, Durban.

KAULFUSS, G.F. 1824. *Enumeratio filicum*: [i]-vi, [1]–300, pls 1,2. Cnobloch, Leipzig.

KAULFUSS, G.F. 1831. Lycopodiaceae et Filices in plantae Ecklonianae. *Linnaea* 6: 181–187.

KÜMMERLE, J.B. 1909. A *Ceterach* génusz üj faja. (Species nova generis *Ceterach*). *Botanikai Közlemények* 8: 286–290.

KUNTZE, O. 1898. *Revisio genera plantarum*, volume III, parts II, II. Felix, Leipzig.

KUNZE, G. 1836. Plantarum acotyledonearum Africae australis recencio nova. *Linnaea* 10: 480–570.

MCNEILL, J., BARRIE, F.R., BURDET, H.M., DEMOULIN, V., HAWKSWORTH, D.L., MARHOLD, K., NICOLSON, D.H., PRADO, J., SILVA, P.C., SKOG, J.E., WIERSEMA, J.H. & TURLAND, N.J. (eds) 2006. International Code of Botanical Nomenclature (Vienna Code) adopted by the Seventeenth International Botanical Congress Vienna, Austria, July 2005. *Regnum Vegetabile* 146: 1–568.

MOORE, T. 1857. *Index filicum: a synopsis, with characters of the genera, and an enumeration of the species of ferns*. W. Pamplin, London.

PAPPE, C.W.L. & RAWSON, R.W. 1858. *Synopsis filicum Africae australis*. Saul Solomon and Co., Cape Town.

PINTER, I., BAKKER, F., BARRETT, J., COX, C., GIBBY, M., HENDERSON, S., MORGAN-RICHARDS, M., RUMSEY, F., RUSSELL, S., TREWICK, S., SCHNEIDER, H. & VOGEL, J. 2002. Phylogenetic and biosystematic relationships in four highly disjunct polyploidy complexes in the subgenera *Ceterach* and *Phyllitis* in *Asplenium* (Aspleniaceae). *Organisms Diversity & Evolution* 2: 299–311.

ROUX, J.P. 1986. A review and typification of some of Kunze's newly described South African Pteridophyta published in his *Acotyledearum Africae Australioris Recensio Nova*. *Botanical Journal of the Linnean Society* 92: 343–381.

ROUX, J.P. 2001. *Conspectus of southern African Pteridophyta*. Southern African Botanical Diversity Network Report No. 13. SABONET, Pretoria.

ROUX, J.P. 2009a. Synopsis of the Lycopodiophyta and Pteridophyta of Africa, Madagascar and neighbouring islands. *Strelitzia* 23. South African National Biodiversity Institute, Pretoria.

ROUX, J.P. 2009b. Pteridophyta: Otto Kuntze's lycopod and fern collections from South Africa. *Bothalia* 39: 227–229.

SCHELPE, E.A.C.L.E. & ANTHONY, N.C. 1986. Pteridophyta. In O.A. Leistner, *Flora of southern Africa*. Botanical Research Institute, Pretoria.

SCHELPE, E.A.C.L.E. 1970. Pteridophyta. In A.W. Exell & E. Launert, *Flora zambesiaca*. Crown Agents for Oversea Governments and Administrations, London.

SCHLECHTENDAL, D.F.L. 1825. *Adumbrationes plantarum*1: [1]–16, pls 1–6. Berlin.

SIM, T.R. 1892. *The ferns of South Africa*, edn 1. Juta, Cape Town.

SIM, T.R. 1915. *The ferns of South Africa*, edn 2. Cambridge University Press, Cambridge.

SWARTZ, O. 1801. *Journal für die Botanik. Herausgegeben von medicinalrath schraber*. 1800. 1–487, t.i–vii. Göttingen, Heinrich Dieterich.

SWARTZ, O. 1806. *Synopsis filicum*, [i]–xviii, [1]–445, t.1–5. Kiel.

TARDIEU-BLOT, M.-L. 1957. The genus *Ceterach* in Madagascar. *American Fern Journal* 47: 108–109.

TARDIEU-BLOT, M.-L. 1958. 5e Famille–Polypodiacées (sensu lato) 5(1) Dennstaedtiacées–(10) Aspidiacées). In H. Humbert, *Flore de Madagascar et des Comores (Plantes Vasculaires)* 1: 181, fig. XVII, 10–12.

THUNBERG, C.P. 1800. *Prodromus plantarum capensium*: [i–viii], [85]–191, Uppsala.

VAN DEN HEEDE, C.J., VIANE, R.L.L. & CHASE, M.W. 2003. Phylogenetic analysis of *Asplenium* subgenus *Ceterach* (Pteridophyta: Aspleniaceae) based on plastid and nuclear ribosomal ITS DNA sequences. *American Journal of Botany* 90: 481–495.

3

The emerging invasive alien plants of the Drakensberg Alpine Centre, southern Africa

C. Carbutt*

Keywords: biological invasions, climate change, disturbance regime, Drakensberg Alpine Centre, early detection, emerging alien plants, expert opinion, life history traits, prioritisation, selection criteria, swift management interventions

ABSTRACT

An 'early detection'-based desktop study has identified 23 taxa as 'current' emerging invasive alien plants in the Drakensberg Alpine Centre (DAC) and suggests a further 27 taxa as probable emerging invaders in the future. These 50 species are predicted to become problematic invasive plants in the DAC because they possess the necessary invasive attributes and have access to potentially suitable habitat that could result in them becoming major invaders. Most of the 'current' emerging invasive alien plant species of the DAC are of a northern-temperate affinity and belong to the families Fabaceae and Rosaceae (four taxa each), followed by Boraginaceae and Onagraceae (two taxa each). In terms of functional type (growth form), most taxa are shrubs (9), followed by herbs (8), tall trees (5), and a single climber. The need to undertake a fieldwork component is highlighted and a list of potential study sites to sample disturbed habitats is provided. A global change driver such as increased temperature is predicted to not only result in extirpation of native alpine species, but to also possibly render the environment more susceptible to alien plant invasions due to enhanced competitive ability and pre-adapted traits. A list of emerging invasive alien plants is essential to bring about swift management interventions to reduce the threat of such biological invasions.

INTRODUCTION

Increased volumes and frequency of trade, travel, and tourism have resulted in an increased spread of invasive and potentially invasive species (Simberloff 2001). The prediction is that such trends are likely to increase in the short and medium-term future in South Africa (Richardson 2001; Le Maitre et al. 2004). One result of increased trade, travel, and tourism is that plants will be moved by humans across geographic barriers far beyond their natural dispersal range. To date, South Africa has been invaded by many species of non-native plants, many of which are already well established and have a negative ecological and economic impact (Wells et al. 1986; Richardson et al. 1997; Van Wilgen 2004), particularly on ecosystem services (Van Wilgen et al. 2008, 2011). The different rates of spread observed in different areas are attributed to synergistic interactions between the basic features of the environment, features of the disturbance regime, and life history traits (Richardson 2001; Thuiller et al. 2006).

Mgidi et al. (2004) predicted that more invasive alien species are likely to reach South Africa in the immediate future. Although upon arrival, such species are still in the infancy of their invasion (either only recently introduced and/or are entering a phase of rapid population growth), they pose an even greater threat than some of the major established invaders because of the large areas they have the potential to invade and the 'unknown factor' associated with the exponential phase of their expansion (Hobbs & Humphries 1995; Nel et al. 2004). Emerging invaders appear to be establishing in areas already heavily invaded by major (well-established) invaders, suggesting that due to certain climatic features, patterns of human settlement, and/or land-use patterns, certain areas are more susceptible and predisposed

to invasive plants in general and that major invaders are also likely to be facilitating invasions of emerging invader species (Nel et al. 2004). Alien plant monitoring and management programmes, historically reactive in nature, should therefore not only target well-established invaders; the 'blacklisting' of emerging invaders as an early warning system will help identify, prioritise, and appropriately manage new invasions (Mgidi et al. 2004) so that the predicted trend of increasing invasions is matched with an ever-increasing ability to nullify the emerging threat (Nel et al. 2004; Olckers 2004). The overall objective should therefore be to proactively curb the threat of invasive alien plant species by stopping the invasion in its tracks, which will afford significant 'savings' in terms of minimising biodiversity losses and minimising overall management costs. Although the established invasive alien plants in the DAC are reasonably well known and their threat to plant biodiversity in the region recognised (Carbutt & Edwards 2004, 2006), no study has focussed explicitly on the emerging invasive alien plants of the DAC and the likely threat that such invaders may pose in the future.

Emerging invasive alien plants of the DAC are here defined as either (i) those alien plant species recorded from the DAC in the past 25 years or less that are currently still in the early stages of invasion (i.e. less than 100 populations with less than 1 000 individuals per population) and given their specific attributes and potentially suitable habitat, could expand further to become major invaders in the future (~ 'current' emerging invasive aliens); or (ii) those alien plant species naturalised in parts of South Africa that do not occur within the confines of the DAC, but in all likelihood will in the future given either their predominantly temperate affinity and/or their current range within 70 km's of the DAC (~ 'future' emerging invasive aliens).

The aims of this study were threefold: (i) outline a basic methodology to identify the emerging invasive alien plants of the DAC; (ii) develop a preliminary list

* Scientific Services, Ezemvelo KZN Wildlife, P.O. Box 13053, Cascades 3202, South Africa. e-mail: carbuttc@kznwildlife.com.

using the desktop component of the methodology identified in (i) above; and (iii) discuss the likely threat posed by these emerging alien invasive plants and the possible relationship between their spread and emerging global change drivers.

METHODS

Scope of study

A desktop approach was adopted to rapidly generate a preliminary list, which is especially helpful if the scope of study encompasses a large geographical area (i.e. where field work would prove costly and time consuming and may take years to complete). The best interim measure is a list to leverage swift management action that can later be fine-tuned with field work.

The geographical scope of the desktop study is the DAC, a temperate region with summer rainfall. Mean annual rainfall varies from ± 640 mm on the more leeward Lesotho side to over 2 000 mm on the main escarpment (Tyson *et al.* 1976; Van Wyk & Smith 2001). The mean temperature of the warmest month is less than 22°C, whilst winter temperatures drop to well below freezing with snow and frost commonplace. Soils of the DAC above 3 000 m therefore have frigid or cryic temperature regimes with mean annual temperatures ranging from 0°C to 8°C (Schmitz & Rooyani 1987). The varied climate is partly responsible for the 11 vegetation types (five grassland types, five shrubland types, and one forest type) occurring in the DAC (see Mucina & Rutherford 2006).

The initial proposed field work component focuses on the South African portion of the DAC although future studies should include representative areas of Lesotho. It is important to note that the term 'DAC' is based on climatological and not floristic grounds (Van Wyk & Smith 2001). The delineated area from 1 800 m a.s.l. to the highest point at 3 482 m a.s.l., encompasses three topo-ecological zones, namely montane (± 1 300 m to ± 1 900 m), sub-alpine (> 1 900 m to ± 2 800 m) and alpine (> 2 800 m to 3 482 m) zones. The DAC is therefore predominantly sub-alpine and alpine in nature, with only the upper limit of the montane zone falling marginally within the DAC. Therefore many of the emerging invasive alien plants being discovered in the montane foothills of the DAC are here listed as 'future' emerging aliens as they currently do not technically occur within the DAC, but in all likelihood will do so in the future once they breach the DAC's climatic envelope.

All sites proposed for the first phase of fieldwork are dispersed along the length of the Free State, KwaZulu-Natal and Eastern Cape Drakensberg (Figure 1; Table 1). These sites are well representative of the eastern DAC, and take into account a range of disturbance regimes and environmental heterogeneity related to altitude (lowest limit to higher altitudes); rainfall (low- to high-altitude gradients as well as north-to-south latitudinal and aspects gradients); temperature (low- to high-altitude gradients as well as north-to-south latitudinal and aspects gradients; see Figure 1) and many of the 11 vegetation types occurring within the DAC.

This study only concerns itself with alien plants introduced from areas outside the borders of South Africa (even though certain species native to South Africa, yet

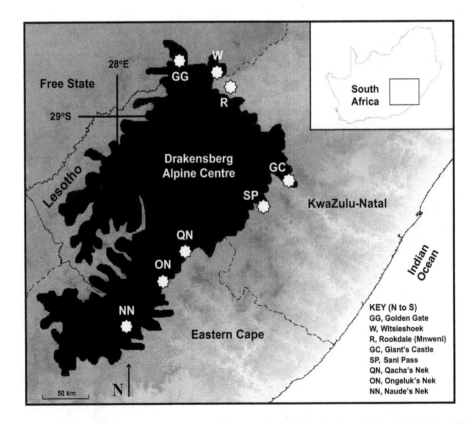

FIGURE 1.—Study sites in the eastern DAC proposed for sampling potential emerging invasive plants in disturbed habitats. This figure should also be interpreted in conjunction with Table 1.

TABLE 1.—Examples of physically disturbed sites which may facilitate the invasion of certain alien plant species in the DAC. Many such physically disturbed sites are also sites of nutrient enrichment (e.g. cattle kraals, heavily grazed or overstocked sites, and rural settlements without on-site or waterborne sanitation).

Sites of disturbance		Explanation	Example(s)	Proposed study site
General	Specific			
1. Paths.	1a. Foot paths.	Paths to homesteads or trading centres.	Qwa Qwa section of Golden Gate HNP.	Golden Gate HNP.
	1b. Tourist trails.	Paths for hiking (recreational).	Main Caves walk, Giant's Castle.	Giant's Castle.
2. Road and rail networks.	2a. Mountain passes.	Trans border access to trading centres.	Sani Pass, Ramatseliso's Gate Pass.	Sani Pass.
	2b. Local service roads.	Roads used by tourists or locals.	Winterton to Cathedral Peak, Barkly Pass, Naudes' Nek Pass, Basutho Gate Pass.	Basuto Gate Pass (Witsieshoek), Naude's Nek.
3. Settlements (habitation).	3a. Rural centres (& cattle kraals).	Disturbance, soil compaction, nutrient enrichment.	Qacha's Nek, Phuthaditjhaba, Rookdale.	Qacha's Nek, Rookdale (Mnweni lowlands, near Bergville).
	3b. Urban centres.			
4. Plantations.	Gum, wattle, pine & poplar.	Immediate sources of alien plants; and potential sites for others.	Boston, Impendhle, Maclear, and Underberg areas.	Sani Pass area.
5. Lands.	5a. Arable lands.	Easily invaded by agrestal & ruderal weeds.	Naude's Nek, Qwa Qwa.	Naude's Nek.
	5b. Pastoral lands.			
	5c. Wastelands.			
	5d. Transformed.			
6. Poorly managed rangelands.	6a. Sub-optimal grazing.	Often leading to erosion dongas.	Eastern Cape Drakensberg, Mnweni, Sani Top.	Lower Mnweni area.
	6b. Sub-optimal burning.			
7. Harvested sites.	Over-collected (exploited).	Harvesting of plants for fuel, medicines, gardens & thatching.	Ongeluk's Nek, Qacha's Nek.	Ongeluk's Nek, Qacha's Nek.
8. Naturally disturbed sites.	8a. Landslides.	Slope instability & disturbance.	Sani Pass.	Sani Pass.
	8b. Rock falls.			
	8c. Terracettes.			
	8d. River beds (flooding).			

naturally absent from the DAC, may be viewed as 'alien' if introduced into the DAC). A classic example is the 'Ermelo' ecotype of *Eragrostis curvula* (Poaceae) from the Highveld. Furthermore, certain native species may even become 'weedy' within their native range. Again such species were excluded [e.g. *Artemisia afra*, *Chrysocoma ciliata*, and *Stoebe vulgaris* (all of the Asteraceae family)]. The study also focuses predominantly on terrestrial invasive plants, the only exceptions being the aquatic herbs *Glyceria maxima* (Poaceae) and *Nasturtium officinale* (Brassicaceae). Plant nomenclature follows Germishuizen & Meyer (2003).

The floristic and threats analysis was based only on the 'current' emerging invasive alien species identified in this study.

Two-tiered 'top-down' and 'bottom-up' approach: introduction

Many alien plant programmes lack objective protocols for prioritising invasive species and areas based on likely future dimensions of spread (Rouget *et al.* 2004) and reliable methods of predicting invasion potential are hard to come by (Nel *et al.* 2004). Criteria using the impact scores of Parker *et al.* (1999) are problematic because they are essentially qualitative, lacking information relating to abundance and rates of spread (Le Maitre *et al.*

2004). Emerging invaders are also not necessarily those most obvious. For example, species that have the greatest available habitat and potential to increase in distribution are sometimes not identified by experts as important invaders (Robertson *et al.* 2004). Many are initially innocuous and restricted within their introduced range (Simberloff 2001). This may be because the dynamics of range expansion and population growth of an invasive alien plant typically include a time lag between its arrival in a new habitat and the start of widespread invasion (Simberloff 2001). Richardson (2001) cited many examples of invaders not showing invasive tendencies for as long as ± 50 to 150 years after introduction.

'Top-down' approach

To avoid 'reinventing the wheel', this study drew in part from the broad-based desktop review of South Africa's emerging invaders from natural and semi-natural habitats (Mgidi *et al.* 2004, 2007; Nel *et al.* 2004). After a screening process, confidence was placed in a final subset of 28 invaders (out of a possible total of 454) because they had been scrutinised a number of times by a number of authors. During this process, Nel *et al.* (2004) applied the expert scoring of four criteria strongly associated with factors that predict the potential invasiveness of plant species ('impact', 'weediness', 'bio-control status' and 'weedy relatives') to 454 emerging alien invaders listed in the Southern African Plant Invaders Atlas (SAPIA) database (Henderson 1998, 2007). In so doing the 454 emerging alien invaders were reduced to 115, and by further filtering were reduced to 84 according to estimates of their potential habitat ('habitat that can potentially be invaded') and current propagule pool size (Nel *et al.* 2004). Further filtering by Mgidi *et al.* (2004, 2007) then reduced the number to 28 during an exercise identifying the areas in South Africa with the greatest likelihood of being invaded. The 28 species were also used by Le Maitre *et al.* (2004) to assess their potential impacts on the biodiversity, water resources and productivity of natural rangelands (bushveld, grasslands, and shrublands) in South Africa.

The list of 28 species was then scrutinised further for the DAC context only. The final sub-set amounted to 23 taxa based on expert opinion by the author [personal field observations, published literature, and specimens lodged in the Natal University Herbarium (NU) of the University of KwaZulu-Natal], and distribution records from Henderson (2001) and the SAPIA database (Henderson 2007). Many of the 23 taxa recognised as 'current' emerging alien species in the DAC are regarded as major invaders in other parts of South Africa and therefore were not part of Mgidi *et al.* (2004) and Nel *et al.* (2004), as these studies focussed solely on emerging invasive alien plants.

Although not the primary focus of this study, a further 27 taxa have been tentatively listed as 'future' emerging invasive alien plants in the DAC. These species were selected on the premise that the next invaders to occupy the DAC in any meaningful way are most likely those of a temperate affinity and are currently located within 70 kms of the DAC's lower altitudinal boundary and therefore stand the greatest chance of breaching the DAC's climatic envelope, or are so poorly known from few localities that their 'emerging' status in South Africa is yet to be investigated in any detail. These 27 'future' emerging taxa should form the basis of future studies lest they become forgotten and their future potential ecological threat overlooked.

'Bottom-up' approach

Land cover monitoring studies have shown that large tracts of South Africa's natural ecosystems are already transformed (Fairbanks *et al.* 2000) and the extent and rate of land transformation will probably increase with time (Macdonald 1989; Tainton *et al.* 1989; D. Jewitt pers. comm.). This trend is in line with other regions of the world (Dale *et al.* 1994; Sala *et al.* 2000). Known for its high plant species richness and high levels of plant endemism, the DAC (~ Eastern Mountain hot-spot) is also characterised by high levels of man-induced habitat transformation and is therefore recognised as one of southern Africa's eight biodiversity 'hot-spots' (Cowling & Hilton-Taylor 1994, 1997).

Given the disturbance factor associated with the study area, the approach takes both disturbed and undisturbed habitats into account because invasive alien plants are able to dominate all stages of succession; early (~ suppression) and late (~ tolerance) successional strategies are contingent upon the specific competitive strategy

employed and can shift in invaded ecosystems over time (see MacDougall & Turkington 2004). Furthermore, disturbed habitats such as mountain pass roads can extend the distribution of alien plants beyond reasonable altitudinal expectations (Kalwij *et al.* 2008). A number of sites have been proposed for the fieldwork component (Figure 1; Table 1). These sites may help to identify additional emerging invasive species undetected in the 'top-down' approach, the focus of which is natural and semi-natural environments.

More specifically, the 'bottom-up' approach should therefore, based on the premise that disturbance is often a critical prerequisite for the invasion of certain alien plants, (i) identify major disturbance nodes (sites) as well as major disturbance types in the DAC (Table 1); (ii) ensure that the suite of disturbance nodes takes all major types of disturbance, environmental heterogeneity, and land tenure of the DAC into account; (iii) document all (potentially invasive) alien plant species at each designated site; and (iv) compare the field list with the SAPIA database, keeping in mind that, if not already data-based, the field species could be an unrecognised emerging invader. Due to the marked environmental gradients that traverse the length of the DAC from north-to-south (and east-to-west), certain alien invasive plants may not be present across all sites (e.g. those of the drier, colder Eastern Cape Drakensberg vs. those of the warmer, wetter northern KwaZulu-Natal Drakensberg).

RESULTS

This study has identified 23 'current' emerging invasive alien plant taxa, represented by 15 families, which pose the most immediate threat to the DAC (Table 2). The families Fabaceae and Rosaceae contribute most of the emerging invasive alien species (four taxa each), followed by Boraginaceae and Onagraceae (two taxa each). All other representative families contribute only a single taxon each (Table 2). In terms of functional type (growth form), most taxa are shrubs (9), followed by herbs (8), tall trees (5), and a single climber (Table 2). Of the 23 'current' emerging invasive alien plants assessed, some 78% are of a northern temperate affinity, and 22% of a tropical affinity (Table 3). Interestingly, none of the assessed taxa were of a southern temperate affinity (*sensu* Henderson 2006). All tall trees and almost all shrubs are of a northern temperate origin, whereas the herbs are equally representative of northern temperate and tropical origins.

This study also highlights a further 27 species as possible 'future' emerging alien invasive plants, as the potential for more recently detected species to invade into the DAC should not be underestimated (Table 4). For example, *Rubus phoenicolasius* (wineberry), locally naturalised in the KwaZulu-Natal Midlands (Stirton 1981) may pose a severe problem in the future given the invasiveness of other *Rubus* species such as *R. cuneifolius* in mesic high-altitude grasslands (O'Connor 2005). Although *R. phoenicolasius* currently does not occur in the DAC, it is regarded as a 'future' emerging invasive alien plant given its temperate affinity and proximity to the DAC (the closest population being ± 70 km's away).

TABLE 2.—The 23 'current' emerging alien plant invaders of the DAC. Taxa are arranged alphabetically.

Taxon	Family	Common name	Functional type (~ growth form)
Argemone ochroleuca subsp. *ochroleuca*	Papaveraceae	White-flowered Mexican poppy.	Herb (forb).
Cirsium vulgare	Asteraceae	Bull/spear thistle.	Herb (forb).
Cotoneaster pannosus	Rosaceae	Silver-leaf cotoneaster.	Shrub.
Cuscuta campestris	Convolvulaceae	Common/field dodder.	Climber (parasitic vine).
Cytisus scoparius	Fabaceae	Scotch broom.	Shrub / short tree.
Echium plantagineum	Boraginaceae	Patterson's curse / purple viper's-bugloss.	Herb (forb).
Echium vulgare	Boraginaceae	Blue echium / viper's-bugloss.	Herb (forb).
Gleditsia triacanthos	Fabaceae	Honey locust.	Tall tree.
Hypericum pseudohenryi	Hypericaceae	St. John's wort.	Shrub.
Juniperus virginiana	Cupressaceae	Eastern red cedar.	Tall tree.
Ligustrum lucidum	Oleaceae	Chinese wax-leaved privet / tree privet.	Tree.
Nasturtium officinale	Brassicaceae	Watercress.	Herb (forb), aquatic.
Oenothera rosea	Onagraceae	Pink evening primrose.	Herb (forb).
Oenothera tetraptera	Onagraceae	White evening primrose.	Herb (forb).
Opuntia ficus-indica	Cactaceae	Sweet prickly-pear.	Shrub / short tree.
Pyracantha angustifolia	Rosaceae	Narrow-leaved / yellow firethorn.	Shrub.
Quercus robur	Fagaceae	Common/English oak.	Tall tree.
Richardia brasiliensis	Rubiaceae	Brazil pusley / tropical Mexican clover.	Herb (forb).
Robinia pseudoacacia	Fabaceae	Black locust.	Tall tree.
Rosa multiflora	Rosaceae	Multi-flora rose.	Shrub.
Rosa rubiginosa	Rosaceae	Eglantine/sweet briar.	Shrub.
Salix fragilis var. *fragilis*	Salicaceae	Crack/brittle willow.	Tall tree.
Ulex europaeus	Fabaceae	European gorse / gorse.	Shrub.

DISCUSSION

The emerging invasive alien plants of the DAC

Conservatively speaking, at least 170 alien angiosperm species (or ± 6% of the DAC's angiosperm flora) have invaded the DAC (Carbutt & Edwards 2004). Poaceae and Asteraceae contribute the most established invaders (Carbutt & Edwards 2004), a reflection of the general success of these two families in the DAC, and together with the legume family, Fabaceae, account for the majority of plant invaders worldwide (Richardson 2001). This study adds a further 23 'current' emerging invasive alien species and recognises a further 27 taxa as potential 'future' invaders of the DAC. The families Fabaceae and Rosaceae that together contributed most of the emerging invasive taxa, also feature as prominent contributors of temperate-affiliated alien invasive species in southern Africa (see Henderson 2006). Of the families mentioned above, only the Fabaceae features prominently in the native angiosperm flora of the DAC (Carbutt & Edwards 2004). A possible reason for the high proportion of alien invasive species of a northern temperate affinity is the long history of cultivating garden ornamentals from Europe and to a lesser extent Asia (i.e. many gardens in the foothills of the DAC have provided the perfect temperate environment for cultivating ornamental alien plants). More broadly speaking, the pattern for southern Africa is an equal representation of taxa from both temperate and tropical origins (Henderson 2006). An interesting trend to monitor is the potential increase in 'future' emerging invasive alien plants of tropical affinity (as a 'barometer' of change), given the

global change predictions regarding warming and displacement of native taxa.

The northern and eastern boundaries of the DAC are estimated to have on average more than 10 alien invaders per quarter-degree grid square, corresponding to areas with the highest levels of transformation, rainfall, and population density (Nel *et al.* 2004). Rouget *et al.* (2004) have predicted that most species of emerging invaders in the DAC will be confined to the sheltered confines of lower altitudes (montane foothills) below the escarpment, particularly in the northern KwaZulu-Natal Drakensberg (warmer and wetter?), the Eastern Cape and Free State Drakensberg (highly transformed through agriculture?), and the warmer, sheltered valleys of the Lesotho Maloti Mountains. Their absence from the alpine summit (Lesotho plateau), particularly at higher altitudes, is attributed to frequent frosts and low mean temperatures of the coldest month (Rouget *et al.* 2004). However, although ecosystem invasibility generally decreases with altitude as fewer alien plants are able to invade high altitude habitats due to the harsh climatic conditions (Keeley *et al.* 2003; Arévalo *et al.* 2005), a study by Kalwij *et al.* (2008) on the distribution of alien plants along Sani Pass has shown that mountain roads (particularly verges) are able to increase the altitudinal limit at which an alien plant is invasive due to the facilitation of a greater propagule pressure, a composite measure encompassing *inter alia* the effects of anthropogenically-induced soil disturbance, increased water runoff, and vehicular traffic (by introducing and spreading propagules). The outcome is the alien species' ability to overcome invasion limiting barriers and hence spread

TABLE 3.—Threats posed by the 23 'current' emerging invasive alien plants of the DAC. Distribution data were derived from Henderson (2001) and supplemented by herbarium records and personal field observations. Additional sources of information, where relevant, are cited under each taxon.

Taxon (and additional references)	Native range	Biogeographical affinity (per Henderson 2006)	Known range in DAC	Potential range in DAC	Threat to DAC (consequences of invasion, including target habitat / predicted niche equivalent and invasive/competitive attributes)
Argemone ochroleuca subsp. *ochroleuca*	Texas (USA) southwards to Mexico.	Tropical.	Invading into DAC from the west and southwest (drier Lesotho lowlands and Eastern Cape).	Probably the western (drier) half of DAC most vulnerable, but may spread eastwards towards the mesic escarpment during episodes of aridification. Disturbed watercourses most susceptible to invasions.	Primarily a weed of waste places (ruderal) and cultivated lands (agrestal), therefore only able to invade and persist in severely and recently disturbed areas. Probably unable to invade into pristine natural vegetation (except perhaps along watercourses subject to episodes of flooding, where it is highly competitive). Produces a large number of seeds that are able to remain dormant during unfavourable conditions for many years.
Cirsium vulgare Hilliard (1977); Streeter (1998).	Europe (including the Mediterranean) and western Asia.	Northern temperate.	Most common along the montane foothills of the KwaZulu-Natal Drakensberg; appears to be scarce in Lesotho (?). NU herbarium localities: Bulwer (1974), Giant's Castle (1966), Katberg (1951), Underberg (1968).	Moist montane sites that have been disturbed and degraded seem most susceptible. Likely to spread across DAC in the absence of episodes of severe aridification.	Primarily a weed of disturbed areas, very competitive in moist, degraded grassland. Probably unable to invade into pristine natural vegetation. Strong competitor in cool, moist, high-altitude conditions. Produces many seeds that remain viable for long periods of time. Seeds (pappus) well adapted for wind dispersal. Its spread is favoured by trampling and soil disturbance. Degrades condition of rangelands.
Cotoneaster pannosus Starr *et al.* (2003).	Sichuan and Yunnan provinces, SW China.	Northern temperate.	Montane region of the central KwaZulu-Natal Drakensberg, eastern Free State, and western Lesotho. A few scattered localities in the Eastern Cape Drakensberg.	Most of DAC is at risk.	Known to invade grassland, forest margins, shrublands, kloofs, riverbanks, rocky outcrops, and roadsides. Has large, aggressive root systems and is known to shade out and smother sun-loving native plants. Dispersed by fruit-eating birds and is able to invade into both disturbed and pristine plant communities. Quick growing; known to quickly dominate a scrub or grassland area. Highly adaptable, can grow in moist or dry soils, and even in thin rocky soils underlying native grasslands. Forest edge and shrubland habitats are most at risk.
Cuscuta campestris Streeter (1998).	North America (Canada, USA) southwards to Mexico and possibly Bahamas, Cuba, and Jamaica.	Northern temperate.	Central KwaZulu-Natal Drakensberg and Lesotho Highlands.	Great potential to spread in DAC, particularly areas with a history of cropping. Appears to be encroaching from the west. Mesic foothills in the Eastern Cape and KwaZulu-Natal Drakensberg seem at greatest risk. Prefers damp soil in full sun.	Smothering parasite on a wide range of host plants; forms dense patches up to 6 m across. Climbs over vegetation. A prolific producer of seeds that are able to remain dormant during unfavourable conditions for long periods; seeds spread by animals and water. Can also spread by stem fragments. Known to invade a wide range of habitats. Disturbed croplands and moist sites are most at risk.
Cytisus scoparius Streeter (1998).	Central and southern Europe and the British Isles (except for Orkney and Shetland Islands).	Northern temperate.	Montane foothills of the KwaZulu-Natal Drakensberg. NU herbarium localities: Highmoor (1959, 1961, 1976), road to Underberg (1961), Van Reenen's Pass (1971).	Has the potential to spread into cool, high-lying areas such as the Eastern Cape Drakensberg.	Invades scrubland, native grassland, forest margins, riverbeds and other waterways. Spread by water and animals and grows rapidly and aggressively into very dense stands. Known to convert open systems into a dense shrubland. Shows signs of drought resistance. Can reproduce vegetatively or by seed. Resprouts after cutting. Can tolerate low soil temperatures. Its invasive success is attributed to its wide tolerance of soil conditions, its ability to fix nitrogen, and its abundant production of hard-coated and long-lasting (up to 80 years) viable seeds.

TABLE 3.—Threats posed by the 23 'current' emerging invasive alien plants of the DAC. Distribution data were derived from Henderson (2001) and supplemented by herbarium records and personal field observations. Additional sources of information, where relevant, are cited under each taxon (continue).

Taxon	Native distribution	Temperate	Distribution in DAC	Threat	Notes
Echium plantagineum Retief & Van Wyk (1998).	Western Europe, including south-western Britain (very rare) and the Mediterranean region.	Northern temperate.	Montane foothills of the Eastern Cape, Free State and KwaZulu-Natal Drakensberg, and Lesotho.	Has the potential to spread throughout the mesic portions of the DAC.	Invades undergrazed or overgrazed rangelands (degrades grasslands); also recorded from rocky slopes in grassland. Persists through a deep taproot and produces ± 5 000 seeds per plant; seeds are dispersed by water and animals and can remain dormant for up to seven years. Can tolerate a wide range of temperatures and can survive periods without rain. Able to germinate at any time of the year, provided conditions are optimal.
Echium vulgare Retief & Van Wyk (1998).	Europe. Distributed throughout the British Isles.	Northern temperate.	Eastern Free State, Lesotho, and Eastern Cape and KwaZulu-Natal Drakensberg.	Has the potential to spread throughout the mesic portions of the DAC.	Primarily a weed of degraded areas, therefore only able to invade and persist in severely and recently disturbed areas. Probably unable to invade into pristine natural vegetation. High threat potential as it has a tendency to invade degraded montane grassland, thereby placing large areas of the DAC at risk.
Gleditsia triacanthos Gilman & Watson (1993a).	North America (mid-western, eastern and east-central USA).	Northern temperate.	Northern and southern KwaZulu-Natal Drakensberg, Eastern Cape Drakensberg, eastern Free State and border of Lesotho on the Free State side.	Highly adaptable; has the potential to spread to almost all parts of the DAC with the exception of forested areas. Favours small stream valleys.	Known to invade grassland and riverbanks. Very adaptable to many adverse conditions (especially heat, drought, poor soils, soils of various pH, soil compaction, and flooding). Seed is viable for many years due to a thick, impermeable seed coat and can germinate under a wide range of conditions. It is a fast-growing member of early and mid-successional stands. Hardy and tolerates both xeric and hydric conditions. Dry *Protea* savanna and *Widdringtonia* fynbos on dry, sunny, low nitrogen sites, and north-facing grasslands burnt too infrequently are most at risk.
Hypericum pseudohenryi.	China	Northern temperate.	Possibly only in the montane foothills of the KwaZulu-Natal Drakensberg. Problematic in the Giant's Castle Game Reserve and Monk's Cowl Forest Station.	Possibly also the wetter parts of the Eastern Cape Drakensberg and Lesotho catchments.	High threat due to its affinity for cooler climates and mid- to high elevations. Also its long-flowering potential and ability to self-fertilise are further threats. Poses the greatest threat to native mid- to high-elevation shrublands, stream scrub and forest margins where plants can readily germinate and form large, dense stands. Capable of altering and displacing native plant communities in areas where they invade by forming monotypic thickets. Will probably degrade stream catchment quality and diminish runoff. May outcompete plants from stream catchments and shade out hillside plants accustomed to full sun. Hardy to about -5°C but plants can resprout from the base if they are damaged by cold.
Juniperus virginiana Gilman & Watson (1993b).	Widespread in north-eastern North America, from south-eastern Canada to the Gulf of Mexico, east of the Great Plains.	Northern temperate.	Eastern Free State, Western Lesotho, and the Eastern Cape Drakensberg.	Most of the DAC is at risk, particularly degraded, open and eroded areas, and those areas being burnt too infrequently.	High threat in overgrazed, fire-excluded rangelands, particularly those of western Lesotho, eastern Free State and the drier parts of the Eastern Cape Drakensberg. However, its ability to tolerate more mesic conditions also places equivalent areas in the KwaZulu-Natal Drakensberg at risk. Known to invade grassland, riverbanks, and rocky outcrops. Frequent pioneer and invader of old fields and other open, often eroded areas. Most competitive in dry, exposed sites, and in disturbed areas. Does not establish well in more competitive, denser vegetation cover that occurs later in succession. Known to invade into rangelands in the absence of fire. Can tolerate extremely xeric (to mesic) conditions. Tolerates a wide range of soil types and soil depths. Resistant to air pollution and is frost hardy.

TABLE 3.—Threats posed by the 23 'current' emerging invasive alien plants of the DAC. Distribution data were derived from Henderson (2001) and supplemented by herbarium records and personal field observations. Additional sources of information, where relevant, are cited under each taxon (continue).

Taxon (and additional references)	Native range	Biogeographical affinity (per Henderson 2006)	Known range in DAC	Potential range in DAC	Threat to DAC (consequences of invasion, including target habitat / predicted niche equivalent and invasive/competitive attributes)
Ligustrum lucidum	Eastern Asia (China and Korea).	Northern temperate.	Western Lesotho and the eastern Free State.	Watercourses and forests across the mesic parts of the DAC are most at risk.	High threat. Fast growing evergreen tree that has the potential to replace mid-canopy trees in forests and can completely dominate an area of forest. Seeds are distributed by frugiferous birds. Prolific producer of seed. Also spreads by means of root suckers. Can tolerate dry and wet conditions.
Nasturtium officinale, Streeter (1998).	Europe (abundant throughout Britain but rare in central and northern Scotland).	Northern temperate.	Confined mostly to the Lesotho highlands.	May spread more widely throughout Lesotho (provided streams remain pristine) and into the fast-flowing clear streams of the Eastern Cape and KwaZulu-Natal Drakensberg (provided these areas don't experience heavy, regular frosts during winter).	Known to invade rivers, riverbanks, and wetlands. Great threat to the pristine streams of the Lesotho Highlands (because it favours cold, clear flowing water). Propagates by rooting stem fragments and seeds.
Oenothera rosea, Goldblatt & Raven (1997).	The New World (Central and South America).	Tropical.	Most common along the montane foothills of the KwaZulu-Natal and Eastern Cape Drakensberg.	Most of the DAC is at risk, particularly degraded and disturbed areas.	Invades riverbanks, moist sites, roadsides and waste places. Has a great ability to utilise suitable light conditions for germination. Autogamous (self-pollinates) and can tolerate aridity.
Oenothera tetraptera, Goldblatt & Raven (1997).	The New World, from Texas in the USA to northern South America.	Tropical.	Eastern Free State, northern KwaZulu-Natal Drakensberg and northern Lesotho.	Most of the DAC is at risk, particularly degraded and disturbed areas.	Invades riverbanks, moist sites, roadsides and waste places. Has a great ability to utilise suitable light conditions for germination. Autogamous (self-pollinates).
Opuntia ficus-indica	Central Mexico.	Tropical.	Not well established in the DAC (unlike most of South Africa), but is invading into the DAC mostly from the north and west (drier Free State and Lesotho lowlands).	Encroaching into Lesotho from the west; if unchecked will continue to invade into the drier reaches of central Lesotho. Should not invade wetlands and hygrophilous grasslands.	Fast-growing, may spread quickly under conditions of aridification and poor range management. It is an aggressive invader of natural vegetation, especially dry and rocky places. Can regenerate from seed, cladode fragments (any broken fragment is capable of regeneration) and underground tubers. A recognised transformer of natural or semi-natural ecosystems, thereby altering ecosystem structure, integrity and functioning. Drought tolerance is enhanced by high water-use efficiency and a large water-holding capacity.
Pyracantha angustifolia	South-western China.	Northern temperate.	Eastern Free State, KwaZulu-Natal Drakensberg, and Eastern Cape Drakensberg.	May also spread into eastern Lesotho.	Very high threat to DAC. Favours high-altitude grassland and cool climates with moderate water availability. Forms dense thickets that exclude other plants and makes access difficult due to its thorns. Known to invade high-altitude grasslands, erosion channels, rocky ridges, and riparian areas. Known to smother and displace native species, particularly in grasslands. Prolific producer of seeds but can also resprout. Fast growing. Favours full sun or part shade (not full shade). Can tolerate a wide range of soil conditions.

TABLE 3.—Threats posed by the 23 'current' emerging invasive alien plants of the DAC. Distribution data were derived from Henderson (2001) and supplemented by herbarium records and personal field observations. Additional sources of information, where relevant, are cited under each taxon (continue).

Taxon	Origin	Biome	Distribution	Potential	Notes on threat
Quercus robur Gilman & Watson (1994a).	Great Britain, Europe (including the Mediterranean) and western Asia (Asia minor).	Northern temperate.	Eastern Cape Drakensberg, southern KwaZulu-Natal Drakensberg, and eastern Free State / western Lesotho.	May also spread into the northern KwaZulu-Natal Drakensberg.	Known to invade forest margins, woodland, roadsides, and riverbanks in grassland and fynbos. Early invader of woodland. Appears to be drought-tolerant. Long-lived (up to 1 000 years).
Richardia brasiliensis Hall et al. (2005).	Central and South America (Colombia, Ecuador, Brazil, Peru and Bolivia).	Tropical.	Montane foothills of the KwaZulu-Natal Drakensberg (up to ± 2 000 m).	Has the potential to spread throughout the DAC.	High threat to DAC. Tolerates a range of environmental conditions. Regarded as subdominant under communal grazing in Highland Sourveld grasslands in the southern KwaZulu-Natal Drakensberg (O'Connor 2005). Highly invasive; spreads rapidly. Nature of invasiveness unknown. Blooms in any month that lacks frost. Is drought tolerant; able to retain moisture in fleshy stems ('semi-succulent'). Very hardy. Prolific seeder. Produces a deep taproot.
Robinia pseudoacacia Gilman & Watson (1994b).	North America (central and south-eastern USA).	Northern temperate.	Eastern Free State, southern KwaZulu-Natal Drakensberg and Eastern Cape Drakensberg.	Has the potential to invade into the drier parts of Lesotho, particularly disturbed areas characterised by inappropriate management practices.	High threat to DAC. Is an early successional species, able to establish and grow quickly. Has a highly invasive root system (vigorous root suckering). Tolerates poor soils and other adverse conditions. Known to invade riverbanks, dongas, roadsides, agricultural areas, disturbed areas, upland natural forest edges, degraded woodland, as well as rangelands and grasslands. Once established, it expands readily into areas where, through shading, it outcompetes sun-loving plants. Fast-growing. Requires little water once established. Drought tolerant.
Rosa multiflora	Eastern Asia (China, Japan and Korea).	Northern temperate.	Upper montane region of the KwaZulu-Natal Drakensberg.	Mesic escarpment most at risk.	Forms impenetrable thickets in grassland, scrub, and forest edge. It restricts the movement of wildlife and displaces native vegetation.
Rosa rubiginosa Streeter (1998).	Europe (including the Mediterranean and Britain).	Northern temperate.	Southern KwaZulu-Natal Drakensberg, eastern Free State, Eastern Cape Drakensberg and the central and western portion of Lesotho. NU herbarium locality: Pitlochrie, Barkly East district (1981).	Most of the DAC is suitable habitat.	High threat to DAC. It is an early successional species capable of rapidly invading open areas. Its rapid growth rate, rapid seed production, efficient seed dispersal aided by animals, and its potential to form a dense shrubland (and hence alter vegetation physiognomy) all contribute to its high threat status. Known to invade high-altitude grassland (especially moist valleys), watercourses, rocky outcrops, roadsides, and overgrazed land around human habitation (including wasteland). Capable of invading into dryland environments. In its invasive range, is known to facilitate the re-establishment of native woody species in disturbed forests by reducing grazing herbivory on native seedlings growing beneath the thorny shrubs. Suckering occurs freely from the crown; bushes therefore often exceed 1 m in diameter. Prevalent in high and low rainfall areas. Spreads through seed dispersal; birds eat the red fruits (hips). Seeds are also spread by run-off from waterways. Plants may also regenerate from root and crown fragments left after disturbance.

TABLE 3.—Threats posed by the 23 'current' emerging invasive alien plants of the DAC. Distribution data were derived from Henderson (2001) and supplemented by herbarium records and personal field observations. Additional sources of information, where relevant, are cited under each taxon (continue).

Taxon (and additional references)	Native range	Biogeographical affinity (per Henderson 2006)	Known range in DAC	Potential range in DAC	Threat to DAC (consequences of invasion, including target habitat / predicted niche equivalent and invasive/competitive attributes)
Salix fragilis var. *fragilis* Cremer (1999).	Western Europe and western Asia.	Northern temperate.	Widespread in the montane foothills of the Eastern Cape Drakensberg, northern and southern KwaZulu-Natal Drakensberg and eastern Free State. Also in Lesotho. NU herbarium locality: Garden Castle / Drakensberg Gardens (1980).	Already widespread in the DAC; now also spreading throughout the watercourses of Lesotho.	High threat to the riparian vegetation in water catchments of the DAC. Dense infestations, if unmitigated, will result in reduced stream flow (and therefore reduced water availability), as well as altering biomass and biodiversity along watercourses. May also threaten wetland environments. Known to invade watercourses, especially riverbanks and mid-stream gravel bars. Shading effects may also be detrimental. Will outcompete and therefore displace terrestrial, semi-aquatic, and aquatic native vegetation. Known to spread its roots into the bed of a watercourse, slowing the flow of water and reducing aeration. It forms thickets which divert water outside the main watercourse or channel, causing flooding and erosion where the stream banks are vulnerable. Its leaves create a flush of organic matter when they drop in autumn, reducing water quality and available oxygen, and directly threatening aquatic plants and animals. This, together with the vast volume of water it uses, impacts negatively on stream health. Characterised by brittle branches which are easily broken (with a 'crack'), providing material for vegetative spread (e.g. can spread prolifically from broken twigs taking root downstream). Can also reproduce prolifically from viable wind-borne seed.
Ulex europaeus Streeter (1998).	Central and western Europe, including the British Isles.	Northern temperate.	Central and southern KwaZulu-Natal Drakensberg and its foothills. NU herbarium localities: Giant's Castle (1965), Highmoor (1957, 1961), and Underberg (1974).	High potential for range expansion, throughout DAC, especially in the DAC's mesic habitats.	High threat to DAC. Its dense growth form results in impenetrable monotypic thickets that rapidly smother and outcompete native vegetation. Invades infertile and disturbed areas, but can also invade undisturbed open areas. Its ability to reach reproductive maturity in two years or less, its production of seeds annually that can remain viable for many years, its quickly spreading vegetative structures that resprout readily following cutting, grazing, or burning, all contribute to its invasiveness. While gorse prefers a cool, moist habitat, this plant has characteristics that allow it to occupy areas of drought or sites that are sunny, exposed, and dry. The characteristics include: spiny leaves covered with thick cuticles; grooved hairy stems; large roots on young plants that allow high water uptake and help anchor plants in exposed, windy sites. Its ability to: (i) fix nitrogen; (ii) acidify and (at least temporarily) impoverish soils by taking up bases; (iii) survive on a variety of soil types; (iv) produce copious amounts of heat-tolerant seeds with long-term viability; and (v) regenerate rapidly from seeds and stumps after disturbances such as brush clearing or fires are all qualities that make it even more problematic. Known to invade grassland, shrubland, vleis, and valleys, mostly in moist mountainous regions. Colonises nitrogen-poor soils, which allows it to outcompete native plants. Can spread quickly by seed or by vegetative growth from stumps after mechanical injury caused by brush clearing or fire.

to elevations either higher than expected or previously recorded (Kalwij *et al.* 2008).

Which approach is least fallible?

Previous studies have shown that the prioritisation of invasive species using a ranking system of criteria is subjective and fallible (see Nel *et al.* 2004; Rouget *et al.* 2004) because there are no objective criteria determining when a score is sufficient to qualify a species for high-priority management action (Nel *et al.* 2004). Comparisons are also difficult between species that occupy a wide range of different habitats with varying levels of disturbance and impact. Robertson *et al.* (2003) reported difficulty in ranking priority species requiring management action at a local scale, compared to more widespread species (perhaps also less abundant across their range) requiring intervention over large areas. Rankings given to species should therefore be viewed as approximate, rather than absolute (Thorp & Lynch 2000). For many of these reasons, the two-tiered methodology using both well-scrutinised scoring systems for invaders of natural and semi-natural ecosystems and a proposed fieldwork component to determine the emerging invaders of disturbed habitats has been proposed.

An alternative approach is a predictive one that makes use of Climate Envelope Models (CEMs). This approach, however, is also potentially fallible as predicting the spread of invasive plants is not a perfect science because predictions are often subject to numerous uncertainties (Schneider & Root 2001). A major flaw in using CEMs is that the role of climate in controlling distributions is not the same for all species, and other factors such as disturbance regime and biotic interactions may sometimes override climatic factors (Richardson & Bond 1991; Hulme 2003; Le Maitre *et al.* 2004; Rouget *et al.* 2004), particularly when recent introductions are not in equilibrium with their environment because their geographic ranges may still be expanding from 'refugia' into larger ranges (Rouget *et al.* 2004). Furthermore, CEMs assume that the current distribution of species provides a good indication of their potential range and the process of averaging climate suitability values assumes that the mean values represent the location where the species occurs. This assumption is likely to be flawed in areas of complex topography (Rouget *et al.* 2004). CEMs therefore appear to give the best correlations with invasive plant species distributions only at a national scale (see Rouget & Richardson 2003; Rouget *et al.* 2004). Even more disconcerting is the lack of congruency between species selected by expert ratings and those determined by CEMs (Nel *et al.* 2004).

Despite the drawbacks of each method, the need to attempt a list is of paramount importance and the value of the proactive approach in identifying emerging invasive alien plants cannot be overstated even if there are no generally accepted ways of quantifying when an area is 'invaded' or when a species is 'invasive' (Richardson 2001), and despite the blurry line dividing the emerg-

TABLE 4.—The 27 'future' emerging alien plant invaders of the DAC. Taxa are arranged alphabetically.

Taxon	Family	Common name	Functional type (~ growth form)
Acacia elata	Fabaceae	Peppertree wattle.	Tall tree.
Achillea millefolium	Asteraceae	Common yarrow/milfoil.	Herb (forb).
Anredera cordifolia	Basellaceae	Bridal wreath / Madeira vine.	Climber.
Arundo donax	Poaceae	Giant reed.	Grass/reed (graminoid).
Campuloclinium macrocephalum	Asteraceae	Pompom weed.	Herb (forb).
Coreopsis lanceolata	Asteraceae	Lance-leaved tickseed.	Herb (forb).
Cortaderia selloana	Poaceae	Pampas grass.	Tall grass (graminoid).
Eucalyptus camaldulensis	Myrtaceae	Red river-gum.	Tall tree.
Glyceria maxima	Poaceae	Reed sweet grass / reed manna grass.	Herb (graminoid), aquatic.
Lythrum hyssopifolia	Lythraceae	Hyssop loosestrife/grass-poly.	Herb (forb).
Nasella tenuissima	Poaceae	White tussock.	Grass/reed (graminoid).
Nasella trichotoma	Poaceae	Nasella tussock.	Grass/reed (graminoid).
Oenothera stricta	Onagraceae	Sweet sundrop.	Herb (forb).
Phytolacca octandra	Phytolaccaceae	Inkberry.	Shrub.
Pinus halepensis	Pinaceae	Aleppo pine.	Tree.
Pinus radiata	Pinaceae	Radiata pine.	Tree.
Pinus taeda	Pinaceae	Loblolly pine.	Tall tree.
Populus alba	Salicaceae	White poplar.	Tree.
Populus deltoides	Salicaceae	Match poplar / cottonwood.	Tall tree.
Populus nigra var. *italica*	Salicaceae	Lombardy poplar.	Tall tree.
Pyracantha crenulata	Rosaceae	Himalayan firethorn.	Shrub.
Richardia stellaris	Rubiaceae	Field madder.	Herb (forb).
Rosa canina	Rosaceae	Dog rose.	Shrub.
Rubus phoenicolasius	Rosaceae	Wineberry.	Shrublet/shrub.
Solanum pseudocapsicum	Solanaceae	Jerusalem cherry / winter cherry.	Shrublet/shrub.
Xanthium spinosum	Asteraceae	Spiny cocklebur.	Herb (forb) / shrublet.
Xanthium strumarium	Asteraceae	Large cocklebur.	Herb (forb) / shrublet.

ing and major invaders. Such a list is useful in (i) helping to select species for modelling their rates of spread; (ii) knowing what species to target and where to focus management action in the future; and (iii) facilitating better trouble-shooting methods for managing biological invasions (Nel *et al.* 2004). The benefits may even be accrued to the management (control) process itself, as South African researchers have shown that bio-control is most effective (Olckers & Hill 1999) and control measures most cost-effective (Hobbs & Humphries 1995; Olckers 2004) during the early stages of invasion.

Role of disturbance

Disturbance, be it naturally occurring or human-induced, is a fundamental driver of plant invasions (Richardson 2001; Simberloff 2001) because it promotes characteristic patterns of environmental heterogeneity and regulates ecosystem processes, population dynamics, species interactions, and species diversity by freeing up limiting resources (Davis & Moritz 2001). Its effect is thought to be so critical that some authors (e.g. Elton 1958) have maintained a view that undisturbed native communities are not susceptible to invasions by introduced species. Irrespective of other factors that facilitate or limit invasions, the susceptibility of communities to invasion by alien plants increases with increasing disturbance up to a threshold after which the disturbance then acts as a barrier—intermediate levels of disturbance are therefore most optimal for invasiveness (Richardson 2001; Woodward 2001). Invasibility resulting from disturbance is also attributed to reduced competition from resident plants through the reduction in standing ground cover (Richardson 2001; Woodward 2001).

The role of disturbance in facilitating the expansion of emerging invasive alien plants has therefore been recognised and incorporated into the proposed methodology of this study. Early successional invaders are confined mostly to post-disturbance environments and are effective at acquiring resources in an environment of high resource availability and relatively low competition (due to traits such as fast growth), and are therefore inherently better at suppressing other species in areas of disturbance (~ suppression-based competition; see MacDougall & Turkington 2004). Fortunately, the fast-growing invasives that dominate post-disturbance environments do not appear to be highly problematic in the long term because they compete poorly in late-successional assemblages. Rather, in the absence of disturbance (i.e. where resources are more limiting in natural or semi-natural environments), the alien flora is able to tolerate reduced resource levels under conditions of intense competition, thus allowing them to dominate in the latter stages of succession. This strategy is termed tolerance-based competition (MacDougall & Turkington 2004).

Opportunities for the spread of invasive plants under climate change

Climate change is predicted to be one of the greatest drivers of ecological change in the coming century (Lawler *et al.* 2009). The DAC is an excellent laboratory for the monitoring of climate change because mountainous regions are highly sensitive to environmental change (e.g. Hill 1996; Midgley *et al.* 2001), especially a change in temperature. For example, a major effect of warming is the tendency of species to track shifting climate and suitable habitat through dispersal and migration in order to remain within their optimal growth environment as present-day plant distributions are determined by their ecological compatibility with present-day climate. When climatic conditions change, plants with a specific set of adaptive characteristics may no longer be suited to the new conditions (Deacon *et al.* 1992; Stock *et al.* 1997). Consequently, species are predicted to move poleward in latitude and upward in elevation (Dunne & Harte 2001). A 3°C change in temperature is equivalent to a move of 250 km of latitude or 500 m of elevation. Alpine species will tend to migrate upslope when cooler, higher elevations begin to warm up (Grabherr *et al.* 1995; Dunne & Harte 2001; Körner 2001). Species already limited to mountaintops (i.e. already at their critical physiological threshold) will be at serious risk of local extinction due to the lack of potentially suitable habitat to migrate to (Dunne & Harte 2001) and because climates are changing more rapidly than species can adapt (Schneider & Root 2001). A further influence of global warming on alpine plant diversity is the lateral migration of species (~ 'niche filling' or 'horizontal reallocation'), with new niches being filled by species and other niches being abandoned (Gottfried *et al.* 1998; Körner 2001). Ultimately, species that are similarly affected will occupy similar habitats (Van Zinderen Bakker & Coetzee 1988; Hill 1996; Midgley *et al.* 2001).

Global climate change may not only result in the direct loss of local native species through climatic incompatibility; native plant communities will also become increasingly susceptible to invasions by alien plants. The impact of alien invasive plants, besides habitat degradation, is the extinction of native species through the effects of competition, parasitism, disease, and hybridisation (Baur & Schmidlin 2007). These invasive plants either (i) arrive pre-adapted because the 'new' local climatic conditions are similar to what they experience in their native ranges elsewhere in the world (Macdonald 1992; Dukes & Mooney 1999), or (ii) because of some form of change, their competitive ability increases in their invasive environment. Studies on plant-climate relationships therefore need to consider both the current selection pressures as well as future ones, as currently non-adaptive traits may pre-adapt taxa to future environmental conditions (Stock *et al.* 1997). A future selection pressure in the DAC is warmer temperature, which may be significant given that the climatic boundaries in the DAC are well defined and may determine the basic distribution limits for plants (Carbutt 2004). The warmer temperatures associated with global climate change may accelerate organic matter decomposition and nitrogen mineralisation, thereby creating a nitrogen environment unsuited to taxa that have evolved in such nitrogen-limited environments where soil inorganic nitrogen availability is heavily constrained by cooler temperatures (Carbutt 2004; Carbutt & Edwards 2008; Contosta *et al.* 2011). Plant communities thriving in the DAC's (currently temperature-mediated) nitrogen-limited soils may be extirpated by future episodes of significant warming because of their inability to cope with nitrogen concentrations far beyond their natural tolerance range. Such communities may therefore be replaced by common nitrophilous ruderal plants (many of which will be invasive alien plants) that are neither conservation-worthy nor native to the DAC.

Management action

Control of invasive alien species is a key operational management function and demands significant financial resources. In the uKhahlamba Drakensberg Park World Heritage Site (UDP WHS), which accounts for a significant area of the KwaZulu-Natal Drakensberg portion of the DAC, some R2 million is spent annually to clear invasive plants. A concern is that no operational funds are allocated to combat invasive alien plants in the UDP WHS; rather all funding is derived from State poverty relief programmes such as 'Landcare', 'Working for Water', 'Working on Fire', and 'Working for Wetlands', placing the Park at high risk should this funding be terminated for whatever reason (Ezemvelo KZN Wildlife 2005).

The 'art and science' of early detection is a pointless exercise unless it is followed up with control and mitigation by the relevant conservation authorities, and the failure to detect and eliminate emerging aliens will add to the financial burden of invasive alien plant control. Whilst it is acknowledged that areas of high biodiversity value will be under constant threat by invasive alien species and therefore the appropriate management action to mitigate this threat is an ongoing need (Richardson *et al.* 2005; Thuiller *et al.* 2008), the smart approach of early detection of new invasions is essential to assist management in the ongoing battle of alien plant threat mitigation. Failure to address emerging invasive alien plants in the short term may double the costs of alien plant control in the long term. The urgent need to act immediately and invest a relatively small amount to control emerging populations before they spread into all available habitats makes good business sense. The benefits of managing emerging alien plants should be calculated to include the stimulation of further job creation in rural communities, improved delivery of ecosystem goods and services, and the safeguarding of native biodiversity and ecological integrity.

CONCLUSION

South Africa has to its credit a long history of alien plant control and research (Macdonald *et al.* 1986; Van Wilgen *et al.* 2011). For example, the Working for Water (WfW) Programme of the Department of Water and Environmental Affairs (DWEA), initiated in 1995, has been widely lauded both locally (Van Wilgen *et al.* 1996; Hobbs 2004; Van Wilgen 2004) and internationally (Mark & Dickinson 2008) for its progressive and proactive approach in eradicating alien plants. Other significant allies in the war against invasive alien plants are the Weeds Research Programme of the Agricultural Research Council's (ARC) Plant Protection Research Institute (PPRI) (which includes the SAPIA database), and the WfW-funded Invasive Alien Plants Early Detection and Rapid Response (EDRR) Programme of the South African National Biodiversity Institute (SANBI).

The critical need for alien plant control and research in the DAC (relating to both major and emerging invaders) is crucial to maintaining the functional ecological integrity of the DAC. Like in many other alpine and subalpine environments, the DAC is a mountainous catchment area which supplies drinking and irrigation water, as well as hydroelectric power. All system functions (including water quality) is inevitably dependent upon a healthy and intact cover of vegetation and a species-rich flora, often likened to an 'insurance policy' in that a species-rich environment has a greater chance of weathering the barrage of human threats and natural environmental changes because there are more likely to be individuals among the many that can withstand a specific threat (see Körner 2001). For example, a species-rich environment has a greater chance of combating invasive plants because the likelihood of an invader encountering a close competitor is higher than if just a handful of native plant species were present (Richardson 2001). Every conceivable effort should ensure that the levels of established (major) invasive alien plants in the DAC are contained to an acceptable minimum and any emerging alien invasive plants are rapidly identified through early detection and eradicated before infestation levels impact negatively on the ecological integrity of the DAC.

FUTURE STUDIES

This desktop-based analysis should be expanded to include the proposed fieldwork component (to both verify and potentially widen the net of possible emerging candidates) and the scope of the assessment enlarged to include representative areas of Lesotho. The value of a proactive approach, both in terms of identifying the invasive alien species and the application of swift counter measures to eliminate emergent populations before they become well established, cannot be overstated.

ACKNOWLEDGEMENTS

The author wishes to thank I. Rushworth, S. McKean and V. Mngomezulu of Ezemvelo KZN Wildlife and Dr. R. Lechmere-Oertel, previously of the Maloti Drakensberg Transfrontier Project, for guidance and sourcing selected literature. Helpful comments and improvements from three reviewers are gratefully acknowledged.

REFERENCES

ARÉVALO, J.R., DELGADO, J.D., OTTO, R., NARANJO, A., SALAS, M. & FERNÁNDEZ-PALACIOS, J.M. 2005. Distribution of alien vs. native plant species in roadside communities along an altitudinal gradient in Tenerife and Gran Canaria (Canary Islands). *Perspectives in Plant Ecology, Evolution and Systematics* 7: 185–202.

BAUR, B. & SCHMIDLIN, S. 2007. Effects of invasive non-native species on the native biodiversity in the River Rhine. In W. Nentwig, *Biological Invasions*: 257–273. Springer-Verlag, Berlin.

CARBUTT, C. 2004. *Cape elements on high-altitude corridors and edaphic islands*. Ph.D. thesis, University of KwaZulu-Natal, Pietermaritzburg.

CARBUTT, C. & EDWARDS, T.J. 2004. The flora of the Drakensberg Alpine Centre. *Edinburgh Journal of Botany* 60: 581–607.

CARBUTT, C. & EDWARDS, T.J. 2006. The endemic and near-endemic angiosperms of the Drakensberg Alpine Centre. *South African Journal of Botany* 72: 105–132.

CARBUTT, C. & EDWARDS, T.J. 2008. *Possible effect of global warming on ecosystem function in the uKhahlamba Drakensberg Park, South Africa: particularly its influence on plant communities through nutrient enrichment.* Proceedings of the launching workshop of a "Global Change Research Network in African Mountains", Kampala (Uganda). 23–25 July 2007. Mountain Research Initiative (MRI) and the Centre for Development and Environment (CDE), University of Bern, Switzerland.

CONTOSTA, A.R., FREY, S.D. & COOPER, A.B. 2011. Seasonal dynamics of soil respiration and N mineralisation in chronically warmed and fertilised soils. *Ecosphere* 2:1–21.

COWLING, R.M. & HILTON-TAYLOR, C. 1994. Patterns of plant diversity and endemism in southern Africa: an overview. In B.J. Huntley, *Botanical diversity in southern Africa*: 31–52. National Botanical Institute, Pretoria.

COWLING, R.M. & HILTON-TAYLOR, C. 1997. Phytogeography, flora and endemism. In R.M. Cowling, D.M. Richardson & S.M. Pierce, *Vegetation of Southern Africa*: 43–61. Cambridge University Press, Cambridge.

CREMER, K. 1999. *Willow management for Australian rivers* (Natural Resource Management Special Issue). Australian Association of Natural Resource Management, Canberra. pp. 2–22.

DALE, V.H., PEARSON, S.M., OFFERMAN, H.L. & O'NEILL, R.V. 1994. Relating patterns of land-use change to faunal biodiversity in central Amazon. *Conservation Biology* 8: 1027–1036.

DAVIS, F.W. & MORITZ, M. 2001. Mechanisms of disturbance. In S.A. Levin, *Encyclopedia of Biodiversity (Volume 2)*: 153–160. Academic Press, San Diego.

DEACON, H.J., JURY, M.R. & ELLIS, F. 1992. Selective regime and time. In R.M. Cowling, *The Ecology of Fynbos: Nutrients, Fire and Diversity*: 6–22. Oxford University Press, Cape Town.

DUKES, J.S. & MOONEY, H.A. 1999. Does global change increase the success of biological invaders? *Trends in Ecology and Evolution* 14: 135–139.

DUNNE, J.A. & HARTE, J. 2001. Greenhouse effect. In S. Levin, *Encyclopedia of Biodiversity (Volume 3)*: 277–293. Academic Press, San Diego.

ELTON, C.S. 1958. *The ecology of invasions by animals and plants.* Methuen, London.

EZEMVELO KZN WILDLIFE. 2005. *Integrated Management Plan (2006–2011): uKhahlamba Drakensberg Park World Heritage Site, South Africa.* Ezemvelo KZN Wildlife, Pietermaritzburg.

FAIRBANKS, D.H.K., THOMPSON, M.W., VINK, D.E., NEWBY, T.S., VAN DEN BERG, H.M. & EVERARD, D.A. 2000. The South African land-cover characteristics database: a synopsis of the landscape. *South African Journal of Science* 96: 69–82.

GERMISHUIZEN, G. & MEYER, N.L. 2003. Plants of southern Africa: an annotated checklist. *Strelitzia* 14:1–1231.

GILMAN, E.F. & WATSON, D.G. 1993a. *Gleditsia triacanthos* var. *inermis* (thornless honeylocust). US Forest Service, Department of Agriculture. Fact sheet no. ST-279. 4p.

GILMAN, E.F. & WATSON, D.G. 1993b. *Juniperus virginiana* (eastern redcedar). USDA Forest Service Fact Sheet ST-327. 4p.

GILMAN, E.F. & WATSON, D.G. 1994a. *Quercus robur* (English oak). US Forest Service, Department of Agriculture. Fact sheet no. ST-558. 4p.

GILMAN, E. F. & WATSON, D. G. 1994b. *Robinia pseudoacacia* 'Frisia' black locust. US Forest Service, Department of Agriculture. Fact sheet no. ST-571. 4p.

GOLDBLATT, P. & RAVEN, P.H. 1997. FSA contributions 9: Onagraceae. *Bothalia* 27: 149–165.

GOTTFRIED, M., PAULI, H. & GRABHERR, G. 1998. Prediction of vegetation patterns at the limits of plant life: a new view of the alpine-nival ecotone. *Arctic and Alpine Research* 30: 207–221.

GRABHERR, G., GOTTFRIED, M., GRUBER, A. & PAULI, H. 1995. Patterns and current changes in alpine plant diversity. In F.S. Chapin & Ch. Körner, *Arctic and Alpine Biodiversity: Patterns, Causes and Ecosystem Consequences*: 167–181. Springer, Heidelberg.

HALL, D.W., VANDIVER, V.V. & FERRELL, J.A. 2005. Brazil pusley, *Richardia brasiliensis* (Moq.). Institute of Food and Agricultural Sciences, University of Florida, Gainesville FL 32611.

HENDERSON, L. 1998. Southern African Plant Invaders Atlas (SAPIA). *Applied Plant Science* 12: 31, 32.

HENDERSON, L. 2001. *Alien weeds and invasive plants: a complete guide to declared weeds and invaders in South Africa.* Plant Protection Research Institute Handbook No. 12. Agricultural Research Council, Pretoria.

HENDERSON, L. 2006. Comparisons of invasive plants in southern Africa originating from southern temperate, northern temperate and tropical regions. *Bothalia* 36: 201–222.

HENDERSON, L. 2007. Invasive, naturalized and casual alien plants in southern Africa: a summary based on the Southern African Plant Invaders Atlas (SAPIA). *Bothalia* 37: 215–248.

HILL, T.R. 1996. Description, classification and ordination of the dominant vegetation communities, Cathedral Peak, KwaZulu-Natal Drakensberg. *South African Journal of Botany* 62: 263–269.

HILLIARD, O.M. 1977. *Compositae in Natal.* University of Natal Press, Pietermaritzburg. 659 pp.

HILLIARD, O.M. & BURTT, B.L. 1987. *The Botany of the Southern Natal Drakensberg.* National Botanic Gardens, Cape Town.

HOBBS, R.J. 2004. The Working for Water Programme in South Africa: the science behind the success. *Diversity and Distributions* 10: 501–503.

HOBBS, R.J. & HUMPHRIES, S.E. 1995. An integrated approach to the ecology and management of plant invasions. *Conservation Biology* 9: 761–770.

HULME, P.E. 2003. Winning the science battles but loosing the conservation war? *Oryx* 37: 178–193.

KALWIJ, J.M., ROBERTSON, M.P. & VAN RENSBURG, B.J. 2008. Human activity facilitates altitudinal expansion of exotic plants along a road in montane grassland, South Africa. *Applied Vegetation Science* 11: 491–498.

KEELEY, J.E., LUBIN, D. & FOTHERINGHAM, C.J. 2003. Fire and grazing impacts on plant diversity and alien plant invasions in the southern Sierra Nevada. *Ecological Applications* 13: 1355–1374.

KÖRNER, Ch. 2001. Alpine ecosystems. In S.A. Levin, *Encyclopedia of Biodiversity (Volume 1)*: 133–144. Academic Press, San Diego.

LAWLER, J.J., SHAFER, S.L., WHITE, D., KAREIVA, P., MAURER, E.P., BLAUSTEIN, A.R. & BARTLEIN, P.J. 2009. Projected climate-induced faunal change in the Western Hemisphere. *Ecology* 90: 588–597.

LE MAITRE, D.C., MGIDI, T.N., SCHONEGEVEL, L., NEL, J.L., ROUGET, M., RICHARDSON, D.M. & MIDGLEY, C. 2004. Plant invasions in South Africa, Lesotho and Swaziland: assessing the potential impacts of major and emerging plant invaders. In *An assessment of invasion potential of invasive alien plant species in South Africa.* CSIR Environmentek Report No. ENV-S-C2004-108. pp. 76–96.

MACDONALD, I.A.W. 1989. Man's role in changing the face of southern Africa. In B.J. Huntley, *Biotic diversity in southern Africa. Concepts and Conservation*: 51–77. Oxford University Press, Cape Town.

MACDONALD, I.A.W. 1992. Global change and alien invasions: implications for biodiversity and protected area management. In O.T. Solbrig, H.M. Van Emden & P.G.W.J. Van Oordt, *Biodiversity and Global Change*: 197–207. International Union of Biological Sciences, Paris.

MACDONALD, I.A.W., KRUGER, F.J. & FERRAR, A.A. 1986. *The ecology and management of biological invasions in southern Africa.* Oxford University Press, Cape Town.

MACDOUGALL, A.S. & TURKINGTON, R. 2004. Relative importance of suppression-based and tolerance-based competition in an invaded oak savanna. *Journal of Ecology* 92: 422–434.

MARK, A.F. & DICKINSON, J.M. 2008. Maximising water yield with indigenous non-forest vegetation: a New Zealand perspective. *Frontiers in Ecology and the Environment* 6: 25–34.

MGIDI, T.N., LE MAITRE, D.C., SCHONEGEVEL, L., NEL, J.L., ROUGET, M. & RICHARDSON, D.M. 2004. Alien plant invasions—incorporating emerging invaders in regional prioritisation: a pragmatic approach for South Africa. In *An assessment of invasion potential of invasive alien plant species in South Africa.* CSIR Environmentek Report No. ENV-S-C 2004-108. pp. 56–75.

MGIDI, T.N., LE MAITRE, D.C., SCHONEGEVEL, L., NEL, J.L., ROUGET, M. & RICHARDSON, D.M. 2007. Alien plant invasions—incorporating emerging invaders in regional prioritization: a pragmatic approach for southern Africa. *Journal of Environmental Management* 84: 173–187.

MIDGLEY, G.F., HANNAH, L., ROBERTS, R., MACDONALD, D.J. & ALLSOPP, J. 2001. Have Pleistocene climatic cycles influenced species richness patterns in the greater Cape Mediterranean Region? *Journal of Mediterranean Ecology* 2: 137–144.

MUCINA, L. & RUTHERFORD, M.C. 2006. *The vegetation of South Africa, Lesotho and Swaziland.* Strelitzia 19. SANBI, Pretoria.

NEL, J.L., RICHARDSON, D.M., ROUGET, M., MGIDI, T., MDZEKE, N., LE MAITRE, D.C., VAN WILGEN, B.W., SCHONEGEVEL, L., HENDERSON, L. & NESER, S. 2004. A proposed classification of invasive plant species in South Africa: towards prioritising species and areas for management action. *South African Journal of Science* 100: 53–64.

O'CONNOR, T.G. 2005. Influence of land use on plant community composition and diversity in Highland Sourveld grassland in the southern Drakensberg, South Africa. *Journal of Applied Ecology* 42: 975–988.

OLCKERS, T. 2004. Targeting 'emerging weeds' for biological control in South Africa: the benefits of halting the spread of alien plants at an early stage of their invasion. *South African Journal of Science* 100: 64–68.

OLCKERS, T. & HILL, M.P. 1999. Biological control of weeds in South Africa (1990–1998). *African Entomology Memoir No. 1.*

PARKER, I.M., SIMBERLOFF, D., LONSDALE, W.M., GOODELL, K., WONHAM, M., KAREIVA, P.M., WILLIAMSON, M., VON HOLLE, B., MOYLE, P.B., BYERS, J.E. & GOLDWASSER, L. 1999. Impact: towards a framework for understanding the ecological effects of invaders. *Biological Invasions* 1: 3–19.

RETIEF, E. & VAN WYK, A.E. 1998. The genus *Echium* (Boraginaceae) in southern Africa. *Bothalia* 28: 167–177.

RICHARDSON, D.M. 2001. Plant invasions. In S.A. Levin, *Encyclopedia of Biodiversity (Volume 4)*: 677–688. Academic Press, San Diego.

RICHARDSON, D.M. & BOND, W.J. 1991. Determinants of plant distribution: evidence from pine invasions. *American Naturalist* 137: 639–668.

RICHARDSON, D.M., MACDONALD, I.A.W., HOFFMANN, J.H. & HENDERSON, L. 1997. Alien plant invasions. In R.M. Cowling, D.M. Richardson & S.M. Pierce, *Vegetation of southern Africa*: 535–570. Cambridge University Press, Cambridge.

RICHARDSON, D.M., ROUGET, M., RALSTON, S.J., COWLING, R.M., VAN RENSBURG, B.J. & THUILLER, W. 2005. Species richness of alien plants in South Africa: environmental correlates and the relationship with indigenous plant species richness. *Ecoscience* 12: 391–402.

ROBERTSON, M.P., VILLET, M.H., FAIRBANKS, D.H.K., HENDERSON, L., HIGGINS, S.I., HOFFMANN, J.H., LE MAITRE, D.C., PALMER, A.R., RIGGS, I., SHACKLETON, C.M. & ZIMMERMANN, H.G. 2003. A proposed prioritisation system for the management of invasive alien plants in South Africa. *South African Journal of Science* 99: 37–43.

ROBERTSON, M.P., VILLET, M.H. & PALMER, A.R. 2004. A fuzzy classification technique for predicting species' distributions: an implementation using alien plants and indigenous insects. *Diversity and Distributions* 10: 461–474.

ROUGET, M. & RICHARDSON, D.M. 2003. Understanding patterns of plant invasion at different spatial scales: quantifying the roles of environment and propagule pressure. In L.E. Child, J.H. Brock, G. Brundu, K. Prach, P. Pyšek, P.M. Wade & M. Williamson, *Plant invasions: ecological threats and management solutions*: 3–15. Backhuys, Leiden.

ROUGET, M., RICHARDSON, D.M., NEL, J., LE MAITRE, D.C., EGOH, B. & MGIDI, T. 2004. Mapping the potential ranges of major plant invaders in South Africa, Lesotho and Swaziland using climatic suitability. In *An assessment of invasion potential of invasive alien plant species in South Africa*. CSIR Environmentek Report No. ENV-S-C 2004-108. pp. 39–55.

SALA, O.E., CHAPIN, F.S. III, ARMESTO, J.J., BERLOW, E., BLOMFIELD, J., DIRZO, R., HUBER-SANWALD, E., HUENNEKE, L.F., JACKSON, R.B., KINZIG, A., LEEMANS, R., LODGE, D.M., MOONEY, H.A., OESTERHELD, M., POFF, N.L., SYKES, M.T., WALKER, B.H., WALKER, M. & WALL, D.H. 2000. Global biodiversity scenarios for the year 2100. *Science* 287: 1770–1774.

SCHMITZ, G. & ROOYANI, F. 1987. *Lesotho: geology, geomorphology, soils.* National University of Lesotho, Roma.

SCHNEIDER, S.H. & ROOT, T.L. 2001. Synergism of climate change and ecology. In S.A. Levin, *Encyclopedia of Biodiversity (Volume 1)*: 709–726. Academic Press, San Diego.

SIMBERLOFF, D. 2001. Effects and distribution of introduced species. In S.A. Levin, *Encyclopedia of Biodiversity (Volume 3)*: 517–529. Academic Press, San Diego.

STARR, F., STARR, K. & LOOPE, L. 2003. *Cotoneaster pannosus—* Rosaceae. United States Geological Survey, Biological Resources Division, Haleakala Field Station, Maui, Hawaii.

STIRTON, C.H. 1981. New records of naturalised *Rubus* in southern Africa. *Bothalia* 13: 333–337.

STOCK, W.D., ALLSOPP, N., VAN DER HEYDEN, F. & WITKOWSKI, E.T.F. 1997. Plant form and function. In R.M. Cowling, D.M. Richardson & S.M. Pierce, *Vegetation of southern Africa*: 376–396. Cambridge University Press, Cambridge.

STREETER, D. 1998. *The wild flowers of the British Isles.* Midsummer Books, London.

TAINTON, N.M., ZACHARIAS, P.J.K. & HARDY, M.B. 1989. The contribution of veld diversity to the agricultural economy. In B. Huntley, *Biotic diversity in southern Africa. Concepts and Conservation*: 107–120. Oxford University Press, Cape Town.

THORP, J.R. & LYNCH, R. 2000. *The determination of weeds of national significance.* National Weeds Strategy Executive Committee, Launceston.

THUILLER, W., RICHARDSON, D.M., ROUGET, M., PROCHES, S. & WILSON, J.R.U. 2006. Interactions between environment, species traits, and human uses describe patterns of plant invasions *Ecology* 87:1755–1769.

THUILLER, W., ALBERT, C., ARAÚJO, M.B., BERRY, P.M., CABEZA, M., GUISAN, A., HICKLER, T., MIDGLEY, G.F., PATERSON, J., SCHURR, F.M., SYKES, M.T. & ZIMMERMANN, N.E. 2008. Predicting global change impacts on plant species' distributions: future challenges. *Perspectives in Plant Ecology, Evolution and Systematics* 9: 137–152.

TYSON, P.D., PRESTON-WHYTE, R.A. & SCHULZE, R.E. 1976. *The climate of the Drakensberg.* The Town and Regional Planning Commission, Pietermaritzburg.

VAN WILGEN, B.W. 2004. Scientific challenges in the field of invasive alien plant management. *South African Journal of Science* 100: 19, 20.

VAN WILGEN, B.W., COWLING, R.M. & BURGERS, C.J. 1996. Valuation of ecosystem services. A case study from South African fynbos ecosystems. *Bioscience* 46: 184–189.

VAN WILGEN, B.W., REYERS, B., LE MAITRE, D.C., RICHARDSON, D.M. & SCHONEGEVEL, L. 2008. A biome-scale assessment of the impact of invasive alien plants on ecosystem services in South Africa. *Journal of Environmental Management* 89: 336–349.

VAN WILGEN, B.W., DYER, C., HOFFMANN, J.H., IVEY, P., LE MAITRE, D.C., MOORE, J.L., RICHARDSON, D.M., ROUGET, M., WANNENBURGH, A. & WILSON, J.R.U. 2011. National-scale strategic approaches for managing introduced plants: insights from Australian acacias in South Africa. *Diversity and Distributions* 17: 1060–1075.

VAN WYK, A.E. & SMITH, G.F. 2001. *Regions of floristic endemism in southern Africa.* Umdaus Press, Pretoria.

VAN ZINDEREN BAKKER, E.M. & COETZEE, J.A. 1988. A review of Late Quaternary pollen studies in east, central and southern Africa. *Review of Palaeobotany and Palynology* 55: 155–174.

WELLS, M.J., POYNTON, R.J., BALSINHAS, A.A., MUSIL, K.J., JOFFE, H., VAN HOEPEN, E. & ABBOTT, S.K. 1986. The history of introduction of invasive plants to southern Africa. In I.A.W. Macdonald, F.J. Kruger & A.A. Ferrar, *The ecology and management of biological invasions in southern Africa*: 21–35. Oxford University Press, Cape Town.

WOODWARD, F.I. 2001. Effects of climate. In S.A. Levin, *Encyclopedia of Biodiversity (Volume 1)*: 727–740. Academic Press, San Diego.

4

Anatomy of myxospermic diaspores of selected species in the Succulent Karoo, Namaqualand, South Africa

H. FOTOUO MAKOUATE*, M.W. VAN ROOYEN*† and C.F. VAN DER MERWE**

Keywords: arid regions, diaspores, dispersal, myxospermy, Namaqualand, scanning electron microscopy, Succulent Karoo

ABSTRACT

Environmental conditions encountered in arid ecosystems differ vastly from those in more mesic ecosystems. Dispersal strategies in arid environments reflect these differences and many mechanisms have evolved that restrict or hinder dispersal. Myxospermy is a trait developed by plant species from arid regions to restrict diaspore dispersal by means of an anchorage mechanism. Several of the abundant plant species in Namaqualand, within the arid Succulent Karoo Biome, display myxospermy. Diaspores of these species produce copious amounts of mucilage when they are moistened and are anchored to the soil once the mucilage dries out again. This study investigated the origin of the mucilaginous layer of 12 species anatomically, using both light and scanning electron microscopy. The mucilage production of the species investigated could best be grouped into three types: 1, epidermal and sub-epidermal cells of seeds and achenes; 2, specialized tissue in wings or the pappus of achenes; and 3, mucilage excreting hairs. Previous systems for classifying the different types of mucilage production did not recognize the mucilaginous nature of wings or a pappus. A short note on the composition of the mucilage is included.

INTRODUCTION

Plants have developed many functional traits that allow them to adapt and survive in different environments. Seed dispersal is an important functional trait that influences the population structure and the spatial and temporal turnover of species within a plant community. Myxospermy is the phenomenon where the epidermis of the diaspore (seed/fruit) contains mucilaginous cells, which swell and become sticky when in contact with water (Gutterman & Shem-Tov 1997; Van Rheede van Oudtshoorn & Van Rooyen 1999). Myxospermy is classified as an antitelechoric dispersal mechanism and is encountered more often in arid than in mesic environments. The original explanation for the prevalence of antitelechory in desert environments was that these mechanisms were adaptive responses to the particularly high mortality of dispersed seeds in deserts and that they had evolved as mechanisms to reclaim the mother site (Murbeck 1919; Zohary 1937; Stopp 1958; Ellner & Shmida 1981; Van Rooyen et al. 1990). Ellner & Shmida (1981) however, argued that antitelechory was a side effect of characters whose adaptive value was not directly related to dispersal. Among the benefits derived from antitelechory are the spreading of germination over time and the provision of suitable conditions for germination and subsequent seedling establishment.

There are many divergent ideas about the ecological importance of myxospermy, and it probably fulfils various functions. When wet, myxospermic diaspores adhere to the soil; and after drying out, remain glued to soil particles and thus resist being carried to unfavourable locations by wind. However, wet diaspores can also adhere to the feet or fur of animals and be dispersed zoochorically. The good contact between the mucilage layer and the soil surface increases water absorption for germination,

and as a result, myxospermic species germinate more successfully on the surface of the soil than non-myxospermic species (Bregman & Graven 1997; Zaady et al. 1997; Van Rheede van Oudtshoorn & Van Rooyen 1999). The water held by the mucilage could also serve as a water reservoir for seedling establishment (Gutterman et al. 1967; Gutterman 1993, 1996), although this function of the mucilage has been challenged (Murbeck 1919; Grubert 1974). Furthermore, the adherence of the seed to the soil by the mucilage prevents massive collection by seed predators (Gutterman 1993), and the repeated wetting of mucilaginous seeds during many nights with dew may affect the repair mechanisms of the DNA, cell membranes, and organelles; thereby enhancing seed viability for many years (Osborne et al. 1980/1981; Leprince et al. 1993; Huang et al. 2008). The mucilage has also been reported to have a stimulatory action on germination (Gat-Tilman 1995) or it may control the germination process by excluding the passage of oxygen when there is excess moisture (Gutterman et al. 1967, 1973; Gutterman 1996; Tamara et al. 2000).

Although several studies have drawn the attention to the ecological importance of myxospermy in the flora of Namaqualand (Rösch 1977; Van Rooyen et al. 1990; Van Rheede van Oudtshoorn & Van Rooyen 1999), the germination behaviour of myxospermic species (Fotouo Makouate 2008) and the response of myxospermic species to grazing pressure (Fotouo Makouate 2008), little attention has thus far been given to the origin and chemical composition of the mucilage of these myxospermic diaspores. The objectives of this study were to investigate the origin and the chemical composition of the mucilage of a representative sample of myxospermic species in Namaqualand, South Africa.

STUDY AREA AND SPECIES

Diaspores of the species investigated in this study were all collected in the Namaqualand Hardeveld Bioregion of the Succulent Karoo (Mucina & Rutherford

* Department of Plant Science, University of Pretoria, 0002 Pretoria.
† gretel.vanrooyen@up.ac.za.
** Laboratory for Microscopy and Microanalysis, University of Pretoria, 0002 Pretoria.

2006). The Succulent Karoo is an arid winter rainfall region stretching along the West Coast of South Africa and Namibia; and is recognized by the IUCN as one of only two entirely arid global hotspots of biodiversity (Conservation International 2009). Many of the biologically unique features of this biome have been attributed to its climate, *i.e.* the effective and relatively predictable seasonal rainfall, and the relatively moderate winter temperatures (Mucina & Rutherford 2006). A prominent feature of the Namaqualand Hardeveld Bioregion is the extravagant spring floral display of winter-growing annuals (Van Rooyen 1999).

Representative species of four families were investigated: within the Lamiaceae, the perennial shrub *Salvia dentata*; within the Acanthaceae, two dwarf perennial shrubs *Acanthopsis horrida* and *Blepharis furcata*; within the Brassicaceae, the annual *Heliophila thunbergii* var. *thunbergii*; and within the Asteraceae, the perennial shrubs *Pentzia incana* and *Othonna cylindrica* as well as the annual species *Cotula barbata, Monoculus (Tripteris) hyoseroides, Senecio arenarius, Ursinia cakilefolia, Oncosiphon grandiflorum* and *Foveolina dichotoma*.

MATERIALS AND METHODS

Scanning electron microscopy (SEM)

Dry seeds/achenes were mounted on a stub and made conductive with RuO_4 vapour (Van der Merwe & Peacock 1999). For wet samples, seeds/achenes of each species were soaked in water for 24 hours before fixation. Seeds/achenes were then fixed in 2.5% glutaraldehyde in 0.075 M phosphate buffer, pH 7.4, for one hour, and rinsed three times (five minutes per rinse) in 0.075 M phosphate buffer and dehydrated in an ethanol series with the respective concentrations of 30%, 50%, 70%, 90%, and 100%. This was followed by critical point drying in liquid CO_2, whereafter the seeds/achenes were mounted on a stub and made conductive with RuO_4 vapour. Samples were viewed with the use of a JSM 840 scanning electron microscope (SEM) (JEOL, Tokyo, Japan), and photographs were taken with the computer program Orion 6.60.4.

Light microscopy

Further studies were done on diaspores where the origin of the mucilage could not be clearly established with the aid of an SEM. Dry samples were fixed in 4% (v/v) formaldehyde in 50% (v/v) ethanol (FAA) to prevent mucilage release. The wet samples were immersed in water for 24 hours. Wet samples were fixed in 2.5% glutaraldehyde in 0.075 M phosphate buffer, pH 7.4, for one hour, and rinsed three times (five minutes per rinse) in 0.075 M phosphate buffer, and dehydrated in an ethanol series with the respective concentrations of 30%, 50%, 70%, 90% and 100%. Samples were subsequently infiltrated over two days with pure LR-White resin and polymerized at 65°C for 24 to 36 hours. Sections were cut with a Reichert Ultracut E microtome. Each section was stained, mounted in immersion oil, and viewed with a Nikon Optiphot light microscope (LM) (Tokyo, Japan). Photographs were taken with a Nikon Digital

camera DXM 1 200 (Tokyo, Japan) and the computer program Nikon ACT-1 (Tokyo, Japan).

Diaspore staining was done according to various manuals of microscopic staining (McClean & Ivimey Cook 1941; O'Brien & McCully 1981; Lawton & Ettridge 1986). Seven different stains were used to test for the presence of various compounds, the details of which are given in Table 1.

TABLE 1.—Stains used for various compounds and expected results

Stain	Compound	Expected results
Aniline blue	Callose	Blue
Biuret solution	Protein	Violet
Methylene blue	Cellulose	Blue
Phloroglucinol	Lignin	Pink
Ruthenium red	Pectic substances	Red
Sudan 4	Lipids	Yellow
Toluidine blue	Polysaccharides	Pink-purple

RESULTS AND DISCUSSION

Anatomical investigation

On the basis of the anatomical origin of the mucilage, Zohary (1937) distinguished seven different types. Grubert (1974) recognized many more types, but his three main categories were based on whether the origin was 1, restricted to epidermal structures, *e.g.* epidermis or hairs; 2, from epidermal and sub-epidermal layers; or 3, restricted to sub-epidermal layers.

Neither the classification systems of Zohary (1937) or Grubert (1974) accommodate diaspores where the mucilage production occurs on wings of achenes, as was reported in the present study. In general, diaspores of most myxospermic species are smooth-coated and do not contain obvious appendages to advance telechory. Contrary to the belief that none of the myxospermic Asteraceae possess a pappus (Grubert 1974), many myxospermic species from Namaqualand are winged or possess a winged pappus. The presence of wings would allow anemochoric dispersal during phase I dispersal (Chambers & MacMahon 1994), but further anemochoric dispersal would be prevented once the achenes were moistened and remained attached to soil particles.

The mucilage production of the species investigated in the present study could best be grouped into three main types:

1 Epidermal and sub-epidermal cells of seeds and achenes:
   1.1 Cells of seeds bursting only in the centre
   1.2 Cells of achenes bursting across the entire surface
2 Specialized tissue in wings or pappus of achenes
3 Hairs:
   3.1 On achenes
   3.2 On seeds

Epidermal mucilage

Heliophila thunbergii var. *thunbergii* (Brassicaceae)

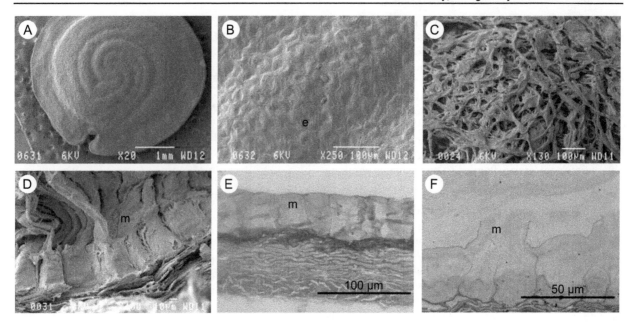

FIGURE 1.—*Heliophila thunbergii* var. *thunbergii*, seeds. A–D, SEM: A, entire dry seed; B, high magnification of dry epidermis; C, high magnification of wet seed surface; D, section of wet seed. E, F, LM: E, section of the dry seed; F, cross section of wet epidermis. e, epidermal layer, m, mucilage.

The scanning electron micrograph of dry seeds of *H. thunbergii* var. *thunbergii* reveals a circular seed (Figure 1A). The epidermal layer (e) consists of circular cells with a raised margin and a cavity in the centre (Figure 1B). Upon wetting, the mucilage swells and the epidermal cells become incapable of containing the excess mucilage and therefore rupture. Once wet, the seed is covered by a gelatinous film (Figure 1C), similar to the mucilaginous seed coat of two species of the Brassicaceae described by Gutterman & Shem-Tov (1997). The thin section of the dry seed also reveals the epidermal cells filled with mucilage (m) (Figure 1E). The section of the wet seed reveals the epidermis with ruptured cells from which filaments of mucilage (m) emerge (Figure 1D, F).

Salvia lanceolata (Lamiaceae)

A similar pattern was observed for the seeds of *S. lanceolata* (Lamiaceae). Once in contact with water, the outer layer of the epidermis absorbs water quickly, the epidermis cells swell, the cuticle ruptures and the contents are excreted as long, continuous threads. Hedge (1970) reported that the contents of the epidermis cells of some Lamiaceae are secreted in the shape of long, continuous, helically coiled threads.

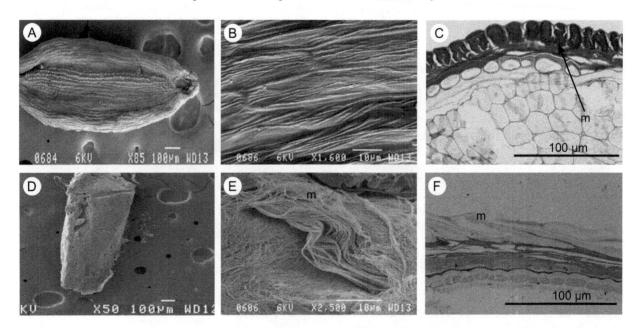

FIGURE 2.—*Cotula barbata,* achenes. A, B, D, E, SEM: A, entire dry achene; B, high magnification of the epidermis; D, entire wet achene; E, high magnification of wet epidermis covered with thick layer of mucilage. C, F, LM: C, cross section of dry achene coat, with mucilage (arrow) retained in epidermal layer; F, cross section of wet epidermis of which outer layer has burst, releasing mucilage from epidermal cells. m, mucilage.

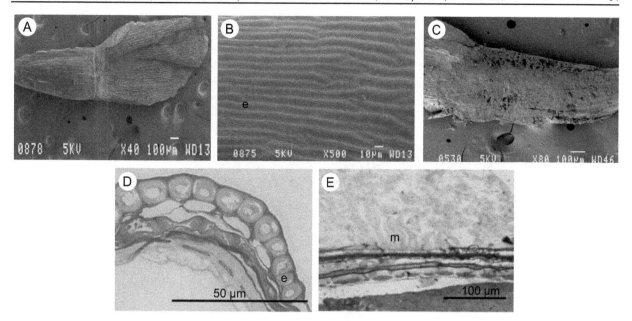

FIGURE 3.—*Foveolina dichotoma,* achenes. A–C, SEM: A, entire achene of *F. dichotoma*; B, high magnification of the epidermis of *F. dichotoma*; C, wet achene. *Oncosiphon grandiflorum*, achenes: D, E, LM: D, thin cross section of dry epidermis; E, thin cross section of wet epidermis with broken epidermal cells that liberate mucilage. e, epidermal layer, m, mucilage.

Cotula barbata (Asteraceae)

In the dry state, the one side of the achene of *C. barbata* is undulated (Figure 2A), whereas the other side of the achene is smooth (Figure 2B). The mucilage structure observed in *C. barbata* is similar to that of *H. thunbergii* var. *thunbergii* with the mucilage contained in the epidermal layer (Figure 2C). When *C. barbata* achenes come into contact with water (Figure 2D), the epidermal cells burst across the entire outer surface (Figure 2E, F) and not only the centre of the cell as in the case of *H. thunbergii* var. *thunbergii*.

Foveolina dichotoma, Oncosiphon grandiflorum and Pentzia incana (Asteraceae)

These three species are taxonomically closely related, and it appears that they have the same way of producing mucilage. The SEMs show that the epidermis of the dry achenes of *F. dichotoma* consists of parallel longitudinal cells (Figure 3A, B), as was also found for *O. grandiflorum* and *P. incana*. Following wetting, fixation, and critical point drying, the achenes are covered with a gelatinous film but the pappus remains unchanged (Figure 3C). Light microscope sections of the dry achene of *O. grandiflorum* show that the epidermal cells are filled

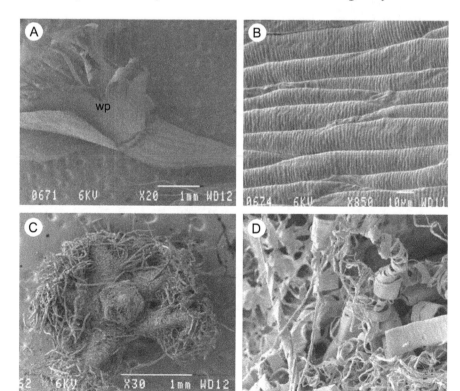

FIGURE 4.—*Ursinia cakilefolia,* SEM of achenes: A, entire dry achene with winged pappus; B, high magnification of dry winged pappus consisting of radially elongated cells; C, entire wet achene with uncoiled winged pappus; D, high magnification of wet winged pappus showing long strands of uncoiled mucilage. wp, winged pappus.

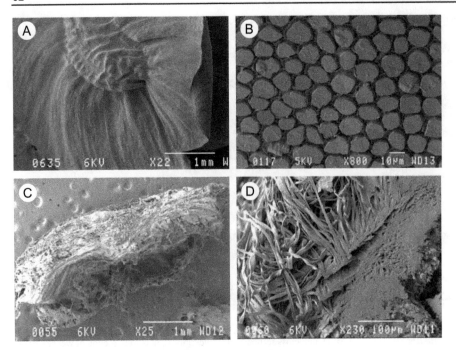

FIGURE 5.—*Monoculus hyoseroides*, SEM of achenes: A, entire dry achene; B, high magnification of a section of the wing; C, entire wet achene with disintegrated wing; D, high magnification of long strands of mucilage.

with mucilage (Figure 3D), which is released when the cells burst open upon swelling in contact with water (Figure 3E).

Mucilaginous wings

Ursinia cakilefolia (Asteraceae)

Dry achenes of *U. cakilefolia* have a winged pappus (wp) (Figure 4A), consisting of a single layer of radially elongated cells with helically thickened cell walls (Figure 4B). When the pappus comes into contact with water, the cells separate at the middle lamellae and the helical thickenings uncoil to form long strands of mucilage (Figure 4C). The different steps through which the pappus uncoils are clearly visible under the SEM (Figure 4D). The same process was observed in *U. nana* subsp. *nana* (Rösch 1977).

Monoculus hyoseroides (Asteraceae)

Achenes of *M. hyoseroides* are three-winged. The epidermis of the seed-bearing part of the achene is undulate and does not become mucilaginous (Figure 5A). The section through the wings shows hexagonally shaped cells completely filled with mucilage and separated by thick cell walls (Figure 5B). Upon wetting, the cells separate and the wings become completely mucilaginous (Figure 5C) and the long mucilage strands remain attached to the achene (Figure 5D).

Mucilaginous hairs

Senecio arenarius (Asteraceae)

In the dry condition, the shiny white hairs (h) have a finger-like shape and lie against the achene surface (Figure 6A, B). The interior of the hair seems to be filled with a coiled thread as seen at high magnification (Figure 6B). In the wet condition, the hairs spread ± at right angles to the seed. Apparently, when the mucilaginous hairs absorb water the threads uncoil and become thin and long (Figure 6C). The same phenomenon was observed for *Othonna cylindrica,* as well as for *O. floribunda* (Rösch 1977).

Acanthopsis horrida and Blepharis furcata (Acanthaceae)

In the dry condition, the hairs on the seed surface of *A. horrida* (Figure 7A, B) and *B. furcata* are tightly appressed to the seed coat and are covered with a thick deposit (Figure 7C). At high magnification it can be seen that these hairs contain coiled spirals (Figure 7B). Once in contact with water the thick deposit expands, the hairs become swollen and spread ± at right angles to the seed

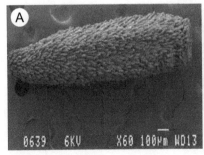
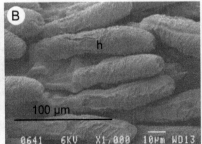
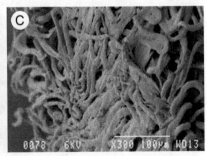

FIGURE 6.—*Senecio arenarius*, SEM of achenes: A, entire dry achene; B, high magnification of dry achene with hair lying against it; C, high magnification of wet hairs. h, hair.

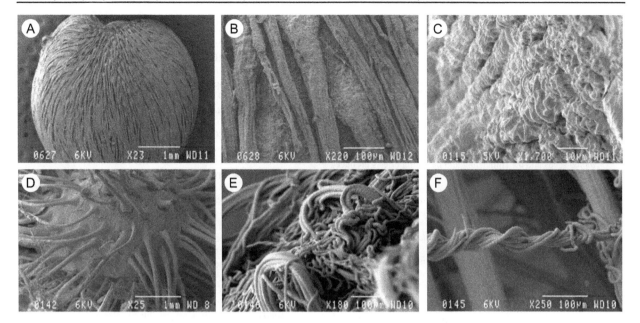

FIGURE 7.—*Acanthopsis horrida,* SEM of seeds: A, entire dry seed; B, hairs; C, dry epidermis covered with thick deposit; D, wet seed; E, swollen hairs; F, uncoiled strands of mucilage.

coat (Figure 7D). The tips of the hairs break open and release a multitude of strands of mucilage (Figure 7E, F).

Chemical composition

Three stains reacted positively with the mucilage of all species: methylene blue, ruthenium red, and toluidine blue; thus confirming the presence of polysaccharides such as cellulose and pectic substances. On the basis of the intensity of the reaction, the epidermal mucilage seemed to contain less cellulose than the mucilaginous wings and mucilaginous hairs. In some species, a slight reaction was also observed with aniline blue, indicating the presence of callose in the mucilage.

Cellulose is the most abundant plant polysaccharide found in the form of microfibrils in cell walls and mucilage. Dry amorphous and fibrous cellulose can absorb considerable amounts of water and becomes soft, flexible, and viscous. Pectin is a heterogeneous grouping of acidic structural polysaccharides with the main molecules being the D-galacturonic acid residues. They do not possess exact structures and are complex. As for cellulose, pectin is a water binder, is highly viscous, and can form a firm gel when in contact with water. In all cases, the mucilage of the diaspores investigated contained both cellulose and pectic substances. This could be due to the presence of microfibres of cellulose embedded in the pectin compounds forming a complex matrix. Both pectic substances and cellulose were also reported in the seed mucilage of *Arabidopsis thaliana* (Brassicaceae) (Windsor *et al.* 2000; Willats *et al.* 2001; Macquet *et al.* 2007). In a detailed chemical and macromolecular study of the composition of *A. thaliana* seed mucilage (Brassicaceae), Macquet *et al.* (2007) found that the mucilage was made up of two layers: a water-soluble layer that could be separated from the seed and an inner layer that remained firmly attached to the seed. The inner layer was itself constituted of two domains,

the internal one of which contained cellulose. The major pectin of both the water-soluble and adherent seed mucilage was rhamnogalacturonan I.

This study has only provided a rough indication of the composition of the mucilage of the diaspores investigated. The chemistry of mucilage is complex and needs a more detailed study to determine the exact type of polysaccharide-forming mucilage in each diaspore.

CONCLUSIONS

The mucilaginous layer of the diaspores of 12 species from Namaqualand was investigated anatomically using both light and scanning electron microscopy. The structural origin of the mucilage produced by the diaspores was diverse. Diaspores of myxospermic plant species might produce mucilage in different ways; but in general, this function resides in the epidermal cells and other external appendages such as hairs and wings. Previous systems to classify the different types of mucilage production did not recognize the mucilaginous nature of wings or pappus. Despite the diversity of the origin and structure of the mucilage, it performs more-or-less the same ecological functions.

ACKNOWLEDGEMENT

The authors gratefully acknowledge the support received from the German Federal Ministry of Education and Research (BMBF) through the BIOTA South Project and the National Research Foundation under grant no. 61277.

REFERENCES

BREGMAN, R. & GRAVEN, P. 1997. Subcuticular secretion by cactus seeds improves germination by means of rapid uptake and distribution of water. *Annals of Botany* 80: 525–531.

CONSERVATION INTERNATIONAL. 2009. http://www.biodiversity-hotspots.org (accessed 2009).

CHAMBERS, J.J. & MACMAHON, J.A. 1994. A day in the life of a seed: movements and fates of seeds and implications for natural and managed systems. *Annual Review of Ecology and Systematics* 25: 263–392.

ELLNER, S. & SHMIDA, A. 1981. Why are adaptations for long-range seed dispersal rare in desert plants? *Oecologia* 51: 133–144.

FOTOUO MAKOUATE, H. 2008. *Dispersal strategies in communal versus privately-owned rangeland in Namaqualand, South Africa.* M.Sc. dissertation, University of Pretoria, Pretoria.

GAT-TILMAN, G. 1995. The accelerated germination of *Carrichtera annua* seeds and the stimulating and inhibiting effects produced by the mucilage at supra-optimal temperatures. *Journal of Arid Environments* 30: 327–338.

GRUBERT, M. 1974. Studies on the distribution of myxospermy among seeds and fruits of Angiospermae and its ecological importance. *Acta Biologica Venezuelica* 8: 315–351.

GUTTERMAN, Y. 1993. *Seed germination in desert plants.* Springer, Berlin.

GUTTERMAN, Y. 1996. Some ecological aspects of plant species with mucilaginous seed coats inhabiting the Negev Desert of Israel. In Y. Steinberg, ed., *Preservation of our world in the wake of change* 6: 492–496. Jerusalem.

GUTTERMAN, Y. & SHEM-TOV, S. 1997. Mucilaginous seed coat structure of *Carrichtera annua* and *Anastatica hierochuntica* from the Negev Desert highlands of Israel and its adhesion to the soil crust. *Journal of Arid Environments* 35: 695–705.

GUTTERMAN, Y., WITZTUM, A. & EVENARI, M. 1967. Seed dispersal and germination in *Blepharis persica* (Burm.) Kuntze. *Israel Journal of Botany* 16: 213–234.

GUTTERMAN, Y., WITZTUM A. & HEYDECKER, W. 1973. Studies of the surfaces of desert plant seeds. II. Ecological adaptations of the seeds of *Blepharis persica*. *Annals of Botany* 37: 1051–1055.

HEDGE, I.C. 1970. Observation on the mucilage of *Salvia* fruits. *Notes from the Royal Botanic Garden Edinburgh* 30: 79–95.

HUANG, Z., BOUBRIAK, I., OSBORNE, D.J., DONG, M. & GUTTERMAN, Y. 2008. Possible role of pectin-containing mucilage and dew in repairing embryo DNA of seeds adapted to desert conditions. *Annals of Botany* 101: 277–283.

LAWTON, J.R. & ETTRIDGE, S.C. 1986. Cytochemical staining of lipids in transition electron microscopy. *Electron Microscopy Society of Southern Africa Proceedings* 16: 115, 116.

LEPRINCE, O., HENDRY, G.A.F. & MCKENZIE, B.D. 1993. Membranes, protection, desiccation and ageing. *Seed Science Research* 3: 272–275.

MACQUET, A., RALET, C.-C., KRONENBERGER, J., MARION-POLL, A. & NORTH, H.M. 2007. *In situ* chemical and macromolecular study of the composition of *Arabidopsis thaliana* seed coat mucilage. *Plant Cell Physiology* 48: 984–999.

MCLEAN, W.C. & IVIMEY COOK, W.C. 1941. *Plant science formulae. A reference book for plant science laboratories (including Bacteriology).* MacMillan, London.

MUCINA, L. & RUTHERFORD, M.C. (eds). 2006. The vegetation of South Africa, Lesotho and Swaziland. *Strelitzia* 19. South African National Biodiversity Institute, Pretoria.

MURBECK, S. 1919. Beiträge zur Biologie der Wüstenpflanzen: Vorkommen und Bedeutung von Schleimabsonderung aus Samenhüllen. *Lunds Universitats Arsskrif* NF Adv 2, 15: 1–36.

O'BRIEN, T.P. & MCCULLY, M. 1981. *The study of plant structure. Principles and selected methods.* Termarcarphi, Melbourne.

OSBORNE, D.J., SHARON, R. & BEN-ISHAI, R. 1980/81. DNA integrity and repair. *Israel Journal of Botany* 29: 259–272.

RÖSCH, M.W. 1977. *Enkele plantekologiese aspekte van die Hester Malan-natuurreservaat.* M.Sc. dissertation. University of Pretoria, Pretoria.

STOPP, K. 1958. Die verbreitungshemmenden Einrichtungen in der südafrikanischen Flora. *Botanische Studien* 8, Jena.

TAMARA, L.W., SKINNER, D.J. & HAUGHN, G.W. 2000. Differentiation of mucilage secretory cells of *Arabidopsis* seed coat. *Plant Physiology* 122: 345–355.

VAN DER MERWE, C.F. & PEACOCK, J. 1999. Enhancing conductivity of biological material for the SEM. *Microscopy Society of Southern Africa—Proceedings* 29: 44.

VAN RHEEDE VAN OUDTSHOORN, K. & VAN ROOYEN, M.W. 1999. *Dispersal biology of desert plants.* Springer, Berlin.

VAN ROOYEN, M.W., THERON, G.K. & GROBBELAAR, N. 1990. Life form and dispersal spectra of the flora of Namaqualand, South Africa. *Journal of Arid Environments* 19: 133–145.

VAN ROOYEN, M.W. 1999. Functional aspects of short-lived plants. In R.W.J. Dean & S.J. Milton (eds), *The Karoo: ecological patterns and processes*: pp. 107–122. Cambridge University Press, Cambridge.

WINDSOR, J.B., SYMONDS, V.V., MENDENHALL, J. & LLOYD, A.M. 2000. *Arabidopsis* seed coat development: morphological differentiation of the outer integument. *Plant Journal* 22: 483–493.

WILLATS, W.G.T., MCCARTNEY, L. & KNOX, J.P. 2001. *In situ* analysis of pectic polysaccharides in seed mucilage and at the root surface of *Arabidopsis thaliana*. *Planta* 213: 37–44.

ZAADY, E., GUTTERMAN, Y. & BOEKEN, B. 1997. The germination of mucilaginous seeds of *Plantago coronopus, Reboudia pinnata,* and *Carrichtera annua* on cyanobacterial soil crust from the Negev. *Plant and Soil* 190: 247–252.

ZOHARY, M. 1937. Die verbreitungsökologischen Verhältnisse der Pflanzen Palästinas. Die antitelechoristischen Erscheinungen. *Beihefte Botanisches Centralblatt* A 56: 1–155.

5

A taxonomic revision of the southern African native and naturalized species of *Silene* L. (Caryophyllaceae)

J.C. MANNING* and P. GOLDBLATT**

Keywords: Caryophyllaceae, new species, seed morphology, *Silene* L., southern Africa, taxonomy

ABSTRACT

The native and naturalized species of *Silene* L. in southern Africa are reviewed, with full synonomy and the description of two new species from the West Coast of Western Cape. Eight native species and three naturalized species are recognized, including the first identification in southern Africa of the Mediterranean *S. nocturna* L. The identity of *S. aethiopica* Burm., which has remained unknown since its description, is established and is found to be the oldest name for *S. clandestina* Jacq. Patterns of morphological variation within each species are discussed and subspecies are recognized for geographically seg-regated groups of populations that are ± morphologically diagnosable. The following new names or combinations are made among the southern African taxa: *S. aethiopica* subsp. **longiflora**; *S. burchellii* subsp. **modesta**, subsp. **multiflora**, and subsp. **pilosellifolia**; *S. crassifolia* subsp. **primuliflora**; *S. saldanhensis*; *S. rigens*; and *S. undulata* subsp. **polyantha**. Each taxon is described, with information on ecology and distribution, and most species are illustrated, including SEM micrographs of the seeds.

INTRODUCTION

Silene L. (tribe Sileneae), with 600–700 species (Greuter 1995b), is distributed mainly through the tem-perate regions of the northern Hemisphere, with its prin-cipal centre of diversity in the Mediterranean and Mid-dle East (Oxelman & Lidén 1995). Over 90 species are recorded in Africa, the vast majority in North Africa, with very few extending southwards into sub-Saharan Africa. Just three native and one introduced species of *Silene* are known from south tropical Africa (Turrill 1956a; Wild 1961), with several additional species cur-rently accepted in southern Africa.

Consensus has still to be reached on the delimita-tion of *Silene*. The circumscription of the genus had been gradually expanded over the last few decades of the twentieth century until it coincided essentially with the delimitation of tribe Sileneae (Greuter 1995b). This consolidation was supported by a preliminary molecu-lar sequence analysis of the tribe (Oxelman & Lidén 1995), which nested all previously recognized genera of Sileneae in *Silene*, with the possible exception of *Agrostemma* L., but had been presaged by the break-down of morphological differences between recognized segregates when these were examined on a global scale. *Cucubalus* L., which is deeply nested in *Silene*, is dis-tinguished primarily by its berry-like fruits, which evi-dently represent a specialized dispersal strategy. *Lychnis* L., which forms a discrete clade within Sileneae, was traditionally diagnosed by having capsules with as many teeth as styles (vs. twice as many in *Silene*) (e.g. Chow-dhuri (1957) and Chater & Walters (1964)) but the two genera were diagnosed only with great difficulty by Bit-

trich (1993) using a combination of trivial characters, and his recognition of the two appears to have been pro-visional until *Silene* could be conserved against *Lychnis*, under which name it had earlier been included by Sco-poli (1771). The generic name *Lychnis* therefore took priority until it was formally rejected in favour of *Silene* (Brummit 1994).

There are currently two opposing views on generic delimitation in Sileneae. The first, advocated by Greuter (1995b), is an inclusive one that recognizes just the two genera *Agrostemma* and *Silene*. In this sense, *Silene* is broadly distinguished within subfamily Caryophylloi-deae (Bittrich 1993) by its actinomorphic flowers with 3 or 5 styles. The alternative view, recently developed by Oxelman *et al.* (2000), proposes the partitioning of *Silene* among seven genera, of which only *Lychnis* (30 spp.) includes more than a handful of species. The clade that constitutes *Silene* in this narrow sense is only weakly supported in the molecular analysis and is also not easily diagnosed morpholgically from its segregates. Although we favour a more inclusive circumscription of the genus, the native and introduced southern African species are all included in *Silene* either way.

We follow the infrageneric classification of *Silene* s. *lat.* proposed by Greuter (1995b), based largely on Chowdhuri (1957) with some rationalization and sev-eral nomenclatural corrections. Although there is mount-ing evidence (Petri & Oxelman 2011) that at least some of these sections are not monophyletic, no satisfactory alternative has yet been offered. Significant characters at sectional level are the type of inflorescence, presence or absence of septa in the ovary, number of styles, shape and venation of the calyx, number of teeth in the dehisc-ing capsule, and the shape of the seeds.

Four global treatments have laid the foundations for our present understanding of the genus (Otth 1824; Rohrbach 1869, 1869–1870; Williams 1896; Chow-dhuri 1957), and several modern regional treatments of the European species have also been published (Greuter 1995b and references therein). Regional floristic studies

* Compton Herbarium, South African National Biodiversity Institute, Private Bag X7, Claremont 7735, South Africa / Research Centre for Plant Growth and Development, School of life Sciences, University of KwaZulu-Natal, Pietermaritzburg, Private Bag X01, Scottsville 3209, South Africa. E-mail: J.Manning@sanbi.org.za.
** B.A. Krukoff Curator of African Botany, Missouri Botanical Garden, P.O. Box 299, St. Louis, Missouri 63166, USA. E-mail: peter.gold-blatt@mobot.org.

exist for parts of North Africa (Wickens 1976; Gilbert 2000) and much of sub-Saharan Africa (Turrill 1956a; Wild 1961) but the taxonomy of *Silene* in southern Africa still rests largely on Sonder's (1860) treatment in *Flora capensis*, and only minor advances in our understanding of the southern African species have been made since then (Rohrbach 1869, 1869–70; Bocquet 1977; Masson 1989; Goldblatt & Manning 2000). The South African species of *Silene* are unique among the flora of the subregion in having been fundamentally misunderstood by Thunberg (1794), who misidentified the majority of his collections as European species (Table 1), leaving it to later authors to recognize that they were indigenous to the country and undescribed. Burman (1768) had earlier made the same mistake, identifying as European all but two (*S. crassifolia* L. and *S. aethiopica* Burm.) of the eight species that he listed from the Cape. Two previous attempts at revising the southern African species, by G. Bocquet in the 1970s and D. Masson in the 1980s, were never completed but many of the southern African herbarium collections bear annotation labels reflecting Masson's unpublished species concepts and names, adding further to the confusion. This revision of *Silene* in southern Africa aims to redress the situation and provide a base for future studies of the genus in the subcontinent. This is especially important as the South African winter rainfall region is a minor centre of diversification for *Silene* in Africa.

Species delimitation in the genus is notoriously difficult and most African species, especially those with a wider distribution, display a complex and often confusing pattern of local variation that defies formal classification (Gilbert 2000). The primary reason for this is almost certainly that the species are often facultatively autogamous, which facilitates the differentiation of local phenotypes, probably independently and repeatedly. Population genetic studies are likely to shed significant light on the nature and origin of these local forms. We have recognized as species only entities that are morphologically (and preferably also ecologically) discontinuous, based on the assumption of ± complete interspecific infertility. Although Stace (1978) observed that interspecific hybridization is generally rare in *Silene*, recent molecular investigations (e.g. Erixon & Oxelman (2008) and Frajman *et al.* (2009)) suggest that it is an important consideration in some species groups. Where possible, we recognize infraspecific taxa for geographically coherent groups of populations that show evident but incomplete morphological differentiation. We have uniformly applied the rank of subspecies to these taxa but experimental investigation might indicate that other

ranks may be useful in some instances, such as in *S. aethiopica*, where subsp. *longiflora* may be better treated at the level of *forma*. The genetics of the maritime forms of *S. aethiopica* and *S. undulata* also require investigation. Our intention, therefore, is to present a preliminary classification that can form the basis for future experimental studies in the genus.

Another intriguing aspect of the genus is its history and diversification in southern Africa. The indigenous species are currently distributed among three sections, all of which are best represented in the northern hemisphere and presumably originated there. *Silene burchellii* (sect. *Fruticulosae*) is widely distributed through the eastern half of the continent, from Ethiopia (and Arabia) to the Cape Peninsula in the extreme south, and the most parsimonious interpretation is that the species migrated from North Africa southwards along the East African Afromontane corridor. *Silene undulata* (sect. *Elisanthe*), in contrast, is largely southern African, extending only as far north as the highlands of eastern Zimbabwe, and its origin in southern Africa remains unclear but long-distance dispersal appears to be more likely. Both sections have subsequently undergone minor adaptive radiation in the coastal regions of the subcontinent. The third lineage is represented by *S. aethiopica* (sect. *Dipterospermae*), the only native southern African annual species. Essentially endemic to the winter-rainfall region in the extreme south and southwest, with no track in tropical Africa, *S. aethiopica* most likely arrived in the subcontinent through long-distance dispersal.

MATERIALS AND METHODS

All relevant types were examined, as well as all herbarium material from BOL, MO, NBG, PRE, and SAM (acronyms after Holmgren *et al.* 1990), the primary collections of species from southern Africa. All species were also studied in the field.

Ripe seeds from dehisced capsules were mounted directly on aluminium stubs and coated with gold-palladium before viewing at the SEM Unit, University of Cape Town. Seeds were sampled from at least two different collections per species (Figures 1, 2).

TAXONOMY

Silene *L.*, Species plantarum: 416 (1753), nom. cons. against *Lychnis* [Brummit: 272 (1994)]. Lectotype:

TABLE 1—South African *Silene* species in the Thunberg Herbarium (UPS-THUNB). Native South African taxa in bold.

Thunberg (1794) and Thunberg Herbarium name	Current name	Specimen (UPS-THUNB no.)
S. bellidifolia Juss. ex Jacq.	**S. ornata** Aiton	*10705*
S. cernua Thunb.	**S. aethiopica** Burm.	*10713, 10714, 10715*
S. crassifolia L.	**S. crassifolia** L.	*10726*
S. gallica L.	*S. gallica* L.	*10744*
S. noctiflora L.	**S. rigens** J.C.Manning & Goldblatt	*10774, 10775*
S. nutans L.	**S. burchellii** Otth	*10780*
S. ?pendula L.	*S. nocturna* L.	*10790* (left hand plant)
S. ?pendula L.	**S. undulata** Aiton	*10790* (right hand plant)

FIGURE 1.—Seed morphology and testa sculpturing in non-native southern African *Silene* sect. *Behen* and sect. *Silene*. A, B, *S. vulgaris, Moffet 613* (NBG): A, side; B, back. C, D, *S. nocturna, Goldblatt & Porter 13551* (NBG): C, side; D, back. E, F, *S. gallica, Bohnen 6351* (NBG): E, side; F, back. Scale bar: 100 μm.

Silene anglica L., nom. rej. in favour of *S. gallica* L., designated by Britton & Brown: 62 (1913) = *S. gallica* L., nom. cons.

Cucubalus L.: 414 (1753). Type: *Cucubalus baccifer* L. = *Silene baccifera* (L.) Roth.

Lychnis L.: 436 (1753), nom. rej. in favour of *Silene* L. [Brummit: 272 (1994)]. Lectotype: *Lychnis chalcedonica* L., designated by Britton & Brown: 62 (1913) = *Silene chalcedonica* (L.) E.H.L.Krause

Viscaria (DC.) Röhl.: 37 (1812). *Lychnis* sect. *Viscaria* DC.: 761 (1805). Type: *Viscaria vulgaris* Röhl. = *Silene viscaria* (L.) Jess.

Annual, biennial or perennial herbs, rarely geophytes or small shrubs, sometimes caespitose, rarely scandent, glabrous or variously pubescent, sometimes glandular-pubescent, rarely gynodioecious or dioecious. *Leaves* opposite, sometimes slightly connate at base, sessile or shortly petiolate, exstipulate. *Inflorescence* a dichasium or monochasium, rarely 1-flowered or sometimes ± capitate. *Flowers* erect, spreading or pendulous, usually bisexual, petals and stamens basally connate with floral axis forming a short anthophore [*fide* Greuter 1995b]. *Calyx* tubular or dilated, sometimes strongly inflated, 5-toothed, 10(–60)-nerved, usually longitudinally ribbed and ± plicate. *Petals* 5(0), usually white or pink, clawed, limb entire to deeply lobed, mostly variously bifid, often with paired coronal scales at base of limb, claw glabrous or ciliate. *Stamens* 5 + 5 in male or bisexual flowers. *Ovary* shortly stipitate, unilocular but usually 3(–5) septate in lower part, styles 3(6 or 10). *Carpel* subglobose to ovoid, borne on persistent carpophore [anthophore], cartilaginous and dehiscing apically by usually twice as

many (rarely as many) teeth as styles, rarely berry-like. *Seeds* numerous, ± reniform, compressed, sometimes disc-like and peripherally winged, variously scupltured, often ± colliculate.

600–700 spp., mainly temperate Eurasia (± 600 spp.), Africa (± 90 spp.) and N America (± 65 spp.).

Key to species [introduced taxa marked with an asterisk]

1a Inflorescence a ± divaricately branched dichasium; plants perennial:
2a Plants entirely glabrous, ± glaucous; calyx inflated, papery, loosely investing capsule, equally 20-veined with veins not raised; petal limbs divided to base, coronal scales obsolete, claw abruptly narrowed and filiform in lower half; ovary distinctly septate (sect. *Behen*) .... 1. *S. vulgaris**
2b Plants pubescent with eglandular and glandular hairs; calyx not inflated, unequally veined with median veins thickened and prominent; petal limbs bifid, coronal scales well-developed, claw gradually tapering to base; ovary unilocular (sect. *Elisanthe*):
3a Lower portion of stem villous with spreading or shaggy, eglandular hairs 1–2 mm long, sometimes mixed with shorter eglandular or glandular hairs, upper part of stem partly or entirely glandular-haired; calyx vesture partly or entirely eglandular (rarely entirely glandular), lobes always fringed with eglandular hairs; stamens and styles often ultimately well-exserted; carpophore (2–)4–10(–15) mm long; widespread through southern Africa .... 2. *S. undulata*
3b Stem either entirely glandular-haired or with a mix of gland-tipped hairs and short, eglandular hairs 0.25–0.50 mm long; calyx vesture entirely glandular and lobes fringed with glandular hairs; stamens and styles included or very shortly exserted; carpophore 2–4 mm long; coastal sands and calcrete in extreme southwestern Western Cape:
4a Stems stiffly erect, vestiture entirely of gland-tipped hairs, without admixture of acute, eglandular hairs; flowers relatively small, claws 8–10 mm long; cap-

FIGURE 2.—Seed morphology and testa sculpturing in southern African *Silene* sect. *Elisanthe*. A, B, *S. undulata*: A, side; B; back, *Goldblatt 13549* (NBG). C, D, *S. saldanhensis, Goldblatt & Manning 13646* (NBG): C, side; D, back. E, F, *S. ornata, Goldblatt & Porter 13254* (NBG): E, side; F, back. G, H, *S. rigens, Goldblatt & Porter 13291* (NBG): G, side; H, back. Scale bar: 100 μm.

sules broadly urn-shaped, 9–10 mm diam. at maturity, causing calyx to split longitudinally at least midway but often to base; seeds echinate . . . . 5. *S. rigens*

4b Stems sprawling, vestiture a mix of gland-tipped hairs and short, eglandular hairs; flowers larger, claws 15–18 mm long; capsules urn-shaped, 6–8 mm diam., not causing calyx to split; seeds colliculate or tuberculate:

  5a Plants mostly laxly sprawling, rarely compact in exposed situations, with flowering stems straggling; calyx without prominent, thickened ribs; petals deep carmine, limbs not overlapping; seeds tuberculate; growing on limestone outcrops in rocky pavement . . . . 3. *S. ornata*

  5b Plants compact, ± cushion-like with flowering stems decumbent or erect; calyx with prominent, thickened ribs; petals mauve, limbs overlapping; seeds colliculate; growing in deep, calcareous sands and consolidated dunes . . . . 4. *S. saldanhensis*

1b Inflorescence a raceme-like monochasium, rarely only 1- or 2-flowered; plants annual or perennial:

  6a Annuals; upper stems and inflorescence axes glandular-pubescent; calyx urn-shaped with triangular or awl-shaped lobes, ribs bristly; carpophore < 1 mm long; seeds compressed and peripherally winged with back flat or shallowly grooved, testa colliculate or tuberculate (sect. *Silene*):

    7a Calyx glandular-pubescent and with glassy bristles 2–3 mm long on veins, without anastomising venation; petals entire . . . . 11. *S. gallica**

    7b Calyx appressed-puberulous and eglandular, with only slightly longer hairs to 0.5 mm long on veins, anastomising venation developed in distal half at least; petals bilobed . . . . 10. *S. nocturna**

  6b Perennials or annuals; stems and inflorescences eglandular-puberulous; calyx clavate with ovate lobes, uniformly shortly puberulous; carpophore 2–12 mm long; seeds compressed and disc-like, peripherally winged with back deeply and acutely grooved between wings, testa striate, cells radially fusiform:

    8a Annuals (sect. *Dipterospermae*) . . . . 9. *S. aethiopica*

    8b Geophytic perennials (sect. *Fruticulosae*):

      9a Plants densely and closely branched, forming small prostrate mats or loose tangled mounds with the

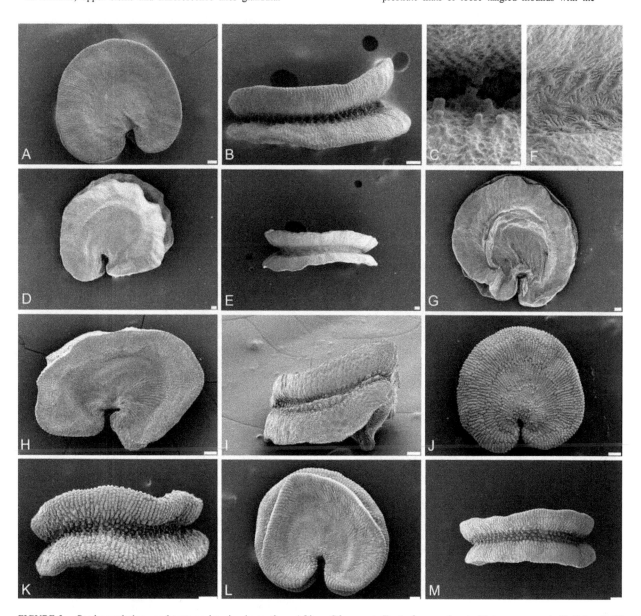

FIGURE 3.—Seed morphology and testa sculpturing in southern African *Silene* sect. *Fruticulosae* and sect. *Dipterospermae*. A–C, *S. burchellii* subsp. *burchellii*, *Penfold 111* (NBG): A, side; B, back; C, detail of dorsal groove. D–F, *S. crassifolia* subsp. *crassifolia*, *Goldblatt et al. 13447* (NBG): D, side; E, back; F, detail of dorsal groove. G, *S. crassifolia* subsp. *primuliflora*, *Middelmost 2046* (NBG): G, side. H, I, *S. mundiana*, *Van der Merwe 1871* (NBG): H, side; I, back. J, K, *S. aethiopica* subsp. *aethiopica* 'typical form', *Goldblatt et al. 13469* (NBG): J, side; K, back. L, M, *S. aethiopica* subsp. *aethiopica* 'maritime form', *Goldblatt & Porter 13569* (NBG): L, side; M, back. Scale bar: A, B, D, E, G–M, 100 μm; C, 10 μm; F, 30 μm.

numerous inflorescences extending only shortly above the vegetative growth; inflorescence axis ± filiform, 0.2–0.5 mm diam.; leaves 7–15 × 1–3 mm; racemes 1- or 2-flowered . . . . 8. *S. mundiana*

9b Plants not forming tangled mounds or mats and with few inflorescences extending well above vegetative growth; inflorescence axis terete, 0.5–1.5 mm diam.; leaves 10–60 × (1–)2–12 mm; racemes mostly more than 2-flowered:

10a Stems mostly erect or suberect, rarely decumbent, mostly annual, 0.5–1.5 mm diam. at base of inflorescence; leaves linear to oblanceolate, rarely obovate, thin-textured, clustered in lower half of stem and well separated from inflorescence; calyx weakly pleated; capsule ovoid, 8–10 × 5–7 mm; plants inland or coastal but never on foredunes . . . . 6. *S. burchellii*

10b Stems ± prostrate or decumbent, relatively robust, perennial and often branching, mostly 1.5–2.0 mm diam at base of inflorescence; leaves oblanceolate to suborbicular, ± succulent, running into inflorescence; calyx strongly pleated; capsule broadly urn-shaped to subglobose, (8–)10–12 × 7–10 mm; plants coastal on sandy foredunes . . . . 7. *S. crassifolia*

Silene sect. **Behen** *Dumort.*, Florula belgica: 107 (1827). Sect. *Inflatae* (*Boiss.*) *Chowdhuri*: 241 (1957), superfluous name. *Silene* [unranked] *Inflatae* Boiss.: 573 (1867). Type: *Silene vulgaris* (Moench) Garcke

± 8 spp., Europe and extratropical Asia, N Africa and Atlantic islands; naturalized elsewhere.

Perennials or biennials, glabrous to sparsely eglandular-pubescent, often glaucous. *Leaves* lanceolate to orbicular. *Flowers* solitary or in few-flowered dichasia, half-pendent at anthesis, vespertine (sometimes indistinctly so); anthophore glabrous. *Calyx* glabrous, inflated, 10- or 20-veined, with conspicuous reticulate venation. *Petals*: claw glabrous, auriculate, limb bifid to base; coronal scales linear or reduced. *Stamens*: filaments glabrous. *Ovary* partially septate. *Capsule* subglobose. *Seeds* semicircular-reniform, with flat to convex flanks and back, radially colliculate-echinate.

1. **Silene vulgaris** *(Moench) Garcke*, Flora von Nord- and Mittel-Deutschland, ed. 9: 64 (1869). *Cucubalus behen* L.: 414 (1753), non *Silene behen* L. (1753). *Behen vulgaris* Moench: 709 (1794). *Silene cucubalus* Wibel: 241 (1799), nom. illegit. superfl. *Silene inflata* Sm.: 467 (1800), nom. illegit. superfl. Type: 'Habitat in Europae septentrionalis pratis siccis', *Herb. Linn. 582.4* [LINN, lecto.!, designated by Aeschimann & Bocquet: 204 (1983)].

[see Chater & Walter (1964) and Greuter (1997) for heterotypic synonyms]

Gynodioecious perennial to 60 cm tall, softly woody at base, stoloniferous; stems ascending or erect, 1–3 mm diam., usually branched, glabrous, with or without axillary non-flowering shoots. *Leaves* ovate-lanceolate or the lower oblanceolate, 30–90 × 6–15 mm, acute to apiculate, base cuneate or narrowed and petiole-like in lower leaves, glabrous, glaucous, with evident side veins. *Inflorescence* a laxly divaricate dichasium, usually several-flowered; bracts smaller than leaves, ovate-attenuate, up to 15 × 5 mm, thinly herbaceous with submembranous margins or entirely ± scarious; pedicels mostly 10–15 mm long at anthesis but up to 30(–50) mm

long in fruit. *Calyx* inflated, urn-shaped and bladder-like, 10–12 × 7–8 mm at anthesis but accrescent to 13× 10 mm in fruit and loosely investing capsule, base intrusive, whitish or pale green with darker veining, equally 20-veined with reticulate venation throughout, not plicate, lobes broadly triangular, 1.5 × 2.0 mm, densely ciliolate along margins and inner face toward apex. *Flowers* spreading or ± pendent at anthesis, bisexual or functionally female, lily-scented at night only but remaining open during day. *Petals* white or pink to purple, claw 10–11 mm long, cuneate and conspicuously auriculate in upper half, abruptly narrowed and filiform in basal half, limb deeply bilobed almost to base, 4–6 × 4–5 mm, coronal scales obsolete. *Stamen* filaments ± 18 mm long and exserted ± 7 mm, or in functionally female flowers 5–6 mm long and deeply included, glabrous. *Ovary* ovoid, ± 3 mm long; styles 18 mm long, exserted 7–8 mm. *Capsule* urn-shaped, ± 10 × 7 mm, usually ± 3 × longer than carpophore; carpophore 2.5–3.0 mm long, glabrous. *Seeds* reniform with flat or convex face and convex back, 1.0–1.5 mm, tuberculate, grey. *Flowering time:* Oct.–Dec.(–Jan.). Figures 1A, B, 4.

Vernacular name: bladder campion.

Distribution and ecology: native to Europe, where it is widely distributed across the continent through the Middle East into temperate Asia; introduced into Ethiopia, Australia, South Africa (where it is cultivated as an ornamental) and also North America, where it is considered a weed.

Diagnosis: readily distinguished by the complete lack of vestiture and by the pale greenish or whitish, inflated, bladder-like calyx with 20 equal longitudinal veins joined by extensive reticulated veining throughout. The petals, deeply divided to the base of the limb with the claw abruptly narrrowed and filiform in the basal half, are also diagnostic. Flowers are either bisexual, with long, well-exserted filaments ± 18 mm long, or functionally female, with relatively short filaments, 5–6 mm long, deeply included within the calyx.

The species is broadly circumscribed by Chater & Walters (1964), who recognize eight subspecies for the European material, most of which have been variously treated by other authors. The South African material is stoloniferous and is thus treated as subsp. *macrocarpa* Turrill (1956b).

subsp. **macrocarpa** *Turrill*, Hooker's Icones plantarum 36: t. 3551 (1956b). Type: Cyprus, Chionistra, 1 620 m, 4 Jun. 1940 [grown in the Herbarium Experimental Ground, Kew, from seeds collected 5 Jul. 1938], *Kennedy s.n. K2312* (K, holo.).

Distribution and ecology: widely distributed through the Mediterranean and long established in southwest Britain, whence it was presumably introduced to South Africa, probably as an ornamental. In South Africa the species has been recorded from mesic situations in the Western Cape winter-rainfall region (Tygerberg Hills, around Stellenbosch and Oudtshoorn, and in the Langkloof) and from the Hogsback in Eastern Cape, as a garden escape in waste places, along roadsides, around old lands, and in vineyards. It was first recorded as an

FIGURE 4.—*Silene vulgaris*, ex hort., no voucher. A, flowering stem; B, petal; C, gynoecium; D, ovary l/s; E, seed. Scale bar: A, 10 mm; B, C, 2.5 mm; D, 2 mm; E, 0.5. Artist: John Manning.

escape around Stellenbosch in the early twentienth century (Figure 5).

In South Africa, *Silene vulgaris* subsp. *macrocarpa* can form dense stands, propagating through stolons but always in transformed habitats and there is no indication that it is invasive. The extensive, stoloniferous rootstock may however become a problem for cultivation if the plants are allowed to establish unchecked over several years in fallow lands, and subsp. *macrocarpa* has become a weed of agronomic importance in Europe (Aeschimann 1983).

Additional specimens

WESTERN CAPE.—**3318** (Cape Town): Tygerberg Nature Reserve, (–DC), 12 Dec. 1975, *Loubser 3476* (NBG); Rustenberg

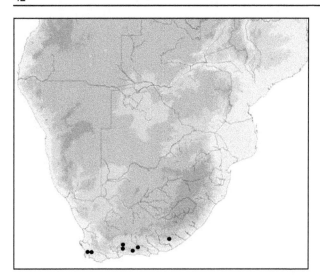

FIGURE 5.—Recorded distribution of *Silene vulgaris* in southern Africa.

Farm, (–DD), 7 Oct. 1973, *Taylor 8608* (NBG, PRE); Idas Valley, Schoongesicht Farm, (–DD), 26 Oct. 1973, *Neser s.n.* (NBG); Stellenbosch, Blaauwklip, (–DD), 10 Oct. 1928, *Gillett 582* (NBG); Stellenbosch, Devon Valley, (–DD), 21 October 1968, *Du Toit s.n.* (NBG); Jonkershoek, (–DD), 19 Oct. 1943, Adamson 3514 (BOL). **3322** (Oudtshoorn): Prince Albert, (–AA), 8 Nov. 1988, *Dean 748* (PRE); Oudtshoorn Experimental Farm, (–AC), 29 Jan. 1970, *Maree 40* (NBG); Boomplaas, Cango Valley, (–AC), 5 Feb. 1975, *Moffett 613* (NBG). **3323** (Willowmore): Longkloof, (–BC), 13 November 1970, *Botha s.n.* (NBG, PRE); Avontuur, (–CA), 2 Jan. 1970, *Wiid s.n.* (NBG)

EASTERN CAPE.—**3226** (Fort Beaufort): Hogsback, Risingham, old cultivated land, (–DB), 10 Jan. 1989, *King s.n. PRE60819* (PRE).

Silene sect. **Elisanthe** *(Fenzl. ex Endl.) Ledeb.*, Fl. Ross. 1: 314 (1842). *Saponaria* sect. *Elisanthe* Fenzl. ex Endl., Gen. Pl.: 972 (1840). *Silene* sect. *Melandriformes* (Boiss.) Chowdhuri: 244 (1957), superflous name, *Silene* [unranked] *Melandriformes* Boiss., Fl. Or. 1: 568 (1867). Lectotype, designated by Pfeiffer: 1186 (1871–1875): *Silene noctiflora* L.

Annuals, biennials and perennials, pubescent with eglandular and glandular hairs, monoecious or dioecious. *Leaves* ovate-lanceolate; bracts herbaceous. *Flowers* in few-flowered dichasia, bisexual or unisexual, spreading to ± erect, mostly vespertine; anthophore glabrous or pubescent. *Calyx* pubescent, clavate, 10-veined, with conspicuous reticulate venation. *Petals*: claw glabrous to ciliate, auriculate, limb bifid; coronal scales present. *Stamens*: filaments glabrous or pubescent. *Ovary* unilocular at maturity; styles 3 or 5. *Seeds* semi-circular-reniform, with convex flanks and back, radially colliculate-echinate.

± 20 spp., Eurasia and N Africa, tropical and S Africa.

The large-flowered southern African species were not treated by Chowdhuri (1957) in an inexplicable omission, but have traditionally been placed in sect. *Elisanthe* (Sonder 1860; Chater & Walters 1964), with which they accord in their dichasial inflorescence, reticulate calyx venation, completely unilocular ovaries, and colliculate-echinate seeds without peripheral wings.

2. **Silene undulata** *Aiton*, Hortus kewensis 2: 96 (1789); Sond.: 126 (1860). *Silene tristis* Salisb.: 301

(1796), nom. superfl. illegit. *Melandrium ornatum* var. *undulatum* (Aiton) Rohrb.: 234 (1869). *Melandrium undulatum* (Aiton) Rohrb.: 245 (1869–70). Type: South Africa, without precise locality, cultivated at Kew from seed introduced by Francis Masson, *Masson s.n. BM000593534* (BM, holo.!)

Tufted perennial, 0.5–2.0 m, producing new vegetative shoots from basal axils at end of flowering season, often developing a gnarled, softly woody crown; rootstock a carrot-like taproot. *Stems* erect to sprawling, rarely prostrate in exposed situations, simple to highly branched, 1.5–5.0(–8.0) mm diam., sparsely or densely pubescent with lower portions of stems bearing acute, eglandular hairs sometimes mixed with scattered, short, gland-tipped hairs, eglandular hairs either all longer or a mix of longer, spreading or shaggy hairs 1–2 mm long and shorter, decurved hairs, upper parts of stems usually with progressively more gland-tipped hairs, sometimes almost entirely glandular-haired. *Leaves* mostly basal, suberect or spreading, lower leaves oblanceolate-spathulate, sub-petiolate, up to 200 × 25 mm, sometimes dead or dying at flowering, cauline leaves elliptical to oblanceolate, mostly 50–80 × 10–20 mm, acute or acuminate, base tapering, margins plane or undulate, usually pubescent, either with mixture of erect, gland-tipped hairs and acute, eglandular hairs or entirely eglandular, sometimes subglabrous or rarely glabrous except on margins, margins always ciliate, with longer hairs to 1 mm long at extreme base, with evident side veins. *Inflorescence* a lax or more compact, symmetrically or asymmetrically branched cyme, axis 1–2 mm diam. below primary flower; bracts similar to upper leaves but smaller, subequal, suberect or spreading; primary pedicels 8–40 mm long, secondary pedicels 5–20 mm long. *Flowers* half-nodding, nocturnal, clove- or soapy-scented at night. *Calyx* cylindrical or narrowly urn-shaped in flower, (18–)20–35(–45) mm long, 10-ribbed, with 1–4 reticulate veins in distal half, densely or sparsely puberulous, hairs usually a mix of acute, eglandular and spreading, gland-tipped hairs but sometimes entirely glandular or entirely eglandular, lobes narrowly triangular to awl-shaped, (3–)5–10 mm long, densely ciliolate. *Petals* white to pink with yellowish reverse, limbs spreading, cuneate, not overlapping with adjacent petals, 9–15 × 8–15 mm, bifid ± halfway, claws strap-shaped, (12–)15–25 × 2.5–3.0 mm, glabrous or pubescent along abaxial midline and margins, exserted up to 10 mm beyond calyx, auriculate apically, auricles of adjacent tepals locking together, coronal scales spreading or suberect, 0.7–1.5 mm long, denticulate. *Stamen* filaments unequal, shorter series 11–25 mm long, longer series 16–30 mm long, reaching to top of claws or exserted; anthers ± 2 mm long, reaching top of claws and included or exserted up to 6 mm. *Ovary* narrowly pyriform, shortly stipitate, 5–7 mm long, stipe ± 1 mm long; styles ± 9–15 mm long at anthesis and included but usually elongating, ultimately reaching up to 25 mm long and exserted up to 10 mm. *Capsule* ovoid, (10–)15–25 × 5–8 mm, 1.5–6.0 × longer than carpophore, minutely granular; carpophore (2–)4–10(–15) mm long, glabrescent. *Seeds* 1.2–1.5 mm diam., reniform with hilum recessed, face flat, back flat, reddish brown, testal cells radial-concentric, colliculate-tuberculate. *Flowering time*: Aug.–Dec. in the winter rainfall region; Nov.–Mar. in the summer rainfall region but as late as June along the subtropical coast. Figures 2A, B, 6, 7.

Vernacular names: Cape catchfly, wild tobacco, wildetabak.

Distribution and ecology: widely distributed through the more temperate parts of southwestern and eastern southern Africa northwards into eastern Zimbabwe (Figure 8), occurring in a variety of stony or rocky habitats, in coastal and inland scrub, often in ravines, along riverbanks and forest margins, and in moist grassland, from just above the high tide mark to well over 2 000 m. The species is absent from the more arid regions, notably Bushmanland and the Kalahari in Northern Cape and the Great Karoo in Western and Eastern Cape.

Growth and flowering of the species in vegetated communities such as thicket margins is stimulated by clearance of the overburden by fire. The flowers are strictly nocturnal and strongly scented at night but although they may remain open for several hours on overcast mornings, sometimes until midday, they are then often no longer scented. The species is clearly pal-atable, being extensively browsed by both wild animals and livestock, but is also used traditionally as a medicine in many diseases, particularly fevers and delirium (Watt & Breyer-Brandwijk 1932).

Diagnosis and relationships: as circumscribed here, *Silene undulata* is a highly variable species with numerous local forms that have in common the development of relatively long, eglandular hairs, 1–2 mm long, on the lower portion of the stem, giving the base of the stems a characteristic, shaggy appearance (Figure 7C). In some instances the eglandular hairs are mixed with scattered, short, gland-tipped hairs, which typically predominate in the upper parts of the stems (Figure 7B), which are sometimes entirely glandular-haired. The vesture of the calyx is similarly variable, usually including a mix of longer, eglandular hairs and short, gland-tipped hairs, but sometimes either entirely eglandular or entirely glandular. The lobes are, however, invariably fringed with eglandular hairs. Other sub-Saharan species in the section have a ± entirely glandular pubescent calyx, including the margins of the lobes. Another characteristic of the species is the relatively long, narrowly triangular to

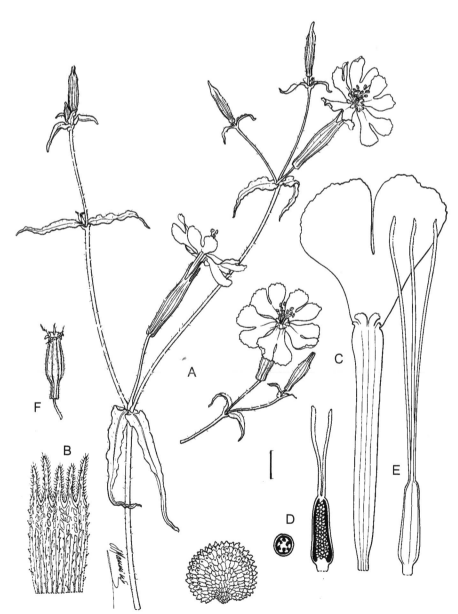

FIGURE 6.—*Silene undulata* subsp. *undulata*, Clanwilliam, *Goldblatt & Porter 13523*. A, flowering stem and flower; B, calyx laid out; C, petal; D, gynoecium before style elongation, l/s and t/s; E, gynoecium after style elongation; F, capsule after dehiscence; G, seed. Scale bar: A, F, 10 mm; B, 6 mm; C–E, 2.5 mm; G, 0.5. Artist: John Manning.

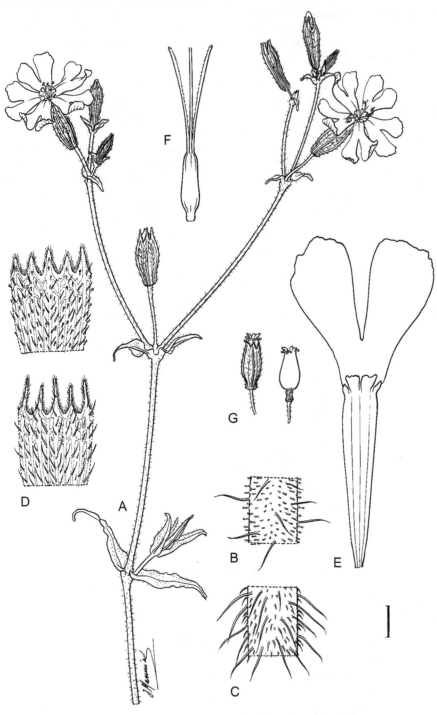

FIGURE 7.—*Silene undulata* subsp. *undulata*, Langebaan, without voucher. A, flowering stem; B, peduncle segment; C, lower stem segment; D, calyces laid out; E, petal; F, gynoecium before style elongation; G, capsules after dehiscence (left hand illustration with calyx removed). Scale bar: A, G, 10 mm; B, C, 1 mm; D, 5 mm; E, 2.5 mm. Artist: John Manning.

subulate calyx lobes, mostly 5–8 mm long (Figures 6B, 7D).

The species varies also in the length and exsertion of the stamens and styles. Style length is dependent on the age of the flowers, and the styles are always included at anthesis. In many populations, however, the styles elongate markedly after anthesis, becoming well exserted by the second or third day.

Silene undulata was described from cultivated plants grown at Kew from seeds collected in the Western Cape by Francis Masson, who visited the region from 1772 to 1775 and again from 1786 to 1795. The earliest collection of the species, however, appears to be an unlocalised specimen at the British Museum collected in 1772 by Franz Oldenburg (*Oldenburg 1260*). Oldenburg, who visited the Cape in1771–1773, accompanied both Masson and Carl Thunberg on collecting trips (Glen & Germishuizen 2010). His fragmentary specimen at the British Museum (*BM001010634*) was initially ignored, then mistakenly identified as the Sicilian species *Silene calycina* C.Presl., before finally being correctly determined as *S. undulata*. It represents the form with a proportionally longer carpophore that was initially recognized at species level under the name *S. capensis* Otth. An equally fragmentary collection of the same form made by Thunberg, possibly from the same population that yielded seeds to Masson since the two men certainly collected together on the Cape Peninsula and along the West Coast where *S. undulata* occurs, was mistaken by him for the Mediterranean *S. pendula* (Table 1).

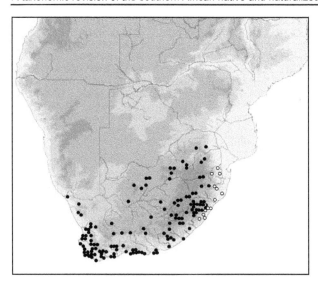

FIGURE 8.—Distribution of *Silene undulata* in southern Africa: subsp. *undulata*, ●; subsp. *polyantha*, ○.

Silene undulata exhibits numerous local forms, varying in habit, degree and nature of vestiture, length of calyx and carpophore, length and shape of calyx lobes, size of petals, degree of exsertion of stamens and styles, and shape and size of the capsule. Some of these have formed the basis of separate species but we are unable to identify any of these character states or combination of states that is consistent with the recognition of more than a single, variable species. The relative length of carpophore and capsule, in particular, has been used to distinguish several species in the complex, viz. *S. bellidioides* (carpophore one quarter as long as capsule), *S. undulata* (carpophore one third as long as capsule), and *S. capensis* (carpophore at least half as long as capsule) (Sonder 1860). A further two species were recognized on the basis of their relatively hairless leaves, *S. eckloniana* from Cape Recife with eglandular stems, and leaves glabrous above but hairy on the margins and along the midrib beneath, and *S. caffra*, also from Eastern Cape, with subglabrous stems and leaves that are ciliolate along the margins but otherwise glabrous on both surfaces. With numerous collections now to hand, we are unable to correlate any of these differences with other character states, nor with geography, and conclude that they represent independent and local variants within a single species.

Some of these names were consolidated by Rohrbach (1869), who treated *S. caffra*, *S. capensis*, and *S. diurniflora* as synonyms of *S. undulata*. Of the remaining names, only *S. bellidioides* and *S. undulata* have been applied in recent years, always with a great deal of uncertainty reflecting the ± continuous and often local variation in the relative lengths of carpophore and capsule.

There is, however, a distinctive morphotype characteristic of the populations from the coastal and near-inland parts of KwaZulu-Natal and Swaziland. Plants from this region consistently produce stiffly erect stems, with closely branched, flat-topped cymes of flowers with short calyces, mostly 12–15 mm long, and unusually small capsules, 10–15 mm long. Although all of these states are found elsewhere in the species, their consistent

association among these populations is unique and suggests some degree of genetic differentiation. We recognize these plants as subsp. *polyantha*.

Key to subspecies

1a Stems erect or sprawling; cymes mostly without tertiary branching; calyx (15–)20–35(–45) mm long; capsules (12–)15–25 mm long, 1.5–4.0 × longer than carpophore; plants widespread but not from coastal KwaZulu-Natal or Swaziland . . . . 2a. subsp. *undulata*
1b Stems erect; cymes with tertiary and often quarternary branching; calyx 12–15(–20) mm long; capsules 10–15 mm long, 4–6 × longer than carpophore; plants from coastal KwaZulu-Natal or Swaziland . . . . . 2b. subsp. *polyantha*

2a. subsp. **undulata**

Silene capensis Otth in DC.: 379 (1824); Sond.: 125 (1860). *Melandrium capense* (Otth.) Rohrb.: 232 (1869). *Melandrium undulatum* var. *capense* (Otth.) Rohrb.: 246 (1869–70). Type: South Africa, 'ad cap. B.-Spei.', without date or collector, *Prodr. 1 p. 379 no. 144* (G-DC, holo.–microfiche!).

Silene diurniflora Kunze: 578 (1844) [*fide* Sond. (1860)]. Type: South Africa, 'Prom. b. sp.', cultivated at University of Leipzig 1842–1843, *Guenzius s.n.* (LZ, holo.†).

Silene bellidioides Sond.: 125 (1860), syn. nov. *Melandrium bellidioides* (Sond.) Rohrb.: 247 (1869–70) [as '*bellidioide*']. Type: South Africa, 'fields near the Zwartkops River, Nov. without year, *Ecklon & Zeyher 1964 S-G-5677* (S, lecto.!, designated by Nordenstam: 279 (1980); SAM!, isolecto.].

Silene eckloniana Sond.: 126 (1860), syn. nov. *Melandrium ecklonianum* (Sond.) Rohrb.: 232 (1869). Type: South Africa, [Eastern Cape], 'sea shore near Cape Recife, Feb. 1830/2', *Ecklon s.n. S-G-8711* (S, holo.!).

Silene caffra Fenzl ex C.Muell. in Walp.: 276 (1868). *Melandrium caffrum* (Fenzl ex C.Muell.) Rohrb.: 232 (1869). *Melandrium undulatum* var. *caffrum* (Fenzl ex C.Muell.) Rohrb.: 246 (1869–70). Type: South Africa, [Eastern Cape], 'Caffraria', *Drège 5342* (W, holo.†).

Silene bellidifolia var. *stricta* Fenzl in Drège: 222 (1844), nom. nud.

Silene bellidifolia var. *foliosa* Fenzl in Drège: 222 (1844), nom. nud. [*fide* Sond. (1860)].

Silene caffra Fenzl in Drège: 222 (1844), nom. nud.

Silene capensis var. *flaccida* E.Mey. in Drège: 222 (1844), nom. nud.

Silene meyeri Fenzl in Drège: 222 (1844), nom. nud. [et herb. *Drège 730*, *fide* Rohrb.: 245 (1869–70)].

Silene ornata lus. *flaccida* Fenzl in Drège: 222 (1844), nom. nud. [*fide* Sond. (1860), as '*florida*'].

Silene thunbergii E.Mey. in Drège: 222 (1844), nom. nud., non *S. thunbergiana* Eckl. & Zeyh. ex Sond. (1860) [*fide* Sond. (1860)].

Melandrium capense var. *strictum* [as '*stricta*'] Fenzl ex Rohrb.: 232 (1869), nom. nud.

Silene noctiflora sensu Thunb., in part [excluding *UPS-THUNB 10774 & 10775* = *S. rigens* J.C.Manning & Goldblatt]: 81 (1794), non. L. (1753).

Stems erect or sprawling. *Cymes* relatively few-flowered, mostly without quarternary branching, rounded. *Calyx* (15–)20–35(–45) mm long. *Capsules* ovoid to narrowly ovoid, (12–)15–25 mm long, 1.5–4.0 × longer than carpophore. *Carpophore* 4–10(–15) mm long. Figures 6, 7.

Distribution and ecology: throughout most of the range of the species but replaced along the eastern seaboard by subsp. *polyantha* (Figure 8).

Diagnosis and relationships: a variable taxon, typically with a lax, rounded inflorescence of relatively large flowers with calyx mostly 20–35 mm long, often with subulate lobes. The nature of the stem and calyx vesiture is very variable, with some plants entirely eglandular and others predominantly glandular, sometimes even within a single collection (e.g. *Taylor 5180*). Variation in the nature and degree of vesiture is especially evident among the populations along the southern Cape coast. Populations between Knysna and Plettenberg Bay often, but not always, have entirely eglandular stems and calyces.

Most populations have the anthers exserted beyond the mouth of the floral tube at anthesis, followed by the styles, which at anthesis are always included in the floral tube. This feature is thus evident only in older flowers. Also diagnostic of many populations are the long, subulate calyx lobes, 5–7 mm long but this can vary within populations. Plants with short calyces, ± 15 mm long, and carpophores ± 4 mm long, are typical of the West Coast, whereas specimens from western Lesotho (e.g. *Schmitz 8095, Hilliard & Burtt 12065*) have among the longest calyces (35–45 mm long) and carpophores (± 15 mm long) recorded for the species.

Plants are very variable in habit, those from coastal and alpine habitats typically more compact and prostrate whereas individuals from sheltered or shaded situations are more lax. Some of this variation is certainly ecological and linked to the habitat but some may well be genetic, since plants from the cliffs at Hermanus evidently retain their compact habit in cultivation [*Davis s.n.* (BOL)].

Separation from subsp. *polyantha* is not always clear in the Kwazulu-Natal Midlands, especially around Pietermaritzburg, where plants have erect stems and well-branched, flat-topped inflorescences but calyces ± 20 mm long.

Additional specimens

LIMPOPO.—**2330** (Tzaneen): Woodbush, (–CC), Jan. 1923, *Wager s.n. TRV23094* (PRE). **2429** (Zebediela): Donkerkloof near Chuniespoort, 1 700 m, (–BA), 14 Mar. 1974, *Vahrmeijer 2414* (MO, PRE).

NORTH-WEST.—**2527** (Rustenburg): Rustenburg, (–CC), 5 Nov. 1940, *Lanham s.n.* (PRE). **2624** (Vryburg): Vryburg, Aarbosvlakte, (–DC), 25 Mar. 1921, *Mogg 8435* (PRE); Vryburg, (–DD), Apr. 1912, *Rogers s.n. BOL45405* (BOL).

GAUTENG.—**2527** (Rustenburg): Witwatersrand, Hekpoort, (–DC), 18 Apr. 1936, *Phillips 33* (PRE). **2528** (Pretoria): Pretoria, Arcadia, E of Union Buildings, (–AC), 12 Jul. 1955, *Smith 323* (PRE); Pretoria, (–CA), 21 Nov. 1904, *Leendertz 4148* (PRE); Waterkloof, (–CA), 25 Oct. 1920, *Verdoorn 131* (PRE); Wonderboom Poort, (–CA), 24 Nov. 1917, *Pole-Evans 230* (PRE); Fountains Valley, (–CA), 14 Mar. 1930, *Verdoorn 820* (PRE); Moreletta spruit, (–CA), 1 Mar. 1980, *Germishuizen 1260* (PRE); Irene, (–CC), Apr. 1924, *Smuts 1318* (PRE); Doornkloof, (–CC), 8 Nov. 1928, *Gillett s.n.* (NBG). **2627** (Potchefstroom): Klifdrif, E of Potchefstroom, (–AD), 8 Jan. 1935, *Theron 1208* (PRE); Parys, (–CC), Apr. 1907, *Grey College Herbarium 576* (BOL).

MPUMALANGA.—**2530** (Lydenburg): Verlorenvallei Nature Reserve, NE of Dullstroom, (–AC), 16 Dec. 1932, *Galpin s.n. BOL45397* (BOL); 10 Apr. 2000, *Van Slageren & Van Wyk 971* (PRE); Mokobulaan Plantation on Skurweberg, 2 100 m, (–BA), 3 Dec. 1985, *Kluge 2663* (PRE). **2629** (Bethal): Ermelo, Nooitgedacht, (–DB), 5 Feb. 1927, *Henrici 1467* (PRE). **2630** (Carolina): Carolina, (–AA), Dec. 1905, *Bolus 11701* (BOL).

FREE STATE.—**2827** (Senekal): Gumtree, (–DD), 13 Jan. 1948, *Reardon 4* (NU). **2828** (Bethlehem): Golden Gate Highland Park, (–DA), Jan. 1966, *Liebenberg 7495B* (PRE); Qwa Qwa National Park, Qwa Qwaberg, (–DB), 2 Feb. 1995, *Zietsman 3112* (PRE). **2829** (Harrismith): Harrismith, Drakensberg Botanical Garden, (–AC), 16 Jan. 1974, *Jacobsz 2015* (MO, NBG, PRE, S); foot of Platberg, (–AC), 2 Feb. 1968, *Van der Zeyde 227* (NBG). **2926** (Bloemfontein): Bloemfontein, Gen. de Wet School, (–AA), 3 Jun. 1967, *Hanekom 910* (PRE); O.F.S. Botanic Garden, (–AA), 27 Mar. 1968, *Müller 236* (NBG); Mazelspoort, (–AA), 7 May 1949, *Steyn 50* (NBG); Dewetsdorp, (–DA), 15 Apr. 1950, *Steyn 942* (NBG). **2927** (Maseru): Ladybrand, (–AB), Nov. 1906, *Rogers 5241* (PRE). **3025** (Colesberg): 5 km from Colesberg on road to Steynsburg, (–CA), 1 Apr. 1981, *Herman 465* (PRE). **3027** (Lady Grey): Zastron campsite, (–AC), 27 Mar. 1980, *Reid 202* (PRE).

KWAZULU-NATAL.—**2730** (Vryheid): Utrecht Dist., Groenvlei, Balele Mountain Lodge, (–CA), 18 Nov. 1997, *Ngwenya 1617* (NH); Hlobane, (–DB), 5 Nov. 1950, *Johnstone 517* (MO, NU). **2828** (Bethlehem): Witzieshoek, (–DB), Dec. 1905, *Thode 5664* (NBG); Royal Natal National Park, near Basuto Gate, (–DB), 2 Feb. 1982, *Stewart & Manning 2254* (NU). **2829** (Harrismith): Van Reenen, (–AD), Jun. 1914, *Bews 965* (NU). **2929** (Underberg): Cathedral Peak Forest reserve, (–AB), 1 Jan. 1983, *Noel 2820* (NU); Giant's Castle Game Reserve, near Dinosaur footprints, (–AD), 27 Dec. 1967, *Trauseld 907* (NU); Lotheni Valley, vicinity of Ash Cave, (–AD), 7 Feb. 1985, *Hilliard & Burtt 18188* (E, NU, PRE); Spring Grove Farm, (–BB/BD), 9 Dec. 1996, *Greene 906* (NH); Kamberg, (–BC), 28 Nov. 1974, *Wright 1985* (NU); 10 Jan. 1990, *Williams 687* (NH); Mpendle Dist., Mulangane Ridge, (–BC), 3 Feb. 1984, *Hillard & Burtt 17507* (E, K, NU, PRE, S); Chameleon Cave area, (–CB), 1 Dec. 1984, *Hilliard & Burtt 17769* (E, K, NU, PRE, S); Sani Pass, (–CB), 18 Feb. 1973, *Hilliard 5339* (NU); 23 Mar. 1977, *Hilliard & Burtt 9801* (E, K, MO, NU); 6 Jan. 1984, *Hilliard, Burtt & Manning 17284* (E, NU); Jan. 2000, *Edwards 1749* (E, K, NU); Cobham State Forest, Polela River, (–CB), 2 Dec. 1987, *MacDevette 2036* (NH); Vergelegen Nature Reserve, (–CB), 2 Jan. 1978, *Hilliard & Burtt 11182* (E, MO, NH); Underberg, Watermead farm, (–DC), 16 Dec. 1989, *Williams 645* (NH); Bulwer, Sunset farm, (–DC), 24 Feb. 1990, *Vos 31* (NU); Polela Dist., Hlabeni Mtn near Creighton, (–DC/DD), 21 Nov. 1994, *Wirminghaus 1310* (NH). **2930** (Pietermaritzburg): Mount West Dist., 3 km N of intersection, (–AA/AC), 13 Nov. 1988, *Greene 609* (NH); Kunhardt's Farm, 21 km from Merrivale on Boston Road, (–AC), Jan.–Apr. 1982, *Kunhardt 10* (NH); Karkloof Mtn Range, summit of Mt Gilboa, (–AD), 18 Nov. 2000, *Johnson & Neal 29* (NU); Pietermaritzburg, World's View, (–CB), 30 Oct. 1964, *Tunnington s.n.* (NU); Richmond/Ixopo area, Hella Hella, (–CC), 1 Nov. 1997, *Swanepoel & Porter 14* (NH). **3029** (Kokstad): Mzimkhulu Dist., Nsikeni Nature reserve, (–AB), 19 Feb. 1992, *Williams 901* (NH); Newmarket, (–AD), 1 Feb. 1895, *Krook s.n. S10-9935* (S); Mt Currie, (–AD), 20 Jan. 1957, *Taylor 5483* (NBG); between Flagstaff and Kokstad, (–DA), 4 Dec. 1928, *Hutchinson 1798* (BOL, K); Ngele, (–DA), Mar. 1883, *Tyson 1286* (SAM); 2 Jan. 1966, *Strey 6365* (NU); 6 Jan. 1990, *Abbott 4838* (NH); 13 Jan. 1990, *Abbott 4925* (NH).

LESOTHO.—**2828** (Bethlehem): Leribe, (–CC), Dec. 1912, *Dieterlen 6791* (SAM). **2927** (Maseru): Maseru, (–BD), 25 Jan. 1951, *Compton 22536* (NBG); between Molimo Nthuse and Blue Mtn Pass, 2 250 m, (–BD), 15 Feb. 1978, *Schmitz 8095* (PRE); slopes behind Malimo Nthuse Hotel, (–BD), 12 Jan. 1979, *Hilliard & Burtt 12065*

(NU). **2929** (Underberg): Mokhotlong, (–AC), 25 Feb. 1949, *Compton 21498* (NBG); 23 km from Mokhotlong on road to Sani Top, 2 250 m, (–AC), 12 Feb. 1987, *Killick 4578* (MO, NH); Sehlabaethebe Reserve, 8 900′ [2 700 m], (–CC), 3 Feb. 1975, *Bayliss 1317* (PRE). **3028** (Matatiele): ridge between Orange River and Maqaba Peak, near Quachasnek, (–BA), 13 Mar. 1936, *Galpin s.n. BOL31667* (BOL).

NORTHERN CAPE.—**2723** (Kuruman): Carrington, (–CB), Apr. 1940, *Esterhuysen 2172* (BOL). **2724** (Taung): Klein Boetsap, (–CC), 1910, *Pagan s.n. PRE54332* (PRE); W of Harz River near Taung, (–DA), 12 Feb. 1948, *Rodin 3646* (BOL, MO, PRE). **2816** (Oranjemund): Richtersveld, Ploegberg S of Khubus, (–DB), 3 Sept. 1977, *Oliver, Tölken & Venter 508* (MO, NBG, PRE). **2917** (Springbok): O'kiep, (–DB), Sept.–Oct. 1926, *Pillans 4983* (BOL). **2922** (Prieska): Niekerkshoop, (–BB), 26 Nov. 1935, *Bryant 1165* (BOL). **3017** (Hondeklipbaai): Bowesdorp, (–BB), Sept. 1941, *Stokoe s.n.* (MO, PRE, SAM); Brakwater on Arakoop, (–BB), 25 Aug. 1977, *Thompson & le Roux 48* (MO. NBG); Grootvlei Pass, (–BB), 3 Sept. 1980, *Le Roux 2786* (BOL, NBG); Kamiesberg Pass, (–BB), 7 Sept. 1981, *Van Berkel 430* (NBG); 12 Sept. 1993, *Strid & Strid 37769* (PRE); Skilpad Wild Flower Reserve, (–BB), 10 Aug. 1993, *Grobler 39* (PRE). **3018** (Kamiesberg): Kamiesberg, Leliefontein, (–AC), 16 Dec. 1936, *Adamson 1454* (PRE); Studer's Pass, (–AC), 11 Sept. 2007, *Snijman 2149* (NBG); Farm Damsland, (–AC), 29 Oct. 2007, *Snijman 2213* (NBG); Langkloof, N of Farm Doringkraal, (–AC), 23 Sept. 2010, *Goldblatt & Porter 13573* (MO, NBG). **3119** (Calvinia): Nieuwoudtville Waterfall, (–AC), Sept. 1930, *Lavis s.n. BOL19622* (BOL); Nieuwoudtville Reserve, (–AC), 7 Sept. 1983, *Perry & Snijman 2312* (NBG, PRE); 13 Sept. 2010, *Goldblatt & Porter 13526* (NBG); Oorlogskloof Nature reserve, (–AC), 20 Sept. 1995, *Pretorius 307* (NBG); between Oorlogskloof and Papkuilsfontein, (–CA), Sept. 1939, *Leipoldt 3076* (BOL); Calvinia, Farm Driefontein, (–DA), 23 Sept. 2009, *Goldblatt, Manning & Porter 13431* (MO, NBG). **3121** (Fraserburg): Platkoppies Farm, 45 km N of Williston, (–AA), 29 Mar. 1993, *Germishuizen 6358* (PRE). **3122** (Loxton): 17 km from Loxton on road to Victoria West, Taaibosfontein, (–BC), 13 May 1976, *Thompson 3053* (NBG, PRE). **3124** (Hanover): Richmond Dist., Elandspoort, (–CB), Oct. 1935, *Thorne s.n. SAM51851* (SAM). **3221** (Merweville): Ezels Kom, (–BA), 12 Apr. 1986, *Shearing 1263* (PRE). **3222** (Beaufort West): Karoo National Park, (–AB), 3 Jan. 1985, *Shearing 866* (PRE). **3223** (Rietbron): Nelspoort, (–AA), without date, *Pearson 2049* (SAM).

WESTERN CAPE.—**3118** (Vanrhynsdorp): Gifberg, (–DC), 15 Oct. 1953, *Esterhuysen 22132* (BOL); Gifberg Pass, (–DD), 11 Sept. 2009, *Goldblatt & Porter 13319* (MO, NBG). **3218** (Vredenburg): Witteklip, (–DD), 8 Sept. 2009, *Goldblatt & Porter 13281* (MO, NBG, PRE). **3218** (Clanwilliam): between Leipoldtville and Eland's Bay, (–AC), Oct. 1947, *Zinn s.n. SAM63437* (SAM); Pakhuis Pass, (–BB), 4 Sept. 1948, *Morris 20922* (BOL); 4 Sept. 1948, *Compton 20922* (NBG); N7 S of Clanwilliam at Kransvlei turnoff, (–BB), 12 Sept. 2010, *Goldblatt & Porter 13523* (MO, NBG); ± 4 km NE of Redelinghuys, (–BC), 9 Sept. 2009, *Helme 6262* (NBG); Citrusdal, Modderfontein, (–CB), 27 Aug. 1968, *Hanekom 1163* (NBG, PRE). **3219** (Wuppertal): Brakfontein, (–AC), 20 Oct. 1983, *Viviers 1245* (NBG); Algeria Forest Station, (–AC), 20 Nov. 1996, *Van Rooyen, Steyn & de Villiers 253* (NBG); Matjiesrivier, (–AC), 6 Apr. 1944, *Wagener 359* (NBG); Dasklip Pass, (–CC), 30 Sept. 1972, *Thompson 1517* (NBG, PRE). **3318** (Cape Town): Langebaan, Lynch point, (–AA), 5 Sept. 1971, *Axelson 478* (NBG); near Darling, (–AD), Sept. 1905, *Bolus s.n.* (BOL); hollow S of Strand Railway, Stickland, (–CD), Oct. 1932, *Acocks 959* (S); Camps Bay, (–CD), Oct. without year, *Ecklon & Zeyher s.n.* (SAM); Table Mtn, (–CD), 29 Jul. 1846, *Prior s.n.* (PRE); Sept. 1897, *Thode 5828* (NBG); Lion's Head, (–CD), 17 Oct. 1897, *Wolley Dod 3511* (BOL); Sept. 1913, *Kensit & Teague s.n BOL45388* (BOL); Devil's Peak, (–CD), Oct. 1877, *Bolus 3844* (BOL); 31 Oct. 1915, *Pillans 2808* (BOL); Table Mtn, Kasteelpoort, (–CD), without date, *Marloth 8999* (PRE); Malmesbury Dist., Helderfontein Farm, (–DA), 11 Sept. 1979, *Boucher 4684* (NBG, PRE); Langverwacht above Kuil's River, (–DC), 1 Oct. 1973, *Oliver 4700* (NBG); Botlaryberg, Farm Koopmanskloof, (–DD), 20 Sept. 1980, *Beyers 86* (NBG); above Jonkershoek, Haelkop, (–DD), 23 Oct. 1928, *Gillett 1784* (BOL, NBG); Jonkershoek, Bosboukloof, (–DD), May 1967, *Kerfoot 5795* (NBG, PRE); Jonkershoek State Forest, (–DD), 31 Oct. 1975, *Haynes 1113* (NBG, PRE). **3319** (Worcester): Witzenberg, Inkruip, (–AA), 1 Oct. 1954, *Esterhuysen 23419* (BOL); 5 km N of Tulbagh on way to Winterhoek State Forest, (–AA), 8 Oct. 1981, *Mauve & Hugo 59* (MO, NBG, PRE); Ceres Dist., Gydouw Pass, (–AB), 12 Aug. 1986, *Van Wyk 2530* (PRE); road from Hottentot's Kloof to karoopoort, (–BA), 29 Nov. 1908, *Pearson 4831* (BOL); Dal Josafat Forest Reserve, (–CA), 19 Oct. 1963, *Taylor 5446* (NBG); Du Toit's Peak, (–CA), 21

Dec. 1975, *Esterhuysen 34165* (BOL); Fonteinjiesberg, (–CB), 1 Jan. 1976, *Esterhuysen 34171* (BOL); Wemmershoek, (–CC), Jan. 1921, *Andreae 778* (PRE); Robertson Dist., Vrolijkheid Nature Reserve, (–DD), 8 Apr. 1974, *Theron & Students 3152* (PRE). **3320** (Montagu): Montagu Div., Keurkloof, (–CC), 20 Sept. 1935, *Lewis s.n. BOL31666* (BOL); Montagu–Barrydale Road, (–CD), Sept. 1923, *Levyns 526* (BOL). **3321** (Ladimsith): Swartberg, Towerkop, (–AC), 16 Dec. 1956, *Esterhuysen 26789* (BOL); Seweweekspoort, (–AD), 27 Dec. 1928, *Andreae 1225* (PRE); Huis River Pass, (–BC), 31 July 1955, *Van Niekerk 536* (BOL); Meiringspoort, (–BC), Feb. 1932, *Thorne s.n. SAM50187* (SAM); Rooiberg, (–DA), 1 Nov. 1931, *Compton 3827* (BOL). **3322** (Willowmore): Swartberg Pass, (–AC), Dec. 1904, *Bolus s.n.* (BOL); Dec. 1943, *Stokoe 9061* (BOL); Cango Valley, Bassonsrus, (–AC), 4 Nov. 1974, *Moffett 401* (PRE); Swartberg, Blouberg Peak, (–AC), 29 Nov. 1987, *Bean & Viviers 1987* (BOL); Wilderness, Fairy Knowe, seashore dunes, (–DC), 13 Dec. 1928, *Mogg 11896* (PRE); Wilderness, (–DC), 14 Nov. 1952, *Van Niekerk 196* (BOL); Ebb and Flow Nature Reserve, Touw River Valley, (–DC), 19 Oct. 1971, *Taylor 7996a* (NBG, PRE). **3323** (Willowmore): Louterwater, (–CC), 18 Dec. 1933, *Compton 4573* (BOL); Herold's Bay, (–CD), without date, *Hugo 2118* (NBG); 17 Sept. 2010, *Goldblatt & Porter 13546* (NBG). **3418** (Simonstown): Sandy Bay, (–AB), 26 Aug. 1978, *Van Jaarsveld 3464* (MO, NBG); Constantiaberg, (–AB), 17 Dec. 1939, *Compton 8209* (NBG); Hout Bay, (–AB), 16 Aug. 1928, *Gillett 411* (NBG); Chapman's Peak, (–AB), 7 Dec. 1943, *Van Niekerk 469* (NBG, PRE); Nov. 1944, *Levyns 57806* (SAM); Silvermine, (–AB), 4 Oct. 1967, *Bayliss 4006* (NBG); Kalk Bay Mtns, (–AB), 30 Oct. 1931, *Adamson 2230* (BOL); 22 Nov. 1943, *Levyns 3553* (BOL); cliffs beyond Simonstown, (–AB), 4 Oct. 1928, *Hutchinson 646* (BOL, PRE); S of Simonstown, above Froggy Pond, (–AB), 16 Nov. 1975, *Esterhuysen 34107* (BOL); Buffelsbaai, (–AD), 28 Nov. 1939, *Adamson 2710* (BOL); Somerset West, (–AB), 31 Oct. 1948, *Parker 4378* (BOL, MO, NBG, S); Kogel Bay, (–BD), 5 Sept. 1946, *Parker 4111* (BOL, NBG); Rooi Els, (–BD), 15 Oct. 1971, *Boucher 1670* (NBG); Pringle Bay, (–BD), 23 Dec. 1982, *O'Callaghan 481* (NBG); Betty's Bay, Jackass Penguin Colony, (–BD), 16 Feb. 1997, *Mucina 50359/1* (PRE). **3419** (Caledon): Houhoek, (–AA), Apr. 1892, *Guthrie 224* (NBG); Caledon, Warmbaths Hotel, (–AB), 5 Oct. 1928, *Gillett 1114* (NBG); 10 miles [16 km] from Caledon on Bot River Road, (–AB), 17 Sept. 1938, *Thorns s.n.* (NBG); Hermanus [cult. at Kirstenbosch], (–AC), Jan. 1932, *Davis s.n. NBG20/31* (BOL); Hermanus, cliffs opposite Marine Hotel, (–AC), 28 Aug. 2010, *Manning 3304* (NBG); Buffeljagsbaai, (–DA/DC), 7 Sept. 1995, *Paterson-Jones 617* (NBG). **3420** (Bredasdorp): Rietfontein, 20 km S Stormsvlei, (–AA), 23 Sept. 1982, *Bayer 3071* (NBG); ± 25 miles [40 km] from Swellendam on Bredasdorp Road, (–AC), 30 Aug. 1962, *Taylor 3847* (NBG); Windhoek, Potberg Nature Reserve, (–AD), Jul./Aug. 1968, *Van der Merwe 870* (PRE); 11 Oct. 1978, *Burgers 1243* (NBG). **3421** (Riversdale): near Kafir Kuils River, (–AB), 14 Oct. 1923, *Muir 2916* (PRE); Stillbay rubbish dumps, (–AC), 9 Sept. 1978, *Bohnen 4091* (NBG); Stillbay Rifle Range, (–AD), 2 Oct. 1978, *Bohnen 4292* (NBG). **3422** (Mossel Bay): Goukamma Nature Reserve, (–BB), 1969, *Heinecken 245* (PRE); Buffalo Bay, (–BB), 26 Nov. 1955, *Taylor 4908* (NBG). **3423** (Knysna): between Georgetown [George] and Swellendam, (–AA), Jul. 1856, *Castelnau 77* (BOL, PRE); Bitou Bridge, (–AA), 15 Dec. 1941, *Fourcade 5522B* (BOL, NBG); Knysna Heads, (–AA), 7 Nov. 1928, *Gillett 2171* (BOL, NBG); foot of Prince Alfred's Pass, (–AA), Oct. 1932, *Fourcade 4853* (BOL, NBG); Keurboomstrand, (–AB), 14 Nov. 1949, *Morris 433* (NBG); 25 Nov. 1955, *Taylor 4917* (NBG); 17 Sept. 1956, *Theron 2099* (PRE); Ratels Bosch, (–BA), Aug. 1905, *Fourcade 16* (BOL); Aug. 1908, *Fourcade 317* (BOL). **3424** (Humansdorp): Groot River, (–AA), Oct. 1942, *Fourcade 5784* (BOL).

EASTERN CAPE.—**3027** (Lady Grey): Witteberg, Joubert's Pass, (–BC), 18 Jan. 1979, *Hilliard & Burtt 12206* (NU); Witteberg, Beddgelert, (–DA), 2 Dec. 1981, *Hilliard & Burtt 14637* (E, NU); Ben McDhui, (–DB), 9 Feb. 1983, *Hilliard & Burtt 16558* (E, K, NU). **3028** (Matatiele): Naude's Nek, (–CA), 27 Dec. 1977, *Bigalke 23* (NU); 13 Feb. 1983, *Hilliard & Burtt 16601* (E, NU). **3126** (Queenstown): Jamestown, Vogelfontein Farm, (–BB), 16 Dec. 1942, *Barker 2246* (NBG); Dordrecht, (–DB), Apr. 1929, *Smuts s.n.* (NBG); 22 Mar. 1964, *Bayliss 2119* (NBG). **3127** (Lady Frere): Maclear, Bastervoetpad, Farm Snowy Side, (–BB), 17 Nov. 1993, *Bester 1713* (NH). **3128** (Umtata): Farm Comarty ± 18 km W of Ugie, (–AA), 19 Apr. 1994, *Bester 2712* (NH); Libode, Misty Mount, S of Port St. Johns, (–DB), 18 Nov. 1991, *Cloete 1344* (NH). **3224** (Graaf-Reinet): 'Waterfurrows' above Graaf-Reinet, (–BA), Nov. 1865, without collector *BOL74044* (BOL). **3226** (Fort Beaufort): Pluto's Vale, (–BA), 29 Oct. 1964, *Bayliss 2481* (NBG); Elandsberg, (–BD), 17 Jan. 1986, *Cadman, Edwards & Norris 3224* (NU); Hogsback, (–DB), May 1932, *Leighton* *s.n. BOL31669* (BOL); June 1961, *Bokelmann 6* (NBG). **3227** (Stutterheim): Kleinemonde, (–CA), ?Nov. 1953, *Taylor 4306* (NBG); King William's Town, (–CD), Jan. 1894, *Sim 1613* (NU). **3228** (Butterworth): Kei Mouth, (–CB), Aug. 1891, *Flanagan 880* (SAM). **3325** (Port Elizabeth): Walmer, (–DC), Sept. 1916, *Paterson 3321* (NBG, PRE); Redhouse, (–DC), Sept. 1914, *Paterson 272* (BOL); Baakens River Valley, Fern Glen, (–DC), 14 Nov. 1974, *Olivier 1241* (MO, NBG). **3326** (Grahamstown): Martindale, Kap River, (–BD), 23 Dec. 1956, *Taylor 5180* (NBG). **3425** (Skoenmakerskop): Sea View, (–AB), 18 Sept. 2010, *Goldblatt 13549* (NBG); Sardinia Bay, (–BA), 18 Sept. 2010, *Goldblatt 13552* (NBG); coast near Cape Recife, (–BA), 18 Sept. 2010, *Goldblatt & Porter 13553* (NBG).

2b. subsp. **polyantha** *J.C.Manning & Goldblatt,* subsp. nov.

TYPE.—KwaZulu-Natal 2930 (Pietermaritzburg): Krantzkloof, (–DD), Oct. 1921, *Haygarth 138* (NBG, holo.; NH, iso.).

Stems erect. *Cymes* relatively many-flowered, with tertiary and often quarternary branching, forming a flat-topped, corymbiform synflorescence. *Calyx* 12–15(–20) mm long. *Capsules* broadly ovoid, 10–15 mm long, 4–6 × longer than carpophore. *Carpophore* 2–4 mm long.

Distribution and ecology: restricted to the coastal and near-inland parts of KwaZulu-Natal and Swaziland between 550–1 000 m, with a single early record from Barberton (Figure 8), occurring along thicket and forest margins and in marshy places.

Diagnosis: distinguished from the typical form by the erect stems and closely branched, flat-topped inflorescence, with tertiary and often quarternary branching, short calyx, mostly 12–15(–20) mm long, and short, broadly ovoid capsules, 10–15 mm long with very short carpophore, 2–4 mm long. This form grades into the typical subspecies in the KwaZulu-Natal Midlands, between Pietermaritzburg and Vryheid.

Additional specimens

MPUMALANGA.—**2531** (Komatipoort): Barberton, (–CC), Dec. 1916, *Pott 5467* (PRE).

SWAZILAND.—**2631** (Mbabane): Mbabane, Forbes Reef, (–AC), 29 Nov. 1963, *Compton 31797* (NBG); Black Mbuluzi Valley, 3500' [1 060 m], 9 Dec. 1958, *Compton 28443* (NBG); 9 Dec. 1964, *Compton 32183* (NBG); Gobolo, (–AC), 23 Nov. 1962, *Dlamini s.n.* (PRE); Hlatikulu, (–CD), Dec. 1910, *Stewart s.n. SAM2523* (SAM); 23 Nov. 1959, *Compton 29502* (NBG).

KWAZULU-NATAL.—**2731** (Louwsburg): Itala Nature Reserve, (–CB), 11 Dec. 1975, *Hilliard & Burtt 8551* (E, K, NU, PRE, S); Nongoma, (–DC), 2 Dec. 1943, *Gerstner 4666* (NBG). **2831** (Nkandla): Eshowe, Reservoir March, (–CD), 9 Nov. 1949, *Lawn 1296* (NH). **2832** (Mtubatuba): Dukuduku, (–AC), 25 Nov. 1965, *Strey 6099* (NU, PRE). **2930** (Pietermaritzburg): ± 25 km NNE Howick, Twin Falls, (–AD), 19 Nov. 1987, *Grové 122* (NU); Baynesfield, Oldfield Farm, (–CD), Oct. 1989, *Edwards 602* (NU); Hammarsdale, 590 m, (–DA), 27 Jun. 1995, *Ward 13189* (NH, NU, PRE); Botha's Hill, (–DB), 10 Dec. 1913, *Medley Wood 12387* (NH, NU). **3029** (Kokstad): Harding, Rooi Vaal, (–DB), 4 Jan. 1957, *Taylor 5275* (NBG). **3030** (Port Shepstone): Alexandra Dist., Umgay [Umgai], (–AD), 9 Oct. 1909, *Rudatis 522* (NBG); *Rudatis 741* (PRE); Campbellton-Dumisa, (–AD), 13 Nov. 1913, *Rudatis 2030* (NBG); near Southport, (–DA), 13 Nov. 1974, *Nicholson 1506* (PRE).

3. **Silene saldanhensis** *J.C.Manning & Goldblatt,* sp. nov.

TYPE.—Western Cape 3317 (Saldanha): Saldanha harbour, SW-facing slopes on ridge at beginning of breakwater, (–BB), calcareous sand, 19 Sept. 2011,

Goldblatt & Manning 13646 (NBG, holo.; K, MO, PRE, S, iso.).

Compact or cushion-forming perennial, 30–60 cm, branching at base with multiple crowns; stems erect or decumbent, 2–4 mm diam., with shoots in lower axils, densely puberulous with gland-tipped hairs mixed with scattered, short, acute eglandular hairs 0.25–0.50 mm long, inflorescence axis often especially densely puberulous. *Leaves* fleshy, densely puberulous with mixture of erect, gland-tipped hairs and acute, eglandular hairs, ciliate at extreme base with longer hairs to 1 mm long, with evident side veins, basal leaves oblanceolate, up to 100 × 35 mm; stem leaves spreading, lanceolate, 35–50 × 12–18 mm, recurved apically, margins weakly revolute and sometimes undulate. *Inflorescence* a lax, few-flowered, assymetrical, paniculate cyme, axis 1.5–2.5 mm diam. below primary flower; bracts similar to upper leaves but smaller, spreading-recurved; primary pedicels 10–20 mm long, elongating to 35 mm in fruit, secondary pedicels 10–20 mm long. *Flowers* half-nodding, nocturnal and faintly gardenia-scented at night. *Calyx* urn-shaped, 17–20 mm long, with ± prominent, thickened ribs, 10-ribbed, with 1–3 reticulate veins in distal half, densely puberulous with erect, gland-tipped hairs, lobes triangular, 3–4 mm long, densely glandular-ciliolate. *Petals* mauve, sometimes with darker corona, limb spreading, broadly cuneate, ± overlapping with adjacent petals, 13–18 × 10–19 mm, bifid ± one third to halfway, sometimes crenulate distally, one or both margins sometimes with small sub-basal tooth, claw strap-shaped, 15–20 × 3–4 mm, auriculate apically, glabrous or sparsely pubescent along abaxial midline and margins, coronal scales suberect, 1.5–2.0 mm. *Stamen* filaments subequal, 15–18 mm long, reaching to top of claws or shortly exserted but not exceeding corona; anthers ± 2 mm long, not exserted beyond corona. *Ovary* narrowly ellipsoid, 6 mm long, stipe ± 1mm long; styles ultimately to 8 mm long, reaching top of corona. *Capsule* ovoid, 17 × 6.5–8.0 mm, ± 5 times longer than carpophore, minutely granular; carpophore 3 mm long, retrosely scabridulous. *Seeds* 1.0–1.2 m diam., subcircular, charcoal grey, with scattered whitish or translucent testal cells, face flat or concave, back flat, testal cells concentrically colliculate. *Flowering time:* Aug.–Oct. Figures 2C, D, 9.

Distribution and ecology: largely restricted to the entrance to Saldanha Bay, where it is known from the breakwater and from across the lagoon in Postberg Reserve (Figure 10), but also known from shortly inland on the Farm Waterboerskraal east of Hopefield. The species is restricted to calcareous sands and consolidated calcrete dunes.

Diagnosis: a compact or cushion-forming perennial with erect or decumbent stems, distinguished by the mauve flowers with broad, overlapping petal limbs. The stamens and styles are included or shortly exserted but never protrude beyond the corona. *Silene saldanhensis* is probably most closely allied to the carmine-flowered *S. ornata* (Figure 11), which it resembles in stem vestiture and in the presence of scattered translucent testal cells in the seeds, but is distinguished by its broader, mauve flowers, colliculate seeds (Figures 2C, D), and in its ecology. Both species have stems with a mix of glan-

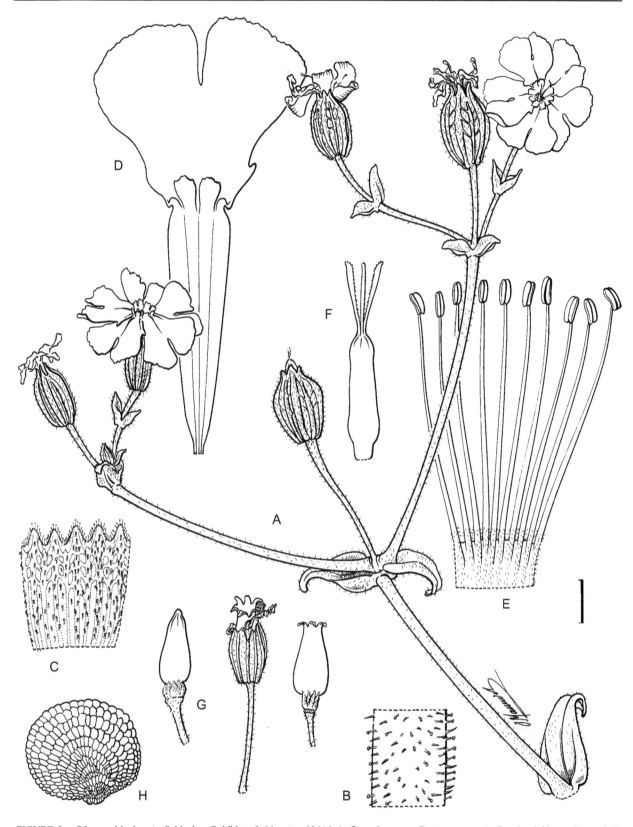

FIGURE 9.—*Silene saldanhensis*, Saldanha, *Goldblatt & Manning 13646*. A, flowering stem; B, stem segment; C, calyx laid out; D, petal; E, androecium laid out; F, gynoecium; G, capsules (left hand illustration before dehiscence with calyx removed; central after dehiscence, right hand after dehiscence with calyx removed); H, seed. Scale bar: A, G, 10 mm; B, 1 mm; C, 5 mm; D–F, 2.5 mm; H, 0.5. Artist: John Manning.

dular hairs and short, scattered eglandular hairs (Figures 9B, 11B) but *S. ornata* is restricted to rocky habitats, and typically has a straggling growth form, even in very exposed situations, with the stems sprawling among bushes or over the rock surface. It typically has slightly smaller flowers, with petal claws ± 15 mm long and the

limbs not overlapping, and tuberculate seeds (Figures 2E, F). *Silene saldanhensis* favours deep, loose sands or compacted dunes, forming neat cushions with the stems either decumbent or erect. It has mostly larger flowers with the claws 15–20 mm long and broad, often overlapping petals, and colliculate seeds, which are distinctive

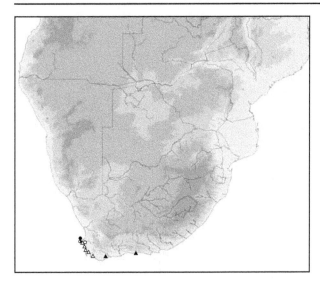

FIGURE 10.—Distribution of *Silene saldanhensis*, ○; *S. ornata*, ●; *S. rigens*, △; *S. mundiana*, ▲.

among the southern African members of sect. *Elisanthe* in lacking papillae or tubercules on the testal cells. The mix of glandular and short, eglandular hairs on the stems, and broad, overlapping petals, with included stamens and styles separate *S. saldanhensis* from coastal forms of *S. undulata* with a short calyx and carpophore.

Conservation note: the restricted occurrence of the species, the low number of plants and development at Saldanha render the species threatened. The Postberg population is protected as part of the West Coast National Park.

Additional specimens

WESTERN CAPE.—**3317** (Saldanha): West Coast National Park, Kraalbaai, (–BB), 13 Sept. 2011, *Manning & Goldblatt 3339* (NBG). **3318** (Cape Town): Hopefield, Waterboerskraal Farm, (–AB), 3 Oct. 1974, *Hugo s.n.* (NBG).

4. **Silene ornata** *Aiton*, Hortus kewensis 2: 96 (1789); Sond.: 126 (1860). *Melandrium ornatum* (Ait.) Asch. ex Rohrb.: 233 (1869). Type: South Africa, without precise locality, cultivated at Kew in 1774 from seed introduced by Francis Masson, *Masson s.n.* (*BM000573045*, holo.!; *BM000573046*, iso.!). Illustrated in Curtis's Botanical Magazine 11: t. 382 (1797).

Silene bellidifolia sensu Thunb.: 81 (1794), non. Juss. ex Jacq. (1777).

Perennial to 1 m, well branched with branches ascending at ± 45 degrees, or plants compact and almost cushion-forming in exposed situations; stems sprawling among vegetation or on rock surface, 1.5–2.5 mm diam., densely puberulous with spreading, gland-tipped hairs mixed with equally short, acute hairs 0.25–0.50 mm long, inflorescence axis often especially densely puberulous. *Leaves* mostly ± cauline, spreading, lower leaves oblanceolate-spathulate, sub-petiolate; stem leaves lanceolate to oblanceolate, mostly 50–80 × 10–20 mm, acute, base tapering or cuneate, margins plane or undulate, puberulous with mixture of erect, gland-tipped hairs and acute, eglandular hairs, ciliate at extreme base with longer hairs to 1 mm long, with evident side veins.

Inflorescence a lax, few-flowered, assymetrical, paniculate cyme, axis 1.0–1.5 mm diam. below primary flower; bracts similar to upper leaves but smaller, subequal, spreading-recurved; primary pedicels 10–40 mm long, secondary pedicels 6–18 mm long. *Flowers* ± horizontally spreading on pedicels flexed sharply apically, ± open day and night, unscented. *Calyx* narrowly urn-shaped in flower, 19–21 mm long, 10-ribbed, with 1–3 reticulate veins in distal half, densely puberulous with erect, gland-tipped hairs, lobes narrowly triangular, 3–5 mm long, densely glandular-ciliolate. *Petals* deep carmine, limb spreading, cuneate, not overlapping with adjacent petals, (7–)10–15 × (7–)12–15 mm, bifid halfway to two thirds, claw strap-shaped, ± 15 × 3 mm, auriculate apically, coronal scales suberect, 2.0–2.5 mm long. *Stamen* filaments subequal, 14–15 mm long, reaching to just below top of claws; anthers ± 2 mm long, reaching top of claws, not exserted beyond corona. *Ovary* narrowly pyriform, shortly stipitate, ± 7 mm long, stipe ± 1mm long; styles 7–12 mm long, included or exserted up to 4 mm beyond petal claw. *Capsule* ovoid, 15–18 × 6–7 mm, 4–5 × longer than carpophore, minutely granular; carpophore ± 4 mm long, glabrescent. *Seeds* 1.0–1.2 mm diam., subcircular with hilum recessed, grey or reddish brown, with scattered translucent cells, face flat or concave, back flat, testal cells radial-concentric, tuberculate. *Flowering time*: (late Aug.) Sept.–Oct. Figures 2E, F, 11).

Distribution and ecology: a local endemic of the limestone hills flanking the mouth of Saldanha Bay (Figure 10), occurring on rocky slopes among coastal scrub. Plants mostly occur on the slopes or summit of hills, with the annual stems straggling through surrounding shrubs, but a population at Stony Head thrives on exposed limestone pavement directly facing the ocean. The plants here are very compact and almost cushion-forming, with the flowering stems sprawling along the rock sheets.

Diagnosis and relationships: *Silene ornata* is readily recognized by the unscented, deep carmine flowers that remain open throughout the day and night. The branching of the flowering stems is largely asymmetrical, and stems and branches are densely puberulous, with a mix of short, glandular hairs and non-glandular hairs. The stamens are always included and the styles mostly so but in some plants they are exserted up to 4 mm beyond the petal claws. This is the form that is illustrated in Curtis's Botanical Magazine 11: t. 382 (1797). Other species in sect. *Elisanthe* have nocturnal, white to pale pink or mauve flowers that are fragrant at night. *Silene ornata* is evidently most closely allied to *S. saldanhensis*, another local endemic of Saldanha Bay, which shares a similar stem vestiture but which is restricted to calcareous sands and consolidated dunes, has slightly larger, mauve flowers with broader tepals, a more tufted habit, and colliculate seeds.

The species, named by Aiton (1789) for its striking, dark red flowers, was introduced to Kew by Francis Masson in 1775. Although the exact provenance of Masson's material is not given, *Silene ornata* is a narrow endemic of Saldanha Bay, which Masson visited for five days in September 1773 in the company of Carl Thunberg (Glen & Germishuizen 2010), and it

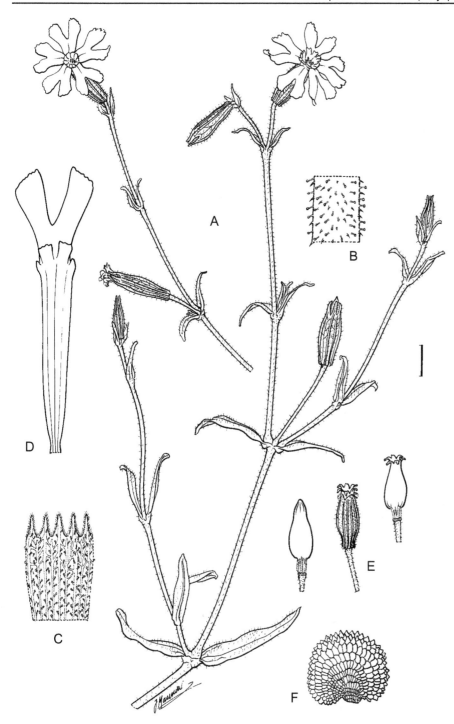

FIGURE 11.—*Silene ornata*, Saldanha, *Goldblatt & Porter 13254*. A, flowering stems; B, stem segment; C, calyx laid out; D, petal; E, capsules (left hand illustration before dehiscence with calyx removed; central after dehiscence, right hand after dehiscence with calyx removed); F, seed. Scale bar: A, E, 10 mm; B, 1 mm; C, 5 mm; D, 2.5 mm; F, 0.5. Artist: John Manning.

is probable that they collected the species at this time. Thunberg (1794) mistakenly associated his collection (*UPS-THUNB 10705*) with the bright pink-flowered Mediterranean *S. bellidifolia*, as correctly realized by Sonder (1860). Masson's brief on the trip was to collect seeds, cuttings and roots for cultivation at Kew and he certainly succeeded with *S. ornata*. Readily propagated by cuttings, the species was widely cultivated in England by 1797, when it was illustrated in Curtis's Botanical Magazine (Curtis 1797) but has since disappeared from cultivation.

Conservation note: the restricted occurrence of the species, the low number of plants and the encroaching urbanisation at Saldanha render the species threatened.

The Postberg population is protected as part of the West Coast National Park.

Additional specimens

WESTERN CAPE.—**3217** (Vredenburg): limestone hill north of Saldanha, (–DD), 17 Sept. 1976, *Goldblatt 4114* (MO); hill N of Saldanha, (–DD), 250′ [76 m], 24 Sept. 2009, *Goldblatt & Porter 13254* (K, MO, NBG, PRE). **3317** (Saldanha): Hoedjies Bay, (–BB), Sept. 1905, *Bolus 12614* (BOL); Postberg, (–BB), 8 Sept. 1957, *Lewis 5273* (NBG); 30 Aug. 1980, *Bond 1715* (MO, NBG); hills between Postberg and Donkergat, (–BB), 9 Sept. 1966, *Rourke 576* (NBG); Konstabelkop and Postberg hills, (–BB), 13 Sept. 2011, *Manning & Goldblatt 3340* (NBG); SAS Saldanha, Bomgat, (–BB), 15 Sept. 2011, *CR15241* (NBG). Without precise locality: hort. Kirstenbosch, Aug. 1937, without collector *BOL31664* (BOL).

5. **Silene rigens** *J.C.Manning & Goldblatt*, sp. nov.

TYPE.—Western Cape 3218 (Clanwilliam): Lange-baanweg, Fossil Park road, (–CC), 127′ [38.4 m], 8 Sept. 2009, *Goldblatt & Porter 13291* (NBG, holo.; MO, PRE, iso.).

Silene noctiflora sensu Thunb., in part [excluding references to Table Mtn and Swartland]: 81 (1794), non. L. (1753).

Tufted perennial, 0.5–1.5 m, producing one or two new vegetative shoots from basal axils at end of flowering season; rootstock a taproot; stems stiffly erect, unbranched except in inflorescence, 3–5 mm diam., densely puberulous with spreading, gland-tipped hairs 0.2–0.5 mm long. *Leaves* mostly basal, suberect or spreading, lower leaves oblanceolate-spathulate, sub-petiolate, 100–140 × 15–25 mm, dead or dying at flowering, stem leaves elliptical to oblanceolate, mostly 60–100 × 10–20 mm, acute, base tapering, margins plane or undulate, pubescent with mixture of erect, gland-tipped hairs and acute, eglandular hairs when young but apparently entirely eglandular at maturity, ciliate at extreme base with longer hairs to 1 mm long, with evident side veins. *Inflorescence* a compact, symmetrically branched cyme with flowers in triads, axis 1.5–3.0 mm diam. below primary flower; bracts similar to upper leaves but smaller, subequal, suberect; primary pedicels 10–40 mm long, secondary pedicels 10–20 mm long. *Flowers* suberect or half-nodding, nocturnal, jasmine-scented or unscented. *Calyx* narrowly urn-shaped in flower, 18–22 mm long, 10-ribbed, with 1–4 reticulate veins in distal half, densely puberulous with erect, gland-tipped hairs, lobes narrowly triangular, 3–6 mm long, densely glandular-ciliolate, splitting almost to base to expose mature capsule. *Petals* whitish or pale dingy pink to mauve, limb spreading, cuneate, not overlapping with adjacent petals, 8–10 × 6–8 mm, bifid ± halfway, claw strap-shaped, 16–18 × 2.5 mm, auriculate apically, coronal scales suberect, ± 1 mm long. *Stamen* filaments unequal, shorter series 9–10 mm long, longer series 12–14 mm long, reaching to 4 mm below top of claws; anthers ± 2 mm long, not reaching top of claws, included. *Ovary* narrowly pyriform, shortly stipitate, 5–6 mm long, stipe ± 0.25 mm long; styles ± 5 mm long, included, ultimately barely exserted. *Capsule* broadly ovoid, 15–20 × 9–10 mm, ± 6 × longer than carpophore, minutely granular; carpophore 2.5–3.0 mm long, glabrescent. *Seeds* 1.2–1.5 mm diam., reniform with hilum recessed, face concave, back flat, reddish brown, testal cells radial-concentric, tuberculate on face but echinate along shoulders and back. *Flowering time*: Sept.–Oct. Figures 2G, H, 12.

Distribution and ecology: restricted to the coastal forelands of the southwestern Cape, from Saldanha Bay along the West Coast to the Cape Flats, and at Vermont near Hermanus (Figure 10), occurring in deep, calcareous sands in strandveld thicket.

Diagnosis and relationships: although overlooked as a distinct species until now, *Silene rigens* is readily distinguished in sect. *Elisanthe* by its characteristic growth form and stem vestiture, flowers, fruits, and seeds. The stems are tall, stiffly erect, and largely unbranched except in the inflorescence, where the branching is symmetrical and the branches relatively short, resulting in distinctive, condensed triads of flowers and fruits. Flowering stems are strictly annual, dying off at the end of the season, to be replaced by axillary shoots from the base. The stem vestiture comprises only gland-tipped hairs, without an admixture of eglandular, acute hairs. Individual flowers are relatively small for the section, with white, dull pink or light mauve petals, the claws 8–10 mm long, and the anthers and styles ultimately barely exserted. The capsules, which are unusually broadly urn-shaped at maturity, 15–18 × 9–10 mm, and with a very short carpophore ± 2.5 mm long, cannot be fully accommodated within the calyx, which splits longitudinally at least midway but often to the base. The distinctly echinate seeds (Figures 2G, H) are unique among the southern African species of sect. *Elisanthe*, which are otherwise colliculate or tuberculate.

Although known for almost 260 years, since the first specimens were collected by Carl Thunberg during his visit to the West Coast in 1773–1774, *Silene rigens* has only now been recognized as a distinct species. Thunberg (1794) mistakenly associated his material (*UPS-THUNB10774 & 10775*) with the European species *S. noctiflora* L. and Sonder (1860) subsequently treated them under *S. undulata*. More recent collections of the species have been variously misidentified as *S. bellidioides*, *S. capensis*, and *S. ornata*.

Conservation note: the restricted occurrence of the species and its local extinction on the Cape Flats render it vulnerable, particualrly as most populations occur in areas undergoing intensive development for housing, industry and recreational activity.

Additional specimens

WESTERN CAPE.—**3317** (Saldanha): Hoedjies Bay, (–BB), Sept. 1905, *Bolus 12613* (BOL). **3318** (Cape Town): Langebaan Peninsula, Stofbergsfontein, (–AA), 28 Nov. 1975 [sterile], *Bucher 2963* (NBG); Elandsfontein West, (–AA), 5 Oct. 1977, 500′ [150 m], *Thompson 3525* (NBG); R27 to Langebaanweg, (–AB), 20 Sept. 2009, *Goldblatt, Manning & Porter 13409B* (MO, NBG); Vygevallei farm, (–AD), 24 Sept. 1996, *Low 2801* (NBG); Duynefontein (Koeberg), (–CB), 31 Oct. 1986, *Bosenberg & Rutherford 177* (NBG); Duynefontein, (–CB), 30 Sept. 2010 [fruiting], *Goldblatt & Manning 13599* (MO, NBG); Melkbosstrand, (–CB), 8 Sept. 1940, *Compton 9329* (NBG); Paarden Eiland, (–CD), 19 Sept. 1942, *Compton 13720* (NBG); damp hollow near Strand Line, Stickland, (–CD), 23 Oct. 1932, *Acocks 1049* (S); Belville, Cape Flats Nature Reserve, University of Western Cape, (–DC), 5 Aug. 1978, *Low 578* (NBG); 31 Aug. 1978, *Low 624* (NBG). **3418** (Simonstown): Cape Flats, Swartklip, along road to military hut in E portion, (–BA), 22 Sept. 1972, *Taylor 8211* (NBG, PRE). **3419** (Caledon): Vermont, near the sea, deep sand, (–AC), 4 Sept. 1986, *S. Williams 1163* (MO).

Silene sect. **Fruticulosae** *(Willk.) Chowdhuri* [as (Rohrb.) Chowdhuri], Notes from the Royal Botanic Garden, Edinburgh 22: 246 (1957). *Silene* [unranked] *Fruticulosae* Willk., Icon. Descr. Pl. Nov. 1: 73 (1854). *Silene* ser. *Fruticulosae* (Willk.) Rohrb., App. Alt. Ind. Sem. Hort. Bot. Berol. 1867: 2 (1867). Lectotype: *Silene ciliata* Pourr., designated by Greuter: 573 (1995b). [Chowdhuri's (1957) lectotypification of the section against *S. burchellii* Otth is treated as an error by Greuter (1995b)].

Perennials or geophytes with woody or tuberous rootstock, eglandular pubescent. *Leaves* linear to oblanceo-

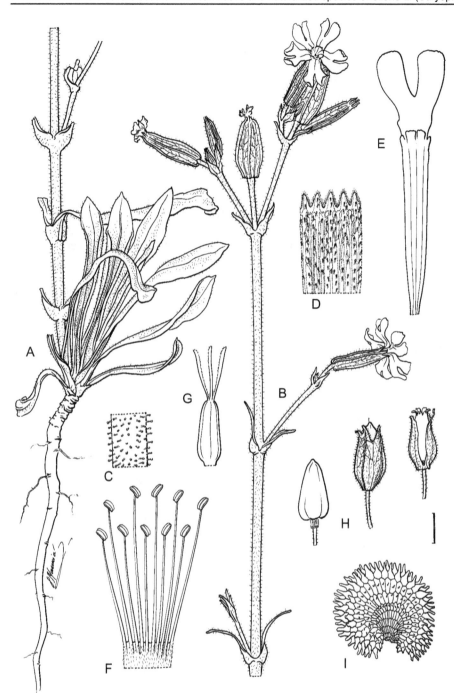

FIGURE 12.—*Silene rigens*, Lange-baanweg, *Goldblatt & Porter 13291*. A, base of plant; B, flowering stem; C, stem segment; D, calyx laid out; E, petal; F, androecium laid out; G, gynoecium; H, capsules (left hand illustration before dehiscence with calyx removed; central before dehiscence, right hand after dehiscence); I, seed. Scale bar: A, B, H, 10 mm; C, 1 mm; D, 5 mm; E–G, 2.5 mm; I, 0.5. Artist: John Manning.

late, lacking apparent lateral veins; bracts herbaceous. *Flowers* in monochasia with axis simple or forked below, spreading or ± erect, vespertine; anthophore well-developed, pubescent. *Calyx* pubescent, clavate, 10-veined, without conspicuous anatomosing venation, lobes ovate, obtuse. *Petals*: claw glabrous or pubescent along veins, weakly auriculate, limb bifid with linear lobes; coronal scales present. *Ovary* partially septate. *Seeds* reniform with deep hilar notch, flanks flat or somewhat concave and smooth or striate, back deeply and narrowly grooved between two undulate peripheral wings.

± 12 spp., S Europe, Africa.

6. **Silene burchellii** *Otth* in DC., Prodromus systematis naturalis regni vegetabilis 1: 374 (1824); Sond.:

128 (1860). Type: South Africa, [Western Cape], kloof between Lion's Head and Table Mt., on the ride towards Camps Bay, without date [probably 1810/1811], *Burchell 271* (G-DC, holo.; K,!, PRE!, iso.). [Although authorship of the species has been attributed either to 'Otth ex DC.' or to 'DC.' alone, De Candolle's note (Candolle 1824: 367) at the foot of the generic treatment makes it clear that the author of the account is Adolphe Otth, and new names published there should thus be attributed to Otth alone.]

Geophytic perennial or suffrutex, mostly 10–50 cm tall; tuber parsnip-like or ovoid; *stems* decumbent or erect, ± suffrutescent below, basal woody portion 1–2 mm diam. but 0.5–1.5 mm diam. at base of inflorescence, ± densely puberulous-scabridulous with short, deflexed or spreading, acute hairs, rarely sub-glabrous.

Leaves cauline but concentrated in lower part of stem, ± abruptly separated from inflorescence, lower sometimes dry or withered, linear to obovate, (10–)15–60 × 1–15(–22) mm, obtuse, apiculate or weakly uncinate, base narrowed, ± adpressed-puberulous, usually more densely so abaxially, rarely glabrous on adaxial or both surfaces, hairs acute, margins ciliate towards base with longer, straggling hairs to 1 mm long, without evident side veins. *Inflorescence* a lax or dense, (2)3–12-flowered monochasium terminating main stems, well separated from foliage, simple or with 1(2) branches from base, rachis 0.5–1.0 mm diam.; bracts smaller than leaves, unequal to subequal, oblong or lanceolate to subulate, puberulous with appressed, acute hairs, margins densely ciliolate; pedicels 3–10(–15) mm long, rarely up to 20 mm long in fruit. *Calyx* clavate in flower, (10–)12–18(–25) × 2.5–4 mm at anthesis, slightly upcurved distal to carpophore when > ± 15 mm long, equally 10-veined, adpressed-puberulous, lobes ovate or triangular, ± 2–4 mm long, densely ciliolate. *Flowers* ± patent or suberect at anthesis, nocturnal, coconut-scented or more medicinal and acrid. *Petals* white to pale yellowish, pale pink or mauve, with darker green or maroon reverse, claw 7–8(–10) mm long, rarely exserted up to 6 mm beyond calyx, narrowed below, without evident auricles, pubescent along midrib and veins on outer surface, limb bifid, 3–7 × 2–5 mm, coronal scales 0.2–0.8 mm long. *Stamen* filaments slightly unequal, 6–10 mm long, shorter series shortly included, longer series reaching top of claw or exserted up to 2 mm. *Ovary* ellipsoid, ± 4 mm long; styles 4–6 mm long, exserted up to 3 mm. *Capsule* ovoid, 8–10 × 5–7 mm, slightly longer than to slightly shorter than carpophore, smooth; carpophore 4–12 mm long, pubescent. *Seeds* 1.0–1.5 mm, reniform with deep hilar notch, flanks flat or somewhat concave, back deeply and narrowly grooved between two undulate wings, reddish brown, testa striate. *Flowering time*: Aug.–Nov. in the winter rainfall region; mainly Oct–May in the summer rainfall region but almost year-round along the coast. Figures 3A–C, 13.

Distribution and ecology: widely distributed through the more temperate parts of southern Africa but absent from the arid interior, through eastern Africa northwards to the highlands of Sudan and Ethiopia, and extending across the Red Sea to Arabia (Figure 13). *Silene burchellii* favours loamy or fine-grained clay soils, often in rocky or stony places, from near sea level up to over 4 000 m. In southern Africa it is mostly a grassland forb but in the winter rainfall region occurs in open, mostly drier, fynbos and succulent shrubland.

Diagnosis and relationships: *Silene burchellii* is distinguished from the other southern African members of sect. *Fruticulosae* by its mostly erect or suberect stems, 0.5–1.5 mm diam. at the base of the inflorescence, and mostly linear to oblanceolate, rarely obovate leaves, (10–)15–60 × 1–16 mm. The leaves are ± concentrated in the lower part of the stem and well separated from the inflorescence by a distinct, leafless upper part ±100–150 mm long bearing one or two pairs of bract-like leaves. The calyx is very variable in length, 10–25 mm long, and is often slightly upcurved. The carpophore ranges from 4–12 mm long, varying from slightly longer than to shorter than the urn-shaped capsule, 8–10 × 5–7 mm.

Silene burchellii is sometimes confused with *S. aethiopica* in the Western Cape, but is distinguished from this annual species by its perennial, geophytic habit, and also never develops the highly branched growth form evident in well-grown plants of *S. aethiopica*. Although the tuber is seldom present on herbarium specimens, the relatively thick stem, terminating abruptly where it has been broken from the tuber, is characteristic. The two species are also ecologically separated, with *S. aethiopica* mainly coastal in deep sandy soils and *S. burchellii* essentially an inland species on finer-grained clay soils. Along the eastern seaboard, especially the Wild Coast and KwaZulu-Natal, broad-leaved *S. burchellii* subsp. *multiflora* can be confused with *S. crassifolia* subsp. *primuliflora*. The latter can usually be identified by its prostrate or decumbent, often branching stems, mostly 1.5–2.0 mm diam. at the base of the inflorescence, leathery or thick-textured leaves, somewhat expanded and strongly pleated calyx, mostly larger coronal scales, 0.8–1.0 mm long, and broadly urn-shaped to subglobose capsules, (8–)10–12 × 7–10 mm. Its strictly coastal distribution, on sandy foredunes, is also diagnostic.

As currently circumscribed, *Silene burchellii* is the most widely distributed of the African species, ranging through the length of the continent, from Arabia and the uplands of Sudan and Ethiopia southwards though East Africa to the Cape Peninsula (Wickens 1976). Although highly variable in leaf shape and width, development of indumentum, number of flowers, and in calyx length, the species has resisted all attempts at segregation into more finely circumscribed taxa. The variation is often broadly correlated with geography and/or ecology but morphological patterns are confounded by the existence of numerous intermediates. The situation was summarised several decades ago by Turrill (1954), who concluded that it was reasonable to assume that the variation in *S. burchellii* is associated both with genetic and with environmental differences. This opinion has not altered since then, and Hedberg (1954, 1957) was unable to maintain any afro-alpine infraspecific taxa on the morphological evidence without supporting evidence from further breeding experiments.

Despite the intractability of the species to satisfactory subdivision, workers on the tropical Africa material have consistently commented on the evident differences between the narrow-leaved tropical plants and the type of the species from the southwestern Cape, which is characterized by ± prostrate or decumbent stems and broad, oblanceolate to obovate leaves (Turrill 1954, 1956a; Wild 1961; Gilbert 2000). There is a general feeling that the tropical African representatives should be distinguished at some level from all or at least some of the southern African material, and Wild (1961) accordingly treated all collections from the *Flora zambesiaca* area as var. *angustifolia* Sond. (1860) to separate it from typical var. *burchellii* from South Africa. With this signal exception, Sonder's (1860) early attempt at segregating the South African material of the species into four varieties, based largely on leaf shape and growth form, has not been implemented. The recent treatment of the species in Ethiopia and Eritrea by Gilbert (2000) avoided all recognition of formal varieties in favour of five informal 'forms'.

FIGURE 13.—*Silene burchellii*: A, subsp. *burchellii*, Signal Hill, no voucher; B, subsp. *pilosellifolia*, Kamiesberg, *Goldblatt & Porter 13574*. Scale bar: 10 mm. Artist: John Manning.

The southern African collections display extensive variation in vegetative and floral morphology but, as with the tropical African material, this is ± continuous, with no clear separation between the extremes, and we are unable to recognize more than a single species. There are, however, some general correlations between morphology and geography that we interpret as demonstrating some level of genetic differentiation within the species. Subsuming all of this variation under a single name obscures its existence and we therefore propose the recognition of four subspecies to highlight the most distinctive variants. The occurrence of intermediates between the subspecies makes the assignment of some specimens difficult, but a combination of morphology and geography is adequate for most material. Although not perfect, this classification provides a basis from which to work.

The type and other collections of *Silene burchellii* from the Cape Peninsula and adjacent coastal regions in the extreme southwestern parts of the Western Cape have a distinctive facies that sets them apart, notably the ± prostrate or decumbent stems and often broad leaves, and we treat these plants as comprising a narrowly circumscribed subsp. *burchellii* that is geographically isolated from other Western Cape populations by the coastal Cape Fold Mountains. A second set of populations from the coastal and near inland parts of the northern Eastern Cape and KwaZulu-Natal and extending into Swaziland is distinctive in its broad leaves and multiflowered inflorescences of nodding or half-pendent flowers with very short calyces, mostly 10–12 mm long. We treat these plants as subsp. *multiflora*. A third set of populations with relatively long calyces, typically 20–25 mm long, is characteristic of the western, southern and central parts of the subcontinent, occurring along the western and southern Escarpment and the mountains of the southwestern Western Cape, inland onto the interior plateau. These populations are treated as subsp. *pilosellifolia*. The remaining southern African plants have moderate-sized flowers, the calyx (12)15–18(–20) mm long, and are treated as subsp. *modesta*. As accepted here, this subspecies is widely distributed, from the Mpumalanga Escarpment southwards into the KwaZulu-Natal Midlands and through the foothills of Eastern Cape and KwaZulu-Natal Drakensberg, and northwards into tropical and North Africa.

Although Carl Thunberg seems to have been the first to collect the species, during his visit to the Western Cape in 1772–1775, he mistakenly associated his collection (*UPS-THUNB* 10780) with the European species *S. nutans* L. (Table 1). Inexplicably, however, this name was omitted from his *Flora capensis* (Thunberg 1794), and the species was only recognized as distinct several decades later (Otth 1824) from material collected on Lion's Head by William Burchell during his visit to the region during 1810–1815.

Key to subspecies

1a Stems prostrate or decumbent, extensively divaricately branched with numerous axillary shoots, not strongly suffrutescent; leaves obovate-spathulate to oblanceolate, ± densely puberulous; calyx 15–20 mm long; coastal regions of the extreme southwestern Western Cape (? also Port Elizabeth) . . . . 6a. subsp. *burchellii*

1b Stems suberect or shortly decumbent, brancing angle acute, ± suffrutescent; leaves variable, upper usually narrower, linear to narrowly oblanceolate, puberulous to subglabrous; calyx 10–25 mm long:

2a Leaves obovate to narrowly oblanceolate, lower 7–20 mm wide; inflorescence mostly (5–)7–12-flowered, usually with well-developed branch, lower internodes 15–20 mm long; flowers patent or slightly deflexed at anthesis; calyx 10–12(–15) mm long; coastal and near inland from E Cape to S Mozambique and Swaziland . . . 6d. subsp. *multiflora*

2b Leaves narrowly oblanceolate to linear-lanceolate or linear (rarely lower leaves oblanceolate), mostly 2–5(–13) mm wide; inflorescence mostly 5–8-flowered, lax or dense, simple or branched, lower internodes 15–40 mm long; flowers mostly suberect, rarely patent; calyx (12–)15–25 mm long, usually arcuate:

3a Calyx (18–)20–25 mm long; western and southwestern coastal and interior southern Africa . . . . 6b. subsp. *pilosellifolia*

3b Calyx (12–)15–18(–20) mm long; southeastern, eastern and northern southern Africa to N Africa and Arabia . . . .
. . . . 6c. subsp. *modesta*

6a. subsp. **burchellii**

Silene thunbergiana Eckl. & Zeyh. ex Sond.: 128 (1860), syn. nov. Type: South Africa, [Western Cape], 'steinige stellen der höhe am Tagelberge und Löwenberge', Sept. [?1832], *Ecklon & Zeyher 253* (SAM, lecto.!, designated by Bocquet & Kiefer: 8 (1978); SAM [2 sheets]!, S!, isolecto.).

Stems decumbent and extensively divaricately branched with numerous axillary shoots, not strongly suffrutescent. *Leaves* obovate-spathulate to oblanceolate, 10–50 × (3–)6–22 mm, ± densely puberulous. *Inflorescence* mostly 5–7-flowered, lax, lower internodes 20–30 mm long. *Flowers* suberect. *Calyx* 15–20 mm long. *Carpophore* 6–10 mm long. Figure 13A.

Distribution: coastal regions of the Western Cape between Darling and Agulhas, and possibly at Port Elizabeth in the Eastern Cape, occurring mainly on shale or loamy soils in renosterveld but also in fynbos, on sandstone or limestone (Figure 14). On the Cape Peninsula, subsp. *burchellii* is found only on Signal Hill and Table Mountain, and is replaced by subsp. *pilosellifolia* on the southern peninsula.

We include *Paterson 2543* (PRE) from Port Elizabeth here although it represents the only collection of subsp. *burchellii* from the Eastern Cape. The plants have prostrate, well-branched stems, oblanceolate leaves, and moderately sized flowers with calyx ± 15 mm long. These features are anomalous for subsp. *pilosellifolia*, the local form from the Port Elizabeth region.

Diagnosis: distinguished by the decumbent, divaricately branching stems with densely pubescent, obovate to oblanceolate leaves rarely more than 4× as long as wide; and moderately long calyx, 15–20 mm long. Plants from Gansbaai and Agulhas are more compact and densely leafy than usual, possibly in response to their proximity to the ocean.

Silene thunbergiana, based on plants collected essentially in the same place as *S. burchellii*, was thought to differ in its broader leaves but this apparent distinction has no merit.

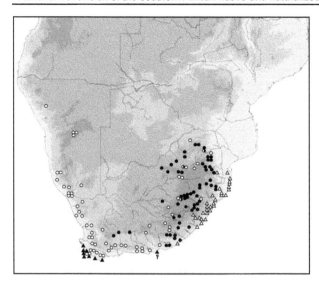

FIGURE 14.—Distribution of *Silene burchellii* in southern Africa: subsp. *burchellii*, ▲; subsp. *pilosellifolia*, ○; subsp. *modesta*, ●; subsp. *multiflora*, △.

Additional specimens

WESTERN CAPE.—**3318** (Cape Town): near Darling, (–AD), Sept. 1905, *Bolus 12615* (BOL); Tygerberg Nature Reserve, 200–400 m, (–DC), 13 July 1975, *Loubser 3320* (NBG); 'steinige lehmige stellen am Löwenstaart [Lion's Rump]', (–DC), July without year, *Ecklon & Zeyher 4748* (BOL, SAM); E slopes of Lion's Head, (–DC), Sept. 1913, *Kensit s.n.* (BOL); Signal Hill, (–DC), Sept. 1897, *Thode STE5830, 5831* (NBG); 28 Aug. 1938, *Penfold 115* (NBG); Signal Hill, Schottsche Kloof, (–DC), Aug. 1938, *Penfold 111* (NBG); above Groote Schuur, (–DC), 12 Sept. 1895, *Wolley Dod 93* (BOL); path to Blockhouse, (–DC), 29 Sept. 1892, *Guthrie 1182* (BOL); Nursery Gorge, (–DC), 24 Oct. 1923, *Compton s.n.* (BOL); E slopes of Table Mtn, (–DC), Sept. 1879, *Bolus 4709* (BOL, PRE); sandy places on Cape Flats, (–DC), without date or collector (NBG). **3418** (Simonstown): Constantiaberg, (–AA), 15 Dec. 1895, *Wolley Dod 472* (BOL); N slope of Orange Kloof, (–AA), 6 June 1897, *Wolley Dod 2616* (BOL); Hangklip, Skilpadsvlei, (–BD), 30' [9 m], 21 Oct. 1969, *Boucher 737* (NBG). **3419** (Caledon): Gansbaai, (–CB), Aug. 1940, *Stokoe 7600* (BOL). **3420** (Bredasdorp); Cape Agulhas, (–CC), 27 Oct. 1940, *Esterhuysen 4411* (BOL).

?EASTERN CAPE.—**3325** (Port Elizabeth): Port Elizabeth, (–DC), Sept. 1914, *Paterson 2543* (PRE).

6b. subsp. **pilosellifolia** *(Cham. & Schltdl.) J.C.Manning & Goldblatt*, stat. nov. *Silene pilosellifolia* Cham. & Schtdl. [as '*pilosellaefolia*'], Linnaea 1: 41 (1862). *Silene burchellii* var. *pilosellifolia* (Cham. & Schltdl.) Sond. [as '*pilosellaefolia*']: 128 (1860). Type: South Africa, [Western Cape], 'prope Plettenbergs-bay ad Doukamma [Goukamma]', Jul. 1821, *Mund & Maire s.n.* (B, holo.† [*fide* Boucquet & Kiefer (1978)]). Neotype: South Africa, Eastern Cape, Uitenhage Dist., sand-hills near the Zwartkopsrivier, Sept. [without year], *Zeyher 240* (NBG, neo.!, designated here; NBG!, SAM [5 sheets]!, iso.). [The holotype appears to have been destroyed (Bocquet & Kiefer 1978) and we accordingly designate a neotype from the same geographical area as the type and that matches both the protologue and the current application of the name. Another collection that has been identified as type material by the Stockholm herbarium, *Mund & Maire s.n. S-G-8699* (S), was collected at Voormansbosch near Swellendam and thus has no direct link to the protologue at all.]

Silene burchellii var. *angustifolia* Sond.: 128 (1860). Type: South Africa, [Western Cape], Caledon Zwartberg,

Dec. [without year], *Ecklon & Zeyher 1959* (S, lecto.!, designated here; PRE!, SAM!, isolecto.).

Silene burchellii var. *cernua* Rohrb.: 121 (1869), nom. illegit. superfl. Type: as for *S. burchellii* var. *angustifolia*.

Silene dinteri Engl.: 383 (1912). Type: [Namibia], 'auf der Granitkuppe bei Aus', 11 Jan. 1910, *Dinter 1141* (SAM, syn.!); 13 Apr. 1911, *Dinter 2235* (SAM, syn.!).

Silene cernua sensu Bartl.: 623 (1832), non Thunb. (1794) (= *S. aethiopica* Burm.).

Silene cernua var. *denudata* Fenzl. in Drège: 222 (1844), nom. nud. [*fide* Rohrb.: 121 (1869)]

Silene cernua var. *hirta* lus. *linifolia* Fenzl. in Drège: 222 (1844), nom. nud. [Specimen: South Africa, 'ad fl. Sonday's River, pr. Nieweweld et in rupestribus as Dutoitskloof, *Drège 550a* [*fide* Rohrb.: 121 (1869)]

Stems ± erect or shortly decumbent, ± tufted with branches subbasal and at an acute angle, ± suffrutescent. *Leaves* oblanceolate to linear-lanceolate or linear but lower leaves sometimes broader, 20–55 × 2–5(–13) mm, puberulous or subglabrous. *Inflorescence* mostly 3–8-flowered, lax or dense, lower internodes 15–40 mm long. *Flowers* mostly suberect, rarely patent. *Calyx* (18–)20–25 mm long, usually flexed slightly upwards. *Carpophore* 9–11 mm long. Figure 13B.

Distribution and ecology: widely distributed through western, southern and the eastern interior of southern Africa, from central Namibia southwards along the western escarpment, through the Cape Fold Mountains into Eastern Cape as far as Zuurberg and inland onto the eastern interior plateau through Free State and Gauteng as far north as Limpopo (Figure 14), occurring in a wide range of open shrubby or grassy habitats, mostly among rocks.

Diagnosis: distinguished from other subspecies by its ± suberect, mostly longer flowers, the calyx (18–)20–25 mm long and curved slightly upwards at the junction between anthophore and ovary. This slight curvature is lost in fruit. Plants are suffrutescent and variable in foliage, usually with oblanceolate to linear-lanceolate or linear leaves although the lower leaves are sometimes broader. Populations from the western edge of the Escarpment in Namibia and Northern Cape southwards into the Cedarberg in Western Cape have sub-glabrous leaves.

The distinction between subsp. *pilosellifolia*, with calyx (18–)20–25 mm long and subsp. *modesta*, with calyx (12–)15–18(–20) mm long is not always a very easy one to draw and even somewhat arbitrary in some cases.

Two collections from the mountains above Simonstown in the southern Cape Peninsula (*Adamson 2333 & Esterhuysen 34106*) are included here on account of their suberect stems and caespitose habit, narrowly oblanceolate leaves, and long calyx, ± 22 mm long. They appear to represent isolated outliers and are the only records of subsp. *pilosellifolia* from the Peninsula—other populations from north of Constantiaberg represent subsp. *burchellii*.

Populations from the summit of Mariepskop [*Van der Schijff 4882* (PRE), *Van der Schijff 5594* (PRE), *Hardy 7000* (PRE)] and from the Wolkberg along the northeastern escarpment [*Müller & Scheepers 128* (PRE)] appear to represent a distinctive ecotype with prostrate or sprawling, highly branching stems, forming tangled mounds, and broad, spathulate leaves. The Mariepskop population was to have formed the basis of the unpublished manuscript name *S. junodii* D.Masson *ms*. At this stage, however, it is unclear how much of this morphological differentiation is phenotypic since these plants invariably occupy sheltered, mostly shaded situations among rocks, where they avoid fires. They are provisionally included in subsp. *pilosellifolia* on account of their calyx length, 20–23 mm long.

Additional specimens

NAMIBIA.—**2014** (Welwitschia): Spitskop, (–BA), 25 Sept. 1981, *Müller 1601* (WIND). **2217** (Windhoek): Grossherzog Friedrichsberg, Farm Regenstein, (–CA), 2 336 m, 19 Mar. 1972, *Giess 11681* (NBG, PRE, WIND); 18 km from Windhoek, (–CB), 27 May 2000, *Zimmermann 239* (WIND). **2616** (Aus): Aus, mountain on Farm Klein Aus, (–CA), 18 Aug. 1963, *Merxmüller 2956* (WIND); Aus, (–CB), 9 Jul. 1922, *Dinter 3579* (BOL, PRE, WIND); 4 Mar. 1929, *Dinter 6109* (BOL, NBG); on road P705, 26 Oct. 1987, *Kolberg 240* (WIND). **2715** (Bogenfels): NW of beacon, (–BD), 9 Sept. 1992, *Oliver 10198* (WIND). **2716** (Bethanie): Karas, E of summit, (–CD), 21 Sept. 1977, *Merxmüller 32193* (WIND); slope of mountain, (–DD), 24 Sept. 1972, *Merxmüller 28811* (WIND). **2718** (Grünau): Klein Karas, (–CA), 9 Apr. 1931, *Örtendahl 50* (PRE, WIND).

LIMPOPO.—**2328** (Baltimore): Blouberg, (–BB), 1 June 1953, *Esterhuysen 21437* (BOL); 20 Feb. 1990, *Stirton, Venter & Edwards 12680* (NU). **2430** (Pilgrim's Rest): Wolkberg, Serala Peak, (–AA), 23 Apr. 1971, *Müller & Scheepers 128* (PRE).

GAUTENG.—**2528** (Pretoria): top of hill behind Louis Botha's home, (–CA), 12 Oct. 1925, *Smuts 870* (PRE); 9 miles [14.4 km], E of Pretoria, The Willows Farm, (–CA), 10 Nov. 1946, *Codd 2139* (PRE); Willowglen, (–CA), Jan. 1951, *Forssman 6* (PRE); Botanical Research Institute, (–CB), 10 Dec. 1975, *Scott 3* (PRE). **2628** (Johannesburg): Doornkloof, (–CC), 25 Nov. 1928, *Gillett 2555* (BOL); 3 Dec. 1928, *Gillett 207* (NBG); 28 Dec. 1928, *Gillett 3246* (BOL).

MPUMALANGA.—**2430** (Pilgrim's Rest): Mariepskop summit, (–DB), 5 Jan. 1960, *Van der Schijff 4882* (PRE); 6 Jul. 1961, *Van der Schijff 5594* (PRE); 3 Sept. 2000, *Burrows 7013* (NU); Mariepskop Radar Station, (–DB), 16 Feb. 1990, *Hardy 7000* (PRE). **2529** (Witbank): Middelberg Dist., Buffelsvlei, (–BC), 1 Dec. 1933, *Rudatis 6* (NBG); Middelburg, (–CD), without date, *Guthrie 4064* (NBG).

FREE STATE.—**2627** (Potchefstroom): 11 miles [17.6 km] from Parys to Potchefstroom, (–DD), 8 Apr. 1967, *Vahrmeÿer 1568* (PRE). **2827** (Senekal): Ficksburg, (–DD), 26 Oct. 1934, *Galpin 13874* (BOL); Strathcona, (–DD), 7 Nov. 1936, *Fawkes 12* (NBG). **2926** (Bloemfontein): Dewetsdorp, (–DA), 15 Apr. 1950, *Steyn 904, 945* (NBG). **3027** (Lady Grey): Zastron Dist., (–AC), Nov. 1926, *Maree 57* (PRE).

KWAZULU-NATAL.—**2829** (Harrismith): foot of Griffin's Hill, (–DD), 2 Apr. 1945, *Acocks 11387* (PRE). **2830** (Dundee): Mpate Mt., (–AA), 26 Nov. 1964, *Shirley s.n.* (NU).

LESOTHO.—**2828** (Bethlehem): Leribe, (–CC), without date, *Dieterlen 362* (PRE, SAM); *Dieterlen 610* (PRE).

NORTHERN CAPE.—**2816** (Oranjemund): Richtersveld, Remhoogte, (–BD), 13 Sept. 1929, *Herre 11778* (NBG); Numees Camping Site, (–BD), 26 Sept. 1981, *Hugo 2787* (NBG, PRE); W side of ridge N of Numees Camp, (–BD), 19 Sept. 1981, *McDonald 688* (NBG); Numees Mountain, (–DB), Sept. 1995, *Williamson & Williamson 5816* (NBG). **2817** (Vioolsdrif): Koeboes [Khubus], (–AC), 17 Sept. 1929, *Herre 11779* (NBG); Rosyntjieberg, neck N of Lelieshoek, (–AC),

1 060 m, 30 Aug. 1977, *Oliver, Tölken & Venter 300* (NBG, PRE); Koeskop, (–AC), Sept. 1995, *Williamson & Williamson 5779* (NBG); Zebrakloof, NE of Rosyntjieberg, (–AC), 9 Oct. 1991, *Germishuizen 5560* (PRE); Ploegberg S of Khubus, (–CA), 600 m, 3 Sept. 1977, *Oliver, Tölken & Venter 509* (NBG, PRE); Ploegberg, (–CA), 25 Sept. 1991, *Van Jaarsveld 11931* (PRE). **2917** (Springbok): Spektakel, (–DA), 26 Aug. 1941, *Compton 11534* (NBG). **2918** (Gamoep): 15 miles [24 km] NE of Springbok, (–AA), 8 Sept. 1950, *Maguire 351* (NBG). **3017** (Hondeklipbaai): Farm 477 Taaibosduin, 20 Aug. 2009, *Bester 9611* (NBG, PRE); Grootvlei, (–BB), 7 Sept. 1945, *Compton 17280* (BOL, NBG); Skilpad Flower Reserve, (–BB), 3 Oct. 1995, *Cruz 132* (NBG). **3018** (Kamiesberg): Langkloof, N of Farm Doringkraal, (–CA), 23 Sept. 2010, *Goldblatt & Porter 13574* (NBG). **3120** (Williston): N of Farm De Hoop, (–CC), 7 Jan 1986, *Snijman 991* (NBG).

WESTERN CAPE.—**3118** (Vanrhynsdorp): Knersvlakte, ± 30 km N of Vanrhynsdorp, Farm Ratelgat, limestone ridge ± 400 m SE of Matjieshuise, (–BC), 7 Aug. 2011, *Koopman s.n.* (NBG). **3119** (Calvinia): upper part of Vanrhyn's Pass, (–AC), 800 m, 13 Sept. 1993, *Strid & Strid 37844* (NBG); parking near top of Vanrhyn's Pass, (–AC), 13 Sept. 2010, *Goldblatt & Porter 13527* (MO, NBG). **3218** (Clanwilliam): Piketberg, Zebra Kop, (–DB), 16 Dec. 1979, *Esterhuysen 35331* (BOL). **3219** (Wuppertal): Kliphuis Gully leading up to Pakhuis Peak, (–AA), 22 Oct. 1987, *Taylor 11863* (NBG); Sneeugat, (–AA), 15 Jan. 1923, *Andreae 922* (PRE); Elandskloof, (–CA), 30 Sept. 1936, *Leipoldt s.n. BOL31661* (BOL); Middelberg Plateau, (–CA), 14 Dec. 1941, *Bond 1351* (NBG). **3318** (Cape Town): Jonkershoek, Guardian Peak, (–DD), [month illegible] 1946, *Adamson 3654* (BOL); Jonkershoek, (–DD), 18 Sept. 1936, *Borchardt 440* (PRE). **3319** (Worcester): Groot Winterhoek, (–AA), 14 Feb. 1934, *Compton 4628* (BOL); Mostertshoek Twins, (–AD), Jan. 1944, *Esterhuysen 9842* (BOL); Hex River Mtns, Sentinel Peak, (–AD), 16 Feb. 1958, *Esterhuysen 27572* (BOL); Groot Drakenstein Mtns, Duiwelskloof, (–CC), 12 Dec. 1943, *Wasserfall 742* (NBG); Haelhoeksneukop, (–CC), 16 Dec. 1975, *Esterhuysen 34161* (BOL). **3320** (Montagu): Laingsburg Dist., Whitehill Ridge, (–BA), 10 Nov. 1935, *Compton 5900* (BOL, NBG); Cabidu, (–BB), 28 Oct. 1950, *Compton 22213* (NBG); Grootvadersbos State Forest, Boosmansbos Wilderness Area, (–DD), 30 Nov. 1988, *Van der Merwe 284* (PRE). **3321** (Ladismith): Gamka Mtn Reserve, (–BC), 25 Nov. 1975, *Boshoff P275* (NBG); Touwsberg, Farm Rietfontein, (–CA), 7 Oct. 1993, *Smook 8705* (PRE). **3322** (Oudtshoorn): Swartberg Pass, (–AC), 7 Aug. 1949, *Steyn 267* (NBG); 6 Dec. 1987, *Vlok 1895* (PRE); near top of Swartberg Pass, (–AC), 21 Mar. 1976, *Thompson 2762* (NBG, PRE); upper Cango valley, Bassonsrus, (–AC), 4 November 1974, *Moffett 433* (NBG); Meiringspoort, (–BC), 12 Nov. 1941, *Thorne s.n.* (NBG). **3418** (Simonstown): hills above Simonstown, edge of swamp, (–AA), 11 Dec. 1938, *Adamson 2333* (BOL); Swartkops Peak S of Simonstown, on firebelt above Froggy Pond, (–AA), 16 Nov. 1975, *Esterhuysen 34106* (BOL); Harold Porter Botanic Garden, above Disa Kloof, (–BD), 17 Oct. 1991, *Forrester 974* (NBG). **3419** (Caledon): Caledon Baths, (–AB), July 1892, *Guthrie 2469* (NBG); between Caledon and Napier, (–BD), 3 Aug. 1940, *Esterhuysen 3051* (BOL); ±12 km NW of Napier, Fairfield Farm, (–BD), 6 Oct. 1994, *Kemper IPC632* (NBG). **3420** (Bredasdorp): Adamskop, (–AC), 200 m, 15 Oct. 1982, *Bayer 3199* (NBG).

EASTERN CAPE.—**3125** (Steynsburg): Middelburg Dist., Bangor Farm, (–AC), Oct. 1917, *Bolus 14040* (BOL). **3126** (Queenstown): Broughton near Molteno, (–AD), Dec. 1892, *Flanagan 1567* (NBG, PRE, SAM); Queenstown, (–DD), Feb. 1896, *Galpin 1988* (PRE); Jan. 1962, *Bokelmann 1* (NBG); 15 km from Dordrecht on road to Queenstown, (–DB), 13 Jan. 1997, *Germishuizen 8891* (PRE). **3323** (Willowmore): between Willowmore and Patensie, Nuweveld Pass, (–BC), 15 Sept. 1982, *Balkwill 451* (NU); Langkloof, near Misgund, (–DC), Oct. 1921, *Fourcade 1718* (BOL). **3224** (Graaf-Reinet): Schimper's Hill near Graaf-Reinet, (–BA), 4 July 1865, without collector *BOL74032* (BOL). **3324** (Steytlerville): 51.1 miles [82 km] from Humansdorp on road to Willowmore, (–CA), 18 Oct. 1934, *Fourcade 5169* (NBG); Zuur Anys Hills [Suuranysberge], (–CC), Dec 1932, *Fourcade 4945* (BOL); kloof on road to Hankey, (–DD), 12 Sept. 1942, *Fourcade 5746* (BOL, NBG). **3325** (Port Elizabeth): Zuurberg, (–BD), 27 May 1965, *Bayliss 2686* (NBG); 8 Jan. 1986, *Van Wyk & Van Wyk 1183* (PRE).

6c. subsp. **modesta** *J.C.Manning & Goldblatt*, subsp. nov.

TYPE.—Eastern Cape, 3227 (Stutterheim): Mount Kemp, (–CB), 13 Jan. 1947, *Compton 19196* (NBG, holo.).

Silene burchellii var. *latifolia* Sond.: 128 (1860), syn nov. Type: South Africa, [Gauteng], shady places at the Crocodile River, *Zeyher s.n. S11-24323* (S, holo.!).

Silene acuta E.Mey. in Drège: 222 (1844), nom. nud. [Specimen: South Africa, [Eastern Cape], between Omtendo [Mtentu] and Omsamculo [uMzimkhulu], *Drège 5340* [*fide* Sond.: 128 (1860)].

Silene cernua var. *hirta* lus. *lanceolata* Fenzl. in Drège: 222 (1844), nom. nud. [Specimen: South Africa, [Eastern Cape], 'iter Gekan et Busche', *Drège 550h* [*fide* Rohrb.: 121 (1869)].

[For synonyms from tropical and North Africa see Turrill (1956a), Wild (1961), Wickens (1976) and Gilbert (2000)].

Stems ± erect or very shortly decumbent, ± tufted with branches subbasal and at an acute angle, ± suffrutescent. *Leaves* oblanceolate to linear-lanceolate or linear but lower leaves sometimes broader, 20–55 × 2–5(–13) mm, puberulous or subglabrous. *Inflorescence* mostly 3–8-flowered, lax or dense, lower internodes 15–40 mm long. *Flowers* mostly suberect, rarely patent. *Calyx* (12–)15–18(–20) mm long, usually flexed slightly upwards. *Carpophore* 6–9(–10) mm long.

Distribution and ecology: widely distributed through eastern southern Africa, Lesotho and Swaziland northwards into tropical and North Africa and Arabia (Figure 14), essentially in temperate or subtropical grassland and savanna, often in rock outcrops.

Diagnosis: subsp. *modesta* constitutes the core of this variable species after the more extreme forms have been segregated and includes ± suffrutescent, tufted plants with mostly oblanceolate to linear leaves and suberect, moderately-sized flowers with calyx (12–)15–18(–20) mm long. This taxon seems to have included the bulk of the collections that were to have been recognized as a new species under the unpublished manuscript name *Silene australis* D.Masson *ms*.

Subsp. *modesta* is replaced in southern Mozambique and coastal KwaZulu-Natal and Eastern Cape by subsp. *multiflorus*, mostly with broader leaves and with generally shorter flowers, the calyx mostly 12–15 mm long, and in the western, southern and eastern interior of the subregion by subsp. *pilosellifolia* with mostly longer flowers, the calyx 18–25 mm long. Subsp. *burchellii*, with procumbent or decumbent stems, is restricted to the extreme southwestern coastal regions.

The tropical African material, all with a calyx length of 12–20 mm, is comfortably included in subsp. *modesta*. Previously aberrant populations from northern Kenya, eastern Uganda, and Ethiopia with exceptionally long calyces, 26–35 mm long, treated as var. *gillettii* Turrill (1954, 1956a), have since been segregated as the distinct species *Silene gillettii* (Turrill) M.G.Gilbert, also distinguished by its broader leaves with broadly cuneate or subcordate bases (Gilbert 2000).

Additional southern African specimens

LIMPOPO.—**2329** (Pietersburg): Farm Duvenhageskraal, (–CD/DC), 3 Dec. 1985, *Venter 11283* (NU). **2330** (Tzaneen): Woodbush Forestry Station, (–CC), Jan. 1923, *Wager 23094* (PRE); 23 Dec. 1928, *Gillett 3205* (NBG). **2428** (Nylstroom): Geelhoutkop, (–AD), Jan. 1918, *Breijer 18089* (PRE); Warmbaths, (–CD), 30 Sept. 1908, *Leendertz 1352* (PRE).

NORTH-WEST.—**2526** (Zeerust): Lichtenburg Dist., Grasfontein, (–CC), 16 Dec. 1929, *Sutton 343* (PRE). **2527** (Rustenburg): Rustenburg Dist., (–BA), without date, *Mudd s.n.* (BOL); Rustenburg Nature Reserve, (–CA), 8 Oct. 1970, *Jacobsen 1067* (PRE). **2626** (Klerksdorp): Ventersdorp, (–BD, Feb. 1932, *Wilman NBG252/30* (BOL).

GAUTENG.—**2528** (Pretoria): Fountains Valley, (–CA), 14 Nov. 1928, *Repton 118* (PRE); Wonderboom Reserve, (–CA), 20 Oct. 1944, *Repton 1887* (PRE); 14 miles [22.4 km] SE of Pretoria on road to Delmas, (–CA), 21 Dec. 1950, *Codd 6302* (PRE). **2627** (Potchefstroom): Northcliffe, (BB), 26 Oct. 1984, *Behr 785* (PRE). **2628** (Johannesburg): ridge above Jeppestown, (–AA), Feb./Mar. 1894, *Galpin 1376* (PRE); Kensington Ridge, (–AA), 31 Oct. 1932, *Burtt Davy 57* (NBG); Melville Koppies, (–AA), 3 Oct. 1960, *Macnae 1265* (BOL); Mondeor, (–AA), 9 Nov. 1961, *Lucas 35146* (PRE); Aasvogels Kop, (–AA), 24 Apr. 1927, *Young 26444* (PRE).

MPUMALANGA.—**2430** (Pilgrim's Rest): Lekgalameetse Nature Reserve, The Downs, (–AA), 21 Jan. 1986, *Stalmans 1005* (PRE); 19 Dec. 1988, *Stahlmans 1811* (PRE); Mt Saheba Nature reserve, (–DC), Jan. 1976, *Forrester & Gooyer 46* (PRE); Graskop, (–DD), 3 Dec. 1887, *Galpin s.n.* (BOL); Stanley Bush Kop, (–DD), 9 Dec. 1986, *Raal & Raal 1016* (PRE). **2530** (Lydenburg): Dullstroom, (–AC), 30 Jan. 1959, *Werderman & Oberdieck 2045* (PRE); 24 Feb. 1982, *Cameron 151* (PRE); Mt. Anderson, (–BA), 25 Dec. 1932, *Smuts & Gillett 2468* (BOL, PRE); Farm de Kuilen, (–BA), 13 Mar. 1985, *Krynauw 313* (PRE); Mac Mac Nature Reserve, (–BB), 7 Mar. 1979, *kluge 1789* (PRE); Belfast, (–CA), 31 Jan. 1929, *Hutchinson 2765* (BOL); 16 Feb. 1964, *Bayliss 2022* (NBG); Dullstroom Dist., Farm Macduff, (–CB), 24 Feb. 1989, *Burgoyne 1053* (PRE); Farm Uitkomst, 18 km from Machadadorp on road to Badplaas, (–CD), 5 Mar. 1986, *Germishuizen 3811* (PRE); Kaapsehoop Asbestos Mine, (–DB), 1 Mar. 1987, *Morrey & Cadman 3679* (NU). **2531** (Komatipoort): Barberton Dist., Lomati Valley, (–CC), Feb. 1922, *Thorncroft 1123* (PRE); summit of Saddleback Mtn, Farm Dycedale, (–CC), 18 Apr. 1987, *Brusse 5053* (PRE). **2630** (Carolina): Carolina Dist., (–AA), 18 Nov. 1909, *Rademacher 8195* (PRE); Ermelo Dist., Farm Nooitgedacht, (–AC), 9 Mar. 1937, *Henrici 1601* (PRE); Songimvelo Game Reserve, between Hooggenoeg and Lochiel, (–BB), 11 Dec. 1992, *Jordaan 2520* (PRE). **2729** (Bethal): Amersfoort, Welverdacht Farm, (–DB), 15 Mar. 1985, *Turner 692* (PRE); Ermelo, (–DB), 1906, *Burtt Davy, 5453* (PRE); Feb. 1910, *Leendertz 7807* (PRE). **2730** (Vryheid): E of Wakkerstroom, (–AD), 13 Jan. 1986, *Glen 1589* (PRE); near Piet Retief, (–BB), 13 Jan. 1951, *Compton 22339* (NBG).

FREE STATE.—**2828** (Bethlehem): Clarens, (–CB), Nov. 1917, *Van Hoepen 18185* (PRE); Golden Gate, (–DA), 15 Jan. 1976, *Fenn 6* (NU); Golden Gate National Park, Mt Pierre, (–DA), 8 Jan. 1989, *Groenewald 8711* (PRE); Golden Gate Highland Park, (–DA), Jan. 1966, *Liebenberg 7495A* (PRE); Witzieshoek, (–DB), Dec. 1905, *Thode STE5663* (NBG). **2829** (Harrismith): Van Reenen, (–AB), Jan. 1914, *Bews 962* (NU); Harrismith, Queen's Hill, (–AC), 22 Feb. 1969, *Jacobsz 1073* (PRE); Harrismith, Botanic Garden, 5 500' [1 667 m], (–AC), 18 Jan. 1970, *Van der Zeyde s.n.* (NBG); Drakensberg Botanic Garden, (–AC), 23 Jan. 1975, *Jacobsz 2093* (NBG, PRE); Swinburne, Farm Grootvlei, (–AC), 30 Jan. 1961, *Jacobsz 9* (PRE); Sterkfontein Dam, (–AC), 28 Nov. 1974, *Jacobsz 1858* (PRE); 11 Dec. 1985, *Blom 333* (PRE). **2926** (Bloemfontein): Wintervalley, N of Bloemfontein, (–AA), 6 Dec. 1968, *Muller 386* (PRE); Bloemfontein, O.F.S. Botanic Garden, (–AA), 6 Dec. 1968, *Müller 386* (NBG); Grant's Hill near Oranje School, (–AA), 23 Oct. 1925, *Potts 3526* (PRE).

SWAZILAND.—**2631** (Mbabane): hilltop near Mbabane, (–AC), 20 Feb. 1961, *Dlamini s.n.* (NBG, PRE).

KWAZULU-NATAL.—**2729** (Volksrust): Newscastle Dist., Blue Ridge, (–DD), 25 Nov. 1989, *Smit 1315* (PRE). **2730** (Vryheid): Hlobane, (–DB), 5 Mar. 1950, *Johnstone 389* (NU). **2731** (Louwsburg): Itala Game Reserve, (–CB), 17 Jan. 1978, *McDonald 475* (NU); 19 Oct. 1982, *Germishuizen 2235* (PRE); Ngome, (–CD), 3 Apr. 1977, *Hilliard & Burtt 9961* (NU); Ngome, Ntendeka Wilderness area, (–CD), 4 dec. 1985, *Jordaan 580* (PRE). **2828** (Bethlehem): Royal Natal National Park, 200 m along path from Basuto Gate to Tendele, (–DB), 5 Feb. 1982, *Manning 87* (NU); Mont aux Sources, (–DD), Apr. 1920, *Allsop 34* (PRE). **2829** (Harrismith): Little Switzerland, (–CB), 21 Apr. 1969, *Anderson 227* (PRE). **2929** (Underburg): Giant's Castle Game

Reserve, (–AD), 26 Jan. 1966, *Trauseld 542* (NU); Mooi River, (–BB), Dec. 1942, *Fisher 419* (NU); Mooi River Dist., Hidcote, (–BB), 15 Jan. 2003, *Potgieter 876* (NU); Kamberg, (–BC), 31 Dec. 1974, *Wright 2090* (NU); Mpendle Dist., Mulangane Ridge above Carter's Nek, (–BC), 3 Feb. 1984, *Hilliard & Burtt 17523A* (NU, PRE); Loteni Nature Reserve, (–BC), 13 Dec. 1978, *Phelan 223* (NU); Sani Pass, (–CB), 21 Mar. 1977, *Hilliard & Burtt 9743* (NU); headwaters of Mhlahl-angubo River, (–CB), 23 Jan. 1982, *Hilliard & Burtt 15343* (E, NU); Bulwer Dist., Mahwaqa Mt., (–DC), 11 Jan. 1992, *Feltham 124* (NU); Polela Dist., Farm Sunset, (–DC), 30 Dec. 1973, *Rennie 445* (NU); 8 Feb. 1985, *Rennie 1616* (NU). **2930** (Pietermaritzburg): Karkloof, Mt. Gilboa, (–AD), 17 Dec. 2000, *Johnson & Neal 33* (NU); Noodsberg, Lager Farm, (–BD), 14 Oct. 1989, *Williams 558* (PRE); Table Mt., (–CB), 7 Jan. 1949, *Killick 220* (NU); Byrne Valley, (–CC), 31 Oct. 2002, *Potgieter 841* (NU); Hella Hella, 3 800′ [1 160 m], (–CC), 1 Nov. 1997, Ingomankulu Hill, (–CD), 15 Dec. 2004, *Young 208* (NU).

LESOTHO.—**2927** (Maseru): Molimo-Nthuse on way to Pass, (–BC), 20 Oct. 1975, *Schmitz 6239* (PRE). **2828** (Bethlehem): Butha Buthe, (–CC), 9 500′ [2 878 m], 2 Feb. 1954, *Coetzee 409* (NBG, PRE); Malibamatzo Valley, (–DC), Dec. 1971, *Schmitz 1545* (PRE); New Oxbow Lodge, (–DC), 13 Jan. 2003, *bester 3937* (PRE). **2928** (Marakabei): from Ha Lephoi Village along Lesobeng River, (–CB), 5 Mar. 1990, *Smook 7271* (PRE). **2929** (Underberg): Sehlabathebe National Park, (–CC), 7 Jan. 1979, *Hoener 2121* (PRE).

WESTERN CAPE.—**3221** (Merwewille): Roggeveld, Farm Uitkyk, (–AD), Oct. 1920, *Marloth 9704* (PRE); Beaufort West Dist., Layton, (–BB), 24 Nov. 1983, *Shearing 400* (PRE). **3222** (Beaufort West): Beaufort West Dist., Nieuweveld Mtns, (–BD), July 1895, *Marloth 2136* (PRE); Nieuweveld Mtns, Mountain View Farm, (–BD), 18 Apr. 1978, *Gibbs Russel, Robinson & Herman 482C* (PRE).

EASTERN CAPE.—**3027** (Lady Grey): Witteberg, (–CB), 11 Mar. 1904, *Galpin 6583* (PRE); Rhodes, Carlisle's Hoek, (–DB), 27 Dec. 1977, *Bigalke 22* (NU). **3127** (Lady Frere): 25 km from Cala at turnoff to Engcobo, (–DB), 12 Jan. 1997, *Germishuizen 8819* (PRE). **3128** (Umtata): Ugie Dist., Pomona, (–AA), Feb. 1928, *Gill 178* (NBG); Elandslaagte, (–AA), Mar. 2000, *Edwards & Potgieter 1922* (NU); Umtata, (–DB), 18 Jan. 1895, *Schlechter 6326* (PRE). **3226** (Fort Beaufort): top of Katberg Pass at Devil's Bellows, (–BC), 24 Apr. 1995, *Victor 1183* (PRE); ± 5 km above Nico Malan Pass, (–BD), 7 Feb. 1995, *Victor & Hoare 324* (PRE). **3227** (Stutterheim): Thomas River, (–AD), 15 Jan. 1947, *Compton 19309* (NBG); *Leighton 2797* (BOL); King William's Town, (–CB), Nov. 1892, *Sim 966, 967* (NU); East London Dist., Amalinda Nature Reserve, (–DD), 22 Oct. 1984, *Currie 9* (PRE). **3326** (Grahamstown): Grahamstown, (–BC), Oct. 1888, *Galpin 224* (PRE); Martindale, (–BD), 30 Nov. 1955, *Taylor 4962* (NBG); 7 Jan. 1965, *Bayliss 2600* (NBG).

6d. subsp. **multiflora** *J.C.Manning & Goldblatt*, subsp. nov.

TYPE.—KwaZulu-Natal, 2930 (Pietermaritzburg): Krantzkloof, (–DD), Nov. 1921, *Haygarth 9437* (NBG, holo.; NH, PRE, iso.).

Stems ± erect or very shortly decumbent, ± tufted with branches sub-basal and at an acute angle, ± suf-frutescent. *Leaves* obovate to narrowly oblanceolate, 20–55 × (2–)7–15(–20) mm, upper leaves usually nar-rower, puberulous or subglabrous adaxially. *Inflores-cence* mostly (5–)7–12-flowered, dense, lower inter-nodes 15–20 mm long (exceptionally lowermost 40 mm long). *Flowers* nodding or slightly deflexed at anthesis. *Calyx* 10–12(–15) mm long. *Carpophore* 3–6 mm long.

Distribution: coastal and near inland from the Fish River Mouth in Eastern Cape northwards through Kwa-Zulu-Natal into southern Mozambique and inland to near Mbabane in Swaziland, occurring mainly in sub-tropical grasslands on loamy or sandy soils (Figure 14).

Diagnosis: distinguished by the shortly decumbent or ± tufted habit with relatively broad, oblanceolate to obovate lower leaves, and the dense, multi-flowered

inflorescence (typically with a well-developed branch), of small, nodding or slightly pendent flowers with short calyces, 10–12(–15) mm long.

The distinction between subsp. *multiflora* and subsp. *modesta* is blurred in the KwaZulu-Natal midlands, where the transition between subsp. *multiflora* with broad-leaves and subsp. *modesta* with narrowly oblan-ceolate leaves, mostly less than 5 mm wide, lies along the Drakensberg foothills at an altitude of ± 1 500 m.

Coastal forms of subsp. *multiflora* with more decum-bent branches may be difficult to separate from *Silene crassifolia* subsp. *primuliflora* in southern KwaZulu-Natal and Eastern Cape where the distribution of the two taxa overlaps. Introgression between the two species is one possible cause. An alternative hypothesis is that *S. crassifolia* is a coastal dune derivative of *S. burchellii*, with the two forms becoming progressively more dis-tinct to the south as ecological and geographical isola-tion between them increases. Plants with the following combination of characters are treated as subsp. *multi-flora*: stems not strongly prostrate, branching angles acute; inflorescence branched and multi-flowered with more than five flowers, well separated from foliage by leafless stem ±100–150 mm long bearing one or two pairs of bract-leaves; calyx 10–12(–15 mm) long; cap-sules ovoid, 8–10 × 5–7 mm.

South African material of this taxon has tradition-ally been identified as var. *latifolia* Sond. but this name is typified by a specimen from Gauteng and correctly applies to broad-leaved forms of subsp. *pilosellifolia*. It was to have been recognized at species level by Da-niel Masson, and several herbarium specimens bear the unpublished name *Silene natalensis* D.Masson *ms*.

Additional specimens

SWAZILAND.—**2631** (Mbabane): Mbabane, (–AC), 20 Dec. 1952, *Compton 23795* (NBG); Ukutula, (–AC), 29 Oct. 1954, *Comp-ton 24597* (NBG); 13 Nov. 1955, *Compton 25249* (NBG).

MOZAMBIQUE.—**2632** (Bela Vista): road from Ponta Molan-gane to Ponta Do Puro, (–BC), 25 Nov. 2001, *Govender 72* (NH); Maputo Elephant Park, (–BC), 29 Nov. 2001, *Govender 90* (NH).

KWAZULU-NATAL.—**2632** (Bela Vista): Kosi Bay, (–DD), 28 Sept. 1961, *Meyer s.n.* (NU); road from Kosi Estuary to Kosi Lake Campsite, 50 m, (–DD), 18 Nov. 1982, *Balkwill 571* (NH, NU); Kosi Bay, Natal Parks Board Camp, (–DD), 27 Nov. 1967, *Strey & Moll 3939* (PRE); Kosi Mouth, (–DD), 24 Apr. 1995, *Lubbe 646* (NU). **2731** (Louwsburg): Nhloenkulu Mission Station, (–DC), July 1927, *Markotter s.n. STE8682* (NBG). **2732** (Ubombo): NW of Sibaya, (–BC), 21 Sept. 1995, *Lubbe 717* (NU); Lake Sibaya, (–BC), 16 Sept. 1965, *Vahrmeijer 1122* (PRE); Mazengwenya, ± 6 km W of Mabibi, (–BC), 17 Sept. 1994, *Lubbe 273* (NH); Mazengwenya, E of Vazi Swamp, (–BC), 27 Nov. 1969, *Moll 4725* (NH); Mbazwane/Sodwana Rd., (–DA), 28 Nov. 1971, *Pooley 1516* (NU); Mbazwana, (–DA), 27 Sept. 1977, *Balsinhas 3270* (PRE); St. Lucia Bay, (–DC), 26 Jul. 1939, *Schweickerdt 1371* (NH, PRE); 1 km N of Lake St. Lucia, (–DC), 24 Sept. 1987, *MacDevette 1892* (NH). **2831** (Nkandla): Melmoth, Mooiplaas, (–CA), ± 930 m, 10 Oct. 1995, *Hutchings & Williams 3554* (NH). **2832** (Mtubatuba): Hluhluwe Game Reserve, (–AA), 5 Oct. 1953, *Ward 1519* (NU); W of Charter's Creek, (–AC), 5 Dec. 1955, *Ward 2840* (NH, NU, PRE); Hlabisa, Mfolozi flood plain, (–AC), 20 oct. 1985, *Steyn 11* (NH); Dukuduku, (–AC), 25 Nov. 1965, *Strey 6098* (NH, NU, PRE); 24 Jul. 1986, *Steyn 55* (NH); Lake St Lucia, East Shores, (–BA), 11 Oct. 1974, *Taylor 238* (NU); Richard's Bay, (–BD), 14 Jul. 1929, *Rump s.n.* (NH, NU); 6 July 1974, *Ward 8670* (NU, PRE); 16 Jul. 1974, *Ward 2490* (NU). **2930** (Pietermaritzburg): Greytown, Blinkwater Trails area, (–BA), 1 dec. 2001, *Potgieter 560* (NU); Pietermaritzburg, (–CB), Nov. 1939, *Thomas 14* (NBG); Pietermaritzburg, Town Hill, (–CB), Oct. 1942, *Fisher 362* (NU); Nov. 1944, *Fisher 741* (NH); 30 Oct. 1952, *Compton 23738* (NBG); Thornville, (–CB), 7 Nov. 1964, *Shirley s.n.* (NU); Drummond, (–CC), Oct. 1929, *Rump s.n.* (NU); Mid Illovo, (–CC), 10 Dec, 2008, *Young 863* (NU); Inanda, (–DA), Nov. 1884, *Medley Wood 393* (NBG, NH); Aug. without year, *Medley Wood 158* (NH); Hammarsdale, (–DA), 4 May 1995, *Ward 13088* (NH, NU), Assagay, (–DC), 2 Nov. 2003, *Wragg 375* (NU); Sept. 1921, *Rogers 24423* (NU); Westville, (–DD), 22 Sept. 1965, *Moll 2340* (NU, PRE). **2931** (Stanger): Tugela Beach, (–AB), 17 Jan. 1952, *Johnson 385* (NBG); Groutville, (–AC), 14 Oct. 1965, *Moll 2559* (NU); The Bluff, Treasure Beach, (–CC), 6 Nov. 1982, *Ellery 3* (NU); Durban, (–CC). Oct. 1883, *Medley Wood 122* (PRE); 25 Oct. 1889, *Medley Wood 4765* (NH); Isipingo, (–CD), May 1948, *Ward 363* (NH, NU); 9 Feb. 1967, *Ward 6054* (NU). **3029** (Kokstad): Kokstad, (–AD), Dec. 1881, *Tyson 1816* (SAM); Mt Currie, (–AD), 25 Nov. 1930, *Goossens 341* (PRE); Weza, Ingeli slopes, (–DA), 2 Jan. 1966, *Strey 6366* (NH, PRE); Weza State Forest, (–DA), 20 Nov. 1986, *Jordaan 857* (NH); Harding, Rooi Vaal, (–DB), 7 Jan. 1957, *Taylor 5359* (NBG). **3030** (Port Shepstone): Alexandra Dist., Umgay [Umgai], (–AD), 7 Nov. 1908, *Rudatis 461* (NBG, PRE); Dumisa, Fairfield, (–AD), 21 Oct. 1997, *Arkell 324* (NH); 22 Oct. 1997, *Ngwenya 1573* (NH); 10 km to Highflats from Umtentweni, (–AD), 6 Jan. 1981, *Schrire 573* (NH); 10 km from Highflats on road to Umzinto, (–AD), 7 Jan. 1981, *Germishuizen 1812* (PRE); Dududu, The Cedars, (–BA), 5 Nov. 1992, *Williams 956* (NH, PRE); Amanzimtoti, (–BB), 27 Sept. 1959, *Wilson 29* (NU); Scottburgh, (–BC), May 1954, *Garbutt 15* (NU); Pennington, (–BC), Apr. 1950, *Gower 17* (NU); 20 Dec. 1960, *Mauve 4098* (PRE); Ifafa, (–BC), 12 Jan. 1943, *Hardley 73* (NU); Park Rynie, (–BC), 8 Apr. 1967, *Baijnath 381* (NU, PRE); Vernon-Crookes Nature Reserve, (–BC), 11 Nov. 2001, *Styles 116* (NU); Umtamvuna Nature Reserve, (–CC), 27 Feb. 1983, *Abbott 874, 875* (NH); Oribi Gorge, (–CD), 11 Jan. 1971, *Glen 328* (NU); Margate, (–CD), Feb. 1931, *Rump s.n.* (NH). **3130** (Port Edward): Port Edward, Farm Blencathra, (–AA), 26 Nov. 2003, *Singh & Baijnath 847* (NH).

EASTERN CAPE.—**3129** (Port St. Johns): Lusikisiki, (–BC), 24 Nov. 1964, *Bayliss 2519* (NBG). **3130** (Port Edward): Mzamba, (–AA/AB), 19 Oct. 1993, *Arkell & Abbott 151* (NH). **3327** (Peddie): Fish River Mouth, (–AC), 22 Oct. 1964, *Bayliss 2458* (NBG).

7. **Silene crassifolia** *L.*, Species plantarum, ed. 2, 1: 597 (1762); Sond.: 129 (1860). Type: South Africa, [Western Cape], Strandfontein, 21 Dec. 1941, *Compton 12781* [NBG, neo.!, designated by Cupido in Cafferty & Jarvis: 1053 (2004)].

Geophytic perennial 7–30 cm tall, forming loose or more compact mats; tuber cylindrical or ovoid; *stems* prostrate or straggling, decumbent, basal woody portion 2–5 mm diam. but (1.0–)1.5–2.0 mm diam. at base of inflorescence, ± densely puberulous-scabridulous with short, deflexed or spreading, acute hairs, rarely subglabrous. *Leaves* cauline, lower mostly dry or withered, oblanceolate to obovate or suborbicular, (10–)15–50 × 5–18 mm, obtuse, apiculate or weakly uncinate, base narrowed in lower leaves, leathery to sub-succulent, ± densely adpressed-puberulous to thickly felted, rarely glabrous on adaxial or both surfaces, hairs acute, margins ciliate towards base with longer, straggling hairs to 1 mm long, without evident side veins. *Inflorescence* a lax or dense, (2)3–8(–10)-flowered monochasium terminating main stems and branches, not evidently separated from leafy stem, rachis 1.0–1.5 mm diam.; bracts smaller than leaves, unequal to subequal, ovate or obovate to oblong or linear-lanceolate, puberulous to thickly felted with appressed or erect, acute hairs, margins densely ciliolate; pedicels 3–10(–15) mm long, rarely up to 20 mm long in fruit. *Calyx* clavate or loosely funnel-shaped in flower, (10–)12–18(–20) × 3–5 mm at anthesis, strongly plicate, equally 10-veined, densely adpressed-puberulous/felted, lobes ovate or triangular, ± 2–3 mm long, densely ciliolate. *Flowers* patent at anthesis, nocturnal. *Petals* white to pale yellowish or

pale pink, claw ± 7(–10) mm long, rarely exserted up to 4 mm beyond calyx, narrowed below, without evident auricles, pubescent along midrib and veins on outer surface, limb bifid, 4–6 × 4–6 mm, coronal scales 0.8–1.0 mm long. *Stamen* filaments slightly unequal, 6–7 mm long, shorter series shortly included, longer series reaching top of claw or exserted up to 1.5 mm. *Ovary* ellipsoid, ± 4 mm long; styles 4–5 mm long, exserted up to 2 mm. *Capsule* broadly ovoid or subglobose, (8–)10–12 × 7–10 mm, subequal to slightly longer than carpophore, smooth; carpophore 4–8 mm long, pubescent. *Seeds* 1.0–1.5 mm, reniform with deep hilar notch, flanks flat or somewhat concave, back deeply and narrowly grooved between two undulate wings, reddish brown, testa striate. *Chromosome number* (subsp. *primuliflora*): $2n = 24$ (Masson 1989). *Flowering time:* mainly Sept.–Mar. but ± throughout the year along the east coast. Figures 3D–G, 15.

Distribution and ecology: distributed along the southwestern, southern, and southeastern coast of South Africa, from Saldanha Bay in the Western Cape to Park Rynie in southern KwaZulu-Natal (Figure 16). The species is strictly coastal, growing in deep sandy soils, often just above the high level waterline but up to 20 m a.s.l., typically on stablized or partially stabilized foredunes at the edge of coastal scrub or grassland, rarely on cliffs but evidently never on limestone or other calcareous outcrops.

Diagnosis and relationships: Silene crassifolia is usually readily identified by its prostrate or decumbent, relatively robust, perennial and often branching stems, mostly 1.5–2.0 mm diam. at the base of the inflorescence, that are leafy almost to the base of the inflorescence, oblanceolate to obovate or sub-orbicular leaves, somewhat inflated and strongly pleated calyx 3–5 mm diam., and broadly urn-shaped to subglobose capsules, (8–)10–12 × 7–10 mm. The strictly coastal distribution, on sandy foredunes, is also diagnostic.

Collections of subsp. *primuliflora* from the eastern coast, especially the Wild Coast and KwaZulu-Natal, may be confused with broad-leaved *Silene burchellii* subsp. *multiflora* and have sometimes been identified as a coastal form of this species. *S. burchellii*, which generally occurs on finer-grained soils inland of the coast, is a more slender plant, with essentially annual, mostly ± erect stems 0.5–1.5 mm diam. at the base of the inflorescence, with the foliage leaves concentrated towards the base of the stem and well separated from the inflorescence by an upper portion of leafless stem, and smaller capsules, 8–10 × 5–7 mm. This is the only region where both species occur, and the possibility of introgression in the coastal grasslands here cannot be discounted.

The two subspecies of *Silene crassifolia* recognized here have traditionally been regarded as distinct species but the greatly increased collections now available have led us to reconsider this interpretation. When Sonder (1860) last monographed the genus, he knew *S. crassifolia* only from the Cape Flats (Blouberg and Riet Valley [Gordon's Bay]), and *S. primuliflora* from a few collections east of Knysna (Tsitsikamma, Port Elizabeth, and southern KwaZulu-Natal). The distinction between the two taxa was quite clear from this sparse material:

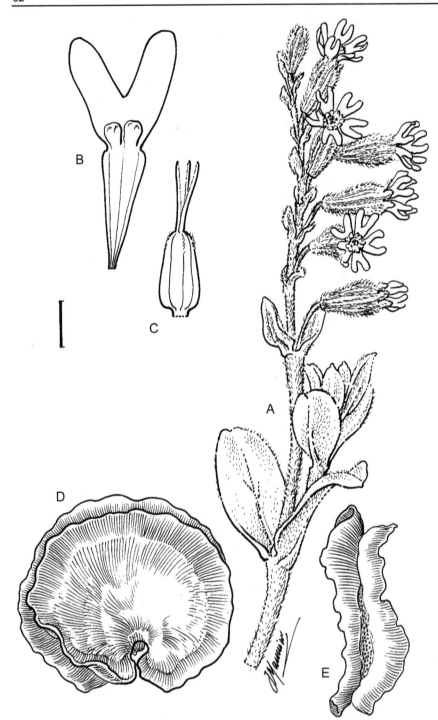

FIGURE 15.—*Silene crassifolia* subsp. *crassifolia*, Saldanha, *Goldblatt & Porter 13568*. A, flowering stem; B, petal; C, gynoecium; D, seed side view; D, seed dorsal view. Scale bar: A, 10 mm; B, C, 2.5 mm; D, E, 0.5 mm. Artist: John Manning.

S. crassifolia was readily diagnosed by its suborbicular or obovate, hairy leaves, ovate bracts, hirsute calyx and relatively shorter carpophore half as long as the capsule, and *S. primuliflora* by its narrower, spathulate, scabrid leaves, ovate-lanceolate bracts, pubescent calyx, and relatively longer carpophore ± as long as the calyx. Later workers (Bocquet & Kiefer 1878; Masson 1989) have maintained these distinctions but the additional collections, especially those from the intervening southern Cape coast, show that these differences are not consistently maintained and that the two extremes grade into one another, especially between Knysna and Mossel Bay, where plants with the broad leaves and short carpophore of *S. crassifolia* develop the short pubescence characteristic of *S. primuliflora*. Further west, at Betty's

Bay, plants have the narrow leaves of *S. primuliflora* but the woolly stem and calyx of *S. crassifolia*, while at Blouberg Strand some collections have the broad leaves and bracts of *S. crassifolia* but the short pubescence of *S. primuliflora*. Other purported differences between the taxa, in calyx length and capsule shape, are similarly variable across the range, with no clear discontinuity between them. This general but incomplete association of morphology with geography is consistent with the rank of subspecies (Stuessy 1990).

Key to subspecies

1a Plants mostly ± thickly felted on stems and leaves, hairs on stem usually spreading; leaves mostly obovate to suborbicular; calyx 12–15 mm long; carpophore 4–5 mm

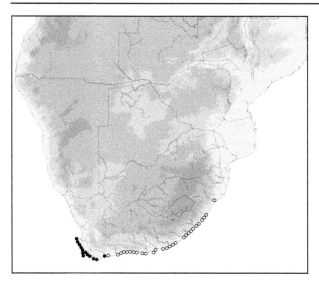

FIGURE 16.—Distribution of *Silene crassifolia*: subsp. *crassifolia*, ●; subsp. *primuliflora*, ○.

long, ± half as long as capsule; west of Breede River mouth . . . . 7a. subsp. *crassifolia*
1b Plants mostly puberulous on stems and leaves, hairs on stem ± deflexed; leaves oblanceolate to obovate; calyx mostly (10–)15–18(–20) mm long; carpophore 5–8 mm long, usually more than half as long as capsule; east of Breede River Mouth . . . . 7b. subsp. *primuliflora*

7a. subsp. **crassifolia**

Plants mostly ± thickly felted with hairs on stem deflexed or spreading. *Leaves* obovate to sub-orbicular, rarely oblanceolate, ± densely pubescent adaxially and thickly felted abaxially, rarely subglabrous adaxially and pubescent abaxially, often sub-succulent. *Calyx* 12–15 mm long. *Carpophore* 4–5 mm long, ± half as long as capsule. Figures 3D–F, 15.

Diagnosis: distinguished by its mostly longer, shaggy or felted pubescence, broader leaves, and often slightly shorter calyx, 12–15 mm long

Distribution: restricted to the southwestern and southern coasts of Western Cape, from the coast near Saldanha to Struisbaai (Figure 16). Plants from the Cape Peninsula exhibit the most extreme phenotypic adaptations to a maritime environment, with obovate to suborbicular leaves that are densely felted, especially on the lower surface.

Additional specimens

WESTERN CAPE.—**3217** (Vredenburg): Jacobsbaai, (–DD), 25 Sept. 2009, *Goldblatt, Manning & Porter 13447* (MO, NBG); Saldanha, Tabacbaai, (–DD), 25 Sept. 2009, *Goldblatt & Porter 13568* (MO, NBG). **3318** (Cape Town): Langebaan Peninsula, Schrywershoek, ± 20 m, (–AA), 26 Nov. 1975, *Boucher 2954* (NBG); Yzerfontein, (–AC), 1 Nov. 1953, *Levyns 10024* (BOL); Blouberg Strand, (–AD), Oct. 1832, *Zeyher 12* (SAM); 5 June 1940, *Bond 415* (NBG); 3 Oct. 2010, *Goldblatt & Manning 13600* (MO, NBG); Rietvalley [Rietvlei], (–AD), Nov. without year, *Ecklon & Zeyher 255* (SAM). **3418** (Simonstown): Fish Hoek, (–AB), Sept. 1882, *Bolus 4763* (BOL); Kalk Bay, (–AB), Jan. 1880, *Bolus 4763* (BOL); Muizenberg, (–AB), 16 May 1903, *Pearson 35* (NBG); 2 Dec. 1938, *Adamson 2320* (BOL); Strandfontein, (–AB), 18 Sept. 1942, *Compton 13706* (NBG); 22 Oct. 1962, *Taylor 4189* (NBG, PRE); Nov. without year, *Lamb 1951* (SAM); Cape Peninsula, Platboom, (–AD), 6 Nov. 1941, *Compton 12297* (NBG); Seekoe River Mouth, (–BA), 16 Sept. 1980, *Parsons 5* (PRE); Somerset West, Swartklip, (–BA), 15 Aug. 1953, *Leistner*

1113 (NBG); Macassar, dunes, (–BA), 23 Nov. 1955, *Van der Merwe 42* (NBG); N of Shuster River, foredunes, 15 m, (–BA), 12 Nov. 1986, *O'Callaghan 1365* (NBG); Betty's Bay, (–BD), 21 Nov. 1952, *Parker 4833* (BOL, NBG); 28 Jan. 1970, *Boucher 1099* (NBG, PRE); 25 Sept. 1996, *Forrester 1143* (NBG); Rooiels, (–BD), 27 Jan. 1962, *Walsh s.n.* (NBG, PRE). **3419** (Caledon): Kleinmond, (–AC), without date, *De Vos 255* (NBG); Bot River Vlei, sandy dunes, 20 m, (–AC), 12 Aug. 1982, *O'Callaghan 206* (NBG); Pearly Beach, (–DA), 4 Nov. 1969, *Taylor 7425* (NBG); 9 March 1980, *Raitt 447* (NBG); Agulhas, beyond lighthouse, (–DD), 16 Nov. 1979, *Taylor 10157* (NBG, PRE). **3420** (Bredasdorp): Lekkerwater, Farm Hamerkop, (–BC), 28 Nov. 1978, *Taylor 9903* (NBG, PRE).

7b. subsp. **primuliflora** *(Eckl. & Zeyh.) J.C.Manning & Goldblatt*, comb. et stat. nov. *Silene primuliflora* Eckl. & Zeyh. [as '*primulaeflora*']: 32 (1834); Sond.: 129 (1860). Type: South Africa, [Eastern Cape], 'campo marino arenoso sinus Algoabay prope Port Elisabeth osque fluvii Zwartkopsrivier', Oct. [without year], *Ecklon & Zeyher 618* (SAM, lecto.!, designated by Bocquet & Keifer: 8 (1978); SAM, isolecto.!).

Silene primuliflora var. *ciliata* Fenzl. ex Sond.: 129 (1860), syn. nov. Type: not designated.

Silene vlokii D.Masson: 485 (1989). Type: South Africa, [Western Cape], 22 Oct. 1987, *D. Masson 1225* (G, holo.!, iso.!).

Silene colorata var. *ciliata* Fenzl. in Drège: 222 (1844), nom. nud.

Plants mostly puberulous with hairs on stem ± deflexed. *Leaves* oblanceolate to obovate, ± puberulous, more densely so abaxially, rarely glabrous on adaxial or both surfaces except along margins. *Calyx* (10–)15–18(–20) mm long. *Carpophore* 5–8 mm long, mostly more than half as long as capsule. Figure 3G.

Diagnosis: distinguished by its shorter pubescence, generally narrower leaves and often slightly longer calyx, (10–)15–18(–20) mm long.

Distribution: distributed along the southern and eastern coast of South Africa, from Vermaaklikheid just east of the Breede River Mouth along the southern and southeastern coast into KwaZulu-Natal where it is mainly recorded south of Durban, with one collection from Shaka's Rock on the North Coast northeast of Durban (Figure 16). Populations between East London and Mazeppa Bay tend to have ± glabrous leaf blades and were distinguished as var. *ciliata* by Sonder (1860) but this is not consistent, *e.g. Maguire 3675* from Kei River Bridge and *Wisura 2655* from Mazeppa Bay, which include individuals that are either glabrous or variously puberulous on one or both leaf surfaces. Plants from the cliffs at Herold's Bay near George described as *S. vlokii* (Masson 1989) represent a dwarf form differing only in their more compact habit, typical of plants from exposed situations such as this, and were synonymized under *S. primuliflora* by Goldblatt & Manning (2000).

Additional specimens

KWAZULU-NATAL.—**2931** (Stanger): Chaka's [Shaka's] Rock, (–CA), without date, *Hillary 156* (NU). **3030** (Port Shepstone): Ifafa, (–BC), 12 Jul. 1974, *Huntley 206* (NU); Park Rynie, (–BC), May 1906, *Thode 5513* (NBG); Uvongo Beach, (–CB), 4 Aug. 1967, *Strey 7602* (NU, PRE); Ramsgate, (–CB), 7 July 1989, *Vos & Gormley s.n.* (NU); 16 Sept. 2003, *Styles 1590* (NU). **3130** (Port Edward): Port Edward, (–AA), 16 Nov. 1963, *Lennox s.n* (NU).

WESTERN CAPE.—**3421** (Riversdale): Farm Koensrus near Vermaaklikheid, (–AC), 3 Mar. 2000, *Goldblatt & Nänni 11292A* (MO, NBG). **3422** (Mossel Bay): Great Brak River Mouth, (–AA), 13 Nov. 1981, *Parsons 375* (NBG); Herold's Bay, (–AB), 7 Jan. 1995, *Victor 864* (PRE); Kleinkrans just outside Wilderness, (–BA), 19 Jan. 1943, *Compton 14333* (NBG); 6 Feb. 1944, *Compton 15573* (NBG); 7 Nov. 1979, *Hugo 1915* (NBG, PRE); Wilderness, beach near river mouth, (–BA), 23 Dec. 1987, *Vlok 1900* (PRE); Wilderness, dunes E of river bridge, (–BA), 19 Sept, 2010, *Goldblatt & Porter 13548* (NBG); Sedgefield, dunes at mouth of vlei, (–BB), 1 Dec. 1959, *Middlemost 2046* (NBG). **3423** (Knysna): Knysna Heads, (–AA), 7 Nov. 1928, *Gillett 2168* (NBG); *Fourcade 4144* (NBG); Buffalo Bay, (–AA), 4 dec. 1962, *Taylor 4442* (PRE); sand dunes E of Robberg, (–AB), 1 Dec. 1943, *Fourcade 6311* (NBG); Keurbooms River, (–AB), 9 Feb. 1936, *Gillett 1417* (BOL); 28 Feb. 1948, *Price s.n. BOL24957* (BOL); 8 Mar. 1983, *O'Callaghan 847* (NBG).

EASTERN CAPE.—**3129** (Port St Johns): Mkambati Nature Reserve, between Daza and Msikaba Rivers, (–BD), 11 Dec. 1986, *Jordaan 986* (NH); Msikaba, (–BD), 5 Aug. 1972, *Coleman 598* (NH); Umgazi River Mouth, (–CB), 9 Dec. 1975, *Taylor 8991* (PRE); Port St Johns, (–DA), Dec. 1943, *Brueckner & Allsopp 200* (NU); 1 Oct. 1962, *Strey 4325* (NH, PRE); Coffee Bay, (–CC), Mar. 1947, *Lewis s.n. SAM63435* (SAM); Dec. 1960, *Van der Shijff 5448* (PRE). **3228** (Butterworth): Bashee River Mouth, The Haven, (–BB), 17 Oct. 1966, *Gordon-Gray 58977* (NH, NU); Mendwana River Mouth, (–BC), 26 July 1965, *Wood 81* (NU); Mazeppa Bay, (–BC), 15 Jun. 1973, *Wisura 2655* (NBG); Kei River, (–CB), 19 Oct. 1951, *Taylor 3675* (NBG); Kei Mouth, (–CB), Jul. 1889, *Flanagan 204b* (PRE); 29 Mar. 1973, *Strey 11237* (MO, NU); Morgan's Bay, (–CB), 15 Jan. 1951, *Wilman 1057* (BOL); Morgan's Bay, Double Mouth, (–CB), 19 Jan. 1979, *Hilliard & Burtt 12453* (NU); Kentani Dist., coast along edge of scrub in sea sand, (–CB), 4 Dec. 1905 & Jul. 1906, *Pegler 1311* (BOL); N side of Nxaco [River] Mouth, landward side of dune, (–CB), 12 July 1966, *Ward 5732* (NH, NU, PRE). **3323** (Willowmore): Nature's Valley, (–DC), 22 Jan. 1978, *Taylor 476/2* (NBG). **3325** (Port Elizabeth): Port Elizabeth, Swartkops River Mouth, (–DC), 23 Jan. 1973 [fruiting], *Dahlstrand 3004* (PRE); Markhan Industrial Area, (–DC), 29 May 1973, *Dahlstrand 3050* (MO, NBG, PRE). **3326** (Grahamstown): Boknes, (–DA), Jan. 1949, *Leighton 3097* (BOL, PRE); 20 oct. 1979, *Botha 2630* (PRE); Kenton-on-Sea, (–DA), Dec. 1949, *Leighton 3096* (BOL, PRE); 27 Jun. 1955, *Acocks 18325* (PRE); Kasouga [Kasuka] River Mouth, (–DA), 21 Sept. 1920, *Britten 2304* (PRE); 22 Sept. 1920, *Britten 2343* (PRE); Kariega River Mouth, (–DA), 6 Dec. 1977, *Hilliard & Burtt 10872* (E, K, MO, NU); Kowie, (–DB), 8 Feb. 1921, *Britten 2689* (PRE); 8 Jul. 1931, *Levyns 3778* (BOL); Port Alfred, (–DB), 25 Nov. 1888, *Galpin 343a* (PRE); Jan. 1907, *Potts 187* (BOL); Jan. 1916, *Tyson s.n. PRE12592* (PRE). **3327** (Peddie): Hamburg, (–AB), 28 Dec. 1931, *Gemmell s.n.* (PRE); [Great] Fish River Mouth, (–AC), Nov. [without year], *MacOwan 737* (BOL); East London, (–BB), Feb. 1888, *Thode STE6598* (NBG); Jan. 1914, *Potts 1783* (BOL); Feb. 1920, *Page s.n. BOL16978* (BOL); Apr.–Jun. 1961, *Bokelmann 2* (NBG); Gonubie Beach, (–BB), Jan. 1953, *Peacock s.n. SAM67466* (SAM); Gonubie Mouth, (–BB), 4 Oct. 1942, *Acocks 9137* (PRE); Shelly Beach, (–BB), Dec. 1952, *Peacock s.n. SAM67467* (SAM); 27 Jan. 1979, *Hilliard & Burtt 12408* (NU). **3424** (Humansdorp): Groot River Mouth, (–AA), Mar. 1910, *Fourcade 6676* (BOL); 29 Dec. 1926, *Phillips 1204* (BOL); Jan. 1927, *Duthie 547* (NBG, SAM); 8 Apr. 1981, *Parsons 195* (NBG); Jeffrey's Bay, (–BA), 17 Dec. 1956, *Taylor 5147* (NBG). **3425** (Skoenmakerskop): Skoenmakerskop, (–BA), 7 Sept. 2002, *Steyn 92* (PRE); Cape Recife, (–BA), 1987, *D.Masson 1230* (PRE).

8. **Silene mundiana** *Eckl. & Zeyh.*, Enumeratio plantarum africae australis extratropicae: 32 (1834); Sond.: 127 (1860) [as '*mundtiana*']. Type: South Africa, [Eastern Cape], 'iter frutices ad Paardekop prope sinum Plettenbergsbay [Plettenberg Bay]', without date, *Mund s.n.* (SAM, lecto.!, designated by Bocquet & Kiefer: 8 (1978); S, isolecto.).

Geophytic perennial 7–30 cm tall, forming loose, tangled mounds or more compact mats; stems sprawling or straggling, decumbent, highly and often closely branched, slender, basal woody portion 1–2 mm diam. but 0.5–1.0 mm diam. at base of inflorescence, closely puberulous-scabridulous with very short, deflexed, acute

hairs. *Leaves* cauline, lower mostly dry or withered, oblanceolate to linear-oblanceolate, 7–15 × 1–3 mm, apiculate or weakly uncinate, base narrowed in lower leaves, puberulous along lower midline and margins, and sometimes also basally, hairs acute, margins ciliate towards base with longer, straggling hairs to 1 mm long, without evident side veins. *Inflorescence* a (1)2-flowered monochasium terminating main stems and branches, rachis filiform, 0.2–0.5 mm diam., not extending much beyond vegetative growth; bracts smaller than leaves, subequal, linear, thinly puberulous with appressed, acute hairs, margins densely ciliolate; pedicels filiform, 3–10(–15) mm long. *Calyx* clavate in flower, 9–11 × 2.5–3.0 mm at anthesis, equally 10-veined, densely adpressed-puberulous/scabridulous, lobes ovate or triangular, ± 1.5 mm long, densely ciliolate. *Flowers* patent at anthesis, ?nocturnal. *Petals* white (?or pale pink), claw ± 6 mm long, narrowed below, without evident auricles, pubescent along midrib and veins on outer surface, limb bifid, 4–5 × 3–4 mm, coronal scales 0.5–0.8 mm long. *Stamen* filaments slightly unequal, 6–7 mm long, shorter series shortly included, longer series shortly exserted. *Ovary* ellipsoid, ± 3 mm long; styles 4–5 mm long, exserted up to 1 mm. *Capsule* broadly ovoid, 6 × 4 mm, slightly longer than carpophore, smooth; carpophore ± 4 mm long, pubescent. *Seeds* 0.8–1.0 mm, reniform with deep hilar notch, flanks flat or somewhat concave, back deeply and narrowly grooved between two undulate wings, reddish brown, testa striate. *Flowering time:* Sept.–Oct.(–Nov.). Figure 3H, I.

Ecology and distribution: originally described from Plettenberg Bay, where it is still known only from the type and thus possibly extinct locally, the species is now also known to occur near Bredasdorp, more than 200 km to the west, in the De Hoop Nature Reserve immediately west of Potberg (Figure 10). Here it is restricted to limestone outcrops near the coast, either fringing seasonal vleis or directly facing the sea. Plants growing on exposed cliffs display the typical maritime adaptations of a more compact habit, forming ± prostrate closely leafy mats, and with smaller, thicker leaves than those sheltered in potholes, which are more diffuse, forming loose cushions in the crevices. The species appears to be locally common where it occurs at De Hoop.

Diagnosis and relationships: one of the least known of the South African species, represented in herbaria by just a handful of specimens.

The highly reduced inflorescences of *Silene mundiana*, not immediately identifiable as monochasia, led Sonder (1860) to place the species provisionally ['doubtfully'] in sect. *Elisanthe* but its geophytic habit, eglandular vestiture, ovate calyx lobes, and especially the winged seeds place it correctly in sect. *Fruticulosae*. Here it is distinguished by its densely and closely branched habit, small leaves 7–15 × 1–3 mm, and ± filiform branches with just 1 or 2 flowers per inflorescence. Plants form small mats or low tangled mounds, with the numerous inflorescences extending only shortly above the vegetative growth.

Conservation note: the restricted occurrence of the species and its possible extinction around Plettenberg Bay render it vulnerable.

Additional specimens

WESTERN CAPE.—**3420** (Bredasdorp): De Hoop farm in mud of dried up vlei, (–AC), 1971 [without month], *Van der Merwe 1871* (NBG); De Hoop, Potberg Nature reserve, limestone cliffs along vlei 2 km NW of De Hoop residence, (–AD), 16 Oct. 1978, *Burgers 1358* (NBG); De Hoop Nature Reserve near Rest Camp, in exposed rock crevices facing coastal winds, (–AD), 22 Oct. 1980, *Mauve, Reid & Wikner 30* (NBG).

Silene sect. **Dipterospermae** *(Rohrb.) Chowdhuri*, Notes from the Royal Botanic Garden, Edinburgh 22: 248 (1957). *Silene* [unranked] *Dipterospermae* Rohrb.: 69 (1869). Lectotype: *Silene colorata* Poir., designated by Chowdhuri: 248 (1957).

Annuals, eglandular pubescent. *Leaves* linear to obovate, lacking apparent lateral veins; bracts herbaceous. *Flowers* in monochasia with axis simple or forked below, ± erect, diurnal, vespertine or cleistogamous; anthophore well-developed, pubescent. *Calyx* pubescent, clavate, 10-veined, without conspicuous anastomosing venation, lobes ovate, obtuse. *Petals*: claw scabrid along midvein above, not auriculate, limb bifid with linear lobes; coronal scales present. *Ovary* partially septate. *Seeds* reniform with deep hilar notch, flanks flat or somewhat concave and smooth or striate, back deeply and narrowly grooved between two undulate peripheral wings.

± 9 spp., circum-Mediterranean to Arabia and W Pakistan, South Africa.

Although Chowdhuri (1957) placed *Silene aethiopica* [as *S. clandestina*] in sect. *Scorpioideae*, it is highly anomalous there in its peripherally winged seeds (Figure 3J–M). These, together with the annual habit, locate it firmly in sect. *Dipterospermae*. Seeds of this type are otherwise characteristic of sect. *Fruticulosae* (Figure 3A–I) and both groups also share a clavate calyx with ovate calyx lobes. The two sections are essentially separated on the basis of an annual vs. a perennial habit, and their exact relationship requires further study. It is especially suggestive that Gilbert (2000) notes that forms of the usually perennial *S. burchellii* (sect. *Fruticulosae*) from the drier parts of Ethiopia are often ephemerals with an evident similarity to the annual *S. colorata* (sect. *Dipterospermae*).

9. Silene aethiopica *Burm.*, Flora indica: cui accedit series zoophytorum indicorum, nec non Prodromus florae capensis: 13 (1768). Type: South Africa, without precise locality or collector, *Burman s.n. Herb. Burm. G00301755* (G, lecto.—digital image!, designated here). [Of the two available syntypes, *G00301754* and *G00301755*, the latter is selected as lectotype as it is a closer match to the protologue in its much branched stem and oblanceolate leaves].

Annual, 7–30(–35) cm tall; stems erect or decumbent, mostly branching from near base, 0.75 mm diam. at base but 0.5–1.5 mm diam. at inflorescence, closely puberulous with very short, deflexed hairs, rarely glabrescent. *Leaves* linear or narrowly oblanceolate to spathulate-oblanceolate, becoming narrower distally, 15–60 × 2–5(–10) mm, apiculate or obtuse, base narrowed in lower leaves, ± pubescent with spreading or ± appressed hairs, rarely glabrescent or glabrous apart

from margins, margins towards base ciliate with hairs to 1 mm long, without evident side veins. *Inflorescence* a secund, spike-like monochasium terminating main stems and branches, 2–8-flowered; bracts smaller than leaves, unequal, fertile bract linear to awl-shaped, densely appressed-puberulous with acute, eglandular hairs; pedicels up to 25 mm long in lower flowers and elongating to 35 mm long in fruit, (2–)6–10 mm long in upper flowers, erect in fruit. *Calyx* clavate in flower, 10–15(–18) × 2–5 mm at anthesis, equally 10-veined, densely adpressed-puberulous, lobes ovate to lanceolate, ± 2 mm long, densely cilio-late. *Flowers* suberect at anthesis, unscented or sweetly scented of coconut-vanilla, nocturnal. *Petals* white to pink with maroon reverse, claw 6–8 mm long, narrowly oblanceolate, without auricles, limb bifid or emarginate, (1–)3–8 mm × 2–6 mm, coronal scales 0.5–1.0 mm long. *Stamen* filaments slightly unequal, 4.5–7.0 mm long, shorter series included up to 2 mm, longer series ± reaching top of claw. *Ovary* ellipsoid, ± 3 mm long; styles 3–5 mm long, ± reaching top of claw. *Capsule* ovoid, (6–)8–9 × 3–4 mm, ± as long as to 2× as long as carpophore; carpophore (3–)4–9 mm long, pubescent. *Seeds* 0.8–1.5 mm, reniform with deep hilar notch, flanks flat or somewhat concave, back deeply and narrowly grooved between two undulate wings, reddish brown, testa striate. *Chromosome number* (subsp. *maritima*): $n = 12$ (Bocquet 1977). *Flowering time*: (Jul.)Aug.–Oct.(–Nov.). Figures 3J–M, 17.

Distribution and ecology: widely distributed through the southern African winter rainfall region, mainly on the coastal lowlands, from the southern Richtersveld southwards to Stilbaai and inland to the Tankwa Karoo and Laingsburg, with a disjunction to Port Elizabeth in the east but evidently absent from the Little Karoo (Figure 18). The species is mostly encountered in well-drained, coarse-grained soils of a wide variety of types, ranging from acidic sands and gravels derived from sandstone and granite to basic limestones and calcareous sands, very rarely in fine-grained clays, occurring on sandy flats, in dry river courses, around salt pans and on limestone, or rarely dolerite, ridges.

Diagnosis and relationships: the only annual species of *Silene* native to southern Africa, *S. aethiopica* is distinguished from the introduced annual species, *S. gallica* and *S. nocturna*, by its wholly eglandular pubescence— the stems with short, deflexed hairs and the clavate calyx covered with short, appressed hairs—and by its discoid, peripherally winged seeds. Above ground it is superficially similar to the perennial *S. burchellii* (sect. *Fruticulosae*), with similar vesture, calyx, and seeds and the two species can be confused in the southwestern parts of the Western Cape where their ranges overlap. Well-grown specimens of *S. aethiopica* with branched stems are unmistakable but unbranched specimens without roots may pose problems. Calyx length is a useful discriminating feature in the region of overlap, with the calyx of *S. aethiopica* typically 10–15 mm long, except in subsp. *longiflora* from the Cold Bokkeveld, where it is 15–18 mm long. In contrast, the more common forms of *S. burchellii* in the region, subsp. *pilosellifolia*, have a calyx 18–25 mm long, although clayx length is shorter, 15–20 mm long, in subsp. *burchellii* in the extreme southwestern coastal areas. Ecologically, *S. aethiopica*

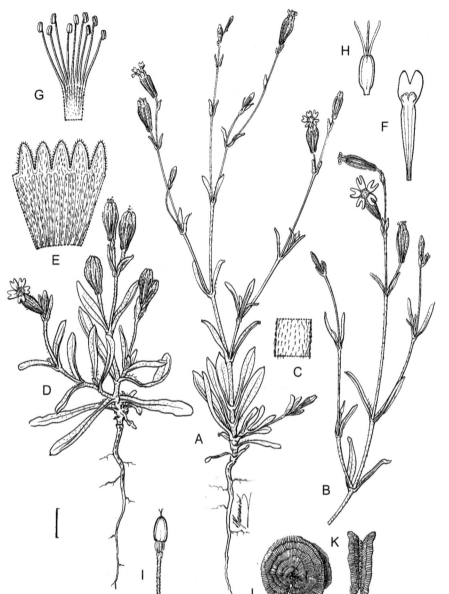

FIGURE 17.—*Silene aethiopica* subsp. *aethiopica*: A, typical form from Tulbagh, no voucher; B, typical form from Gordon's Bay, no voucher; C–K, maritime form from Blouberg Strand, *Goldblatt & Manning 13601*. A, D, flowering plants; B, flowering stem; C, stem segment; E, calyx laid out; F, petal; G, androecium laid out; H, gynoecium; I, capsule before dehischence with calyx removed; J, seed side view; K, seed dorsal view. Scale bar: A, B, D, I, 10 mm; C, 1 mm; E, 5 mm; F–H, 2.5 mm; J, K, 0.5. Artist: John Manning.

prefers well-drained, sandy soils, whereas *S. burchellii* is mostly found in stony places or in finer-grained clays. The two taxa are, however, best separated by habit. *S. aethiopica* is an annual species with a slender taproot that is invariably present in herbarium specimens, which are always pulled from the ground with the main root intact. *S. burchellii*, in contrast, is a perennial with annual flowering stems but a persistent, woody stem base, developing a persistent, short or long rootstock that is swollen and tuber-like distally, although the swollen terminal portion is very rarely present in herbarium specimens.

As in other species of *Silene*, *S. aethiopica* is variable in stature, leaf morphology and flower size, notably in the length of petal limbs, carpophore and calyx. Much of the variation is either ecological or recurrent through the range of the species but populations from the southern Knersvlakte and the Olifants River valley are consistently less pubescent than usual, with the leaf blades either glabrous, apart from the margins, or glabrescent, and the stems often glabrescent as well. Large- and

small-flowered forms occur throughout the range of the species, sometimes within single populations.

Strand forms of *Silene aethiopica* from the southwestern coast (Figure 17D), distinguished from more inland plants (Figure 17A, B) by the typical maritime adaptations of a more spreading habit, broader and more fleshy leaves, and fleshy calyx with distinctly bulging nerves, were described as a separate species under the name *S. dewinteri* (Bocquet 1977). The distinction between the two forms is not absolute, however, and some collections comprise material that could be assigned to both taxa (eg. *Goldblatt & Porter 13509*). We visited the type locality of *S. dewinteri* at Table View and confirm Bocquet's observations regarding the texture of the leaves and calyx but not of the habit— all of the plants that we encountered were erect, possibly because our visit to the site was earlier in the season than Bocquet's. More significantly, however, our transect across the dunes passed through a gradation from typical *S. aethiopica* on the landward side to *S. dewinteri* on the seaward side. We encountered similar

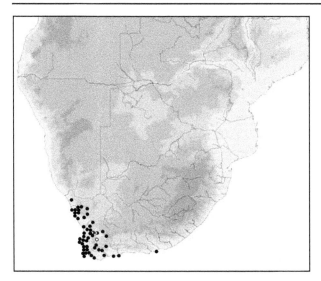

FIGURE 18.—Distribution of *Silene aethiopica*: subsp. aethiopica, ●; subsp. *longiflora*, ○.

morphological clines at Jacobsbaai and at Elands Bay. Although plants at the two extremes of the ecological spectrum are certainly very different in appearance, they grade into one another in the ecotone, leading us to conclude that *S. dewinteri* represents a ± distinct ecotype rather than a separate species. This interpretation is also implicit in Bocquet's (1978) subsequent description of *S. dewinteri* as a 'sand-dune ecotype'.

That there is some genetic basis for the maritime form seems to be borne out by Bocquet's (1977) observation that the greater fleshiness of the plants was retained in cultivated plants grown from seed. It is not clear, however, whether this variant represents recurrent selection for a maritime genotype, a situation that is readily envisaged in a self-fertile species. In any event, we do not recognize this maritime form at any taxonomic rank. An investigation of the population genetics of the species is necessary before the taxonomic and evolutionary status of these variants can be accurately assessed.

A more clear-cut morphological disjunction is evident in populations from the Cold Bokkeveld and adjacent Cedarberg–Bidouw Mtns, which have markedly longer calyces (and carpophores) than usual. We describe this variant as subsp. *longiflora*.

Nomenclatural note: The identity of *Silene aethiopica* Burm. (1768) has remained uncertain until now. Based on unlocalised material from the 'Cape of Good Hope' and one of the earliest names published in African silenes, the published diagnosis *caule ramoso, floribus subspicatus, petalis bifidis obtusis, foliis lanceolatis viscidis* includes two characteristics not combined in any southern African species, viz. 'viscid leaves' and a 'subspicate' inflorescence. Thunberg (1823) treated the species as a synonym of what he called *S. noctiflora* (a European species) but his application of this name in the southern African context actually encompasses the newly described *S. rigens*, a cymose-flowered species in sect. *Elisanthe*, and the basis for his interpretation of *S. aethiopica* is unclear. Others dealing with *Silene* ignored the name, including Otth (1824), Ecklon & Zeyher (1834), and Sonder (1860) in his account of the

genus for *Flora capensis*. Williams (1896), however, in his revision of *Silene*, cited *S. aethiopica* as an uncertain synonym of the later *S. burchellii*, and the African Plants Database (2011 version) follows this lead. This tentative identification of Burman's species is clearly based on the 'subspicate' nature of the infloresence. It is most fortunate, therefore, that Nicolas Fumeaux at the Burman Herbarium has succeeded in locating two sheets bearing the name *S. aethiopica*, which clearly constitute original material of the species and are thus syntypes of the name. One of the sheets (*G00301754*) also bears the inscription *Lychnoides aethiopica angustifolia fl. incarnatis, petalis acutis* from John Ray's (1704) *Histoire plantarum*, and the other (*G00301754*) bears an additional manuscript name and diagnosis for *Lychnis monomotapensis*. Both collections represent the species known until now as *Silene clandestina*, and *G0030175* was in fact identified under this name by Daniel Masson in 1991. The name *S. aethiopica* Burm. (1768) thus takes priority over *S. clandestina* Jacq. (1791), which becomes a later synonym. Burman's description of the leaves of this species as viscid must be assumed to be an error as the type material does not show this feature. Curiously, the same specimen had earlier been determined as *S. dewinteri* Bocquet (1977) by Daniel Bocquet in 1976, who should therefore have taken up the epithet *aethiopica* for this taxon were it to be recognized at species level, which we do not. In any event, we regard the Burman specimen to represent a slightly luxuriant individual (possibly cultivated) of typical *S. aethiopica* and not the maritime '*dewinteri*' form.

Although the source of Burman's specimens is not given, other collections of South African material in the Burman herbarium are known to have been made by [Heinrich] Oldenland, who was was employed by the Dutch East India Company at the Company Gardens in Cape Town in the late seventeenth century. Ray (1704) certainly used material collected by Oldenland in the preparation of his *Historia plantarum*, and his mention of the species is a clear indication that it had reached Europe by the early eighteenth century.

Key to subspecies

1a Calyx 10–15 mm long; carpophore 3–8 mm long, shorter
  than to as long as capsule . . . . 9a. subsp. *aethiopica*
1b Calyx 15–18 mm long; carpophore ± 10 mm long, slightly
  longer than capsule; Cold Bokkeveld . . . . . 9b. subsp. *longiflora*

9a. subsp. **aethiopica**

Silene clandestina Jacq.: 111 (1791 [as '1789']), syn. nov.; Sond.: 127 (1860). Type: South Africa, without precise locality, illustration in Jacq., Collectanea pl. 3, fig.3 (1791 [as '1789'] (lecto., designated here).

Silene cernua Thunb.: 81 (1794), syn. nov. *Silene burchellii* var. *cernua* (Thunb.) Williams: 76 (1896). Type: South Africa, [Western Cape], without precise locality or date, *Thunberg s.n.,* Herb. Thunb. 10714 (UPS-THUNB—microfiche!, lecto., designated here). [This specimen is selected as the most complete of the three under this name in the Thunberg Herbarium (*UPS-THUNB10713, 10714 & 10715*), clearly showing the annual habit of the species]

Silene linifolia Willd.: 473 (1809) [*fide* Sond. (1860)], syn. nov. Type: not designated and identity uncertain.

Silene recta Bartl.: 623 (1832) [*fide* Sond. (1860)], syn. nov. Type: South Africa, [Western Cape], 'auf dem Tafelberge auf dem zweiten Höhe', without date, *Ecklon & Zeyher s.n.* (not located).

Silene crassifolia var. *angustifolia* Bartl.: 623 (1832), syn. nov. Type: South Africa, [Western Cape], 'beim obersten Blockhause am Teufelsberge', without date, *Ecklon & Zeyher s.n.* (not located).

Silene constantia Eckl. & Zeyh.: 32 (1834), syn. nov. Type: South Africa, [Western Cape], 'collium capensium prope Constantiam et Hottentotsholland', *Ecklon & Zeyher 251* (SAM, lecto.!, designated here). [Bocquet's (1977) contention that Ecklon & Zeyher (1834) automatically rendered this name illegitimate and superfluous by citing *S. crassifolia* var. *angustifolia* Bartl. in synonymy is incorrect as names have priority only at their rank: *S. constantia* is therefore a legitimate name at species rank, with its own type (McNeill *et al.* 2006: Art. 11.2)].

Silene clandestina var. *minor* Sond.: 128 (1860), syn. nov. Type: not designated.

Silene clandestina var. *major* Sond.: 128 (1860), syn. nov. Type: not designated.

Silene dewinteri Bocquet, Bothalia 12: 309 (1977), syn. nov. Type: South Africa, [Western Cape], Table View, 9 Nov. 1975, *Bocquet 17774* (ZT, holo.!).

Plants ± lax or compact. *Stems* erect or decumbent. *Leaves* linear to oblanceolate, 2–5 mm wide. *Calyx* 10–15 mm long. *Carpophore* (3–)4–8 mm long, half as long to as long as capsule. Figure 17.

Diagnosis: distinguished from subsp. *longiflora* by the shorter calyx, 10–15 mm long.

Distribution: throughout the range of the species but replaced on the Cold Bokkeveld and Cedarberg–Bidouw Mtns by subsp. *longiflora* (Figure 18). The maritime ecotype, characterized by a compact habit, sub-succulent leaves 3–10 mm wide, ± fleshy calyx, and consistently short carpophore, 4–5 mm long, is restricted to the seaward side of coastal dunes above the high-tide mark in the extreme southwestern Western Cape between Saldanha Bay and Onrus River Mouth, in stabilized or semi-stabilized, calcareous sands. The absence of this variant along the coast to the north and east may be due to the lack of suitable sandy beaches there.

Additional specimens

NORTHERN CAPE.—**2917** (Springbok): between Kawarass and Lekkersing, 250 m, (–AA), 4 Sept. 1925, *Marloth 12437* (NBG); turn-off to Komaggas on Koingnaas-Kleinzee Road, (–CC), 11 Aug. 2007, *Bester 7895* (PRE); Farm Zonnekwa, 200 m, (–CD), 7 Oct. 1986, *Le Roux & Lloyd 519* (NBG); Spektakel, (–DA), Sept. 1883, *Bolus 6667* (BOL); Hester Malan Reserve, (–DB), 3 Oct. 1974, *Rösch & Le Roux 836* (PRE); 26 Aug. 1985, *Struck 145* (NBG); Farm De Draay, (–DB), 29 Aug. 1990, *Le Roux 4122b* (NBG); Groot Vlei, (–DC), 7 Sept. 1945, *Compton 17296* (NBG). **2918** (Gamoep): between Gamoep and Aggenys close to Vaalheuwel Farm, (–DC), 19 Sept. 2002, *Steyn 271* (PRE). **3017** (Hondeklipbaai): Farm Koingaas, 50 m, (–AB), 27 Aug. 1986, *Le Roux & Lloyd 354* (NBG); Hondeklip Bay, (–AD), Oct. 1924, *Pillans 17942* (BOL); 11 miles [17.6 km] E of Hondeklipbaai on road

to Garies, (–AD), 11 Sept. 1970, *Thompson 1089* (NBG); Kookfontein, (–BA), Aug. 1883, *Bolus 31659* (BOL); ± 2 km W of Sarisam homestead, (–DA), 2 Aug. 2006, *Bester 7029* (PRE). **3018** (Kamiesberg): Platbakkies, (–BC), 10 Sept. 1976, *Thompson 2866* (PRE); Langkloof, N of Doringkraal, (–CA), 23 Sept. 2010, *Goldblatt 13573A* (MO, NBG). **3119** (Calvinia): Niewoudtville Waterfall, (–AC), Sept. 1930, *Lavis s.n. BOL19818* (BOL); Nieuwoudtville Reserve, (–AC), 7 Sept. 1983, *Perry & Snijman 2279* (NBG). **3220** (Sutherland): Farm Bergsigt, (–DA), 7 Sept. 1988, *Crosby 965* (PRE).

WESTERN CAPE.—**3118** (Vanrhynsdorp): Draaihoek Farm, 4 km SW of Vredendal, 70 m, (–CB), 27 Aug. 1986, *Hilton-Taylor 1152* (NBG); Troe-troe Farm, 3 km from Vanrhysndorp on Nieuwoudtville road, (–DB), 3 Aug. 1977, *Le Roux 2007* (NBG); Widouw Farm, 15 km S of Van Rhynsdorp, 500' [150 m], (–DB), 9 Sept. 1976, *Thompson 2826* (NBG, PRE); 22 Sept. 1985, *Zietsman & Zietsman 1157* (PRE); Windhoek Farm at foot of Gifberg, (–DC), 22 Sept. 191, *Pearson 6760* (BOL); Doorn River Mouth, (–DC), 22 Jul. 1941, *Compton 11037* (BOL, NBG); Brandewyn River, (–DD), 26 Aug. 1950, *Barker 6588* (NBG); Nardouw Mtns S of Bulshoek Barrage, (–DD), 2 Sept. 1977, *Goldblatt & Manning 10719* (MO, NBG). **3217** (Vredenburg): limestone hill N of Saldanha, (–DD), 30 Sept. 2009, *Goldblatt, Manning & Porter 13469* (MO, NBG); dunes N of Jacobsbaai, (–DD), 9 Sept. 2010, *Goldblatt & Porter 13509* (MO, NBG); Saldanha Peninsula, Tabacbaai, dunes along coast, (–DD), 22 Sept. 2010, *Goldblatt & Porter 13569* (MO, NBG, PRE). **3218** (Clanwilliam): Wadrif Soutpan, (–AB), 14 Sept. 1984, *O'Callaghan & Van Wyk 96* (NBG, PRE); 9–10 km S of Lambert's Bay, (–AB), 14 Sept. 2010, *Goldblatt & Porter 13533, 13534* (MO, NBG); between Leipoldtville and Elands Bay, (–AD), Oct. 1947, *Zinn s.n. SAM63443* (SAM); Elands Bay, sandy coast below bushman cave, (–AD), 14 Sept. 2010, *Goldblatt & Porter 13535* (MO, NBG); hills around Clanwilliam, (–BB), Jul. 1897, *Leipoldt 594* (NBG, SAM); Olifants River Dam, (–BB), 10 Sept. 1949, *Barker 5736* (NBG); near Paleisheuwel, (–BD), 5 Sept. 1954, *Levyns 10162* (BOL); Kriedouwkrantz on road to Algeria Forest Station, (–BD), 7 Sept. 1976, *Hugo 429* (NBG); N of Velddrift, 10 m, (–CC), 15 Oct. 1986, *O'Callaghan 1220* (NBG); Salt pans near Zoutkloff, (–CD), 10 Sept. 1949, *Barker 5803* (BOL, NBG); Piketberg Dist., Het Kruis, (–DA), 22 Sept. 1940, *Compton 9519* (NBG). **3219** (Wuppertal): Driefontein Farm, 4 km SW of Doringbos, 200 m, (–AA), 2 Oct. 1986, *Hilton-Taylor 1858* (NBG); Boontjieskloof Farm E of Pakhuis Pass, ± 1 500' [450 m], (–AA), 4 Sept. 1976, *Taylor 9319* (NBG, PRE); Heuningvlei, Groot Koupoort, ± 3150' [945 m], (–AA), 11 Oct. 1975, *Kruger 1696* (NBG); Alpha Farm, (–AA), 20 Jul. 1941, *Compton 10934* (NBG); Algeria Forest Station, (–AC), 8 Sept. 1997, *Van Rooyen, Steyn & de Villiers 710* (NBG); Cedarberg Forest reserve, Langrug, 1 000' [300 m], (–AC), 12 Sept. 1982, *Viviers 581* (NBG); Tankwa Karoo National Park, W end, (–BA), 2 Aug. 2007, *Bester 7682* (NBG). **3220** (Sutherland): Syferwater Farm, 1 212 m, (–AA), 18 Sept. 2006, *Rosch 558* (NBG). **3318** (Cape Town): Langebaan, (–AA), 5 Sept. 1971, *Axelson 486* (NBG); around pan SE of Jakkalsfontein, 60 m, (–AD), 24 Sept. 1996, *Low 2831* (NBG); Melkbosstrand, (–AD), 24 Sept. 1966, *Dahlstrand 1066* (PRE); 30 Sept. 2010, *Goldblatt & Manning 13598* (MO, NBG); Bok Point, (–CB), 14 Sept. 1940, *Barker 779* (NBG); Bok Bay, (–CB), 14 Sept. 1940, *Compton 9389* (NBG); Robben Island, (–CD), Oct. 1932, *Adamson s.n.* (BOL); 20 Aug. 1943, *Walgate 512* (NBG); 19 Nov. 1987, *Van Jaarsveld & De Lange 9546* (NBG); Blaauwberg [Blouberg], (–CD), Sept. 1954, *Stokoe s.n. SAM67560* (SAM); Blouberg Strand, dunes S of Dolphin Hotel, (–CD), 3 Oct. 2010, *Goldblatt & Manning 13601* (MO, NBG); Paarden Eiland, Klein Zoar Vlei, 1 m, (–CD), 8 Oct. 1979, *Linder s.n.* (BOL); Milnerton, (–CD), 17 Sept. 1913, *Phillips s.n. NBG103292* (NBG); 'in leter montis tabularis', (–CD), Oct. 1873, *Bolus 2718* (BOL); 'prope Rondebosch', (–CD), Oct. 1878, *Bolus 2718a* (BOL); Bergvliet Farm, (–CD), 1–2 Nov. 1915, *Purcell s.n.* (SAM); Van Kampsbay [Camps Bay], without date, *Zeyher s.n.* (BOL); Camps Bay, (–CD), 13 Oct. 1956, *Cassidy 51* (BOL, NBG); W slopes of Table Mtn, (–CD), 5 Jul. 1936, *Adamson 970* (PRE); Malmesbury Dist., near Pella, Burgers Post Farm, 200 m, (–DA), 9 Oct. 1979, *Boucher & Shepherd 4764* (NBG, PRE); Cape Flats Nature Reserve, University of Western Cape campus, ± 200 m, (–DC), 9 Oct. 1980, *Low 1095* (NBG); Cape Flats near Tygerberg Station, (–DC), 15 Oct. 1927, *Smuts 2490* (NBG). **3319** (Worcester): Tulbagh Waterfall, (–AC), Oct. 1928, *Levyns 2566* (BOL); Laken Vlei, (–BC), 19 Oct. 1941, *Compton 12076* (NBG); Worcester, Riverside, (–CB), 11 Sept. 1962, *Walters 878* (NBG); Robertson, (–DD), 25 Sept. 1935, *Lewis s.n. BOL31663* (BOL). **3320** (Montagu): Laingsburg Dist., [Matjiesfontein], Fisantekraal Valley, (–BC), 7 Nov. 1948, *Compton 21152* (NBG); Montagu, (–CC), 5 Jul. 1938, *Levyns 6457* (NBG). **3418** (Simonstown): Llandudno, (–Ab), 26 Sept. 1928, *Hutchinson 535* (BOL, K); Hout Bay, (–AA), 20 Sept. 1941, *Compton 11749*

(NBG); Red Hill, (–AB), 6 Oct. 1962, *Taylor 4128* (NBG); Clovelly, (–AB), 22 Oct. 1940, *Adamson s.n.* (BOL); Noordhoek, (–AB), 1 Nov. 1944, *Barker 3293* (NBG); Muizenburg, (–AB), Dec. 1894, *Guthrie s.n.* (BOL); Gifkommetjie turnoff, Cape of Good Hope Nature Reserve, (–AD), 6 Oct. 1971, *Taylor 7952* (NBG); Buffels Bay, (–AD), 8 Oct. 1966, *Taylor 6949A* (PRE); Cape Flats, Swartklip, ± 170′ [51 m], (–BA), 6 Sept. 1972, *Taylor 8171* (NBG, PRE); Penhill, Eersterivier, 50 m, (–BA), 24 Oct. 1984, *Raitt 5431* (NBG); Gordon's Bay, (–BB), Dec. 1901, *Bolus s.n.* (BOL); Strand/Macassar, Somchem, 5 m, (–BB), 28 Aug. 1995, *Low 2462* (NBG). **3419** (Caledon): Onrus River, (–AC), 28 Sept. 1958, *Williams 57* (NBG). Kleinmond, (–AC), 15 Oct. 1949, *De Vos 1489* (NBG); Hermanus, Riviera, (–AC), 5 Oct. 1916, *Purcell s.n.* (SAM); road from Stanford to Gansbaai, (–CB), 21 Sept. 1938, *Gillett 4294* (BOL); Baviaansfontein Farm, ± 3.5 miles [5.6 km] E of De Kelders, (–CB), 27 Sept. 1962, *Taylor 4106* (NBG); Uilkraals River, 2 m, (–CB), 3 Nov. 1987, *O'Callaghan 3/11/2e* (NBG). **3420** (Bredasdorp): De Hoop, Potberg Nature Reserve, (–AD), 2 Aug. 1979, *Burgers 2145* (NBG). **3421** (Riversdale): Jongensfontein Farm, Stillbay, (–AD), 23 Aug. 1978, *Bohnen 3995* (NBG); Melkhoutfontein, (–BD), 7 Oct. 1897, *Galpin 3775* (PRE).

EASTERN CAPE.—**3325** (Port Elizabeth): Zwartkopsrivier, (–DC), Sept. without year, *Zeyher 1961 S11-24180* (S); Swartkops River near Port Elizabeth, (–DC), 28 Aug. 1947, *Rodin 1064* (BOL).

9b. subsp. **longiflora** *J.C.Manning & Goldblatt*, subsp. nov.

TYPE.—Western Cape, 3219 (Wuppertal): Swartruggens, Knolfontein Farm, 60 km NE of Ceres, 1 184 m, (–DC), 25 Oct. 2006, *I. Jardine & C. Jardine 562* (NBG, holo.).

Plants ± lax. *Stems* erect. *Leaves* narrowly oblanceolate, 3–5 mm wide. *Calyx* 16–18 mm long. *Carpophore* ± 10 mm long, slightly longer than capsule.

Diagnosis: distinguished by its longer calyx, 16–18 mm long.

Distribution and ecology: restricted to the Cold Bokkeveld and Cedarberg–Bidouw Mtns, occurring in sandy soils on sandstone slopes at mid-altitudes (Figure 18).

Additional specimens

WESTERN CAPE.—**3219** (Wuppertal): pass into Biedouw Valley, 6.2 miles [10 km] from Clanwilliam-Calvinia road, 1 500′ [450 m], (–AA), 27 Aug. 1967, *Thompson 351 pp.* (NBG); Driehoek Valley, (–CA), 3 Oct. 1952, *Esterhuysen 20573* (BOL); Swartruggens, Knolfontein Farm, 60 km NE of Ceres, 1 185 m, (–DC), 7 Oct. 2005, *Jardine & Jardine 230* (NBG).

Silene sect. **Silene**

Annuals, with eglandular and mostly glandular hairs. *Leaves*: lowermost obovate-spathulate. *Flowers* in monochasia, with axis simple or 1–3-forked below, spreading or erect, vespertine, diurnal or cleistogamous; anthophore short, pubescent. *Calyx* pubescent to hirsute, with or without glandular hairs, 10-veined, with or without anastomising venation, tubular to fusiform in flower, tapering apically in fruit. *Petals*: claw glabrous, not auriculate, sometimes wanting, limb entire or bifid; coronal scales present. *Stamens*: filaments glabrous or barbellate in proximal half. *Ovary* partially septate. *Capsule* oblong or ovoid-oblong. *Seeds* reniform, flanks excavate, back flat or furrowed, radially colliculate-echinulate.

± 20 spp. Circum-Mediterranean, extending to Macronesia and Pakistan; introduced elsewhere.

10. **Silene nocturna** *L.*, Species plantarum 1: 416 (1753). Type: 'Habitat in Italia, Pensylvania', *Herb. Linn. 583.8* [LINN, lecto.—digital image!, designated by Ghafoor: 91 (1978)].

Annual to 50 cm tall; stems erect, wand-like, simple or branched, 0.75–1.50 mm diam., lower parts closely puberulous with very short, spreading hairs mixed with longer hairs up to 1 mm long; upper parts of stems and inflorescences closely puberulous with very short, spreading hairs mixed with short, spreading, gland-tipped hairs. *Leaves* mostly basal, oblanceolate to spathulate-oblanceolate becoming oblong-lanceolate distally, 15–50 × 5–15 mm, apiculate or obtuse, base narrowed in lower leaves, ± densely pubescent with ± appressed hairs, mixed on upper leaves with gland-tipped hairs, margins towards base ciliate with hairs to 1 mm long, without evident side veins. *Inflorescence* a secund, spike-like monochasium terminating main stems and branches, (3–)5–8-flowered, lower flowers remote; bracts smaller than leaves, unequal, fertile bract linear to awl-shaped, densely puberulous with gland-tipped hairs mixed with acute, eglandular hairs; pedicels up to 12 mm long in lower flowers and elongating to 35 mm long in fruit, mostly 3–4 mm long in upper flowers or more rarely upper flowers subsessile, lowermost pedicels sometimes patent or deflexed in fruit. *Calyx* narrowly urn-shaped in flower, 9–12 × 2.0–2.5 mm at anthesis, equally 10-veined, reticulately veined in distal half, densely adpressed-puberulous with slightly longer hairs 0.5 mm long on veins, lobes narrowly triangular-lanceolate, ±2 mm long, densely ciliolate. *Flowers* suberect at anthesis, unscented, nocturnal or cleistogamous. *Petals* pink, claw 6–8 mm long, oblong but narrowed below, without auricles, limb bifid or emarginate, 1.5–4.0 × 1.5–4.0 mm, coronal scales 0.25–0.50 mm long. *Stamen* filaments unequal, 5–7 mm long, shorter series shortly included, longer series shortly exserted. *Ovary* pyriform, ± 4 mm long; styles ± 2 mm long, deeply included. *Capsule* ovoid, 6–8 × 4 mm, ± 4–5 × longer than carpophore, minutely transversely rugulose; carpophore ± 1.5–2.0 mm long, pubescent. *Seeds* 0.5–1.0 mm, reniform with hilar notch, reddish brown, face excavate-auriculate, testal cells radially elongated-fusiform, colliculate, back broad with narrow, tuberculate groove. *Flowering time:* (Jun.–)Aug.–Nov. Figures 1C, D, 19.

Vernacular name: Mediterranean catchfly.

Ecology and distribution: widespread through the Mediterranean, extending to Portugal and Northern Spain; introduced and established in western Australia and an occasional introduction in the southwestern Cape, where it has been recorded mainly from the Cape Peninsula but also from Bredasdorp, with a recent record from Sardinia Bay west of Port Elizabeth (Figure 20). The species appears to be restricted to disturbed ground and roadsides.

The species was first collected in South Africa under the name *Silene ?pendula* L. by Carl Thunberg (Table 1), who travelled in the country from 1772 to 1774. Another early South African record of *Silene nocturna* that we have seen was made by Carl Zeyher (*Zeyher 1960*) in the mid-nineteenth century, somewhere in the vast tract of country between Malabarshoogde [near

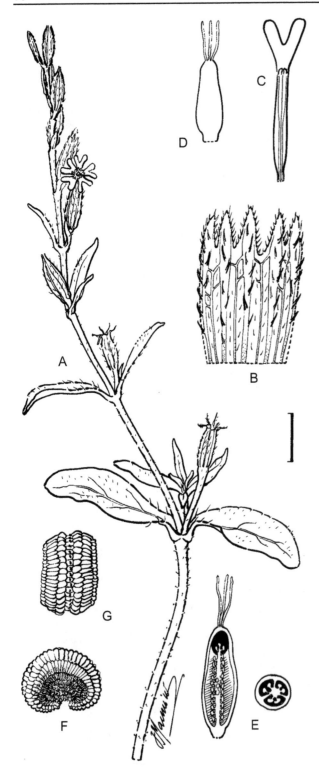

FIGURE 19.—*Silene nocturna*, Sardinia Bay, *Goldblatt & Porter 13661*. A, flowering stem; B, calyx laid out; C, petal; D, gynoecium; E, gynoecium l/s and t/s; F, seed side view; G, seed dorsal view. Scale bar: A, 10 mm; B, 5 mm; C, D, 2.5 mm; E, 2 mm; F, G, 0.5 mm. Artist: John Manning.

Queenstown in Eastern Cape] and Hessaquaskloof [near Riviersonderend in Western Cape]. The next collection, from Wynberg Hill, dates from the end of the nineteenth century. Certainly, within two decades of this latter collection, the species was being found at several other localities around the Peninsula but has so far failed to establish itself as invasive. These early collections were

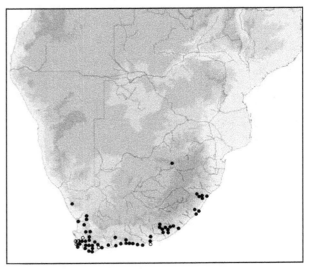

FIGURE 20.—Recorded southern African distribution of *Silene gallica*, •; *S. nocturna*, ○.

misidentified as either *S. gallica* or *S. clandestina* [now *S. aethiopica*] or were left unnamed.

Diagnosis and relationships: the occurrence of *Silene nocturna* in southern Africa has been overlooked until now due to confusion with *S. gallica*, another introduced annual. The two species are superficially similar vegetatively and can be confused as herbarium specimens, although *S. nocturna* usually has more slender and lax flowering stems, typically with the lowermost flowers on long pedicels up to 35 mm long in fruit. They are best distinguished by details of stem vestiture, and especially by their flowers and seeds. Both species have a mixture of short and long eglandular hairs on the lower portions of the stem but in *S. gallica* the longer hairs are distinctly shaggy and 1–2 mm long, whereas in *S. nocturna* they are shorter and at most up to 1 mm long. The flowers of *S. nocturna*, as the name suggests, are typically nocturnal with distinctly bilobed petals (Figure 19C), and the calyx is appressed-puberulous, without gland-tipped hairs and with only slightly longer hairs on the veins (Figure 19B). *S. gallica*, in contrast, has diurnal flowers with distinctive, unlobed petals (Figure 21C), and a glandular-haired calyx with characteristic glassy bristles 2–3 mm long on the veins (Figure 21B). Both species have seeds with deeply excavated flanks but differ in detail: seeds of *S. nocturna* have spreading, almost wing-like shoulders with a narrowly grooved back (Figure 1C, D, 19F, G), and those of *S. gallica* have rounded, unwinged shoulders and a flat or slightly concave back (Figure 1E, F, 21G, H).

Additional specimens

WESTERN CAPE.—**3318** (Cape Town): mountain side above Clifton, (–CD), Aug. 1913, *Kensit s.n.* (BOL); Cecilia boundary, (–CD), 16 Sept. 1945, *Levyns 17426* (BOL); Rondebosch, (–CD), Sept. 1902, *Anon STE 13264* (NBG); Cape Peninsula, between Bishopscourt and Kirstenbosch, (–CD), 3 Oct. 1951, *Salter 9048* (BOL); Bishopscourt, (–CD), Nov. 1951, *Pillans 10646* (MO); Paradise Estate, N slopes of Wynberg Hill, (–CD), Sept. 1918, *L. Bolus s.n.* (MO); Suider Paarl, (–DB), Aug. 1917, *Roberts & Adendorf 17671* (PRE); Tygerberg Nature Reserve, (–DC), 6 Sept. 1975, *Loubser 3380* (NBG); Sept./Oct. 1976, *Loubser 3384* (PRE). **3418** (Simonstown): Wynberg Hill, (–AB), 8 Nov. 1896, *Wolley Dod 1817* (BOL); E slopes of Vlaggenberg [Vlakkenberg], 19 Sept. 1915, *Pillans 2832* (BOL). **3420**

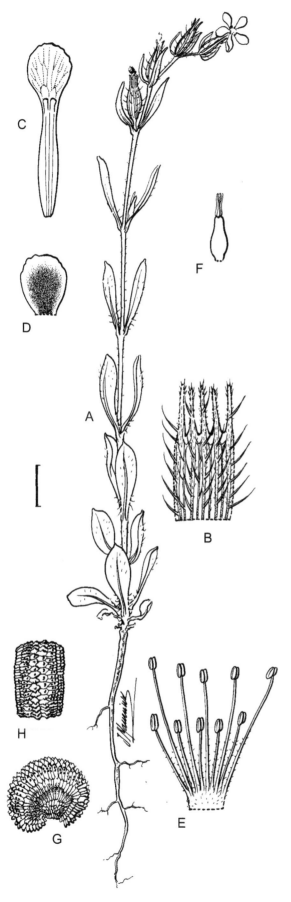

FIGURE 21.—*Silene gallica*, Kleinmond, without voucher. A, flowering plant; B, calyx laid out; C, petal; D, petal limb [var. *quinquevulnera*]; E, androecium laid out; F, gynoecium; G, seed side view; H, seed dorsal view. Scale bar: A, 10 mm; B, 5 mm; C–F, 2.5 mm; G, H, 0.5 mm. Artist: John Manning.

(Bredasdorp): De Hoop, flats in Grootwildkamp, (–AD), 2 Aug. 1979, *Burgers 2144* (NBG).

EASTERN CAPE.—**3425** (Skoenmakerskop): Sardinia Bay, road side at Loerie road intersection, (–BA), 18 Sept. 2010, *Goldblatt & Porter 13661* (MO, NBG). *Locality uncertain*: South Africa, between Malabarshoogde and Hessaquaskloof, Sept. [without year], *Zeyher 1960 S11-24179* [left hand specimen] (S).

11. **Silene gallica** *L.*, Species plantarum: 417 (1753), nom cons.; Sond.: 127 (1860). Type: 'Habitat in Gallia', *Herb. Linn. 583.11* [LINN, lecto.—digital image!, designated by Greuter: 102 (1995a)].

Silene anglica L.: 416 (1753), nom. rej. in favour of *S. gallica* L. *S. gallica* var. *anglica* (L.) W.D.J.Koch: 109 (1843). Type: 'Habitat in Anglia, Gallia', *Herb. Linn. 583.1* [LINN, lecto., designated by Talavera & Muñoz Garmendia: 498 (1989)].

[see Chater & Walter (1964) and Greuter (1997) for additional heterotypic synonyms].

Annual to 40 cm tall; stems ascending or erect, simple or well-branched from base and above, 0.75–3.00 mm diam., lower parts closely puberulous with very short, ± deflexed or spreading hairs mixed with longer, shaggy, spreading or upcurved hairs 1–2 mm long but later glabrescent; upper parts of stems and inflorescences closely puberulous with very short, ± deflexed or spreading hairs mixed with short, spreading, gland-tipped hairs, inflorescence axis especially densely gladular-pubescent. *Leaves* oblanceolate to spathulate-oblanceolate becoming oblong-lanceolate distally, 15–50 × 4–15 mm, apiculate, base narrowed in lower leaves, sparsely or densely pubescent with erect or ± appressed hairs, mixed on upper leaves with gland-tipped hairs, margins towards base ciliate with long, straggling hairs to 3 mm long, without evident side veins. *Inflorescence* a secund, spike-like monochasium terminating main stems and branches, (4–)8–15(–20)-flowered, lower flowers often remote; bracts smaller than leaves, unequal, fertile bract linear, sterile bract larger and more leaf-like, densely puberulous with gland-tipped hairs mixed with acute, eglandular hairs; pedicels mostly up to 3 mm long in lower (rarely up to 12 mm long in lowermost flower developed in axil of upper most leaf) but upper flowers subsessile. *Calyx* sub-cylindrical in flower, 8–10 × 1.5–2.0 mm at anthesis, equally 10-veined without reticulate venation, densely puberulous with mix of gland-tipped and acute hairs and with conspicuous, glassy bristles 2–3 mm long on veins, lobes awl-like, 2–3 mm long. *Flowers* suberect at anthesis, unscented, diurnal. *Petals* longitudinally twisted and propellar-like, white or pink, sometimes with large crimson blotch or stain, claw 6–8 mm long, oblong, without auricles, limb obovate or oblong, minutely crenulate, unlobed, 1.5–4.0 × 1.0–2.5 mm, coronal scales ± 1 mm long. *Stamen* filaments unequal, shorter series ± 4–6 mm long, included, longer series ± 7–9 mm long, exserted up to 2 mm, puberulous. *Ovary* pyriform, ± 2 mm long; styles ± 1.5 mm long, deeply included. *Capsule* ovoid, 6–8 × 4–5 mm, ± 6–8 × longer than carpophore, minutely transversely rugulose; carpophore ± 1 mm long, pubescent. *Seeds* 0.7–1.0 mm, reniform, greyish brown, face concave-excavate, testal cells radially elongated-fusiform and mostly 2–4 granulate, back flat or weakly concave, tuberculate. *Flowering time:* (Jun.–)Aug.–Nov. in the southwestern and southern Cape; Sept.–April in the summer-rainfall region. Figure 1E, F, 21.

Vernacular names: small-flowered catchfly, gunpowder weed (Sonder 1860), joppies (*Hanekom 1005*), hardebolkeiltjies [hardebolletjies] (*Wagner s.n. STE16912*).

Distribution and ecology: native through Europe and western Asia; now introduced as a weed into many parts of the world, and well established in South Africa, mainly in the winter-rainfall region, where it is most common in the southwestern Cape, ranging northwards onto the Bokkeveld Escarpment and the Kamiesberg in central Namaqualand, and eastwards along the Eastern Cape coast to Grahamstown and East London, with isolated records further north, in southern and central KwaZulu-Natal and Potchefstroom (Figure 20); it is also recorded from the the highlands of east tropical Africa in Kenya, Tanzania, and Uganda (Turrill 1956a). *S. gallica* typically occurs as a weed of agricultural lands and in waste places, along roadsides, and in other disturbed sites, but in some places has also invaded more open native vegetation that has been subjected to light disturbance, such as trampling and grazing.

The species was already established in South Africa by the late eighteenth century, when it was encountered here by Carl Thunberg, who travelled in the country from 1772 to 1774. Another early collection made near Port Elizabeth by Ecklon in the 1830s was misidentified as *S. cernua* Thunb., a later synonym of *S. aethiopica* Burm.

Diagnosis and relationships: an erect or somewhat spreading, rarely ±prostrate annual with obovate-spathulate lower leaves, densely glandular-puberulous on the upper parts of the stem and on the inflorescence, and highly characteristic flowers with ± entire petals. The calyx is ± tubular at anthesis with distinctive glassy bristles 2–3 mm long on the veins, and the urn-shaped capsule, 6–8 mm long, is carried on a very short carpophore, ± 1 mm long. The styles are very short, ± 1.5 mm long.

The seed testal cells are radially elongated on the flanks, which are deeply excavated, and each cell is ornamented with a single series of 2–4 granules (Figure 1E). A double series of cells along the midline demarcates the flat or weakly convex seed back (Figure 1F).

Minor variants have been distinguished in the past, based on branching, flower colour, petal shape, and differences in capsule orientation (Greuter 1995a) but only the striking colour form, var. *quinquevulnera*, is recognized here. This variant cannot be distinguished from the typical form on any other grounds and identification of herbarium specimens depends on colour notes on the collecting labels or on residual coloration of the petals. Both varieties are sometimes represented on a single herbarium sheet. A second variety, var. *anglica*, was listed by Sonder (1860) as applying to more highly branched and spreading plants but as far as we can tell this is purely a manifestation of growing conditions and has no taxonomic value at all.

11a. var. **gallica**

Petals uniformly white to pink. Figure 21C.

Distribution: as for species (Figure 20).

11b. var. **quinquevulnera** *(L.) W.D.J.Koch*, Synopsis florae germanicae et helveticae, ed. 2: 109 (1843); Sond.: 127 (1860). *Silene quinquevulnera* L.: 416 (1753). Type: 'In Lusatia, Italia, Gallia', *Herb. Burser XI: 72* [UPS, lecto., designated by Talavera & Muñoz Garmendia: 409 (1989)].

Petals with large crimson spot or staining in centre of petal limb. Figure 21D.

Distribution: this variant occurs mainly in the southwestern Cape, extending into the Eastern Cape as far as Grahamstown (Figure 20), but has not been recorded further east or north in southern Africa, nor from tropical Africa (Turrill 1956a; Wild 1961). It occurs mixed with the typical variety, usually in the minority.

Additional specimens

*Collections comprising or including individuals of var. *quinquevulnera*.

NORTH-WEST.—**2627** (Potchefstroom): Potchefstroom, (–AC), 10 Oct. 1903, *Burtt Davy 1761* (NH).

KWA-ZULU NATAL.—**2930** (Pietemaritzburg): Lidgetton, (–AC), 1 April 1917, *Mogg 624* (BOL); Lions River Dist., Umgeni above Midmar, (–AC), 13 Oct. '964, *Moll 1246* (NU); Pietermaritzburg, (–CB), Sept. 1946, *Huntley 51* (NH, NU); 1968 [without month], *Garrett 76* (NU); Inchanga, (–DA), Sept. 1955, *Alexander 14* (NU); Umlaas Road, (–DA), 4 Sept. 1981, *Manning 18* (NU); Botha's Hill, (–DC), 30 Sept. 2003, *Styles 1597* (NU). **2931** (Stanger): Durban flat, (–CC), Sept. 1883, *Wood 2245* (NH). **3029** (Kokstad): Harding, Farm Bedford, (–DB), 28 Sept. 1963, *Lennox s.n.* (NU).**3030** (Port Shepstone): St Michaels-on-Sea, (–CD), 10 Oct. 1973, *Mogg 38284* (NH). **3130** (Port Edward): Port Edward, (–AA), without date or collector, *NU2490/3* (NU). Imprecise locality: Natal, received Feb. 1884, *Wood 1932* (BOL).

NORTHERN CAPE.—**3017** (Hondklipbaai): Kamiesberg, Skilpad Wildflower Reserve, (–BB), 29 Sept. 1995, *Cruz 120* (NBG). **3119** (Calvinia): top of Vanrhyns Pass, (–AC), 1 Oct. 1947, *Compton 2867* (NBG); Oorlogskloof Nature Reserve, (–CA), 16 Oct. 1996, *Pretorius 394* (NBG).

WESTERN CAPE.—**3118** (Vanrhynsdorp): Gifberg, (–CD), 16 Sept. 1911, *Phillips 2490* (NBG). **3218** (Clanwilliam): Clanwilliam, Olifants River Dam, (–BB), 10 Sept. 1949, *Barker 5737* (BOL, NBG). **3219** (Wuppertal): Cold Bokkeveld, Ondertuin, (–CC), 20 Dec. 1980, *Hanekom 2610* (PRE); Cold Bokkeveld, Skoongesig, (–DC), 4 Nov. 1967, *Hanekom 1005* (NBG, PRE). **3318** (Cape Town): Signal Hill, (–CD), Sept. 1887, *Thode 9236* (NBG); Cape Town, slopes above De Waal Drive, (–CD), 26 Oct. 1928, *Gillett 1819** (NBG, PRE); slopes of Table Mt above Cape Town, (–CD), 26 Oct. 1928, *Hutchinson 997** (BOL); E slopes of Lions Head, (–CD), Sept. 1913, *Kensit s.n. BOL45322* (BOL); Raapenberg, (–CD), Dec. 1890, *Guthrie 840* (BOL); near Rondebosch, (–CD), 8 Sept. 1895, *Wolley Dod 96** (BOL); Rondebosch, (–CD), 7 Aug. 1938, *Adamson 1926* (BOL); Kuils River, Langverwacht, (–DC), 26 Oct. 1971, *Oliver 3708* (NBG); Jonkershoek, Bosboukloof, (–DC), Oct. 1967, *Kerfoot 6080* (PRE); Bonterivier Farm, SW of Stellenbosch, (–DD), 6 Oct. 1989, *Buys 56** (NBG); Stellenbosch, Voëltjiesdorp, (–DD), 10 Sept. 1978, *Boucher 3928** (NBG); Stellenbosch, (–DD), 19 Sept. 1966, *Taylor 6888** (PRE, NBG). **3319** (Worcester): Ceres, Bokkerivier Farms, (–AD), 19 Nov. 1963, *Booysen 95* (NBG); Worcester High School, (–CB), 18 July 1980, *Walters 2078* (NBG); Worcester, Langerug Koppie, (–CB), 24 Aug. 1977, *Walters 1880* (NBG). **3320** (Montagu): Cogmans Kloof, (–CC), 18 Oct. 1964, *Bayliss 2429* (NBG). **3418** (Simonstown): Fishhoek, (–AB), July 1918, *Pahe s.n. BOL45321* (BOL); Noordhoek, (–AB), Sept. 1937, *Eames s.n. BOL45325* (BOL); Kommetjie vlei, Farm Imhoffs Gift, (–AB), *Davies 21* (NBG, PRE); 5 km from Hout

Bay, (–AB), 14 Oct. 1980, *Davies 40** (NBG); Faure, (–BA), 14 Sept. 1946, *Jordaan s.n. STE2490** (NBG); Somerset West, (–BA), 10 Aug. 1944, *Parker 3898** (BOL, NBG); Betty's Bay, Harold Porter Botanic Reserve, (–BD), 11 Sept. 1968, *Ebersohn 52/68** (NBG); Kogelberg State Forest, along road near 2nd dwelling, (–BD), 29 Nov. 1991, *Kruger 102* (NBG). **3419** (Caledon): Caledon, Zwartberg [Swartberg], (–AB), 30 Sept. 1980, *Hilliard & Burtt 13068* (NU); Palmietrivier near Kleinmond, (–AC), Sept. 1930, *Rossouw s.n. NBG11240* (NBG); ± 3 km from Greyton on Genadendal Road, (–BA), 29 Sept. 1997, *Meyer 1402** (PRE); 5 miles [8 km] NW of Riviersonderend, (–BB), 18 Sept. 1949, *Heginbotham 100* (BOL, NBG); Napier, (–BC), Jan. 1937 [fruiting], *Jordaan 834* (NBG); Caledon/Napier road ± 1 km E of Stanford/Rietpoel crossroads, (–BD), 25 Aug. 1995, *Paterson-Jones 565* (NBG); Fairfield Farm, (–BD), 3 Oct. 1994, *Kemper IPC657* (NBG); Groot Hagelkraal, (–DA/DC), 25 July 1995, *Paterson-Jones 470* (NBG); Bredasdorp, Anyskop, (–DD), Jan. 1937 [fruiting], *Jordaan 137* (NBG). **3420** (Bredasdorp): Swellendam, (–AB), 25 Aug. 1956, *Theron 2052* (BOL); Potberg, (–BC), 20 Aug. 1980, *Burgers 2484* (NBG).

EASTERN CAPE.—**3226** (Fort Beaufort): Cold Springs, (–BA), 9 Oct. 1983, *Jacot Guillarmon 9245* (PRE); Grahamstown, (–BC), Oct. 1888, *Galpin 227* (PRE); Grahamstown Memorial Garden, (–BC), 9 Oct. 1988, *Jacot Guillarmod 10120** (PRE); Port Alfred, (–DB), 28 Sept. 1918, *Britten 806* (PRE); Fort Hare Farm, (–DD), 11 Aug. 1978, *Gibbs Russell 4382* (PRE). **3227** (Stutterheim): Albany, (–AC), 6 Nov. 1965, *Bayliss 2998* (NBG); Albany Dist., Howison's Poort, (–AC), 16 Oct. 1966, *Bayliss 3645* (NBG); Fort Cunnyngham, (–AC), Sept. 1897, *Sim 968* (NU), *2181* (PRE); cultivated ground near Komgha, (–BD), Nov. 1891, *Flanagan 1181** (BOL, PRE); Grahamstown, Settler's Monument, (–BC), 5 Oct. 2004, *Ramdhani 496* (NH); Stutterheim Dist., Kologh State Forest, (–CB), 7 Dec. 2001, *Klein 78** (PRE); King Williamstown, (–CD), Nov. 1891, *Sim 971* (NU); East London, (–DD), Oct. 1963, *Bokelmann 8* (NBG). **3228** (Butterworth): Kentani, (–CB), 19 Oct. 1906, *Pegler 1412* (PRE). **3322** (Oudtshoorn): Bassonsrus, upper Cango Valley, (–AC), 4 Nov. 1974, *Moffett 434* (NBG). **3323** (Willowmore): near Avontuur, (–CA), Nov. [without year], *Bolus 2259* (BOL). **3325** (Port Elizabeth): Zuurberg Mountains, (–BC), 10 Oct. 1975, *Bayliss 7114** (NBG); Uitenhage Dist., sandy hills near the Zwartkop River, (–DC), Aug.–Oct. [without year], *Ecklon 431* (BOL, NBG); Port Elizabeth, (–DC), July 1942, *Cruden 448* (BOL); 15 Sept. 1982, *Immelman 353* (PRE); Red House, (–DC), Sept. 1914, *Paterson 16232* (PRE). **3421** (Riversdale): Reisiesbaan siding, (–AB), 28 Aug. 1979, *Bohnen 6351* (NBG). **3422** (Mossel Bay): road from Kleinbrakrivier to Gannakraal along Moordkuilrivier valley, (–AA), 19 Oct. 1990, *Joffe 878* (NBG); Pinedew farm, E of Wilderness, (–BA), 7 Nov. 1979, *Hugo 1926* (NBG). **3423** (Knysna): Knysna, Belvedere, (–AA), 9 June 1921, *Duthie s.n. STE29219* (NBG); Plettenberg Bay, (–AB), Sept. 1921, *Smart sub Rogers 15459* (PRE); Ratels Bosch, (–BA), June 1908, *Fourcade 282* (BOL). **3424** (Humansdorp): Witte Els Bosch [Witelsbos], (–AA), Nov. 1922, *Fourcade 2411** (BOL, NBG); Humansdorp, (–BB), June 1932, *Wagner s.n. STE16912* (NBG); Humansdorp Dist., Modderfontein, (–BB), 19 Feb. 1932 [cult. Ondersterpoort], *Steyn s.n.* (PRE); Jeffrey's Bay, (–BB), 17 Dec. 1956, *Taylor 5154* (NBG).

ACKNOWLEDGMENTS

The following colleagues kindly located type specimens in their collections and provided us with digital images: Nicolas Fumeaux of the Conservatoire et Jardin botaniques de la Ville de Genève (G); Jens Klackenberg (Curator) and Mia Ehn of the Swedish Museum of Natural History (S); and John Hunnex from the British Museum of Natural History (BM). Thanks to Dr. Bruno Wallnöfer, Curator, Vascular Plant Collections, Naturhistorisches Museum, Vienna (W), for confirming the loss of the type of *Silene caffra* Fenzl ex C.Muell.; Martin Callmander, Missouri Botanical Garden for assistance with types; Roy Gereau, Missouri Botanical Garden, for advice on nomenclatural matters; Mary Stiffler, Missouri Botanical Garden, for help with literature searches; Anne-Lise Fourie of the Mary Gunn Library for allowing us access to the De Candolle Herbarium microfiche; Esmerialda Klaassen of the Windoek Herbarium (WIND) for providing some critical measurements; and Miranda Waldron, SEM Unit, University of Cape Town for technical assistance. Ingrid Nänni provided material of *Silene vulgaris* for illustration, and Stephen Boatwright prepared the electronic figures. We are also grateful to all curators for allowing us access to their collections, and to Bengt Oxelman for valuable comments on the manuscript.

REFERENCES

AESCHIMANN, D. & BOCQUET, G. 1983. Etude biosystématique du *Silene vulgaris* s.l. (Caryophyllaceae) dans le domaine alpin. Notes nomenclatures. *Candollea* 38: 203–209.

AESCHIMANN, D. 1983. Le *Silene vulgaris* s.l. (Caryophyllaceae), évolution vers une mauvaise herbe. *Candollea* 38: 575–617.

AITON, W. 1789. *Hortus kewensis*, vol. 2. George Nicol, London.

BARTLING, T. 1832. Plantae ecklonianae. *Linnaea* 7: 620–652.

BITTRICH, V. 1993. Caryophyllaceae. In K. Kubitzki, J.G. Rohwer & V. Bittrich, *The families and genera of vascular plants* II. Flowering plants. Dicotyledons. Magnoliid, Hamamelid and Caryophyllid families. Springer-Verlag, Berlin.

BOCQUET, G. & KIEFER, H. 1978. The Ecklon & Zeyher collection at the Compton Herbarium with special reference to the species of *Silene* (Caryophyllaceae). *Bericht der Schweizerischen Botanischen Gesellschaft* 88: 7–19.

BOCQUET, G. 1977. *Silene dewinteri*, a new species of the Caryophyllaceae from the south-western Cape. *Bothalia* 12: 309–311.

BOISSIER, P.E. 1867. *Flora orientalis*, vol. 1. Georg, Basel.

BRITTON, N.L. & BROWN, A. 1913. *An illustrated flora of the Northern United* states, ed. 2, vol. 2. Charles Scribner's Sons, New York.

BRUMMIT, R.K. 1994. Report of the Committee for Spermatophyta: 41. *Taxon* 43: 271–277.

BURMAN, N.L. 1768. *Flora indica: cui accedit series zoophytorum indicorum, nec non Prodromus florae capensis*. Cornelius Haak, Leiden.

CAFFERTY, S. & JARVIS, C.E. 2004. Typification of Linnaean plant names in Caryophyllaceae. *Taxon* 53: 1049–1054.

CHAMISSO, A. & SCHLECHTENDAL, D. 1826. De plantis in expeditione spectulatoria Romanzoffiana: Caryophylleaca. *Linnaea* 1: 37–54.

CHATER, A.O. & WALTERS, S.M. 1964. *Silene* L. In T.G.T. Tutin, V.H. Heywood, S.M. Walters & D.A. Webb, *Flora europaea* 1: Lycopodiaceae to Platanaceae. Cambridge University Press.

CHOWDHURI, P.K. 1957. Studies in the genus *Silene*. *Notes from the Royal Botanic Gardens Edinburgh* 22: 221–278.

CURTIS, W. 1797. *Silene ornata*. Dark-coloured catchfly. *Curtis's Botanical Magazine* 11: t. 382. W. Curtis, London.

DE CANDOLLE, A.P. 1805. *Flore francaise* 3. Desray, Paris.

DE CANDOLLE, A.P. 1824. *Prodromus systematis naturalis regni vegetabilis* 1. Treuttel & Wurz, Paris.

DRÈGE, J.F. 1844. Zwei pflanzengeografische Dokumente. *Flora* 1843.

DUMORTIER, B.C.J. 1827. *Florula belgica*. Casterman, Tournay.

ECKLON, C.F. & ZEYHER, K.L. 1834. Enumeration plantarum Africae Australis extratropicae, 1. Hamburg.

ENDLICHER, S.L. 1840. *Genera plantarum*. Beck, Vienna.

ENGLER, A. 1912. Beiträge zur Flora von Afrika XL. Caryophyllaceae africanae. *Botanische Jahrbücher für Systematik, Pflanzengeschichte und Pflanzengeographie* 38: 380–384.

ERIXON, P. & OXELMAN, B. 2008. Reticulate or tree-like chloroplast DNA evolution in Sileneae (Caryophyllaceae). *Molecular Phylogenetics and Evolution* 48: 131–325.

FRAJMAN, G., EGGENS, F. & OXELMAN, B. 2009. Hybrid origins and homoploid reticulate evolution within *Heliosperma* (Sileneae, Caryophyllaceae)—a multigene phylogenetic approach with relative dating. *Systematic Biology* 58: 1–18.

GARCKE, C.A.F. 1869. *Flora von Nord- and Mittel-Deutschland*, ed. 9. Wiegandt, Berlin.

GHAFOOR, A. 1978. Caryophyllaceae. In S.M.H. Jafri & A. El-Gadi, *Flora of Libya* 59. Al Faateh University, Tripoli.

GILBERT, M.G. 2000. Caryophyllaceae. In S. Edwards, M. Tadesse, S. Demissew & I. Hedberg, *Flora of Ethiopia and Eritrea* 2, 1: 196–228. Addis Ababa University, Addis Ababa and Uppsala University, Uppsala.

GLEN, H.F., & GERMISHUIZEN, G. 2010. Botanical exploration of southern Africa, edn. 2. *Strelitzia* 26. South African National Biodiversity Institute, Pretoria.

GOLDBLATT, P. & MANNING, J.C. 2000. Cape plants: a conspectus of the Cape Flora. *Strelitzia* 9. National Botanical Institute & Missouri Botanical Garden.

GREUTER, W. 1995a. Proposal to conserve the name *Silene gallica* L. (Caryophyllaceae) against several synonyms with equal priority. *Taxon* 44: 102–104.

GREUTER, W. 1995b. *Silene* (*Caryophyllaceae*) in Greece: a subgeneric and sectional classification. *Taxon* 44: 543–581.

GREUTER, W. 1997. Caryophyllaceae. In Strid, A. & Tan, K., *Flora hellenica* 1. Koeltz, Königstein.

HEDBERG, O. 1954. Caryophyllaceae. *Svensk Botanisk Tidskrift Utgifven af Svenska Botaniska Foreningen* 48: 199–210.

HEDBERG, O. 1957. Afroalpine vascular plants: a taxonomic revision. *Symbolae botanicae Upsalienses* 15:1–411.

HOLMGREN, P.K., HOLMGREN, N.H. & BARNETT, L.C. 1990. *Index Herbariorum. Part. 1: The Herbaria of the World.* New York Botanical Garden, New York.

JACQUIN, N.L. 1777. *Hortus botanicus vindobonensis*, vol. 3. Published by the author, Vienna.

JACQUIN, N.L. 1791 ['1789']. *Collectanea*, vol. 3. Wappler, Vienna.

KOCH, W.D.J. 1843. *Synopsis florae germanicae et helveticae*, ed. 2. Wilmans, Frankfurt.

KUNZE, G. 1844 ['1843']. Pugillus secundus plantarum adhuc ineditarum s. minus cognitarum, quas ann. 1842 et 1843 praeter alias alio loco descriptas v. describendas coluit hort. bot. Univers, litt. Lipsiensis. *Linnaea* 17: 567–580.

LEDEBOUR, C.F. VON. 1842. *Flora rossica*, vol. 1. Schweizerbart, Stuttgart.

LINNAEUS, C. 1753. *Species plantarum*. Salvius, Stockholm.

LINNAEUS, C. 1762. *Species plantarum*, Edition 2. Salvius, Stockholm.

MASSON, D. 1989. *Silene vlokii* D.Masson sp. nov., new species of Caryophyllaceae for South Africa. *Candollea* 44: 485–491.

MCNEILL, J., BARRIE, F.R., BURDET, H.M., DEMOULIN, V., HAWKSWORTH, D.L., MARHOLD, K., NICOLSON, D.H., PRADO, J., SILVA, P.C., SKOG, J.E., WIERSEMA, J.H. & TURLAND, N.J. 2006. International Code of Botanical Nomenclature (Vienna Code) adapted by the 17th International Botanical Congress, Vienna, Austria, July 2005. *Regnum Vegetabile* 146: i–xviii, 1–568. Koeltz, Königstein.

MILLER, P. 1768. *The gardeners dictionary,* ed. 8. London.

MOENCH, C. 1794. *Methodus*. Cattor, Marburg.

MUELLER, K. 1868. *Annales botanices systematicae*, vol. 7. Hofmeister, Leipzig.

NORDENSTAM, B. 1980. The herbarium of Lehmann and Sonder in Stockholm, with special reference to the Ecklon and Zeyher collection. Taxon 29: 279–291.

OTTH, A. 1824. *Silene* L. In Candolle, A.P. de, *Prodromus systematis naturalis regni vegetabilis* 1: 367–385. Paris, Strasburg & London.

OXELMAN, B. & LÍDEN, M. 1995. Generic boundaries in the tribe Sileneae (Caryophyllaceae) as inferred from nuclear rDNA sequences. *Taxon* 44: 525–542.

OXELMAN, B., LIDÉN, M., RABELER, R.K. & POPP, M. 2000. A revised generic classification of the tribe Sileneae (Caryophyllaceae). *Nordic Journal of Botany* 20: 743–748.

PETRI, A. & OXELMAN, B. 2011. Phylogenetic relationships within *Silene* (Caryophyllaceae) section *Physolychnis*. *Taxon* 60: 953–968.

PFEIFFER, L.K.G. 1871–1875. *Nomenclator botanicus*, vol. 1 Kassel.

RAY, J. 1704. *Historia plantarum*, vol. 3. Smith & Walford, London.

RÖHLING, J.C. 1812. *Deutschlands Flora,* ed. 2, part 2. Wilmans, Bremen.

ROHRBACH, P. 1869 ['1868']. *Monographie der Gattung* Silene. Engelmann, Leipzig.

ROHRBACH, P. 1869–70. Beiträge zur systematik der Caryophyllinen. *Linnaea* 36: 170–270.

SALISBURY, R.A. 1796. *Prodromus stirpium in horto as Chapel Allerton vigentium*. London.

SCOPOLI, J.A. 1771. *Flora carniolica*, Edition 2. Krauss, Vienna.

SMITH, J.E. 1800. *Flora brittanica*. London.

SONDER, O.W. 1860. Caryophylleae. In W. Harvey & Sonder, O.W. *Flora capensis* 1: 120–151. L. Reeve, London.

STACE, C.A. 1978. Breeding systems, variation patterns and species delimitation. In H.E. Street, *Essays in Plant Taxonomy*: 57–78. Academic Press, London.

STUESSY, T.F. 1990. *Plants taxonomy: the systematic evaluation of comparative data.* Columbia University Press, New York.

TALAVERA, S. & MUÑOZ GARMENDIA, F. 1989. Sinopsis del género *Silene* L. (Caryophyllaceae) en la Península Ibérica y Baleares. *Anales del Jardin Botanico de Madrid* 45: 407–460.

THUNBERG, C.P. 1794. *Prodromus plantarum capensium*, vol. 1. Edman, Uppsala.

THUNBERG, C.P. 1823. *Flroa capensis*, Edition 2. Cotta, Stuttgart.

TURRILL, W.B. 1954. New varieties of *Silene burchellii. Kew Bulletin* 1954: 57, 58.

TURRILL, W.B. 1956a. *Silene*. In W.B. Turrill & E. Milne-Redhead, *Flora of tropical East Africa: Caryophyllaceae*: 31–34. Crown Agents for Oversea Governments and Administrations, London.

TURRILL, W.B. 1956b. *Silene vulgaris* (Moench) Garcke subsp. *macrocarpa* Turrill. *Hooker's Icones Plantarum* 36: t. 3551. Royal Botanic Gardens, Kew.

WALPERS, W.G. 1868. *Annales botanices systematique*, vol. 7,2. Hofmeister, Leipzig.

WATT, J.M.W. & BREYER-BRANDWIJK, M.G. 1932. *The medicinal and poisonous plants of southern Africa*. Livingstone, Edinburgh.

WIBEL, A.W.E.C. 1799. *Primitiae florae werthemensis*. Goepferd, Jena.

WICKENS, G.E. 1976. *Flora of Jebel Marra (Sudan Republic) and its geographical affinities.* Her Majesty's Stationery Office, London

WILD, H. 1961. *Silene*. In A.W. Exell & H. Wild, *Flora zambesiaca* 1: 350–355. Crown Agents for Oversea Governments and Administrations, London.

WILLDENOW, C.L. 1809. *Enumeratio plantarum*. Berlin.

WILLIAMS, F.N. 1896. A revision of the genus *Silene* Linn. *Journal of the Linnean Society, London* 32: 1–196.

WILLKOMM, M. 1854. *Icones et descriptiones plantarum novarum*, vol.1,8. Payne, Leipzig.

6

Systematics of the hypervariable *Moraea tripetala* complex (Iridaceae: Iridoideae) of the southern African winter rainfall zone

P. GOLDBLATT* and J.C. MANNING**

Keywords: Chromosome cytology, Iridaceae, Iridoideae, *Moraea* Mill., new species, southern Africa, taxonomy

ABSTRACT

Field and laboratory research has shown that the *Moraea tripetala* complex of western South Africa, traditionally treated as a single species, sometimes with two additional varieties, has a pattern of morphological and cytological variation too complex to be accommodated in a single species. Variation in floral structure, especially the shape of the inner tepals, degree of union of the filaments, anther length and pollen colour form coherent patterns closely correlated with morphology of the corm tunics, mode of vegetative reproduction, and in some instances capsule and seed shape and size. The morphological patterns also correlate with geography, flowering time and sometimes habitat. It is especially significant that different variants of the complex may co-occur, each with overlapping or separate flowering times, a situation that conflicts with a single species taxonomy. We propose recognizing nine species and three additional subspecies for plants currently assigned to *M. tripetala*. *M.* **grandis**, from the western Karoo, has virtually free filaments and leaves often ± plane distally; closely allied *M.* **amabilis**, also with ± free filaments and often hairy leaves, is centred in the western Karoo and Olifants River Valley. Its range overlaps that of *M.* **cuspidata**, which has narrowly channelled, smooth leaves, linear inner tepals spreading distally and filaments united for up to 1.5 mm. *M.* **decipiens** from the Piketberg, *M.* **hainebachiana**, a local endemic of coastal limestone fynbos in the Saldanha District, *M.* **ogamana** from seasonally wet lowlands, and early flowering *M.* **mutila** constitute the remaining species of the complex in the southwestern Western Cape. *M.* **helmei**, a local endemic of middle elevations in the Kamiesberg, Namaqualand, has small flowers with short, tricuspidate inner tepals. All but *M. amabilis* and *M. mutila* are new species. We divide *M. tripetala sensu stricto* into three subspecies: widespread subsp. *tripetala*, subsp. **violacea** from the interior Cape flora region, and late-flowering subsp. **jacquiniana** from the Cape Peninsula and surrounding mountains.

INTRODUCTION

The Afro-Eurasian and largely sub-Saharan genus *Moraea* Mill. (Iridaceae: Iridoideae) comprises ± 215 species of cormous geophytes (Goldblatt & Manning 2009, 2010). Although florally diverse, the genus is recognized by a bifacial and channelled (rarely terete) leaf blade (isobilateral leaves are ancestral for Iridoideae) and a corm of a single internode derived from a lateral bud in the axil of the lowermost cataphyll. The majority of *Moraea* species have *Iris*-like flowers with clawed tepals, prominent, spreading outer tepal limbs marked with nectar guides, and flattened, petaloid style branches to which the anthers are appressed. Unlike *Iris* L., the tepals of *Moraea* are normally free and the filaments are at least partially united or rarely secondarily free in a few species (always free in *Iris*). Species of *Moraea* with other flower types (Goldblatt 1986b; 1998), usually with subequal tepals, reduced style branches and, in some species, a perianth tube, are derived and limit the utility of floral characters in circumscribing the genus.

Relatively widespread in the southern Africa winter-rainfall zone, *Moraea tripetala* is one of the more common species of subg. *Vieusseuxia* (D.Delaroche) Goldblatt (± 30 spp.). The group is distinguished by the flowers usually lasting three days (vs. fugaceous and lasting less than a single day, the plesiomorphic condition), a single, channelled foliage leaf, and inner tepals variously modified in most species into trilobed or linear structures, occasionally reduced to short hair-like cusps or rarely absent (Goldblatt 1976a; 1986a). As currently understood (Goldblatt 1976b) *M. tripetala* is recognized by a blue, violet or purple perianth, reduced inner tepals that are usually represented by short, hair-like cusps 2–3 mm long, filaments either free or shortly united for 1–2 mm (1/4 to 1/3 their entire length) and an associated short style with the style branches appressed to the stamens for almost their entire length. Variation in the colour and texture of the corm tunics, length and shape of the leaves, in the size and shape of the inner tepals, and the occasional development of pubescence on the stems and leaves, was documented in some detail by Goldblatt (1976b), who at that stage preferred not to accord any of the variant forms taxonomic recognition. Goldblatt (1976b) also noted that *M. tripetala* was variable cytologically but the three populations sampled were diploid with a basic chromosome number of *x* = 6, this shared with all species of subgen. *Vieusseuxia*.

During extensive field work in western southern Africa over the past decade we have encountered a wide range of plants currently identified as *Moraea tripetala*. The patterns of variation in the flower, especially the form of the inner tepals, and associated differences in corm, capsule and seed morphology, and in the means of asexual reproduction have led us to the conclusion that recognition of a single species for the complex does not reflect biological reality. Particularly difficult to reconcile with a single species treatment for *M. tripetala* is the co-occurrence of two morphological variants, either flowering virtually side-by-side, or in bloom several weeks or months apart in the same locality, and some-

* B.A. Krukoff Curator of African Botany, Missouri Botanical Garden, P. O. Box 299, St. Louis, Missouri 63166, USA. E-mail: peter.goldblatt@mobot.org
** Compton Herbarium, South African National Biodiversity Institute, Private Bag X7, Claremont 7735, South Africa / Research Centre for Plant Growth and Development, School of life Sciences, University of KwaZulu-Natal, Pietermaritzburg, Private Bag X01, Scottsville 3209, South Africa. E-mail: J.Manning@sanbi.org.za.

times differing in their habitat. With this in mind we have undertaken an extensive review of the variation in *M. tripetala*. We conclude that a better treatment of the complex, which we consider monophyletic, is to recognize nine species, with *M. tripetala* subdivided into three subspecies.

MATERIALS AND METHODS

We first made field observations across the entire range of the *Moraea tripetala* complex, noting in addition to floral morphology, details of corm tunics, mode of vegetative reproduction, and whenever possible capsule and seed morphology. These observations were then integrated with an examination of specimens in herbaria with important holdings of southern Africa flora, BOL, K, MO, NBG, PRE, and S (abbreviations following Holmgren *et al.* 1990), plus a study of type material of the named variants in the complex. Abbreviation of author names follows Brummitt & Powell (1992).

Chromosome counts were determined from mitotic squashes using root tips. Material for the original counts reported here was prepared according to the protocol described by Goldblatt & Takei (1993). The vouchers are housed at the Missouri Botanical Garden Herbarium (MO). Counts are based on samples of three to four individuals and, following standard practice in plant cytology, are assumed to represent an entire population.

RESULTS

The lectotype of *Moraea tripetala* (designated by Goldblatt 1976b), is a Thunberg collection from the southwestern Cape, with filaments united for ± 1 mm, inner tepals reduced to linear, acute cusps ± 3 mm long, and a linear, narrowly channelled, glabrous leaf. Corms are lacking in the type material but we are confident that a range of plants recorded from Aurora on the west coast of the Western Cape eastward to Knysna and mainly flowering from August to the end of September accord with the type. These often have slightly sweet-scented flowers and the inner tepals vary from short hair-like vestiges to linear structures up to 5 mm long. Apart from plants matching the type relatively well, there is a range of somewhat to significantly different populations both in the southwestern Cape and in the interior, in the western Karoo and Little Karoo. We describe the variant populations below, dealing first with plants from the near southwestern Cape followed by those from the interior.

Variant 1: the first significant variant in the southwestern Cape is an early flowering taxon, mostly blooming from mid-August to early September, often with relatively broad leaves. These are sometimes plane distally, 3–5(–7) mm wide and laxly twisted, ± half as long to ± as long as the stem, or rarely exceeding it. Plants occur at low elevations on clay or loamy soils. The inner tepals are linear to narrowly lanceolate, up to 12.5 mm long, and differentiated into an erect portion (equivalent to the claw of the outer tepals) and an outspread portion (equivalent to the limb) that is either slightly expanded at the base or, in some populations (e.g. *Goldblatt 2310*

MO), with two obtuse, lateral lobes, rendering the limb ± trilobed. The filaments are united for 1.0–1.5 mm, rarely only ± 0.5 mm. In two collections the leaves and stem of some plants are thinly hairy (e.g. *Nordenstam & Lundgren 1998* MO, NBG, S; *Acocks 1912* S). This variant was first recognized at species rank by C.F. Ecklon (1827) as *Vieusseuxia mutila* and later by J.G. Baker (1904) as *Moraea punctata* but has remained poorly understood. *Moraea monophylla* (Baker 1906) and *M. tripetala* var. *mutila* (Baker 1896) represent the same variant.

Plants from the immediate area of Sir Lowry's Pass, east of Cape Town, flowering in August, have inner tepals ± 10 mm long, consisting of a limb with a short central tapering cusp and short rounded lateral lobes. They are somewhat out of range for variant 1 and have a narrow, channelled, smooth leaf (e.g. *Loubser 872* NBG; *Goldblatt 2506* MO). These plants may belong here or alternatively may be hybrids with *M. unguiculata* or another species: we discuss them below in more detail.

Variant 2: a second southwestern Cape variant is the plant G.J. Lewis (1941) named *Moraea tripetala* var. *jacquiniana* (= *M. jacquiniana* Schltr. ined.). Often shorter in stature, 140–300 mm high and with smaller flowers than in typical *M. tripetala*, it usually has exceptionally slender leaves, mostly 2–3 mm wide, V-shaped in cross section. When dry the blades are closely folded together and appear terete. Flowering in this plant is from mid-November to January on the Cape Peninsula (Lewis 1950), but sometimes in late October elsewhere. Illustrated in *Flora of the Cape Peninsula* (Maytham Kidd 1950), the flowers are often dark purple, but sometimes light mauve to ± blue as in *M. tripetala sensu stricto*, and as far as known, consistently have white nectar guides. The linear inner tepals are 5–6 mm long (e.g. *Pillans 10279* BOL, MO), thus slightly more than 1/2 as long as the outer tepal claws and reaching to between the base and middle of the anthers. The outer tepals have limbs 11–16 mm long and claws 8–11 mm long (vs. limbs 12–18 and claws 10–12 mm long in *M. tripetala sensu stricto*). At several localities, notably in the southern Cape Peninsula, at Jonkershoek and in the Grabouw area, both typical *M. tripetala* and var. *jacquiniana* occur, the former blooming two to four months earlier, late July to September. *M. pulchra* Eckl. (1827) from 'Hottentotshollandkloof' collected in flower on 25 November represents the earliest name for this plant at species rank.

Variant 3: restricted to lowland wetlands of the southwestern Cape and flowering in September, this variant has an unusual, narrow ovary, cylindric rather than ovoid, 10–13 mm long, and darkly lined vertically on locules, and nearly cylindric capsules (15–)20–24 mm long, somewhat angled on the locules. Typical *M. tripetala* has an ellipsoid ovary 6–10 mm long and capsules 8–14 mm long, both round in cross section. The flowers of this variant are also unusual in having the outer tepal limbs pale blue with darker veins radiating from a yellow nectar guide. Plants are always relatively short, rarely exceeding 200 mm, have leaves usually shorter than the stems, and small corms with blackish, wiry tunics. There is also invariably a small cormlet in the axil of the foliage leaf, and plants often grow in small groups, the result of vegetative reproduction from these cormlets. Typical *M. tripetala* does not normally produce cormlets either in the leaf axils or at the corm base, and is always a solitary plant, favouring well-drained habitats. Variant 3 extends from Voëlvlei in the north to Harmony Flats at Strand in the south.

Variant 4: a last variant in the southwestern Cape is a vegetative apomict, restricted to limestone pavement or calcareous sands in the Saldanha Bay area, and blooming in August and early September, rarely later. Relatively short, up to 300 mm high, plants have lilac or pale to mid-blue flowers with pale yellow nectar guides, outer tepals 27–29 mm long and unusual inner tepals ± 4 mm long, spindle-shaped, sometimes oblanceolate and obscurely 3-lobed, tapering distally with the apex curving inward, sometimes the inner distal surface pilose. Unusual for the complex, it has pale grey-blue anthers and off-white or yellow pollen. Capsules never develop to maturity and are shed several days after flowers fade and seeds are not formed. Microscopic examination of pollen confirms that the grains are all malformed. Propagation of new plants is accomplished by the production of several cormlets born in the axil of the foliage leaf and sometimes the lowermost sheathing leaf. Non-flowering individuals produce similar cormlets at the base of the main corm. This variant co-occurs with typical *M. tripetala*, which blooms in the same sites in late September and early October, and has paler blue flowers and dark blue or purple anthers and red pollen typical of the species.

Variant 5: some plants from the Piketberg assigned to *Moraea tripetala* by Goldblatt (1976b) have large inner tepals 7–10 mm long with distally expanded limbs that are ± trilobed and tapering to an attenuate, twisted tip. These plants are also unusually tall, reaching up to 450 mm, but the flowers are relatively small for the complex, with outer tepals 20–23 mm long and claws 8–9 mm long. The filaments, only ± 4 mm long, are united for ± 2 mm. In general appearance these plants resemble *M. unguiculata* Ker Gawl., but the filaments, united for less than 1/2 (slightly more than 1/3) their length, are characteristic of the *M. tripetala* complex. Both typical *M. tripetala* and white-flowered *M. unguiculata* occur locally in the Piketberg, where the former has the short, hair-like inner tepals typical of the species.

Variant 6: outside the immediate southwestern Cape, in the western Karoo, plants included in *Moraea tripetala* usually have free filaments (sometimes united basally for < 0.4 mm), inner tepals reduced to small hair-like cusps mostly 1–2 mm long, a linear leaf, shallowly channelled below but often plane distally, and light brown corm tunics of medium-textured to soft fibres. Plants always form clonal colonies, the result of vegetative propagation by the production of two or more cormlets at the base of the flowering stem that replace the parent corm, and/or by axillary cormlets (Figure 14).

Populations of this variant fall into two main groups. The first of these, from the western Karoo and the Olifants River Valley and surrounding mountains, mostly from lighter, sandstone or shale-derived soils, has smaller flowers with outer tepals 23–25 mm long (claws 10–14 mm), anthers 4.5–7.0(–8.0) mm long and ±

as long as to 1.5 × as long as the filaments, and either white to yellow pollen (eastern populations) or red pollen (western populations) (Figure 14). Pollen colour is lost in older collections but our observations are based on examination of living plants across almost the entire range of this morph. The 5(6)-sided seeds are 1.7–2.0 × ± 1 mm and stand out in having pale, spongy, ridged angles and an extended pale spongy mass at the micropylar end. Seeds of typical *M. tripetala* and related variants are 1.0–1.3 × ± 1 mm, with sharply raised, pale angles and a finely wrinkled surface through which the underlying dark brown seed body is evident. This variant matches the type of *M. amabilis* Diels.

Plants closely resembling this variant growing on sandstone pavement in the Bokkeveld and Gifberg–Matsikamma Mtns are often particularly short, usually < 180 mm high. Capsules and seeds in these plants are consistent with this variant.

A second western Karoo variant comprises large-flowered plants, mostly of dolerite-derived soils, with outer tepals 30–40 mm long (claws 12–18 mm long), anthers 8–11 mm long and ± twice as long as the filaments, and consistently red pollen (Figure 15). The capsules are 16–19 mm long, narrowly ellipsoid-oblong with a markedly thickened apical rim. Seeds in the few fruiting collections that we have been able to examine have exceptionally large, brown seeds, 2–3 × 1.8–2.3 mm.

Variant 7: also in the western Karoo and adjacent interior southern Cape, another variant of the complex has linear inner tepals (8–)10–15 mm long, ascending below and spreading distally, filaments united for 0.5–1.8 mm, and narrowly channelled leaves with the leaf halves often appressed. The relatively large corms have tunics of wiry, usually thickened fibres, and cataphylls that decay into a particularly prominent collar of vertical fibres around the base. The filaments are (3–)4–6 mm long and the anthers, 4–8 mm long, with red pollen. Late flowering, these populations seldom bloom before the last week of September and only in late October at higher elevations in the Swartberg Mtns. Plants are sympatric in the Roggeveld and Klein Roggeveld with the western Karoo morph of variant 6, which has short, hair-like inner tepals and pale corm tunics, and their flowering times often overlap (e.g. *Goldblatt & Porter 13461* and *13462*, MO, NBG, from the Farms Fortuin and Nuwerus). The few plants we have seen with mature capsules have distinctive seeds with a spongy testa slightly thicker on the angles.

Variant 8: plants from the Kamiesberg, Namaqualand, discovered only in 2009 by Cape Town botanist N.A. Helme, and evidently belonging to the *Moraea tripetala* complex, have filaments united in the lower 1.5 mm but differ notably in their short, trifid inner tepals, represent the last significant variant of the complex (Figure 4). Apart from the unique inner tepals, the flowers are small, with outer tepals 21.0–23.5 mm long bearing bright yellow, velvety nectar guides. The outer tepal claws have a pair of marginal teeth just below their apices.

Even with the exclusion of these variants, the remaining populations of *Moraea tripetala* are still variable, and we discuss the major patterns below under that species. Extending from Aurora and the Piketberg on the west coast to the Cape Peninsula and thence across the southern Cape to Knysna, plants have moderately-sized, pale blue, purple or violet flowers with white or yellow nectar guides (many populations are variable for flower colour), inner tepals hair-like and rarely > 4 mm long, and the ovary is ellipsoid and 8–10 mm long. Vegetative reproduction via axillary or basal cormlets is rare. Among these populations, plants from the southern Cape sometimes lack inner tepals entirely and those from the Langeberg foothills may have free filaments. Neither trend is consistent as far as we can determine. From the few fruiting specimens available, seeds of what we consider typical *M. tripetala* are consistently small, with pale, raised, ± winged angles.

DISCUSSION

Species, subspecies or merely ecotypes?: the immediate question our observations raise is whether or not any of the several morphological variants should be recognized taxonomically, thereby reducing the otherwise florally distinctive *Moraea tripetala* to just one of several taxa that are only moderately distinctive florally and not easily distinguished without examination of corms, capsules, and seeds. We conclude that the patterns of variation that we now recognize renders the single species solution unacceptable, particularly because of the correlated character differences in the corm, leaf, flower, and seeds, and the examples of sympatry of two variants at some sites, either flowering at the same time or flowering weeks to months apart. Moreover, the range of habitats and flowering times would be remarkable for a single species and unacceptable for the lack of conformity with any but the broadest morphological species concepts. Similarly, treatment of the major variants as subspecies violates any biological species concept. We conclude that all the distinctive entities that co-occur and have overlapping flowering times should be treated as separate species. Likewise, co-occurring variants that have flowering times separated by weeks or months require taxonomic recognition if they can be readily identified by more than one unique morphological marker.

This framework has been the underlying philosophy in our decision to dismember *Moraea tripetala* sensu Goldblatt (1976b). We also note that Baker (1896) in *Flora capensis* recognized one variety in the species and later added two more (Baker 1904, 1906). Lewis (1950) recognized *M. tripetala* as one species with three varieties in *Flora of the Cape Peninsula* alone. Three of the eight segregate species we recognize already have names in the literature, providing a precedent for our revised taxonomy [*M. amabilis*, *M. tripetala* var. *jacquiniana* (= *Vieusseuxia pulchra*) (Variant 2), *M. punctata* (= *Vieussuexia mutila* = *M. tripetala* var. *mutila*) (Variant 1)]. To these we now add *M. decipiens* (variant 5); *M. hainebachiana* (variant 4); *M. ogamana* (variant 3); *M. cuspidata* (variant 7); and *M. helmei* (variant 8). We also recognize *M. amabilis* and *M. grandis* from the western Karoo for the two subsets of variant 6.

With the recognition of nine species in the *Moraea tripetala* complex, section *Vieusseuxia* now includes 36 species, and the genus *Moraea* ± 215 species (Goldblatt & Manning 2010 and unpublished).

Floral and reproductive biology: as might be expected in species with similar flowers, all members of the complex share the same pollination system, large bodied, mostly anthophorine bees (usually *Amegilla* and *Anthophora* spp.), or sometimes *Apis mellifera* (Goldblatt *et al.* 2005). Our impression is that the species of the complex are remarkably successful and it is rare to see plants later in the season without multiple capsules with full complements of seeds; autogamous selfing can be dismissed because in these iris-type flowers the anthers do not reach the stigmatic lobes (Goldblatt 1998) and we infer that insect mediated pollen transfer is necessary to achieve pollination.

Propagation by seed is complemented in several species either by production of cormlets in the leaf or cataphyll axils and at the base of the corm (*Moraea hainebachiana*, *M. ogamana*) or sometimes by production of two new corms in place of the parent corm (*Moraea amabilis* and often *M. grandis*). Enhanced vegetative reproduction in these species results in plants forming clonal colonies. These two vegetative propagation strategies are uncommon elsewhere in the *M. tripetala* complex. The lowland wetland populations we treat as *M. ogamana* consistently produce a single cormlet in the foliage leaf axil but the vegetative apomict, *M. hainebachiana* produces multiple cormlets in the foliage leaf axil as well as at the corm base, a feature particularly well-developed in non-flowering individuals. This last species produces malformed pollen, and the capsules are invariably shed soon after flowers wilt, without the ovules developing into seeds.

Chromosome cytology: as noted by Goldblatt (1976a), chromosomal karyotypes are variable for populations then included in *Moraea tripetala*. Sampling is limited (Table 1) but we note the following. *M. amabilis*, *M. cuspidate*, and *M. grandis* have the longest three chromosome pairs either metacentric or sub-metacentric (Figures 1A, B) and a small satellite on the fourth longest ± acrocentric pair (Goldblatt 1971 and Table 1, as sat type 1). This karyotype is the common one in subgen. *Vieusseuxia*. The karyotype of *M. mutila* has small satellites on the two longest, submetacentric chromosome pairs (sat type 3) (Figure 1C). In contrast, plants of *M. tripetala* subsp. *tripetala* and subsp. *violacea* (Table 1), have very large satellites on a long, telocentric chromosome pair (sat type 2). In three of four populations of subsp. *tripetala* sampled a second large satellite (Figure 1D) is evident on a relatively short acrocentric chromosome pair (sat type 2a) (poor preparation in the fourth sample makes it uncertain whether the second pair of satellites is present or not). These examples, inadequate though they are, show that chromosomal rearrangements via inversion or translocation occur with some frequency in the complex and potentially provide a genetic basis for reproductive isolation because of meiotic disruption in any inter-populational hybrids. The three karyotypes also provide support of our revised taxonomy in showing different karyotypes, consistent within species. *M. decipiens*, *M. hainebachiana*, *M. helmei*, *M. ogamana*, and *M. tripetala* subsp. *jacquiniana* remain to be examined cytologically.

TABLE 1.—Chromosome counts in *Moraea tripetala* complex with corrected taxonomy. Satellite characteristics are as follows: small on acrocentric chromosome, sat type 1; large on telocentric chromosome, sat type 2, type 2a with second large satellite on short acrocentric pair; small on large submetacentric chromosome, sat type 3. Original counts are in bold type in column 2 and their voucher specimens are at the Missouri Botanical Garden Herbarium (MO).

Species	Diploid number, 2n; satellite position	Reference for previous count or voucher data for new count
grandis	12; sat type 1	Goldblatt (1971) as *M. tripetala, Goldblatt 101*
	12; sat type 1	W Cape, near Nieuwoudtville, *Goldblatt 3948*
cuspidata	12; sat type 1	Goldblatt (1976) as *M. tripetala, Goldblatt 547*
amabilis	**12; sat type 1**	W Cape, Clanwilliam, Boskloof road, *Goldblatt 2547*
	12; sat type 1	W Cape, 70 km N Matjiesfontein, *Goldblatt 3222*
	12; sat type 1	N Cape, Calvinia to Middelpos, *Goldblatt 4381*
	12; sat type 1	W Cape, Nardouw Pass, *Goldblatt 3852*
mutila	**12; sat type 3**	W Cape, north of Piketberg, *Goldblatt 6127*
	12; sat type 3	W Cape, Camphill Village road, *Goldblatt* no voucher
tripetala subsp. *tripetala*	**12; sat type 2a**	W Cape, Pearly Beach, *Goldblatt 2607*
	12; sat type 2a	W Cape, Caledon Swartberg, *Goldblatt 2509*
	12; sat type 2a	W Cape, near Bot River, *Goldblatt 3997A*
	12; sat type 2	W Cape, Mamre Road Station, *Goldblatt 2487*
subsp. *violacea*	12; sat type 2	Goldblatt (1976) as *M. tripetala, Goldblatt 667*
	12; sat type 2	W Cape, Op-de-Tradouw, *Goldblatt 4184*

TAXONOMY

Key to the species of the *Moraea tripetala* complex

1a Inner tepals 7–20 mm long, linear or expanded and lobed in middle or 3-lobed with central lobe tapering and twisted:

2a. Inner tepals 7–9 mm long, 3-lobed with longer, tapering, often twisted central lobe; style crests ± 5 mm long; plants from middle elevations of the Piketberg, flowering mainly November . . . . 1. *M. decipiens*

2b. Inner tepals 7.5–15.0 mm long, linear throughout or expanded and lobed in middle; style crests 5–10 mm long; plants widespread on a variety of soils, flowering mainly August to October:

3a Foliage leaf usually shorter than stem, even half as long, leaf blade distally at most shallowly channelled or ± plane and laxly twisted, mostly 3.5–5.0 mm wide; leaf blades and/or stem or spathes and sheathing leaves sometimes hairy; inner tepals 7.5–12.5 mm long, often expanded in middle, sometimes into small rounded lobes at base of spreading distal half; ovary 7–8 mm long; plants flowering mainly in August(September) 3. *M. mutila*

3b Foliage leaf as long as or more often exceeding stem, sometimes twice as long; leaf blade narrowly channelled with leaf halves often folded together, 1.5–3.0 mm wide; vegetative parts never hairy; inner tepals 8–15 mm long, linear for entire length, spreading in distal half to one third; ovary 11–15 mm long 4. *M. cuspidata*

1b Inner tepals (1–)2–6 mm long or occasionally absent, either 3-lobed in upper third or reduced to hair-like or lanceolate to linear cusps, acute and usually widest at base:

4a Inner tepal ± 4 mm long, 3-lobed in upper third; filament column and lower abaxial surface of inner tepals puberulous; plants of the Kamiesberg, Namaqualand . . . 2. *M. helmei*

4b Inner tepal variable in length (1–6 mm), a hair-like, linear cusp; filament column and lower abaxial surface of inner tepals smooth; plants of the western and southern Cape and western Karoo:

5a Axil of foliage leaf bearing several dark-coloured cormlets breaking through leaf sheath; plants never forming mature capsules or seeds; anthers pale blue and pollen blue or white to yellow; plants restricted to Western Cape west coast near Saldanha on limestone pavement or calcareous sands . . . . 6. *M. hainebachiana*

5b Axil of foliage leaf rarely with a solitary cormlet enclosed in sheath or without cormlets; anthers and pollen orange-red or white (rarely yellow); plants always forming mature capsules and fertile seeds; plants widespread from the Western Cape coast and

Western Karoo on various soils:

6a Filaments usually united for 0.5–1.5 mm (rarely free); plants not forming colonies, occasionally a single cormlet present in axil of foliage leaf; leaves narrowly channelled and leaf halves often closely appressed; stems and abaxial leaf surface smooth:

7a Ovary narrowly ellipsoid, mostly 4–9 mm long; capsules ellipsoid, 8–14 mm long; foliage leaf (or leaves) usually as long as or longer than stem; axillary cormlets absent; pollen usually orange-red, occasionally white . . . . 5. *M. tripetala*

7b Ovary linear-cylindric, 10–13 mm long; capsules ± cylindric (15–)20–24 mm long; foliage leaf often shorter than stem, sometimes half as long; bearing a single cormlet in axil of foliage leaf; pollen white 7. *M. ogamana*

6b Filaments ± free or united basally for up to 0.4 mm; plants usually forming colonies due to vegetative reproduction by axillary cormlets and/or two new cormlets replacing parent corm; leaves widely channeled, often ± plane distally; plants sometimes velvety on stems and abaxial surface of leaves:

8a Flowers with outer tepal limbs (9–)11–18 × 10–14 mm, always as long as or longer than wide; anthers 4.5–7.0(–8.0) mm long and up to one and a half times as long as filaments; corm tunics usually of hard, coarse, light brown (occasionally dark brown) fibres . . . . 8. *M. amabilis*

8b Flowers with outer tepal limbs 15–20(–23) × 15–20 mm, often ± as long as wide; anthers 8–11 mm long and ± twice as long as filaments; corm tunics usually of soft, fine (rarely medium-textured), light to dark brown fibres . . . . 9. *M. grandis*

1. **Moraea decipiens** *Goldblatt & J.C.Manning*, sp. nov.

TYPE.—Western Cape, 3218 (Clanwilliam): Piketberg, Farm Noupoort, south-trending, rocky sandstone slope in shallow pockets of soil, (–DA), 2 Nov. 2011, *Goldblatt & Porter 13709* (NBG, holo.; K, MO, PRE, iso.).

Plants 300–450 mm high. *Corm* 9–12 mm diam. with pale, fibrous tunics. *Stem* usually simple, rarely 1-branched, glabrous. *Foliage leaf* solitary, linear, conduplicate with margins tightly folded together, 0.7–1.5

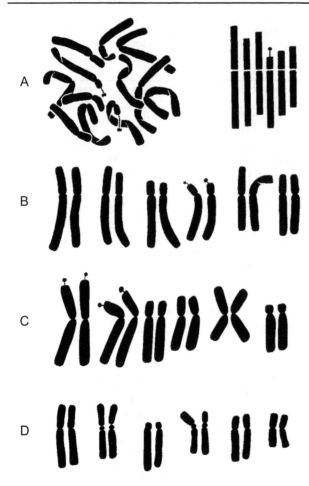

FIGURE 1.—A. Mitotic metaphase and idiogram of karyotype of *Moraea grandis*, *Goldblatt 101*); B–D, karyotypes from mitotic metaphases with chromosomes arranged in pairs for comparison; B. *M. amabilis, Goldblatt 3852*; C. *M. mutila, Goldblatt 6127*). D. *M. tripetala* subsp. *tripetala, Goldblatt 2607B*).

mm diam., exceeding stem and often trailing above, sometimes with axillary cormlet; sheathing cauline leaves 45–55 mm long, green with dry attenuate apices. *Rhipidia* mostly 3–5-flowered; spathes green with dry, brown, attenuate tips, inner 35–40 mm long, outer ± half as long as inner. *Flowers* pale to deep purple, outer tepal limb bases with white, wedge- or V-shaped, minutely papillate nectar guides dark purple in centre; outer tepals 20–23 mm long, claw 8–9 mm long, with wide, dark violet median stripe, limb 12–15 × 6–8 mm; inner tepals 7–10 long, violet, claw suberect, plane, expanded to 2 mm wide at apex, often into rounded lobes, limb 3–4 mm long, tapering to attenuate tip, laxly twisted distally. *Stamens* with filaments 3.5–4.0 mm long, united in lower ± 2 mm; anthers 4.5–6.0 mm long, reaching or shortly exceeding stigma lobe, dark purple-black; pollen orange-red. *Ovary* narrowly oblong-truncate, 7–9 mm long, usually exserted when flower mature; style ± 2 mm long, dividing ± 0.5 mm above top of filament column, branches, 7–9 × 1–2 mm; stigma shallowly bilobed; style crests narrowly wedge-shaped, ± 5 mm long, erect. *Capsules* narrowly ovoid, 8–11 mm long. *Seeds* 5(6)-sided, 1.2–1.5 mm × ± 1 mm, facets slightly wrinkled, brown seed body visible though testa, angles between facets forming narrow, raised, pale, slightly spongy ridges. *Chromosome number* unknown. *Flowering time*: late Oct.–mid-Nov. Figure 2.

Distribution: known only from the western half of the Piketberg in Western Cape (Figure 3), *Moraea decipiens* is evidently restricted to middle elevations of the range, rather than the emergent peaks, and based on our own collection, occurs on stony sandstone slopes in shallow pockets of soil among Restionaceae and low shrubs. None of the other three collections that we have seen include information about the habitat.

Diagnosis: named for the deceptive (*decipiens* = Latin deceiving) appearance of the flowers to those of *M. tripetala*, *M. decipiens*, nevertheless, differs in having the inner tepal claws distinctly expanded distally, sometimes into rounded lobes, with the limb linear-attenuate and curved inward above. The inner tepals recall those of *M. unguiculata* and *M. algoensis* but unlike those species, the filaments of *M. decipiens* are only shortly united for ± 2 mm and the style branches diverge above the fused part of the filaments whereas in *M. unguiculata* and *M. algoensis* the filaments are united almost to their apices in a relatively thick column. The style branches of *M. decipiens* are relatively short, ± 7 mm long, unlike those in *M. tripetala* which are usually 9–12 mm long. The markings on the outer tepals are also notable, consisting of a wedge- or V-shaped white zone surrounding a dark purple basal mark that is continuous as a wide, longitudinal purple streak on the claw. Plants from Pakhuis Pass, flowering in late October (*Helme 5726*, NBG) bear a superficial resemblance to *M. decipiens* in size of the flowers, colour of the nectar guides and the broad inner tepals but the filaments are united for 2.5 mm and 6.5 mm long, thus longer than the anthers, which are just 4 mm long. More material of this plant is needed before an informed decision can be made about its status but we suspect it represents yet one more undescribed species of subgen. *Vieusseuxia*.

Typical *Moraea tripetala* itself also occurs at higher elevations on the Piketberg, e.g. *Linder 414* BOL, from Levant Mtn where *M. decipiens* also occurs, but flowers some two to four weeks later. Early flowering *M. mutila* of the complex occurs in renosterveld on the lower eastern slopes of the range and has inner tepals broadly similar to those of *M. decipiens*, although narrower and longer. The population at the type locality was relatively large and plants showed no variation in critical features, thus unlikely to be hybrid between *M. tripetala* and *M. mutila*. Barker's collection of *M. decipiens* likewise comprises several plants (*Barker 7563*), all uniform for flower features, and the pollen appears normal under the microscope.

Additional specimens

WESTERN CAPE.—3218 (Clanwilliam): hills NW of Mouton's Vlei, (–DA), Nov. 1934, *Pillans 7489* (BOL); Piketberg, Levant Mtn, Farm Avontuur, (–DA), 5 Nov. 1973, *Linder 81* (BOL); Piketberg, without precise locality, 3 Nov. 1951, *Barker 7563* (NBG).

2. **Moraea helmei** *Goldblatt & J.C.Manning*, sp. nov.

TYPE.—Northern Cape, 3018 (Kamiesberg): southern Kamiesberg south of Farm Karas, near top of Langkloof, 1 125 m, (–AB), 28 Oct. 2011, *Goldblatt & Porter 13688* (NBG, holo.; MO, iso.).

Plants 250–350 mm high. *Corm* 9–12 mm diam. with pale, fibrous tunics. *Stem* simple or with 1 or 2

FIGURE 2.—*Moraea decipiens, Goldblatt & Porter 13709*. A, flowering stems and corm; B, inner tepal variation; C, inner perianth whorl plus filaments and style; D. capsules. Scale bar: A, D, 10 mm; B, C, 2 mm. Artist: John Manning.

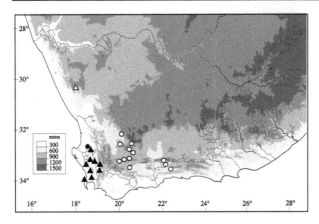

FIGURE 3.—Distribution of *Moraea cuspidata*, ○; *M. decipiens*, ●; *M. helmei*, △; and *M. mutila*, ▲.

short branches, glabrous. *Foliage leaf* solitary, linear, channelled, remaining ± erect, ± 2 mm wide when opened flat; sheathing cauline leaves 2 or 3, 35–45 mm long, green, attenuate, becoming dry at tips. *Rhipidia* 2(3)-flowered; spathes green with dry tips, attenuate, inner 40–45 mm long, outer ± two thirds as long, entirely sheathing. *Flowers* pale blue to violet, outer tepal limbs darkly veined, with narrowly triangular, shortly velvety, yellow nectar guides at base of outer tepal limbs outlined pale blue to dark violet; outer tepals 21.0–23.5 mm long, claw 8–10 mm long, with a shallow raised tooth either side just below apex, limb 12.5–13.5 × 7–8 mm; inner tepals short, ± 4 mm long, 3-forked in upper 1 mm, inner lobe slightly larger than laterals, ± puberulous. *Stamens* with filaments ± 4 mm long, united in lower 1.5 mm, column puberulous; anthers ± 6.5 mm long, dark violet; pollen yellow. *Ovary* narrowly ellipsoid, ± 7 mm long, initially partly included, ultimately fully exserted; style ± 2 mm long, dividing at top of filament column, branches linear, 9 × 1 mm; style crests ± 7 mm long, erect. *Capsules* and *seeds* unknown. Flowering time: late Oct.–mid-Nov. Figure 4.

Distribution and habitat: known only from the Kamiesberg in central Namaqualand (Figure 3), *Moraea helmei* has been collected just twice, from a "somewhat marshy site the top of the Langkloof south of the Farm Karas" and "from a seasonal seep south of Karas at the upper end of the Langkloof". Plants occur at an altitude of about 1 125 m. Conservation status is impossible to assess from the limited data available about the species, but the species is undoubtedly rare.

Diagnosis and relationships: the filaments, united for 1.5 mm and slightly more than one third their length, and the short style, ± 2 mm long, are consistent with the *M. tripetala* complex but the inner tepals of *M. helmei* are short and expanded distally into three short, subequal lobes (vs. linear and usually cusp-like in *M. tripetala*). The puberulous filament column appears to be unique in the complex. The trilobed inner tepals of *M. helmei* recall those of *M. unguiculata*, which has a longer, narrow, coiled central lobe and rounded lateral lobes, but the filaments of this species are united in a prominent column for more than half their length. A peculiarity of the outer tepal claws of *M. helmei* is the pair of shallow tooth-like projections just below the apices, a feature not known elsewhere in *Moraea*: the tepal claws in all other

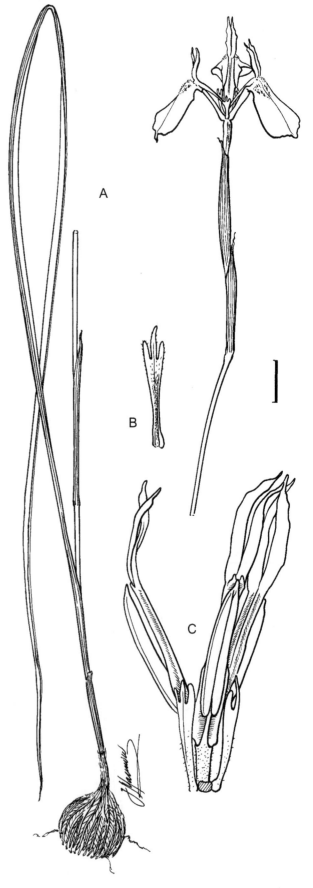

FIGURE 4.—*Moraea helmei*, *Goldblatt & Porter 13688*. A, flowering stem and corm; B, inner tepal; C, inner perianth whorl plus filaments and style. Scale bar: A, 10 mm; B, C, 2 mm. Artist: John Manning.

species are straight or slightly bowed outward and have plane margins.

Additional specimen

NORTHERN CAPE.—**3018** (Kamiesberg): Kamiesberg, ± 1–2 km south of Farm Karas [Welkom], in seep area west of Langkloof road, 1 125 m, (–AC), 28 Oct. 2009, *Helme 3269* (NBG, photo).

3. **Moraea mutila** (*C.H.Bergius ex Eckl.*) Goldblatt & J.C.Manning, comb. nov. *Vieusseuxia mutila* C.H.Bergius ex Eckl.*:* 12 (1827). Type: [Western Cape], vicinity of Camps Bay, *Bergius 15 S7238* (S, lecto.!, designated by Goldblatt: 715 (1976b) [Syntype: [Western Cape], 'Löwenrücken' [Lion's Rump], *Ecklon s.n.* (S, syn.!, two collections made on different dates, also probably *M. mutila*)].

Iris tripetala Thunb.: 13 (1782, Dec.), hom. illegit. et nom superfl. pro *M. tripetala* L.f. (1782, Apr.). Type: South Africa, without precise locality, *Thunberg s.n.* (UPS: Herb. Thunberg, syn.).

M. tripetala var. *mutila* Baker: 23 (1896), as nom. nov. pro *Vieusseuxia mutila* Licht. ex Klatt: 621 (1866), hom. illegit. non Eckl. (1827). Type: [Western Cape], vicinity of Camps Bay, *Bergius 15 S7238* (S, holo.!).

M. punctata Baker: 1003 (1904). Type: [Western Cape], 'Piketberg Road' [Gouda], 17 Aug. 1897, *Schlechter 4851* (K, lecto.!, designated by Goldblatt: 751 (1976); B!, GRA!, PRE!, isolecto.).

M. monophylla Baker: 24 (1906). Type: [Western Cape], 'Olifants River, Clanwilliam', Aug. 1894, *Penther 685* [K, lecto.!, designated by Goldblatt: 751 (1976)].

Plants 15–30 cm high. *Corm* mostly 10–15 mm diam., evidently without cormlets at base, tunics of medium to coarse fibres, vertical members often thickened below into prominent claws. *Stem* simple, rarely 1-branched (1 of 28 plants seen), glabrous or occasionally pilose, sheathed below by brown, fibrous cataphylls, these often accumulating. *Foliage leaf* solitary, linear, shallowly channeled below, often becoming flat and slightly twisted in distal third, ± half to as long as stem, 3–5(–7) mm wide, glabrous or sometimes pilose abaxially or on margins; sheathing cauline leaves 20–40 mm long. *Rhipidia* mostly 2- or 3-flowered; spathes green, inner 35–40(–45) mm long, outer ± half as long as inner, rarely pilose. *Flowers* pale blue or white, nectar guides triangular, white to pale yellow with dark blue dots and often edged with darker color, velvety (or smooth?); outer tepal claws 9–12 mm long, limbs dipping at 45°, ovate, widest in distal third, 14–15 × ± 10 mm, margins recurving later, inner tepals ± linear to slightly expanded in middle, sometimes as rounded lobes, 7.5–12.5 mm long. *Stamens* with filaments 3.0–3.5 mm long, united for 0.5–1.5 mm; anthers 4.5–6.5(–7.5) mm long, pollen red or yellow. *Ovary* narrowly ellipsoid, 7–8 mm long; style 1–2 mm long, style branches 7–9 mm long, crests linear, 8–10 mm long. *Capsules* ellipsoid, 15–17 × ± 5 mm. *Seeds* ± 1.8 × 1.3–1.4 mm, pale straw-coloured, testa spongy, angles narrowed into wings. *Chromosome number* 2*n* = 12 (Table1). *Flowering time*: late Aug.–late Sept., depending on subspecies. Figure 5.

Distribution: largely a species of renosterveld, *Moraea mutila* is centred in the western lowlands of Western Cape, extending from the Cape Peninsula to Piketberg, and extending locally inland to the Tulbagh Valley (Figure 3). A collection said to be from the Olifants River Valley near Clanwilliam made by Arnold Penther is more likely mislabelled and was perhaps gathered en route from Porterville to Clanwilliam. At the very least the Olifants River Valley record requires confirmation. Soils on which *M. mutila* has been collected are described as shale and clay, consistent with renosterveld but probably also on granites, and plants we have seen grew on loamy clay. *M. mutila* is evidently rare on the Cape Peninsula and as far as we have been able to determine was first collected there by C.H. Bergius in 1817, then by Ecklon and by Zeyher in the 1820s, and later by Friedrich Wilms, Rudolf Marloth, and Harry Bolus but there are no recent records. Nevertheless, it probably persists on clay soils, its preferred habitat, on the lower slopes of Lions Head and Signal Hill above Cape Town.

Possibly belonging here are pale blue-flowered plants from the western lower slopes of Sir Lowry's Pass (*Loubser 872* NBG, *Goldblatt 2506* MO) with the inner tepals trilobed as in *M. mutila* but the filaments are united for ± 2.5 mm and the style is 4 mm long. Alternatively these plants may be hybrids between *M. tripetala* and *M. unguiculata*.

Diagnosis: early blooming *Moraea mutila* (the name means mutilated, presumably due to the reduced inner tepals) is recognized by the pale blue, rarely white flowers with linear to narrowly lanceolate inner tepals 7.5–12.5 mm long, erect below and spreading distally. Often difficult to see in preserved specimens, the inner tepals are usually slightly wider in the midline and sometimes expanded into rounded lobes near the base of the spreading distal half. The flowers otherwise seem little different from those of *M. tripetala* but the pilose nectar guides are usually white or palest yellow and dotted with dark blue to purple, whereas the nectar guides of *M. tripetala* are more often yellow, edged with dark blue

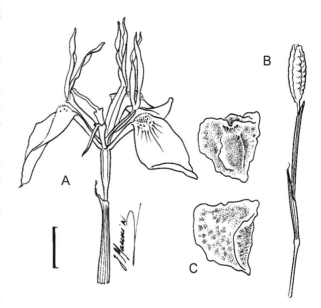

FIGURE 5.—*Moraea mutila, Goldblatt 2310*. A, flower; B, capsule; C, seeds. Scale bar: A, B, 10 mm; C, 2 mm. Artist: John Manning.

or purple. The leaf is typically relatively short, in most collections shorter than the stem, and widely channelled, sometimes plane in the distal half. Several specimens of two collections (*Acocks 1912* and *Nordenstam & Lundgren 1998*) have leaves velvety on the abaxial surface and velvety spathes. The few collections from the Cape Peninsula also have partly hairy leaves, either on the abaxial surface or on the margins. The pubescence is evidently variable in the populations that show the character, thus of limited value for identification, although we have not seen any collections of true *M. tripetala* with pubescence of any kind on the vegetative parts.

Taxonomic history: the earliest record of *Moraea mutila* is the plant described in detail by C.P. Thunberg (Dec. 1782) under the name *Iris tripetala*. Thunberg's plant had smooth, channelled leaves, a blue flower with yellowish nectar guides, and notably, inner tepals with very narrow claws, convex abaxially, and patent limbs at the base of which were two opposed lobes (*dentibus duobus*). Thunberg's *M. tripetala* has, however, no nomenclatural status or is at best superfluous as he cited *I. tripetala* L.f. (April 1782) in his account. The latter, described incompletely but said to have subulate (awl-shaped or needle-like) inner tepals, is the basionym for *M. tripetala*. Thunberg actually collected multiple specimens of the *M. tripetala* complex from at least three sites but clearly had *M. mutila* in mind in his extended description of *I. tripetala*. Since his description conflicts with that of Linnaeus fil. we treat Thunberg's *I. tripetala* as a separate, but superfluous name.

Moraea mutila appears again as *Vieusseuxia mutila* C.H.Bergius ex Eckl. (Ecklon 1827), published with the brief diagnosis, '*Blumen blassblau*' (flowers pale blue), with citations of a C.H.Bergius collection from Camp's Bay (August 1817), and two Ecklon collections from Lion's Head [as Löwenrücken], Cape Town, one collected in August (year not stated), and the other in September 1826. Nordenstam (1972) accepted the name as valid despite the brief diagnosis that mentions neither the leaf vestiture nor the form of the inner tepals, features critical to identifcation of the species. The epithet 'mutila' was taken from the manuscript name on Bergius's collection, now in the Stockholm Herbarium and annotated '*Iris nov. species mutila mihi*' in Bergius's hand. We wonder whether Ecklon actually saw the Bergius specimen, which was probably already in Berlin when he arrived in Cape Town in 1823 but, nevertheless, cited it. Bergius, incidentally, almost certainly took the epithet 'mutila' from Martin Lichtenstein's manuscript *Spicilegium flora capensis*, which he would have seen when he was in Berlin under Lichtenstein's patronage at Berlin before he departed for the Cape (Gunn & Codd 1981). The Bergius collection was designated the lectotype of *V. mutila* (Goldblatt 1976b), thus maintaining the link between Bergius's epithet and specimen.

Klatt's *Vieusseuxia mutila*, a homonym for Ecklon's *V. mutila*, explicitly excludes the specimens on a sheet then at the Berlin Herbarium (now at Stockholm), collected by C.H. Bergius that has smooth vegetative parts (referring them to *Moraea tripetaloides* DC. = *M. tripetala*). Other specimens on that sheet with hairy leaf margins, a pilose stem and linear inner tepals are the type of the name and are *M. mutila*. Baker (1896) in *Flora*

capensis recognized one variety of his *M. tripetala*, var. *mutila*, with 'leaf pilose' [the entire diagnosis] and references to both *Vieusseuxia mutila* Klatt and *Iris mutila* Licht. ex Roem. & Schult. The latter, described as glabrous on all parts according to the protologue (Roemer & Schultes 1817) (there is no extant type material), is in conflict with Baker's variety. Both name and description of *Iris mutila* were taken directly from Martin Lichtenstein's manuscript, *Spicilegium flora capensis*. Klatt's species, however, matches Baker's var. *mutila*, but as it is illegitimate, we treat Baker's variety as an inadvertent new name but based on the same type as *V. mutila* Klatt. Baker's variety is *M. mutila*, and identical with a second species, *M. punctata*, which he described in 1904, from smooth-leaved specimens collected by Rudolph Schlechter. Baker (1906) clearly did not realize that his *M. monophylla* was a close match for *M. punctata*, which now falls into synonymy under *M. mutila*.

Representative specimens

WESTERN CAPE.—**3218** (Clanwilliam): flats N of Piketberg at Eendekuil turnoff, renosterveld, (–DB), 1 Sept. 1981, *Goldblatt 6127* (K, MO, NBG, PRE); Piketberg, Jobskloof, (–DB), 11 Aug. 1973, *Linder 635* (BOL). **3318** (Cape Town): Dassenberg N of Mamre, W slopes, (–AD), 23 Sept. 1974, *Nordenstam & Lundgren 1998* (MO, S); near Moorreesburg, Klein Swartfontein, renosterveld on shale, (–BA), 25 Aug. 1970, *Acocks 24320* (K, MO, PRE); Porterville, flats, (–BB), 10 August 1959, *Loubser 933* (NBG); slopes of Lions Rump near Tamboerskloof, (–CD), Aug., *Zeyher 5010* (SAM); Rosebank, Cape Town, (–CD), Aug. 1877, *H. Bolus 3732* (BOL); Signal Hill, 100 m, (–CD), Aug. 1894, *Marloth 195* (PRE), 26 Aug. 1883, *Wilms 3670* (K); 29 miles [45 km] from Cape Town to Malmesbury, (–DA), 29 Aug. 1932, *Lewis s.n.* (BOL45082); between Salt River and Kalabas Kraal, (–DA), 4 Sept., *Hutchinson 161a* (K, PRE); damp flats N of Bottelary road, (–DC), 26 Aug. 1933, *Acocks 1912* (S); Stellenbosch, (–DD), 1865, *Sanderson 959* (K). **3319** (Worcester): flats S of Tulbagh Road Station, (–AC), 11 Aug. 1974, *Goldblatt 2310* (MO); Voëlvlei Tortoise Reserve, (–AC), 10 Aug. 1989, *Solomon 16* (NBG); Wolseley, Farm Romansrivier, burned April 2011, (–AC), 11 Aug. 2011, *Boucher 7670* (MO, NBG); near Wellington, (–CC), 15 August 1926, *Grant 2361* (MO).

4. **Moraea cuspidata** Goldblatt & J.C.Manning, sp. nov.

TYPE.—Western Cape, 3220 (Sutherland): near Farm Fortuin, north of Matjiesfontein, near shaded rest stop (co-blooming with *M. amabilis*), (–DC), 28 Sept. 2009, *Goldblatt & Porter 13462* (NBG, holo.; MO, iso.).

Plants solitary, 180–300 mm high. *Corm* 15–25 mm diam., without cormlets at base, tunics of brown to almost black, tough, medium to coarse fibres. *Stem* smooth, simple or 1(2)-branched exceptionally with up to 6 branches, with cataphylls of medium textured, light brown, mostly vertical fibres forming a collar around base, sometimes accumulating in a dense mass. *Foliage leaf* solitary, linear(–linear-filiform), channelled, 1.5–3.0 mm wide, glabrous, V-shaped in section, leaf halves closely appressed in dry conditions; sheathing cauline leaves (30–)40–55 mm long, with dry attenuate tips. *Rhipidia* several-flowered; spathes green with attenuate, dry tips, inner (40–)50–80 mm long, outer ± half as long as inner, glabrous. *Flowers* pale blue, mauve or violet blue, outer tepal limbs lanceolate, with large white, velvety nectar guides dotted and usually edged with dark violet; outer tepal claws (8–)10–12 mm, usually speckled with dark blue to violet dots or purple in midline, limbs (12–)16–22 × 10–16 mm, inner tepals linear,

(8–)10–15 mm long, ascending below, distal third to half usually spreading horizontally (or ± erect when short), often pilose at base. *Stamens* with filaments (3–)4–6 mm long, united basally for 0.5–1.8 mm; anthers 4–8 mm long, pollen red (rarely white). *Ovary* exserted, 11–15 mm long; style vestigial, branches 8–10 mm long; crests 5–8 mm long, linear, arching inward. *Capsules* ovoid-ellipsoid, 14–16 mm long. *Seeds* 1.8–2.0 × 1.3–1.4 mm, pale straw-coloured, testa spongy, thicker on angles. *Chromosome number* 2n = 12 (Table 1) *Flowering time*: mid-Sept. and Oct. Figure 6.

Distribution: a species of semi-arid habitats, *Moraea cuspidata* grows in both mountain renosterveld and dry,

marginal fynbos, usually in sandy or sandy loam soils derived from sandstones of the Cape System or from the Beaufort Series of the Karoo System. It flowers late in the season—the earliest record is in mid-September but mid- to late October is usual. It is relatively poorly collected, not (we suspect) because it is rare, but because it grows in areas not much botanized at that time of year. In the late spring of 2009 we found the species at several sites we visited between Touw's River and Komsberg Pass in the Klein Roggeveld. The recorded range (Figure 3) extends from the Bonteberg and Voetpadsberg near Touw's River through the Klein Roggeveld and southern Roggeveld to the Swartberg and higher mountains of the

FIGURE 6.—*Moraea cuspidata, Goldblatt & Porter 13462*. A, flowering stems and corm; B, inner perianth whorl plus filaments and style. Scale bar: A, 10 mm; B, 2 mm. Artist: John Manning.

Little Karoo. An outlying record from Perdekloof, south-east of Oudtshoorn near Camfer (*Goldblatt & Porter 12575*), probably belongs here, although the flowers are exceptionally small, the inner tepals are 8–9 mm long, consistent with *M. cuspidata*. The apparent gap in the range between the Roggeveld and the Swartberg may be an artifact due to incomplete collecting.

Diagnosis: recognized immediately by the linear inner tepals up to 15 mm long (rarely less than 10 mm), spreading in the distal half (Figure 6B), *Moraea cuspidata* (named for the long, cusp-like inner tepals) also usually has a particularly narrow foliage leaf, the leaf halves tightly appressed in dry conditions, and brown, sometimes almost black corm tunics, the vertical fibres of which are usually heavily thickened. In addition, the cataphylls tend to persist as a collar of stiff fibres around the base (not always present in herbarium material). The flowers are pale blue to violet with the large, velvety nectar guides spotted with dark blue on a white background. The filaments are usually united for 1.0–1.5 mm but occasionally less, and ± as long as or slightly shorter than the dark purple anthers that bear red pollen. We have seen only two collections with ripe capsules and these are relatively large, 14–16 mm long. The seeds of these two collections are also relatively large for the complex, 1.8–2.0 mm long and unique in having a pale, spongy testa much thickened into prominent ridges on the angles.

The species is sympatric at some sites and even co-blooming with the smaller-flowered *Moraea amabilis* (also of the *M. tripetala* complex), but this species has a broadly channeled foliage leaf, sometimes plane distally and often pilose on the abaxial surface, and consistently short, hair-like inner tepals typically 1.5–2.0 mm long. At sites where we have seen the two growing together, anthers of *M. amabilis* have white pollen, contrasting with the orange-red pollen of *M. cuspidata*. The range of *M. cuspidata* also coincides in part with *M. tripetala* subsp. *violacea*, which extends from Gydo Pass to the Hex River Valley and Touw's River, but we have not seen them growing near one another.

Representative specimens

NORTHERN CAPE.—**3220** (Sutherland): Roggeveld, Soekop, Huis Kamp, (–AA), 16 Sept. 2006, *Rosch 556* (NBG); Koedoes Pass, W slopes, (–CC), 22 Sept. 1970, *Goldblatt 547* (BOL); Klein Roggeveld, Smousrand, (–DC), Sept. 2009, *Goldblatt & Porter 13459* (MO, NBG); Klein Roggeveld, slopes near Farm Meintjiesplaas, (–DC), 28 Sept. 2009, *Goldblatt & Porter 13456A* (MO, NBG).

WESTERN CAPE.—**3319** (Worcester): Bonteberg, Farm Karrona, (–BD), 15 Sept. 1971, *Thompson 1257* (NBG); **3320** (Montagu): Bonteberg, flats at N entrance to Pienaarspoort, (–AA), 27 Sept. 2009, *Goldblatt & Porter 13453* (MO, NBG, PRE); Cabidu Karoo, Laingsburg, (–AA), 28 Sept. 1951, *Barker 7503* (NBG); Tweedside, (–AB), ex Hort. Kirstenbosch, *Barker 7466* (NBG), Sept. 1932, *Lewis s.n. Nat. Bot. Gard 2708/32* (BOL); Matjiesgoed Mtns, Farm Klein Spreeufontein, ± 990 m, (–BC), 16 Sept. 1987, *Bean, Vlok & Viviers 1938* (MO, PRE). **3322** (Oudtshoorn): Swartberg Pass, rocky slopes near picnic site, (–AC), 23 Oct. 1986, *Goldblatt 8003* (MO, PRE); top of Swartberg Pass, (–AC), Nov. 1896, *H. Bolus s.n.* (BOL); Cango Caves, Oudtshoorn, (–AC), 16 Oct. 1928, *Gillett 1695* (NBG); Perdepoort N of Camfer, sandstone slopes, (–CD), 28 Sept. 2004, *Goldblatt & Porter 12575* (MO).

5. **Moraea tripetala** *(L.f.) Ker Gawl.* in Curtis's Botanical Magazine 19: t. 702 (1803). *Iris tripetala* L.f.: 92 (1782).

Vieusseuxia tripetaloides DC.: 138 (1803), nom. illegit. superfl. pro *I. tripetala* L.f. (1782). *Vieusseuxia tripetala* (L.f.) Voigt: 602 (1845). *Vieusseuxia tripetala* (L.f.) Klatt: 155 (1894), hom. illegit., non (L.f.) Voigt (1845). Type: [Western Cape], without precise locality, *Thunberg s.n. Herb. Thunb. 1186* [UPS-THUNB, lecto.! designated by Goldblatt: 175 (1976)].

Plants (140–)250–450 mm high, glabrous on all vegetative parts. *Corm* mostly 8–15 mm diam., without cormlets at base, tunics of moderately coarse, wiry, usually dark grey fibres. *Stem* simple or 1- or 2(–6)-branched, sheathed below by brown cataphylls and sometimes with collar of brown fibres. *Foliage leaf* solitary (rarely 2 in subsp. *violacea*), leathery, narrowly channeled, C- or V-shaped in section, to 5 mm wide, usually exceeding stem and distally trailing; sheathing cauline leaves mostly 45–60 mm long. *Rhipidia* 3–several-flowered; spathes green with dry attenuate tips, inner 35–65(–70) mm long, outer ± half as long as inner. *Flowers* pale blue, purple, or violet, nectar guides triangular, yellow edged with violet, or white dotted with dark blue and edged with darker blue, lightly honey scented, unscented or strongly scented of carnation (subsp. *violacea*); outer tepals 20–30(–32) mm long, claws 9–12 mm long, limbs obovate, widest in distal third, 11–18 × ± 8–13 mm, plane, spreading at 45°, margins eventually recurved, inner tepals hair-like to ± linear, 1–4(–6) mm long, or lacking. *Stamens* with filaments 4–6 mm long, united for 0.5–1.5 mm, occasionally (southern populations) ± free; anthers 4–7 long, pollen usually orange-red, rarely white. *Ovary* 4–9 mm long; style 1–2 mm long, style branches 9–12 mm long, crests ± linear, 5–12 mm long. *Capsules* ellipsoid, 8–14 mm long. *Seeds* 1.0–1.3 mm × ± 1 mm, flat surfaces slightly wrinkled, brown seed body visible through testa, angles forming raised, golden-brown ridges. *Chromosome number* 2n = 12 (only subspp. *tripetala* and *violacea* counted) (Table 1). *Flowering time*: mainly Aug.–mid-Sept., occasionally in Oct. Figures 7, 8.

Distribution: widespread in Western Cape, *Moraea tripetala* extends from Aurora on the west coast to near Knysna in the east, and inland to Ceres and the Little Karoo (Figure 9). Habitats vary from limestone flats in fynbos, to neutral and acid sands in sandveld, to loamy alluvium in coastal and montane fynbos. Interior populations from the Warm Bokkeveld (Ceres) and western Little Karoo, here segregated as subsp. *violacea*, occur in significantly drier habitats on clay soils in renosterveld. Populations segregated as subsp. *jacquiniana* are mostly montane and occur above 500 m but at lower elevations in the southern Cape Peninsula, and flower significantly later in the year. Although much of its lowland habitat has been lost to agriculture and its original range significantly reduced, *M. tripetala* still occurs extensively in undisturbed low to middle elevation sites. Interior populations have lost little of their original habitat except immediately around Ceres where orchards have replaced much of the original vegetation.

Diagnosis: in our revised, narrower circumscription, *Moraea tripetala* includes plants with reduced hair-like to linear inner tepals mostly 2–4, rarely up to 6 mm long, and filaments united for up to 1.5 mm and occasionally free. The outer tepal limbs are broadly ovate,

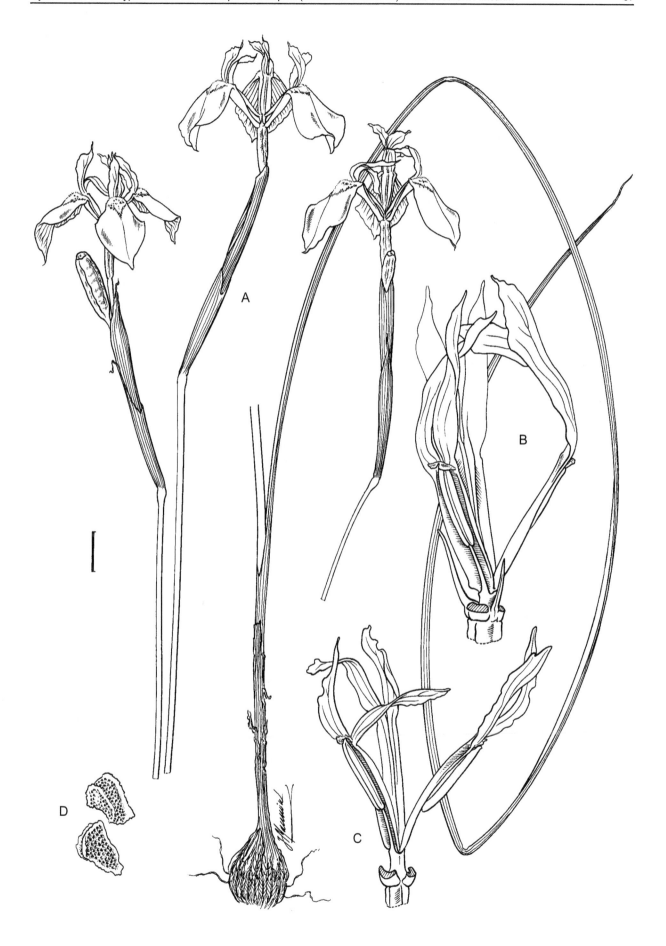

FIGURE 7.—*Moraea tripetala* subsp. *tripetala*, A, B, D, *Goldblatt & Porter 13467* (MO); C, *Goldblatt & Porter 12374*. A, flowering stems and corm; B, C, inner perianth whorl plus filaments and style; D, seeds. Scale bar: A, 10 mm; B–D, 2 mm. Artist: John Manning.

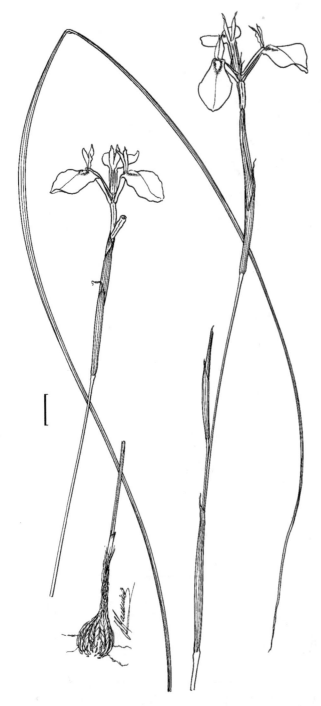

FIGURE 8.—*Moraea tripetala* subsp. *jacquiniana*, Red Hill, no voucher. Scale bar: 10 mm. Artist: John Manning.

longer than wide and oriented ± 45° below horizontal. The corms have dark brown to grey tunics, usually composed of medium-textured fibres, and do not produce cormlets at the base. The linear leaves, normally exceeding the stem, are narrowly channeled, and in dry conditions the leaf halves lie parallel to one another. As far as we can determine, vegetative reproduction by production of cormlets does not occur in *M. tripetala* as now circumscribed. Flower colour ranges from pale to deep blue or violet or rarely light purple, sometimes even in the same population. Nectar guides are often yellow edged with dark blue or purple, but occasionally white, then dotted with blue to purple. The seeds are small, and typical of the complex in being 5(6)-sided with the fac-

ets slightly wrinkled and showing the brown color of the seed body (Figure 7D). The angles are raised into narrow wings of pale straw colour. In contrast, *M. amabilis*, *M. cuspidate*, and *M. grandis* have larger seeds of more complex structure.

Variation: there are a number of notable variants among the plants we include in *Moraea tripetala*, one of the most important of which includes populations from the interior Western Cape on clays and shales from Ceres to the western Little Karoo. These plants have dark violet outer tepals with yellow nectar guides, are usually relatively short in stature, and have short stiff, hair-like inner tepals, mostly ± 2 mm long. The flowers we have examined alive have a strong clove or carnation-like scent. Unusually for the complex, and for subg. *Vieusseuxia* as a whole, some populations of this morph have two foliage leaves (e.g. *Dymond s.n.* from near Ouberg Pass; *Goldblatt 4184, 11438* from Op-de-Tradouw near Barrydale; *Van Wyk 61* from the Voetpadsberg; *Goldblatt & Snijman 6976A* from Worcester). We recognize these populations of the interior Western Cape as subsp. *violacea*.

Elsewhere, *Moraea tripetala* has pale or deep blue, or sometimes purple outer tepals, either with yellow or white nectar guides, in the latter case usually speckled with dark blue or purple, and flowers are faintly sweet-scented or odourless. It is also relatively tall (allowing for reduced stature in years of poor rainfall) and usually has hair-like to linear inner tepals 2–4(–6) mm long (occasionally the inner tepals are absent in southern Cape populations). Among these populations, those that flower from November to January on the Cape Peninsula and surrounding mountains stand out in their softly fibrous corm tunics, slightly smaller flowers, and usually very narrow leaves. The flowers also often have inner tepals 4–6 mm long. Typical *M. tripetala* in the same area flowers from August to early October. We think it useful to recognize these two sets of populations taxonomically but favour subspecies rank because the morphological differences between them and typical *M. tripetala* are small and not absolute. The taxon was treated by Lewis (1941, 1950) as var. *jacquiniana*.

Variation elsewhere across the wide range of *Moraea tripetala* seems to us less taxonomically significant and less consistent. We discuss this under subsp. *tripetala*.

History: although the type of *Moraea tripetala* was collected by C.P. Thunberg, the basionym *Iris tripetala* was described by Linnaeus fil. (April, 1782), several months before Thunberg (Dec. 1782) published his own, much more extensive description. Goldblatt (1976b) selected as lectotype a specimen that had short, hair-like inner tepals and a long, narrow, channelled leaf, thus fixing the application of the name. Thunberg's plant is obviously a different species, and has the distinctive inner tepals of what is now *M. mutila* (see extended discussion under that species). The reduced inner tepals rendered *M. tripetala* so distinctive that it acquired few synonyms. De Candolle (1804), when transferring *I. tripetala* to *Vieusseuxia*, the genus then used for southern African species originally treated as *Iris*, renamed it *V. tripetaloides*, a morphologically more apt epithet. The name is superfluous as De Candolle cited *I. tripetala* L.f.

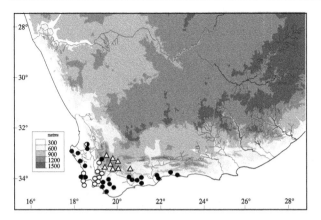

FIGURE 9.—Distribution of *Moraea tripetala* subsp. *tripetala*, •; *M. tripetala* subsp. *violacea*, △; *M. tripetala* subsp. *jacquiniana*, ○.

and *I. tripetala* Thunb. in synonymy. *I. tripetala* was transferred to *Moraea* in 1803 by Ker Gawler but the confusion over the circumscription of *Moraea* persisted and Voigt (1847) and then Klatt (1866) independently transferred *I. tripetala* to *Vieusseuxia*.

Key to subspecies

1a Plants mostly of montane habitats in rocky sandstone soils, flowering mainly November to January, occasionally in October; ovary 4–6 mm long; inner tepals often 5–6 mm long, reaching bases of anthers; tepals mostly dark violet
.... 5c. subsp. *jacquiniana*
1b Plants of clay, granitic or sometimes rocky limestone or sandstone soils, flowering mainly August to early October; ovary 6–9 mm long; inner tepals 1–3(–5) mm long (or absent), rarely reaching base of anthers; tepals pale to dark blue to violet:
  2a Ovary 6–8 mm long; outer tepals 24–26 mm long and inner tepals 1–2 mm long; tepals violet with yellow nectar guides; foliage leaves 1 or 2; plants of clay soils in renosterveld in the Warm Bokkeveld, Bonteberg, and Little Karoo .... 5b. subsp. *violacea*
  2b Ovary 8–9 mm long; outer tepals 23–30(–32) mm long; and inner tepals absent or 1–3(–5) mm long; tepals pale blue or deep blue or purple usually with white (occasionally yellow) nectar guides; foliage leaf always solitary; plants of sandy flats and slopes, sometimes in limestone, extending from Aurora to Knysna .... 5a. subsp. *tripetala*

5a. subsp. **tripetala**

?*Iris mutila* Licht. ex Roem. & Schult.: 447 (1817). *Vieusseuxia mutila* (Licht. ex Roem. & Schult.) A.Dietr.: 494 (1833), hom. illeg. non Eckl. (1827). Type: unknown, assigned here based on description alone.

Plants 250–450 mm high, stem sometimes sheathed below by collar of brown fibres. *Corm* mostly 12–15 mm diam. *Foliage leaf* always solitary, linear. *Inner spathes* 45–65(–70) mm long. *Flowers* pale blue or purple with yellow or white nectar guides, faintly honey-scented or unscented; outer tepals 23–30(–32) mm long, claws 10–12 mm long, limbs 12–18 mm long,; inner tepals usually hair-like, 1–3(–5) mm long, or absent. *Stamen filaments* 4–6 mm long, united for 1.0–1.5 mm, occasionally (southern populations) ± free; anthers 4.5–7.0 long; pollen usually red or white. *Ovary* 8–9 mm long; style branches ± 12 mm long, crests mostly 8–12 mm. *Capsules* 12–14 mm long. *Flowering time*: late Aug.–mid-Sept., occasionally in Oct. (exceptionally July, e.g. *Wurts 212*). Figure 7.

Distribution: largely a plant of coastal forelands, subsp. *tripetala* extends from Aurora on the west coast of Western Cape to near Knysna in the east. It also occurs on lower mountain slopes, occasionally up to ± 700 m, and has been recorded on the Piketberg, the Hottentots Holland–Jonkershoek mountain complex, the Witzenberg at Ceres, and the lower southern slopes of the Langeberg (Figure 9). Populations along the southern Cape, notably in the De Hoop area, the Agulhas Peninsula (*Goldblatt 2607*), and some from near Caledon (*Goldblatt 6196*) occasionally lack inner tepals, or individual flowers may lack one or two of the inner tepals (e.g. *Goldblatt & Manning 12248*, east of Swellendam). Plants from the southern foothills and lower slopes of the Langeberg sometimes have free filaments (e.g. *Thorne s.n.* SAM, from Garcia's Pass; *Goldblatt 3741* from near Suurbraak). Subsp. *tripetala* is usually found in sandy habitats or on coastal limestone, thus in sandveld, strandveld and fynbos vegetation.

Representative specimens

WESTERN CAPE.—**3218** (Clanwilliam): Piketberg, plateau on Kapteins Kloof Mtn, (–DA), 21 Oct. 1935, *Pillans 7796* (BOL); Piketberg, plateau E of Levant Mt., edge of dam, (–DD), 20 September 1974, *Linder 414* (BOL). **3318** (Cape Town): R27 to Langebaanweg, sandveld, (–AC), 29 Sept. 2009, *Goldblatt & Porter 13467* (MO); Dassenberg, N of Mamre, (–AD), 23 Sept. 1974, *Nordenstam & Lundgren 1999* (MO, S); near Mamre Road Station, sandy waterlogged ground, (–BC), 3 Sept. 1974, *Goldblatt 2487* (MO); sandy flats near Wynberg, (–CD), Sept. 1884, *Macowan 215* (BOL); Camps Bay, (–CD), Sept. 1886, *Thode 8531* (NBG). **3319** (Worcester): Ceres Nature Reserve, top of Michell's Pass, rocky sandstone slopes, (–CD), 29 Sept. 2009, *Goldblatt & Porter 13463* (MO, NBG). **3320** (Montagu): Swellendam, below Langeberg range at Clock Peaks, (–CD), 2 July 1952, *Wurts 212* (NBG). **3419** (Caledon): Caledon Swartberg and the Baths, (–AB), Aug., *Ecklon & Zeyher s.n.* (MO); 18.4 km SW of Greyton, shale ground in renosterveld, (–BA), 28 August 1970, *Acocks 24337* (K, MO, PRE); Rotary Drive, Hermanus, (–AC), 10 October 1974, *Goldblatt 3009* (MO, PRE); hill above, Pearly Beach, (–CB), 11 Sept. 1974, *Goldblatt 2607* (K, MO, PRE). **3420** (Bredasdorp): Suurbraak to Heidelberg, W of Strawberry Hill, (–BA), *Goldblatt 3741* (MO); clay bank E of Swellendam, (–BB), 7 Sept. 2003, *Goldblatt & Manning 12248* (MO, NBG); 10 km S of Bredasdorp, entrance to Die Poort, limestone, (–CA), 13 Sept. 1978, *Goldblatt 4860* (MO). **3421** (Riversdale): Riversdale, Garcias Pass, (–AA), Oct. 1926, in *Thorne s.n.* (SAM 43214); 2 km from Herbertsdale to Cloete's Pass, (–BA), 26 Sept. 2003, *Goldblatt & Porter 12374* (MO). **3422** (Mossel Bay): Pacaltsdorp, George, (–AB), 22 Aug. 1978, *Moriarty 339* (MO); Goukamma Nature Reserve, (–BB), 1968, *Heinecken 206* (PRE).

5b. subsp. **violacea** Goldblatt & J.C.Manning, subsp. nov.

TYPE.—Western Cape, 3320 (Montagu): Op-de-Tradouw, Barrydale to Montagu, south facing clay slope in renosterveld, (–DC), 26 Aug. 2000, *Goldblatt 11438* (NBG, holo.; K, MO, PRE, iso.).

Plants mostly 200–300 mm high, stem sometimes with collar of fine fibres around base. *Corm* mostly 8–10 mm diam., tunics of medium-textured, usually pale fibres, sometimes thickened below. *Foliage leaf* 1 or 2, linear, mostly 1.5–2.5 mm wide, leaf halves often folded together and blade appearing terete. *Inner spathes* 35–45(–50) mm long. *Flowers* violet with yellow, velvety nectar guides, strongly scented of cloves (?always); outer tepals 24–26 mm long, claws ± 10 mm long, limbs 14–16 mm long, inner tepals hair-like, 1–2(3) mm long. *Stamen filaments* 4.0–6.5 mm long, united for 0.5–1.2 mm; anthers 4.0–6.5 mm long, pale violet, pollen white.

Ovary 6–8 mm long; style branches ± 12 mm long, crests 6–8(10) mm long. *Capsules* 8–11 mm long. *Flowering time*: late Aug.–late Oct.

Distribution: subsp. *violacea* extends from Gydo Pass and the Warm Bokkeveld though the southern foothills of the Hex River Mtns near Worcester inland to the Bonteberg, Touws River and the western Little Karoo near Barrydale (Figure 9). Populations with two foliage leaves appear to occur randomly across its range. Plants almost invariably occur on clay soils in renosterveld but we have also seen the species growing among low succulent shrubs.

Representative specimens

WESTERN CAPE.–**3319** (Worcester): Gydo Pass, (–AB), *Barker 6836* (NBG); Hottentots Kloof, (–BA), 28 Sept. 1944, *Barker 3031* (NBG); E slopes of Theronsberg Pass, (–BC), 31 Oct. 1974, *Goldblatt 3228* (MO); Ceres, Lakenvlei, (–BC), 10 Oct. 1941, *Barker 1336* (NBG); 2 km east of De Doorns, Hex River Valley, (–BC), 27 Sept. 1974, *Nordenstam & Lundgren 2056* (MO, S); ± 12 km N of Worcester at Brandwagt turnoff, Worcester West, (–CB), 26 Sept. 1983 (fr.), ex hort. Missouri, Mar. 1986, *Goldblatt & Snijman 6976A* (MO); Doringkloof, Voetpadsberg, (–DB), 24 Aug. 1985, *Van Wyk 61* (NBG); Koo, foot of Naudesberg, (–DD), 12 Sept. 1962, *Lewis 6054* (NBG). **3320** (Montagu): Ouberg to Touws River, (–CB), 29 Sept. 1932, *Dymond s.n.* (BOL21234); Keurkloof, off Cogman's Kloof, (–CC), 24 Sept. 1935, *Lewis s.n.* (BOL); Op-de-Tradouw, Barrydale to Montagu, S facing clay slope in renosterveld, (–DC), 23 Sept. 1976, *Goldblatt 4184* (MO), 12 Sept. 1994, *Goldblatt & Manning 9990* (K, MO, NBG).

5c. subsp. **jacquiniana** *(Schltr. ex G.J.Lewis)* Goldblatt & J.C.Manning, stat. et comb. nov. *Moraea tripetala* var. *jacquiniana* Schltr. ex G.J.Lewis: 54 (1941). Type: [Western Cape], Constantiaberg, Dec. [without year], *Wolley Dod 1919* (BOL, lecto.!, designated by Goldblatt: 751(1976b); K!, iso.).

Vieusseuxia pulchra Eckl.: 13 (1827). Type: [Western Cape], 'Hottentotshollandkloof', 25 Nov. [without year], *Ecklon & Zeyher s.n.* (S, holo.!).

Plants 140–300 mm high, stem usually sheathed below by collar of fine, loosely netted fibres. *Corms* ± 10 mm diam., tunics of relatively fine fibres, not usually accumulating. *Foliage leaf* solitary, narrowly channelled or leaf halves appressed together thus appearing terete, exceeding stem and up to twice as long, 2–3(–5) mm wide (when dry often apparently ± 1.5 mm diam.). *Inner spathes* 40–50 mm long. *Flowers* usually purple or blue with triangular, white nectar guides spotted with dark blue and edged darker color, unscented, outer tepals 20–25 mm long, claws 9–11 mm long, limbs lanceolate, 11–16 × 8–11 mm, inner tepals hair-like, mostly 5–6 mm long, reaching or exceeding anther bases. *Stamen filaments* 4–6 mm long, united for 1.0–1.5 mm; anthers 5–6 long, pollen usually (always?) red. *Ovary* 4–6 mm long; style branches 10–11 × 2 mm long, crests linear, 5–8 mm long. *Capsules* mostly 10–14 mm long. *Flowering time*: mid-Nov.–Jan., rarely in late Oct. Figure 8.

Distribution: occurring entirely within the range of subsp. *tripetala*, subsp. *jacquiniana* extends from the Cape Peninsula to the Hottentots Holland and Houw Hoek Mtns and north to the mountains around Franschhoek, Paarl, and Wellington (Figure 9). There are also isolated records from the Piketberg, the mountains near Ceres, and in the vicinity of Citrusdal (*Ecklon & Zey-*

her Irid 20), this last in need of confirmation. Although largely montane, subsp. *jacquiniana* occurs at quite low elevations in the southern Cape Peninsula. Plants grow in fynbos on stony sandstone slopes. Flowering is very late in the season, mostly November and December, but plants have been collected in January and sometimes in late October. Subsp. *tripetala* flowers in August and September, usually at lower elevations, sometimes in similar habitats.

Diagnosis: subsp. *jacquiniana* is poorly differentiated from subsp. *tripetala* but apart from the later flowering time it can be distinguished by the finely fibrous corm tunics, shorter rhipidial spathes 40–50 mm long, and slightly smaller flowers, usually deep violet, consistently with inner tepals 4–6 mm long (usually shorter or absent in subsp. *tripetala*). The outer tepals are typically 20–25 mm long, the ovary 4–6 mm long and capsules, 10–14 mm long, all smaller than in subsp. *tripetala*. As far as is recorded, the violet flowers consistently have white nectar guides whereas *M. tripetala* has yellow or less often white nectar guides, then speckled with blue dots. The colour illustrations of flowers of the two taxa can readily be compared in *Wild Flowers of the Cape Peninsula* (Maytham Kidd 1950).

The taxon was named in honor of N.J. Jacquin, whose monumental volumes of coloured illustrations of plants, many of them from the Western Cape and Namaqualand, were an inspiration and major resource for accurate plant depictions, including his painting of three variants of *Moraea tripetala* (*Icones plantarum rariorum* vol. 2, t. 211), one of which perhaps was thought by Schlechter to represent what is now subsp. *jacquiniana*. The epithet was first used, without description, by Bolus & Wolley Dod (1903) in a checklist of plants of the Cape Peninsula as if validly published by Schlechter. The epithet was taken from the collection labels of *Schlechter 7222* (collected in 1896) and was only validated at varietal rank by G.J. Lewis in 1941. A collection from 'Hottentots Holland Kloof' was named *Vieusseuxia pulchra* by Ecklon in 1827. Ecklon's descriptive phrase, 'flowers darker blue' [than *M. setacea* the name Ecklon used for *V. tripetala*] barely qualifies as a diagnosis but Nordenstam (1972) regarded the name as validly published.

Representative specimens

WESTERN CAPE.—**3218** (Clanwilliam): along the Olifants River, Clanwilliam to Citrusdal, (–?BB), Nov., *Ecklon & Zeyher Irid 20* (MO); Piketberg, plateau on Kapteins Kloof Mtn, (–DA), 21 Oct. 1935, *Pillans 7796* (BOL). **3318** (Cape Town): Cape Peninsula, Table Mt., (–AD), Dec. 1950, *Pillans 10279* (BOL, MO); Table Mt., lower plateau, Nov. 1944, *Lewis 946* (SAM). **3319** (Worcester): Michell's Peak, Ceres, (–AD), 16 Dec. 1948, *Esterhuysen 14767* (BOL); Upper Wellington Sneeukop, moist slope below shale band, flowers dark violet-blue, (–CA), 23 Jan. 1972, *Esterhuysen 32799a* (BOL); Groot Drakenstein Mtns, Devil's Tooth, (–CC), 12 Dec. 1943, *Esterhuysen 9549* (BOL); mountain at top of Franchhoek Pass, (–CC), 18 Nov. 1974, *Nordenstam & Lundgren 2259* (MO, NBG); Kaaimansgat, neck above High Noon Estate, (–CD), 6 Jan. 1980, *Goldblatt 5422* (MO); Onklaarberg, 20 miles [± 30 km] S of Worcester, (–DC), Dec. 1924, *Stokoe 1106* (PRE). **3418** (Simonstown): Cape Peninsula, Vlakkenberg, (–AB), 1 Jan. 1896, *Wolley Dod 457* (BOL); burned lower slopes of Klaasjagersberg adjacent to Cape Point Reserve, (–AB), 26 Nov. 1979, *Goldblatt 5258* (MO, PRE); Paulsberg slopes, (–AB), 6 Nov. 1939, *Lewis 675* (SAM). Muizenberg Mt., 1 500 ft [465 m], (–AB), 29 Nov. 1938, *Wall s.n.* (S); Sir Lowry's Pass, 1 219 m, '*Moraea jacquiniana* Schltr., n. sp.', (–BB), 14 Jan. 1896, *Schlechter 7222* (GRA, PRE). **3419** (Caledon): Haasvlakte, Houwhoek road to Highlands, (–AA),

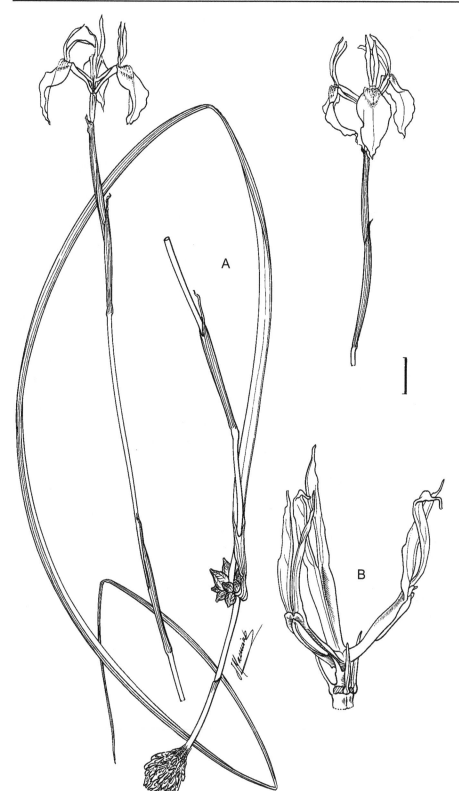

FIGURE 10.—*Moraea hainebachiana, Goldblatt & Porter 13262*. A, flowering stems and corm; B, inner perianth whorl plus filaments and style. Scale bar: 10 mm. Artist: John Manning.

14 Oct. 1988, *Boucher & Stindt 5452* (NBG, PRE); Lebanon Forest Reserve, Jakkalsrivier Catchment 1a, (–AA), 23 Nov. 1970, *Haynes 476* (NBG, PRE).

6. **Moraea hainebachiana** *Goldblatt & J.C.Manning*, sp. nov.

TYPE.—Western Cape, 3217 (Vredenburg): Jacobsbaai, 1 km E of town on calcrete ridge, Erf 890, (–DD), 31 Aug. 2011, *Claassens 95* (NBG, holo.; MO, iso.).

Plants 180–280 mm high. *Corm* mostly 7–12 diam., usually with cormlets at base, tunics of coarse, almost black, hard, wiry fibres, thickened below into claw-like ridges. *Stem* usually simple, rarely 1-branched, glabrous, bearing several, dark-coloured cormlets just below ground level in foliage leaf axil. *Foliage leaf* solitary, linear, channelled, inserted well above top of cataphyll, trailing above, (1.5–)2.0–4.0 mm wide, exceeding stem by 100–150 mm; sheathing cauline leaves ± 55 mm long, green with dry attenuate apices, lowermost sometimes with axillary cormlets. *Rhipidia* usually 2-flow-ered, spathes green with dry attenuate tips, inner 60–65 mm long, outer ± half as long as inner. *Flowers* pale vio-let to deep blue, outer tepal limbs darkly veined, nectar guide white with lines of dark blue radiating from limb base, scented of vanilla, outer tepals 24–25 mm long, claw ± 10 mm long, white-hairy on adaxial surface, with prominent basal, yellow nectary ± 1.5 mm long, limb 14–15 × 9 mm, usually ultimately recurving, inner tepals spindle-shaped, sometimes oblanceolate and obscurely 3-lobed, 5–6 × 0.5 mm, tapering distally with apex curving inward, sometimes inner distal surface pilose. *Stamens* with filaments ± 6 mm long, united in lower 1.0–1.5 mm; anthers 5–6 mm long, blue-grey, pollen off-white or yellow. *Ovary* narrowly oblong-truncate, included or exserted, 7–9 mm long; style ± 2 mm long, dividing ± 1 mm above top of filament column, branches wedge-shaped, 7–8 × ± 1.5 mm at apex; stigma shal-lowly bilobed; style crests narrowly wedge-shaped (or sublinear), ± 10 × 1 mm. *Capsules* not developed, shed soon after flowers wilt. *Seeds* not produced. *Chromo-some number* unknown. *Flowering time*: Aug.–mid Sept. Figure 10.

Distribution: restricted to the Saldanha District in Western Cape (Figure 11), *Moraea hainebachiana* is a narrow edaphic endemic of rocky, limestone flats and slopes and calcareous sands along the coast and adja-cent hills. Plants grow in humus-rich pockets of loam between fractured limestone as well as in coarse calcare-ous sand. Populations extend from the Farm Trekoskraal north of Jacobsbaai through the limestone hills north of Saldanha to the southern end of the Donkergat Peninsula in West Coast National Park. Distinctive as the habitat is, typical *M. tripetala* grows in the same places, but blooms four to five weeks later. *M. hainebachiana* was evidently first collected in 1932 near Langebaan by the late G.J. Lewis, expert on the systematics of southern African Iridaceae. Her collection was referred with-out question to *M. tripetala* at the time. The probabil-ity that the species was distinct from *M. tripetala* was brought to our attention by Koos (Jakobus) Claassens of Jacobsbaai, who has made a thorough study of the local and strongly endemic flora of the Saldanha limestone areas. Given its very narrow range, *M. hainebachiana* must be considered threatened by coastal development although it is probably secure at the southern end of its range in the West Coast National Park under current low wildlife stocking levels.

Diagnosis: named for the Hainebach family of Cape Town for their generous contribution to conservation of the Cape flora, *Moraea hainebachiana* at first seems to be fairly typical *M. tripetala* except for its low stat-ure and slightly smaller flowers. On close examination, however, the species exhibits several unusual features. The first of these is the cluster of small, dark grey or blu-ish cormlets in the foliage leaf axil and at the base of the corm, the latter feature especially pronounced in non-flowering individuals. The foliage leaf is always inserted on the stem well above the top of the cataphyll and clearly separated from it whereas in other species in the complex, the insertion of the foliage leaf is concealed by the cataphyll. The flowers of *M. hainebachiana* are con-sistently pale violet to deep blue and have inner tepals 5–6 mm long, somewhat longer than is typical for *M. tripetala* and unusual in being expanded in the mid-dle, thus spindle-shaped or sometimes oblanceolate and obscurely 3-lobed, and occasionally pilose on the inner (adaxial) surface. More significantly, the filaments are as long as or slightly longer than the anthers, whereas the anthers are slightly longer than the filaments in *M. trip-etala* and in some species of the complex considerably so. The anthers are pale grey-blue and contain whitish or yellow pollen. Orange-red pollen is more frequent in the complex. The nectar guides of *M. hainebachiana* are also somewhat unusual, consisting of lines of dots on a whitish background with the edges not clearly defined as they are in *M. tripetala* in which the yellow or white nectar guides are typically edged in darker blue to violet. Vegetative reproduction does not occur in *M. tripetala* as circumscribed here and the production of multiple corm-lets in leaf axils and at the base of the corm is unknown elsewhere in the *M. tripetala* complex.

Visiting the type locality at Jacobsbaai after flowering to examine capsules and seeds, we found all capsules poorly developed and ready to be shed, if not already fallen, before developing any seeds. We confirmed the condition for two more populations. Microscopic exami-nation of pollen grains shows them to be malformed and evidently infertile. We conclude that *M. hainebachiana* is a vegetative apomict. Propagation is effected by corm-lets that plants liberally produce.

Representative specimens

WESTERN CAPE.—**3217** (Vredenburg): Jakobsbaai, Swart-riet Farm, calcrete ridge to N of road to farm house, 50 m, (–DD), 7 Aug. 1993, *Boucher 5801* (NBG); Jacobsbaai, limestone pavement, (–DD), 1 Sept. 2010 (fr.), *Goldblatt & Manning 13491* (MO, NBG, PRE); limestone hill N of Saldanha, limestone pavement, (–DD), 4 Sept. 2009 (fl.), *Goldblatt & Porter 13262* (MO). **3318** (Cape Town): Langebaan, (–AA), 4 Sept. 1932, *Lewis s.n.BOL45094* (BOL); West Coast National Park, flats SW of Konstabel Kop on calcareous sands, (–AA), 13 Sept. 2011, *Manning & Goldblatt 3338* (NBG).

7. Moraea ogamana *Goldblatt & J.C.Manning*, sp. nov.

TYPE.—Western Cape, 3319 (Worcester): Elands-berg Farm, foot of Elandskloof Mts, seasonally wet allu-vial flats, in fynbos, (–AC), 9 Sept. 2010, *Goldblatt & Manning 13520* (NBG, holo.; MO, iso.).

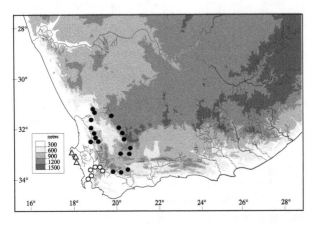

FIGURE 11.—Distribution of *Moraea amabilis*, ●; *M. hainebachiana*, Δ; *M. ogamana*, ○.

Plants 150–200(–380) mm high. *Corm* 6–12 mm diam., tunics of hard, wiry, black fibres. *Stem* usually unbranched, 2 or 3 internodes long. *Foliage leaf* solitary, sometimes with second leaf from axillary cormlet, usually shorter than stem, rarely longer, 2–3 mm wide when opened flat, widely channelled or leaf halves almost folded together, smooth, apple-green, inserted at ground level or shortly below ground, usually bearing small axillary cormlet; sheathing cauline leaves (30–)45 mm long, green with dry attenuate tips. *Rhipidia* mostly 3-flowered; spathes green with dry brown attenuate tips, inner to 55 mm long, outer ± half as long as inner. *Flowers* pale blue, with triangular yellow nectar guides, with dark veins radiating from nectar guide, outer tepal claws 12–14 mm long, limbs ovate, ± 15 mm long, inner tepals hair-like, up to 4 mm long. *Stamens* with filaments ± 5 mm long, united for ± 1 mm; anthers ± 6 mm long, dark bluish black, pollen white. *Ovary* linear-cylindric, 10–13 mm, darkly lined vertically on locules and septa; style ± 2 mm long, branches ± 12 mm long, crests ± linear, 8–12 mm long. *Capsules* ± cylindric, (15–)20–24 × 2 mm. *Seeds* 0.8–1.5 mm diam., golden brown, shortly winged on angles. *Chromosome number* unknown. *Flowering time*: mid-Aug.–mid-Sept. Figure 12.

Distribution: restricted to the Western Cape lowlands between Voëlvlei and Strand and locally in the upper Breede River Valley between Wolseley and Botha, *Moraea ogamana* occurs in waterlogged stony ground

(Figure 11). Judging by the few collections, the species is rare. With rapid agricultural development in lowland Western Cape *M. ogamana* must be regarded as endangered, although its existence is secure in the Elandsberg Nature Reserve at Bo-Hermon immediately south of Voëlvlei, a small part of its once much wider range.

Diagnosis: the narrow, linear-cylindric ovary 10–13 mm long, often partly included in the spathes, and pale blue flower with outer tepal limbs with yellow nectar guides and dark radiating venation immediately set *Moraea ogamana* apart in the *M. tripetala* complex. Associated with these features are the short stature, rarely exceeding 180 mm, fairly small corms, 7–12 mm diam., with wiry, black tunic fibres and a relatively short leaf, usually shorter than, but occasionally exceeding the stem. Unlike most populations of *M. tripetala* in the Western Cape forelands, which have orange-red pollen, the pollen of *M. ogamana* is white. An additional, but somewhat trivial distinction is the style crests, up to 12 mm, which are unusually long for the relatively small flowers. More importantly, almost all specimens we have seen have a single, small cormlet in the foliage leaf axil. This modest level of vegetative reproduction results in plants sometimes growing in small clumps of three or four individuals. Correlated with the long, angular ovary, the capsules are ± cylindric and normally 20–24 mm long. Capsules of *M. tripetala* are ellipsoid-oblong and only 10–14 mm long. The species is named in honour of Ms Naoka Ogama of Japan for her generous donation to the study of southern African Iridaceae.

Representative specimens

WESTERN CAPE.–**3318** (Cape Town): between Paarl and Pont, (–DB), *Drège s.n.* (S); near Groenfontein, Klapmuts, level wet area, (–DD), 15 Sept. 1983, *Van Zyl 3515* (NBG, PRE); damp place near Bottelary Road, (–DD), 9 Sept. 1934, *Acocks 2157* (S). **3319** (Worcester): Bo-Hermon, Elandsberg Farm, seasonally wet alluvial fynbos, (–AC), 20 Oct. 2010 (fr.), *Manning 3307* (MO, NBG); between Bain's Kloof and Wolseley, wet sandy flats, (–AC), 24 Aug. 1974, *Goldblatt 2428* (MO); Worcester, Botha's Halt, (–CB), 14 Sept. 1928, *Gillett 293* (NBG). **3420** (Simonstown): Harmony Reserve, Strand, (–BB), 28 Aug. 2000, *Runnalls 1041* (NBG).

8. **Moraea amabilis** *Diels* in Botanishe Jahrbücher der Systematik 44: 118 (1910). Type: [Northern Cape], Bokkeveld, Calvinia, Oorlogskloof, 13 Sept. 1900, *Diels 626* (B, lecto.!, designated by Goldblatt: 751 (1976), as holotype).

Plants 150–250 mm high, growing in clonal colonies. *Corm* mostly 7–12 mm diam., with tunics of pale, medium to fairly coarse fibres (rarely soft relatively fine fibres), often replaced annually by two daughter corms. *Stem* simple or 1(2)-branched, glabrous or velvety, sometimes with cormlets in below-ground axils, enclosed at base by brown cataphylls, occasionally accumulating in a collar of fine fibres around base. *Foliage leaf* solitary, linear, channeled, usually broadly so, 2–5 mm wide, abaxial surface often velvety-pilose; cauline sheathing leaves ± 40–52 mm long, glabrous or velvety, green with dry apices. *Rhipidia* several-flowered; spathes green, dry and attenuate above, becoming dry entirely, glabrous or velvety, inner mostly 45–55 mm long, outer ± half to two thirds as long as inner. *Flowers* mostly purple or violet to dark blue (rarely pink, dull yellow or buff-brown), nectar guides velvety, white

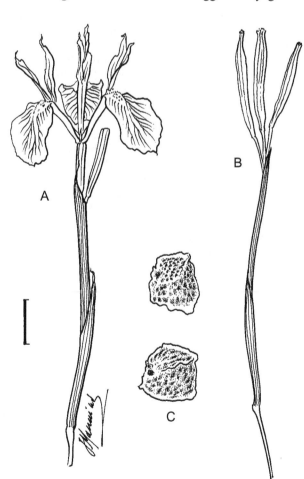

FIGURE 12.—*Moraea ogamana, Goldblatt & Manning 13520.* A, flowering stem; B, capsules; C, seeds. Scale bar: 10 mm. Artist: John Manning.

speckled with purple or dark blue, often with larger dark mark at base, often slightly scented of honey or vanilla, outer tepals 20–31 mm long, claws pale mauve often with darker central line and spotted purple-black, 10–12 mm long, velvety, limbs ± broadly lanceolate, (9–)11–18 × 10–14 mm, apex obtuse-apiculate, inner tepals erect, hair-like, acute, 1.5–2.5 mm long. *Stamens* with filaments ± free, 3.5–6.0 mm long; anthers 4.5–7.0(–8.0) mm long, pollen white (often red in west of range, rarely pale blue). *Ovary* exserted from spathes, 6.0–9.5 mm long, flushed red; style vestigial, < 0.5 mm long, branches ± 10 mm long; crests linear, erect or arching inward, 6–9 mm long. *Capsules* narrowly ellipsoid, 11–16 mm long, sometimes with thickened apical rim. *Seeds* 1.7–2.0 × ± 1 mm, facets with spongy edges and prominent spongy micropylar crest. *Chromosome number 2n* = 12 (Table 1). *Flowering time*: mostly Sept.–mid-Oct. Figure 13.

Distribution: relatively widespread, *Moraea amabilis* extends from the Olifants River Valley to the Bokkeveld Escarpment and in an arc across the northern edge of the Roggeveld Escarpment and south through the Roggeveld and Klein Roggeveld to Worcester and the western end of the Little Karoo (Figure 11). Plants grow in light clay soils or in sandy loam. Locally in the Bokkeveld Mtns and Gifberg, plants grow in shallow sandy ground on sandstone pavement where they are usually dwarfed, we assume because of the shallow, nutrient poor soil. In the west of its range plants flower from late August to mid-September, but at higher elevations to the east, plants bloom from mid September into October, sometimes as late as the end of that month.

Diagnosis: *Moraea amabilis* is recognized in the *Moraea tripetala* complex by the virtually free filaments (united for < 4 mm) and inner tepals reduced to hair-like cusps up to 2.5 mm long, and by the distinctive mode of vegetative reproduction in which the parent corm is replaced annually by two new corms. Plants also often have a cormlet in the leaf axil and as a result form clonal colonies. This unusual characteristic is only evident if the plants are carefully extracted from the ground (see Figure 14A). The species is most likely to be confused with *M. grandis*, which also has virtually free filaments but that species is larger in all features, notably the ovary, 6–9 mm long in *M. amabilis* vs. 10–12(–15) mm in *M. grandis*, and anthers 4.5–7.0(–8.0) mm and up to half again as long as the filaments vs. 9–11 mm long and twice as long as the filaments in *M. grandis*. The corms of *M. amabilis* have pale (rarely dark) brown, coarsely fibrous tunics with prominent vertical elements in contrast to the finer tunic fibres in *M. grandis*. Apart from smaller flowers, *M. amabilis* often has white pollen in contrast to the consistently bright red-orange pollen of *M. grandis* (Table 2).

Capsules of *Moraea amabilis*, typically 11–16 mm long, contain relatively small, sharply angular seeds, ± 2 mm at longest axis, with a prominent, pale, spongy micropylar crest (Figure 14D). In contrast, capsules of *M. grandis* are significantly larger, up to 19 mm long, have a thickened apical ridge and the large, dark brown seeds are up to 3 mm long (Figure 15D. A velvety to pilose pubescence is universal on the leaves, stem and

spathes of western Karoo populations of *M. amabilis* but those from the Bokkeveld Mtns, Gifberg and Cedarberg–Olifants River Mtns usually have smooth leaves although some notable exceptions include *Goldblatt & Porter 13522* from Kransvlei and *Levyns 10155* from Paleisheuwel. We place only moderate confidence in the pubescence character as some other members of the *M. tripetala* complex occasionally have hairy leaves (notably *M. mutila*), but we have seen no collections of *M. grandis* with hairy leaves. On the dolerite flats near Farm Keiskie, southeast of Calvinia, plants of *M. grandis* (*Goldblatt et al. 13366*) were entirely hairless, whereas *M. amabilis* growing nearby had pubescent leaves, stems and sheathing leaves (*Goldblatt et al. 13365*). The possibility that *M. amabilis* and *M. grandis* hybridize in the Nieuwoudtville area where the two co-occur is discussed briefly under the latter species.

Plants from near Worcester match *Moraea amabilis* vegetatively, notably in their corm tunics and velvety leaves and stems but have unusual, buff-yellow, or dusty pink flowers (e.g. *Goldblatt 4089*). Plants from the Rabiesberg, not far distant, also have velvety stems and leaves and were described as having brown (we assume buff-brown) flowers. These colour variants are unique for the *M. tripetala* complex.

Moraea amabilis was described by Friedrich Diels in 1910 based on plants he had collected in September 1900 when he lived in Calvinia. Goldblatt (1976b) erroneously designated *Diels 626* as the holotype, overlooking the second collection cited in the protologue, *Diels 1169* (incidentally no longer extant). The so-called holotype becomes the lectotype. Treated merely as a variant morph of *M. tripetala* by Goldblatt (1976b), we now regard *M. amabilis* as one the most distinctive members of the *M. tripetala* complex in its habit of often producing two new corms in place of the single parent corm. Flowers of the type, evidently somewhat shrunken, have tepal claws 10–12 mm long and limbs ± 12–14 × 11 mm, an ovary ± 7 mm long, and the stamens have filaments ± 3.5 mm and anthers 5–6 mm long. Leaves and corms are lacking, but as these were described they were presumably present in the syntype, now lost, *Diels 1169*. The extant type material most closely matches the morph from the sandstone pavement in the Bokkeveld Mtns close to the edge of the escarpment.

Representative specimens

NORTHERN CAPE.—**3119** (Calvinia): between Oorlogskloof and Papkuilsfontein, (–AC), Sept. 1939, *Leipoldt 3059* (BOL), *Leipoldt 3002* (BOL, PRE); Oorlogskloof Nature Reserve, sandstone pavement near entrance, burned last summer, (–AC), 13 Sept. 2010, *Goldblatt & Porter 13532* (MO, NBG, PRE); 0.5 km S of top of Vanrhyn's Pass, (–AC), 29 Aug. 1974 *Goldblatt 2471* (MO; 15 km SSW of Nieuwoudtville, Oorlogskloof Nature Reserve, (–AC), 31 Aug. 1995, *Pretorius 259* (NBG, PRE); Lokenburg, rugged TMS, (–CA), 28 July 1956, *Acocks* 18900 (K, PRE); Farm Keiskie, dolerite flats SE of Keiskie-se-Poort, plants hairy (co-blooming with *M. grandis*), (–DB), 16 Sept. 2009, *Goldblatt, Manning & Porter 13365* (MO, NBG). **3120** (Williston): ± 70 km SE of Calvinia to Middelpos on Blomfontein road, (–CC), 26 Oct. 1976, *Goldblatt 4381* (MO, PRE). **3220** (Sutherland): Roggeveld, Soekop, Rooipunt Camp, (–AA), 17 Sept. 2004, *Rosch 253* (NBG); Roggeveld Plateau, Farm Blesfontein, sandy flats, (–AD), 4 Nov. 2011, *Goldblatt & Porter 13720* (MO, NBG); Smoushoogte, slopes above road (visited by *Rediviva macgregorii*), (–DC), 17 Sept. 1993, *Steiner 2710* (MO, NBG); top of Verlaten Kloof, (–DC), 28 Oct. 1982, *Bayer 3276* (NBG); Moordenaars Karoo on road to farm Meintjiesplaas, shaded S slope above stream, (–DC), 30 Sept.

FIGURE 13.—*Moraea amabilis*, *Goldblatt & Porter 13461*. A, flowering stems and corm; B, inner perianth whorl plus filaments and style; C, capsules; D, seeds. Scale bar: A, C, 10 mm; B, D, 2 mm. Artist: John Manning.

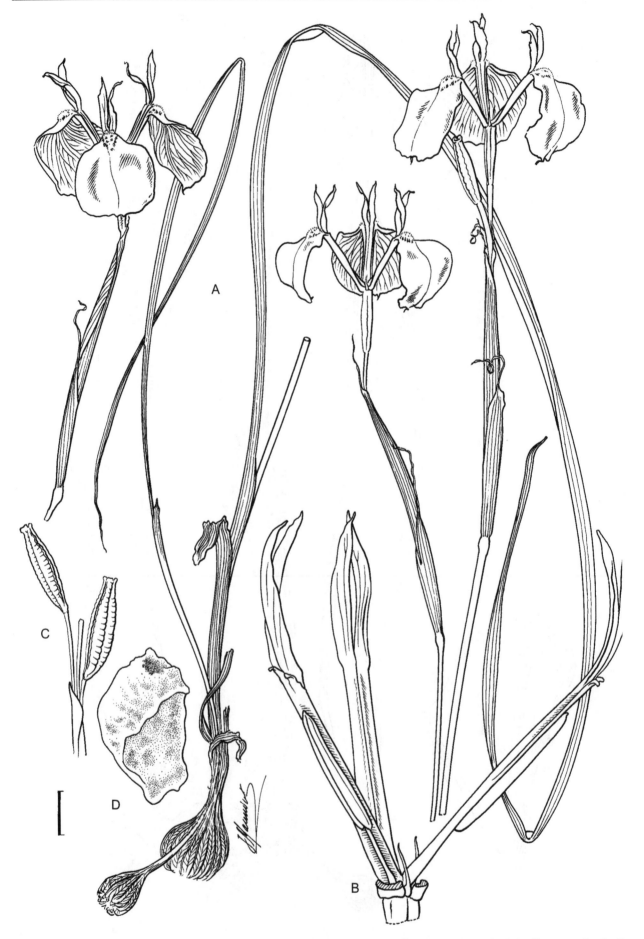

FIGURE 14.—*Moraea grandis, Goldblatt & Manning 13658*. A, flowering stems and corm; B, inner perianth whorl plus filaments and style; C, capsules; D, seeds. Scale bar: A, C, 10 mm; B, D, 2 mm. Artist: John Manning.

TABLE 2.—Comparison of taxonomically important features in *Moraea amabilis* and *M. grandis*. Filaments are virtually free or basally united for less than 0.4 mm in all specimens of both species examined.

Species	Ovary/ capsule (mm)	Outer tepals (mm) Limb	Claw	Inner tepal (mm)	Filaments (mm)	Anthers (mm)	Leaf width (mm)	Inner spathe (mm)
grandis	(10–)12–15/ 16–19	18–23 × 15–20	12–18	1.5–3.0	4.5–5.5	8–11 orange-red,	3.0–5.5	50–80
amabilis	6.0–9.5/ 11–16	11–18 × 8–14	10–14	1.5–2.5	3.5–6.0	4.5–8.0 white or red (blue)	2–5	45–55

2004, *Goldblatt & Porter 12656* (MO); Klein Roggeveld, (–DC), Oct. 1920, *Marloth 9595* (PRE), *9602* (PRE).

WESTERN CAPE.—**3118** (Vanrhynsdorp): Gifberg, plateau, (–DB), 23 Aug. 1984, *Goldblatt 7230* (MO), 13 Sept. 2010 (fr.), *Goldblatt & Porter 13524* (MO, NBG), 23 Jun. 1913, *Phillips 7524* (K); Bulhoek, (–BD), 2 Aug. 1896, *Schlechter 8382* (MO, PRE); top of Nardouw Pass, (–BD), 13 Aug. 1976, *Goldblatt 3853* (MO). **3218** (Clanwilliam): Clanwilliam, at turnoff to Boskloof, (–BB), 29 Aug. 1974, *Goldblatt 2457* (MO, PRE), 11 Sept. 2010 (fr.), *Goldblatt & Porter 13522A* (MO, NBG); W slopes of Pakhuis Pass, 10 km from Clanwilliam, (–BB), [?date], *Nordenstam & Lundgren 1325* (MO, S); sandy slopes near Paleisheuwel, (–BC), 5 Sept. 1954, *Levyns 10155* (BOL). **3219** (Wuppertal): Klipfonteinrand, (–AA), 20 Aug. 1974, *Nordenstam & Lundgren 1347* (MO, S). **3220** (Sutherland): Koedoes Mtns, rocky gully above stream, (–CC), 14 Sept. 2009, *Goldblatt, Manning & Porter 13327* (NBG, MO); 70 km S of Sutherland to Matjiesfontein, (–DC), 30 Oct. 1974 (late fl. & fr.), *Goldblatt 3222* (MO); N of Farms Fortuin and Nuwerus at Komsberg turnoff from Sutherland road (co-blooming with *M. cuspidata*), (–DC), 28 Sept. 2009, *Goldblatt & Porter 13461* (MO). 3319 (Worcester): Rabiesberg, Montagu, (flowers brown), ex hort. Kirstenbosch, (–CA), July 1940, *Van der Gaast s.n.* (Nat. Bot. Gard. 2673/35 in NBG); Rabiesberg, lower slopes, flowers pink, (–CA), 27 Sept. 1935, *Lewis s.n.* (BOL 21733, K); foot of Fonteintjiesberg, Farm Onse Rug, clay soil, flowers pink or buff, (–CB), 12 Sept. 1976, *Goldblatt 4089* (K, MO, PRE). 3320 (Montagu): Memorial Hill, rocky slope above cemetery, (–AB), 29 Sept. 2009, *Goldblatt & Porter 13467* (MO); S-facing bank of stream opposite Pietermeintjies Siding, (–AC), 23 Sept. 1976, *Goldblatt 4175* (MO, PRE); Op-de-Tradouw, Barrydale to Montagu, (–DC), *Goldblatt 4185* (MO).

9. **Moraea grandis** *Goldblatt & J.C.Manning*, sp. nov.

TYPE.—Northern Cape, 3119 (Calvinia): Wild Flower Reserve at Nieuwoudtville, edge of shrubs near top of dolerite slope, scented of vanilla, (–AC), 22 Sept. 2011, *Goldblatt & Manning 13658* (NBG, holo.; MO, PRE, iso.).

Plants 200–400 mm high, usually in small colonies. *Corm* mostly 10–15 mm diam, with tunics of light (rarely dark brown), relatively fine, firm or soft fibres, rarely accumulating in a thick mass; often producing two new corms to replace parent corm. *Stem* glabrous, rarely velvety to pilose, simple or 1(2)-branched, base sheathed by dry light brown cataphylls, these usually persisting as collar of fine fibres around base; usually producing one or more cormlets in leaf or cataphyll axils. *Foliage leaf* solitary, shallowly channelled, usually exceeding stem and up to twice as long, 3.5–5.5 mm wide, glabrous; cauline sheathing leaves green or becoming dry from tips, long-attenuate, 55–70 mm long. *Rhipidia* several flowered; spathes green with dry, attenuate tips, glabrous, inner 55–80 mm long, outer ± half as long as inner. *Flowers* pale blue to pale violet or light purple, nectar guides small, triangular, yellow (rarely cream) on a white background and edged dark blue, short-velvety (rarely glabrous), scented of vanilla,

outer tepals 30–40 mm long, claws 12–18 mm long, velvety, limbs ± orbicular, widest above middle, usually ± as wide as long, 15–20(–23) × 15–20 mm, plane, obtuse-apiculate, spreading at 45–60°, inner tepals hairlike, erect, 1.5–3.5 mm long. *Stamens* with filaments ± free or united < 0.2 mm, 4.0–5.5 mm long; anthers 8–11 mm long, usually not reaching stigma lobes, pollen orange-red. *Ovary* (10–)12–15 mm long, slightly narrowed below apex; style vestigial, < 0.4 mm long, branches ± 15 mm, crests linear, 8–12 mm long. *Capsules* 16–19 mm long, narrowly ovoid-oblong with thickened rim. *Seeds* relatively large, 2–3 × 1.8–2.3 mm, facets rounded, angles with wing-like ridges. *Chromosome number* 2n = 12 (Table 1). *Flowering time*: mainly late Aug.–late Sept. Figure 14.

Distribution: restricted to the northern end of the western (winter rainfall) Karoo, *Moraea grandis* extends from the Langberg west of Loeriesfontein through the higher country between Loeriesfontein and Calvinia and along the Bokkeveld Plateau to the northern end of the Roggeveld Escarpment (Figure 15). Plants favour heavier soils and are most often found in heavy red clay among dolerite rocks but also grow on lighter soils derived from tillite and shale.

Diagnosis: the most striking of the species of the *Moraea tripetala* complex, *M. grandis* has relatively large corms up to 15 mm diam, with tunics usually of fine, soft (rarely firm) fibres, and the largest flowers in the alliance. The outer tepals are 30–40 mm long (Table 2) and have ± orbicular limbs 18–23 × 15–20 mm, thus almost as wide as long, and have unusually small nectar guides, usually yellow (sometimes cream) on a white background, minutely spotted with dark blue, and usually edged with a darker colour than the rest of the tepal limb. The limbs are also widest in the upper third

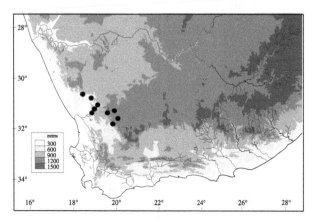

FIGURE 15.—Distribution of *Moraea grandis*, ●.

and usually have an abrupt, comma-like tip. The flowers have a pleasant, vanilla-like odour. Like its relative, *M. amabilis*, it typically grows in colonies resulting from vegetative reproduction by axillary cormlets and the sometimes replacement of the parent corm by two daughter corms. It also resembles *M. amabilis* in having virtually free filaments (united for < 0.2 mm), but the anthers, 8–11 mm long, are at least twice as long as the filaments and sometimes 2.5 × as long, easily the largest in the *M. tripetala* complex. The pollen is consistently bright orange-red, not unusual for the entire group, and contrasting with the white pollen in most western Karoo populations of *M. amabilis*, which has smaller anthers 4.5–7.0 mm long and ± as long as or up to 1.5 × as long as the filaments.

Moraea grandis sometimes grows in close proximity to *M. amabilis*, and the two can be seen flowering together on the farm Keiskie south of Calvinia. Difficulty in distinguishing the two in the immediate vicinity of Nieuwoudtville is probably the result of hybridization and introgression between the two species, which share the same large-bodied bees as pollinators.

Moraea grandis has been illustrated in several wild flower volumes notably in the *Nieuwoudtville, Bokkeveld Plateau and Hantam* wildflower guide (Manning & Goldblatt 1997: 69) and *The color encyclopedia of Cape bulbs* (Manning *et al.* 2002: 19, 307). The photographs show the distinctive, broad outer tepal limbs oriented almost vertically in fully open flowers.

Representative specimens

NORTHERN CAPE.—**3018** (Kamiesberg): Langberg, wet gully on shale, (–DB), 5 Sept. 2006, *Goldblatt & Porter 12765* (MO, NBG). **3019** (Loeriesfontein): slopes of Kubiskou Mtn, clay ground, (–CC), 6 Sept. 2006, *Goldblatt & Porter 12783* (MO); ± 8 km W of Loeriesfontein, slate hills, (–CD), 12 Sept. 1976, *Thompson 2883* (NBG, PRE). **3119** (Calvinia): Nieuwoudtville, Farm Glenlyon, (–AC), 2 Sept. 1959, *Hardy 73* (PRE); 3 miles [4.5 km], W of Nieuwoudtville, (–AC), 28 Aug. 1950, *Lewis 2274* (PRE); Glen Lyon; 2 mi. S of Nieuwoudtville, red clay flats. Flowers blue. Fairly frequent in between bulbinellas and rhenosterbos, (–AC), 11 Sept. 1974, *Oliver & Mauve 42* (K, PRE); foot of Vanrhyns Pass, (–AC), 16 Oct. 1974 (fr.), *Goldblatt 3098* (MO); Calvinia District, 8 km on Rietfontein road, W of Calvinia–Loeriesfontein road, (–AD), 25 August 1976, *Goldblatt 3948* (MO, PRE); Langfontein ± 3 km to Kotzeskolk, (–BB), 2 Sept. 1986, *Burger & Louw 184* (NBG); Farm Perdekraal, S slope on dolerite, (–BC), 12 Sept. 1981, *Goldblatt 6247* (MO); Hantamsberg, upper slopes NW of Calvinia, (–BC), 16 Sept. 1980, *Goldblatt 1980* (MO); road to Toring, 4.1. km N of Loeriesfontein–Calvinia road, dolerite slope with heavy red clay, (–BD), 21 Sept. 2011, *Goldblatt & Manning 13655* (MO, NBG); Bloukrans Pass, near summit in wet gully, (–DA), 31 Aug. 1981, *Snijman 622* (NBG, PRE); Farm Keiskie, dolerite flats SE of Keiskie-se-Poort (co-blooming with *M. amabilis*), (–DB), 16 Sept. 2009, *Goldblatt, Manning & Porter 13366* (MO, NBG).

ACKNOWLEDGEMENTS

We thank Elizabeth Parker and Lendon Porter for their assistance and companionship in the field; Koos Claassens of Jacobsbaai for help in investigating the status of *M. hainebachiana*; Mary Stiffler, Research Librarian, Missouri Botanical Garden, for help with literature searches; Roy Gereau, Missouri Botanical Garden, for assistance with nomenclatural questions; and Clare Archer for lists of exsiccatae from the PRE collection. Rhoda McMaster provided the initial stimulus for this study by questioning the range of seed types in plants then treated as *Moraea tripetala*. Collecting permits were provided by the Nature Conservation authorities of Northern Cape and Western Cape, South Africa. Field work was funded in part by the Mellon Foundation and the B.A. Krukoff Fund, Missouri Botanical Garden.

We acknowledge the generous donation of the Hainebach family of Camps Bay, Cape Town, for conservation of the Cape Flora. Their contribution is remembered in the name *Moraea hainebachiana*. We are also delighted to recognize the valuable support towards our research provided by Ms Naoko Ogama, Miyagi prefecture, Japan, by commemorating her in the name *Moraea ogamana*.

REFERENCES

BAKER, J.G. 1896. Iridaceae *in* W. T. Thiselton-Dyer, *Flora capensis* 6: 7–171. Reeve & Co., London.

BAKER, J.G. 1904. Beiträge zur kenntnis der Afrikanischen-Flora. XVI. Iridaceae. *Bulletin de l'Herbier Boissier* ser. 2, 4: 1003–1007.

BAKER, J.G. 1906. II.—Diagnoses Africanae, XVI. Royal Botanic Gardens, Kew. *Bulletin of Miscellaneous Information* 1906: 15–30.

BOLUS, H. & WOLLEY DOD, F. 1903. List of the flowering plants of the Cape Peninsula. *Transactions of the South African Philosophical Society* 14.

BRUMMITT, R.K. & POWELL, C.E. 1992. *Authors of plant names.* Royal Botanic Gardens, Kew.

DE CANDOLLE, A.P. 1804. Mémoire sur le *Vieusseuxia*, genre de la famille des Iridées. *Annales du Muséum National d'Histoire Naturelle* 2: 136–141.

DIELS, L. 1910. Formationen und Florenelemente im nordwestlichen Kapland. *Botanische Jahrbücher fur Systematik* 44: 92–124.

DIETRICH, A.G.1833. *Species plantarum* edn. 6, 2. Nauck, Berlin.

ECKLON, C.F. 1827. *Topographisches Verzeichniss der Pflanzensammlung von C.F. Ecklon.* Reiseverein, Esslingen.

GOLDBLATT, P. 1971. Cytological and morphological studies in the southern African Iridaceae. *Journal of South African Botany* 37: 317–460.

GOLDBLATT, P. 1976a. Evolution, cytology and subgeneric classification in *Moraea* (Iridaceae). *Annals of the Missouri Botanical Garden* 63: 1–23.

GOLDBLATT, P. 1976b. The genus *Moraea* in the winter-rainfall region of southern Africa. *Annals of the Missouri Botanical Garden* 63: 657–786. 1976.

GOLDBLATT, P. 1986a. The Moraeas of southern Africa. *Annals of Kirstenbosch Botanical Garden* 14: 1–224.

GOLDBLATT, P. 1986b. Convergent evolution of the *Homeria* flower type in six new species of *Moraea* (Iridaceae) in southern Africa. *Annals of the Missouri Botanical Garden* 73: 102–116.

GOLDBLATT, P. 1998. Reduction of *Barnardiella, Galaxia, Gynandriris, Hexaglottis, Homeria* and *Roggeveldia* in *Moraea* (Iridaceae: Irideae). *Novon* 8: 371–377.

GOLDBLATT, P. & MANNING, J.C. 2009. New species of *Moraea* (Iridaceae: Iridoideae), with range extensions and miscellaneous notes for southern African species. *Bothalia* 39: 1–10.

GOLDBLATT, P. & MANNING, J.C. 2010. *Moraea intermedia* and *M. vuvuzela* (Iridaceae: Iridoideae), two new species from western South Africa, and some nomenclatural changes and range extensions in the genus. *Bothalia* 40: 146–153.

GOLDBLATT, P., MANNING, J.C. & BERHNARDT, P. 2005. Pollination mechanisms in the African genus *Moraea* (Iridaceae: Iridoideae): floral divergence and adaptation for pollen vector variability. *Adansonia* 27: 21–46.

GOLDBLATT, P. & TAKEI, M. 1993. Chromosome cytology of the African genus *Lapeirousia* (Iridaceae: Ixioideae). *Annals of the Missouri Botanical Garden* 80: 961–973.

GUNN, M.D. & CODD, L.E. 1981. *Botanical exploration of southern Africa.* Balkema, Cape Town.

HOLMGREN, P.K., HOLMGREN, N.H. & BARNETT, L.C. 1990. *Index Herbariorum. Part. 1: The Herbaria of the World.* New York Botanical Garden, New York.

KER GAWLER, J. 1803. *Moraea tripetala. Curtis's Botanical Magazine* 19: t. 702.

KLATT, F.W. 1866. Revisio Iridearum (Conclusio). *Linnaea* 34: 537–689.

KLATT, F.W. 1894. Iridaceae. Pp. 143–230 *in* T. Durand & H. Schinz, *Conspectus Florae Africae* 5. De Mat, Brussels.

LEWIS, G.J. 1941. Iridaceae. New genera and species and miscellaneous notes. *Journal of South African Botany* 7: 19–59.

LEWIS, G.J. 1950. Iridaceae. Pp. 217–265 *in* R.A. Adamson & T.M. Salter, *Flora of the Cape Peninsula*. Juta. Cape Town.

LINNAEUS, C. fil. 1782 [as 1781]. *Supplementum plantarum*. Orphanotropheus, Brunswick.

MANNING, J.C. & GOLDBLATT, P. 1997. Nieuwoudtville, Bokkeveld Plateau and Hantam. *South African Wild Flower Guide* 9. Botanical Society of S. Africa, Cape Town.

MANNING, J.C., GOLDBLATT, P. & SNIJMAN, D. 2002. *The color encyclopedia of Cape bulbs*. Timber Press, Portland, OR.

MAYTHAM KIDD, M. 1950. *Wild flowers of the Cape Peninsula*. Oxford University Press, Cape Town.

NORDENSTAM, R.B. 1972. Types of Ecklon's 'Topographisches Verzeichniss' in the Swedish Museum of Natural History in Stockholm. *Journal of South African Botany* 38: 277–298.

ROEMER, J.J. & SCHULTES, J.A. 1817. *Systema vegetabilium secundum* 1. Cotta, Stuttgart.

THUNBERG, C.P. 1782. Dissertatio de Iride. Edman, Uppsala.

VOIGT, J.O. 1845. *Hortus suburbanus calcuttensis*. Bishop's College Press, Calcutta.

7

Systematics of the southern African genus *Ixia* (Iridaceae: Crocoideae): 4. Revision of sect. *Dichone*

P. GOLDBLATT* and J.C. MANNING**

Keywords: Iridaceae, *Ixia* sect. *Dichone*, morphology, new species, southern Africa, taxonomy, winter rainfall zone

ABSTRACT

The southern African genus *Ixia* L. comprises ± 90 species from the winter-rainfall zone of the subcontinent. *Ixia* sect. *Dichone* (Salisb. ex Baker) Goldblatt & J.C.Manning, one of four sections in the genus and currently including 10 species and three varieties, is distinguished by the following floral characters: lower part of the perianth tube filiform and tightly clasping the style; filaments not decurrent; upper part of the perianth tube short to vestigial; style branches involute-tubular and stigmatic only at the tips; and so-called subdidymous anthers. We review the taxonomy of the section, providing complete descriptions and distribution maps, and a key to the species. *I. amethystina* Manning & Goldblatt is recognized to be a later synonym of *I. brevituba* G.J.Lewis. Most collections currently included under that name represent another species, here described as *I. rigida*. We recognize five additional species in the section: early summer-blooming *I. altissima* from the Cedarberg; *I. bifolia* from the Caledon District; *I. flagellaris*, a stoloniferous species from the Cedarberg; *I. simulans* from the western Langeberg; and *I. tenuis* from the Piketberg. We also raise to species rank *I. micrandra* var. *confusa* and var. *minor*, as *I. confusa* and *I. minor* respectively. Foliar and associated floral variation in the widespread *I. scillaris* has led us to recognize two new subspecies among its northern populations, broad leaved subsp. **latifolia** and the dwarfed, smaller flowered subsp. **toximontana**; subsp. *scillaris* is restricted to the immediate southwestern Cape, from Darling to Somerset West. Sect. *Dichone* now has 17 species and two subspecies.

INTRODUCTION

Restricted to the winter-rainfall zone of southern Africa, the genus *Ixia* L. comprises some 90 species divided among the four sections *Dichone* (Salisb. ex Baker) Goldblatt & J.C.Manning, *Ixia*, *Hyalis* (Baker) Diels and *Morphixia* (Ker Gawl.) Pax (Goldblatt & Manning 2011). Sect. *Dichone*, the smallest of these, comprised eight species and three varieties when last revised by Lewis (1962). In the later account of *Ixia* for *Flora of southern Africa*, De Vos (1999) followed Lewis's taxonomy closely but included the new species, *I. collina*, described by Goldblatt & Snijman (1985), and altered the arrangement of species to place those with specialized, unilateral stamens at the end of the account. The most recent addition to the section, *I. amethystina*, was described by Manning & Goldblatt (2006), but we have now discovered it to be conspecific with the type of *I. brevituba*. Most plants currently referred to *I. brevituba* in herbaria represent a separate species, described here as *I. rigida*.

Field work conducted since 1999 has resulted in the discovery of three sets of populations belonging to sect. *Dichone* that do not accord with any of the species currently recognized. We describe one of them from the Langeberg and allied to *I. scillaris*, as *I. simulans*, another from the Piketberg as *I. tenuis* and the third, evidently allied to the *I. micrandra* complex, as *I. bifolia*.

*B.A. Krukoff Curator of African Botany, Missouri Botanical Garden, P. O. Box 299, St. Louis, Missouri 63166, USA. E-mail: peter.goldblatt@mobot.org.
** Compton Herbarium, South African National Biodiversity Institute, Private Bag X7, Claremont 7735, South Africa / Research Centre for Plant Growth and Development, School of life Sciences, University of KwaZulu-Natal, Pietermaritzburg, Private Bag X01, Scottsville 3209, South Africa. E-mail: J.Manning@sanbi.org.za.

Examination of living plants and herbarium specimens matching *I. micrandra* var. *confusa* and *I. micrandra* var. *minor* provide evidence that both are better regarded as separate species, which we treat as *I. confusa* and *I. minor* respectively. We identified two further undescribed species from the Cedarberg in herbaria, namely *I. altissima*, an early summer blooming species of uncertain affinities and *I. flagellaris*, allied to *I. scillaris* but the only stoloniferous species in the section.

We provide a complete review of sect. *Dichone*, expanding it to include 17 species, with *I. scillaris* subdivided into three subspecies.

TAXONOMIC HISTORY

As is the case for many Cape plants, the taxonomic history of sect. *Dichone* is complex. The genus *Dichone* (the meaning of the Greek translation, two tubes, might refer to the incompletely dehiscent anther thecae) was proposed without description by R.A. Salisbury in 1812 as 'Dichone crispa Laws. Cat.' for a plant known at the time as *Ixia crispa* L.f. (1782) but now treated as *I. erubescens* Goldblatt. The cryptic reference, 'Laws. Cat.', refers to an unpublished catalogue, perhaps merely a handwritten list, of plants evidently grown in the garden of the Scottish nurseryman, Peter Lawson, about which little appears to be known. C.F Ecklon (1827) likewise treated *Ixia crispa* and its immediate allies in a separate genus but for which he used the name *Agretta*, also without description. His species and combinations in *Agretta* are currently invalid. Ecklon intended to admit five species to his genus, one of which (*A. stricta*) is now *I. stricta* (Eckl. ex Klatt) G.J.Lewis, and another (*A. pallideflavens*), which is now *I. odorata* Ker Gawl. (sect. *Hyalis* according to Goldblatt & Manning 2011). *Agretta stricta* was referred by Klatt (1882) to his genus *Tritonixia* and the species was only transferred to *Ixia* (sect.

Dichone) by Lewis in 1962. We have not identified Eck-lon's *A. grandiflora*.

The name *Dichone* was validated by J.G. Baker (1877) as sect. *Dichone* Salisb. ex Baker of *Tritonia*, at which time it included *T. scillaris* (L.) Baker, *T. trinervata* Baker and *T. undulata* (Burm.f.) Baker (which he erroneously believed was an earlier synonym of *Ixia crispa*). *T. undulata* is correctly a species of *Tritonia* and is currently known by that name (Goldblatt & Manning 2006). *Ixia micrandra* Baker (sect. *Dichone*) remained in *Ixia* on account of its actinomorphic flowers, although it has the filiform perianth tube, short anthers and involute style branches of *Dichone*. To add to the series of misunderstandings, *Ixia retusa* Salisb., described in 1796 by R.A. Salisbury but without a known type, was considered conspecific with what Ker Gawler (in Sims 1801) called *I. scillaris*. Later, Ker Gawler (1803) cited *I. retusa* under what he called *I. polystachia* (sic), which he then considered an earlier name for *I. scillaris*. The plate, t. 629 of *Curtis's botanical magazine*, is not, however, *I. scillaris* but is either *I. confusa* (G.J.Lewis) Goldblatt & J.C.Manning or *I. stricta* (Eckl. ex Klatt) G.J.Lewis. It appears that Ker Gawler believed that *I. retusa* was a member of what we now call sect. *Dichone*. Plate t. 629 was later cited by Klatt (1882) under his combination *Watsonia retusa*, which like its basionym, *I. retusa*, is without a type. Both Lewis (1962) and De Vos (1999) excluded *I. retusa* from *Ixia* on the basis that the protologue is inadequate to identify the species. We concur.

The four species of sect. *Dichone* that occur in the western Karoo remained unknown until the early twentieth century, when collections of the three species *Ixia curvata* G.J.Lewis, *I. rigida* Goldblatt & J.C.Manning, and *I. trifolia* G.J.Lewis were made by Rudolf Marloth in 1920. The first of the western Karoo species to be recognized, *I. trifolia*, was described by Lewis (1934), based on her own collection grown at Kirstenbosch Botanical Garden. In assigning the species to *Ixia*, Lewis evidently did not then realize that *I. trifolia* was closely allied to species of sect. *Dichone*, which at the time was included in *Tritonia*. Although she noted the unusual anther dehiscence via very narrow slits in *I. trifolia*, she regarded the species as belonging in sect. *Euixia* (i.e. sect. *Ixia*) and close to *I. ovata* (Andrews) Sweet, now *I. abbreviata* Houtt.

By 1954 Lewis regarded sect. *Dichone* as more closely allied to *Ixia* than to *Tritonia* and recommended its recognition as a separate genus (Lewis 1954: 100) because of its unique anther dehiscence. Later (Lewis 1962) she formally transferred sect. *Dichone* to *Ixia*, noting that the section was also unusual in the genus in its involute-tubular style branches and its distinctive anthers. She described *I. brevituba* G.J.Lewis and *I. curvata* in 1962, assigning Marloth's collection of what we here describe as *I. rigida* to her *I. brevituba* (Lewis 1962), which is typified by a 1929 collection made by Grant & Theiler.

Time has provided new evidence for Lewis's conclusion that *Dichone* is more closely related to *Ixia* than to *Tritonia*. Species examined cytologically have the same basic chromosome number, $x = 10$, and karyotype as other species of *Ixia* whereas *Tritonia* has $x = 11$ (Goldblatt 1971a; De Vos 1982; Goldblatt & Manning 2011). *Ixia* also has derived pollen grains with a single banded operculum (Goldblatt *et al.* 1991 and unpublished) whereas *Tritonia* has the plesiomorphic pollen grains of subfamily Crocoideae with a two-banded operculum. The two genera are clearly closely allied and available molecular studies (Goldblatt *et al.* 2006, 2008) indicate a sister relationship (with moderate bootstrap support, 80%) between the one species each of *Ixia* (*I. latifolia* D.Delaroche) and *Tritonia* (*T. disticha* (Klatt) Baker) that were sequenced. Although Goldblatt & Manning (1999) initially treated sect. *Dichone* as one of two sections of subg. *Ixia* they later reverted to Lewis's four-section classification of *Ixia*. There have been no molecular studies that shed light on the relationship of *Ixia* and *Dichone* but the unstated assumption implicit in the current taxonomy is that *Dichone* is nested within *Ixia*, hence its status as a section of that genus.

MATERIALS AND METHODS

Using standard methods of taxonomic investigation, we examined the holdings of *Ixia* in herbaria with significant southern African collections, BOL, K, MO, NBG, PRE, and SAM (acronyms following Holmgren *et al.* 1990). Our herbarium studies were accompanied by field investigation to determine variation within populations for some species, and their ecology, especially soil, aspect, and altitude.

MORPHOLOGY

Morphology: the vegetative morphology of sec. *Dichone* corresponds closely to that of other species of *Ixia*, with few exceptions. Leaves are plane and form a basal fan in most species but *I. erubescens* has crisped and undulate leaf margins, and leaves of *I. scillaris* are sometimes undulate, or rarely crisped in a few populations from the Piketberg, Pakhuis–Biedouw Mtns, and near Tulbagh. Leaves of *I. minor* (= *I. micrandra* var. *minor*) are terete to subterete or plane whereas those of *I. micrandra* are filiform, and both species usually have two or sometimes three leaves, the uppermost ± entirely sheathing the lower part of the stem. *I. bifolia*, *I. trinervata*, and usually *I. simulans* also have two leaves, a basal foliage leaf and a second, largely to entirely sheathing, upper leaf. In *I. trinervata* the lanceolate foliage leaf is distinctive in having three ± equally prominent veins. The lowermost leaf of *I. simulans* has one prominent vein and thickened margins, and the upper sheathing leaf is inserted above the stem base.

Flowers are produced in spikes typical of Crocoideae. The floral bracts are typical of *Ixia*, being membranous and ± translucent, becoming dry with age. The outer of each bract pair usually has 3 prominent veins and 3 teeth but is occasionally 1-toothed, sometimes varying within a species but consistently so in *I. rigida*. Flowers of most species are predominantly pale to deep pink, sometimes mauve pink, but with a white to pale yellow throat often edged darker pink to mauve. The perianth is occasionally white in some individuals or populations of *I. scillaris*, notably in the Olifants River Valley populations of subsp. *latifolia*, but is purple in *I. brevituba*, the flowers of which have a dark centre and stamens. Flower orientation ranges from fully upright to half to fully nodding, thus facing to the side. Flowers of several species are scented. *I. bifolia*, *I. confusa*, *I. stricta*, and at least some populations of *I. scillaris* produce a rose-like scent, and *I. collina* and *I. rigida* are also sweetly scented (Goldblatt & Snijman 1985).

The perianth tube is usually described as filiform (or slender throughout) in sect. *Dichone* but the tepals are in fact fused for some distance beyond the lower filiform part of the tube. The length of the upper part of the tube is often so short as to be conveniently ignored but in *Ixia altissima* and *I. rigida* fully half the length of the tube is flared for ± 2 and 1 mm respectively and the tube is thus clearly funnel-shaped. In *I. trifolia* the flared upper part of the tube is also developed to a significant degree. The walls of the slender, lower part of the tube always tightly clasp the style and the tube contains no nectar.

Although radial floral symmetry is the common, and by inference ancestral, condition in the group, flowers of *Ixia collina*, *I. erubescens*, *I. simulans*, *I. scillaris*, *I. tenuis*, and probably also *I. flagellaris* are weakly zygomorphic. The flowers in these species face sideways and are thus nodding, the tepals held more-or-less vertically, with the stamens unilateral and the anthers horizontal or drooping. At dehiscence the stamens in the nodding flowers of *I. curvata* and *I. stricta* are symmetrically arranged but become unilateral with age, evidently through gravity (De Vos 1999).

Filaments are usually free and filiform, and inserted at the mouth of the narrow part of the perianth tube. In some populations of *Ixia trifolia* the filaments are united basally for up to 0.7 mm, or up to one fifth their length. Anthers are oblong to subrotund and slightly to markedly recurved above the base. Species from the Western Karoo typically have oblong anthers, 4–5 mm long, with the thecae ± twice or more as long as wide, whereas those from the southwestern Cape below the Escarpment are characterized by their derived, short, suborbicular to broadly oblong anthers, 2–4 mm long and < twice as long as wide. Several species within this group have unilateral stamens (*I. collina*, *I. erubescens*, *I. simulans*, and *I. scillaris*), with the thecae sharply acute at their bases. The filaments in these species are characteristically sharply recurved apically such that the thecae are held almost horizontally. Anthers in sect. *Dichone* dehisce from the base via narrow slits that do not open fully, a unique feature in the genus and family. Dehiscence is either ± complete to the locule apex or incomplete and restricted to the lower third or half of the anther, as in *I. scillaris*. Anthers have been described as subdidymous (Lewis 1962), a term meant to imply the vestigial nature of the anther connective, which in sect. *Dichone* is a narrow, adaxial strip of tissue that continues from the apex of the filament to shortly below the anther apex. The two thecae are closely appressed without the usual well-developed sterile tissue of the connective on their adaxial surfaces.

The filiform style divides at the mouth of the narrow part of the perianth tube, thus at or shortly beyond the filament bases, and the style branches spread outwards, either ± straight (e.g. *I. bifolia*, *I. micrandra*, *I. trifolia*) or falcate to recurved. The style branches are involute, with the margins rolled upward to form an enclosed, tubular channel open only at the apex, with the stigmatic surfaces restricted to the apices, which are often expanded or slightly bifurcate (e.g. in *I. collina* and *I. micrandra*).

The capsules are, as far as known, typical of *Ixia* (Goldblatt & Manning 2011), as are the seeds, which have a smooth surface and an excluded vascular trace (Goldblatt *et al.* 2006; Goldblatt & Manning 2011). Capsules are known for only a few species and we do not routinely include capsule features in the species descriptions below. Notably, however, the few available capsules of *I. scillaris* are ± obovoid, possibly a distinguishing taxonomic feature.

INFRASECTIONAL CLASSIFICATION

No classification below sectional rank has until now been proposed for sect. *Dichone*. However, with 17 species, it seems useful to us to recognize two species groups, or series. We include species with radially symmetric flowers and either complete or incomplete anther dehiscence in ser. *Euanthera*, thus including the Cedarberg *I. altissima*, all western Karoo species, and several from the southwestern Cape. The anther thecae in ser. *Euanthera* dehisce conventionally along narrow slits for their entire length or almost so and are either ± rounded or acute and recurved at the base. Species with mainly zygomorphic flowers and ± unilateral stamens, *I. stricta* excepted, are included in ser. *Dichone*. The anthers in this group of species are mostly horizontal or pendent, acute and recurved at the base, and dehisce incompletely from below. No vegetative specializations consistently accompany these floral differences. Ser. *Dichone*, which is evidently monophyletic, is centred in the southwestern Cape below the Escarpment, with populations of *I. scillaris* ranging into northern Namaqualand and onto the Gifberg–Matsikamma Mtn complex. Ser. *Euanthera* is comparatively widespread.

SYSTEMATICS

Ixia sect. **Dichone** *(Salisb. ex Baker) Goldblatt & J.C.Manning* in Bothalia 29: 63 (1999). *Tritonia* sect. *Dichone* Salisb. ex Baker: 163 (1877). *Tritonia* subg. *Dichone* (Salisb. ex Baker) Baker: 190 (1892). *Tritonixia* sect. *Dichone* (Salisb. ex Baker) Klatt: 357 (1882). *Ixia* subg. *Dichone* (Salisb. ex Baker) G.J.Lewis: 150 (1962). Type: *I. scillaris* L., lectotype here designated [Lewis: 159 (1962) and Goldblatt & Manning: 63 (1999) listed *I. crispa* L. (or its synonym *I. erubescens* Goldblatt) as the type but only *Tritonia undulata* (with its presumed synonym *I. crispa*) and *I. scillaris* were included by Baker in the section when it was first validly described].

Dichone Lawson ex Salisb.: 320 (1812), nom. nud.

Agretta Eckl.: 23 (1827), nom. nud.

Corm as in *Ixia*; stolons with a terminal cormlet produced in *I. flagellaris*. *Leaves* as in *Ixia* but sometimes only two and then lower with expanded blade and upper largely to entirely sheathing. *Spike*: outer bracts with 3(1) main veins acutely 3-toothed, 3-lobed, or 1-toothed. *Flowers* pink to magenta (rarely white), tepal bases white or yellow edged darker pink (blue-mauve with dark centre in *I. brevituba*), scented or unscented; perianth tube usually relatively short, filiform below and clasping style sometimes funnel-shaped with upper part flared. *Stamens* symmetrically disposed or unilateral and reclinate, filaments inserted at apex of narrow part of tube, filiform or flattened; anthers straight and linear to oblong or short and suborbicular, with vestigial connective, thecae fully or incompletely dehiscent, then splitting from base by narrow slits. *Style* dividing at mouth of tube opposite or just above base of filaments,

branches filiform-tubular and involute, thus stigmatic only at apices, straight or falcate-recurved.

Ser. **Euanthera** *Goldblatt & J.C.Manning*, ser. nov.

Flowers upright or nodding. *Stamens* symmetrically arranged (at least in bud and on opening, sometimes later by gravity unilateral in *Ixia curvata*); anthers straight, rounded at base, or recurved and ± acute at base, dehiscing conventionally by longitudinal slits to reach apex or almost so. *Style* branches ± straight or falcate. Type species: *I. micrandra* Baker.

10 spp., western half of the Cape floral region and western Karoo.

1. **Ixia altissima** *Goldblatt & J.C.Manning*, sp. nov.

TYPE.—Western Cape, 3219 (Wuppertal): Gonna-

Key to species of sect. *Dichone*

1a Plants 0.9–1.2 m high; perianth tube funnel-shaped with flared upper portion ± 2 mm long and ± as long as filiform lower half; style dividing opposite middle 1/3 of anthers . . . . 1. *I. altissima*

1b Plants 0.9–1.5 m high; perianth tube cylindrical almost throughout, flared only in upper 0.5–1.0 mm if at all; style dividing opposite or just above bases of filaments:

  2a Flowers radially symmetric, upright to half nodding; stamens erect with anthers parallel or diverging:

    3a Perianth tube very short, 2.0–2.5 mm long; stem with short, stiffly erect lateral branches held close to main axis, each with up to 8 flowers . . . . 2. *I. rigida*

    3b Perianth tube 2.3–6.0 mm long; stem simple or with 1(–several) suberect to spreading branches:

      4a Leaves ± terete or filiform, rarely more than 1 mm wide; corm tunics of fine fibres extending upward in a collar around the stem base:

        5a Style branches straight and extending outward below anthers, 3–5 mm long; filaments filiform; outer floral bracts with 3 main veins and several secondary veins, acutely 3-toothed . . . . 9. *I. micrandra*

        5b Style branches falcate, recurved at tips, up to 2 mm long; filaments flattened and wider in middle; outer floral bracts with 3 main veins but lacking secondary veins, usually shallowly 3-toothed or 3-lobed . . . . 10. *I. minor*

      4b Basal and sometimes other leaves linear to sword-shaped or lanceolate, usually at least 1.5 mm wide, blades plane with main vein and sometimes multiple veins ± equally prominent; corm tunics fine to coarse-textured, with or without fibrous collar around the stem base:

        6a Anthers 4–5 mm long, oblong and > twice as long as wide; spike nodding in bud, becoming erect as flowers open:

          7a Anthers yellow; perianth pink, tepals darker at base and yellow in throat; foliage leaves (2.5–)5–12 mm wide; style branches 4–5 mm long . . . . 3. *I. trifolia*

          7b Anthers purple; perianth light purple with small dark centre; foliage leaves 1.5–2.0 mm wide; style branches up to 2 mm long . . . . 4. *I. brevituba*

        6b Anthers 1.5–3.0 mm long, suborbicular to broadly oblong and < twice as long as wide; spike erect in bud:

          8a Leaves 2, basal leaf with expanded, ± lanceolate blade, upper leaf entirely sheathing:

            9a Basal leaf with 3 ± equally prominent, thickened veins; style branches falcate to recurved (white drying ± white); tepals 12–15 × 6–8 mm when fully open; spike mostly 7–12-flowered . . . . 6. *I. trinervata*

            9b Basal leaf with 1 prominent vein (translucent when alive, appearing raised when dry); style branches ascending, ± straight (pink, often drying purple); tepals remaining slightly cupped, 12–14 × 5–7 mm; spike mostly 4–7-flowered . . . . 7. *I. bifolia*

        8b Leaves > 2, usually 3 to 5, with no visible veins or only main vein prominent:

          10a Leaves (3)4 or 5, often drying at anthesis; foliage leaf blades lanceolate, mostly 4–9 mm wide . . . . 11. *I. stricta*

          10b Leaves (2)3 or 4; foliage leaf blades linear, mostly 1.5–4.0 mm wide . . . . 8. *I. confusa*

  2b Flowers usually half to fully nodding, facing to side; stamens unilateral with anthers held horizontally or pendent:

    11a Anthers dehiscing completely by longitudinal slits . . . . 5. *I. curvata*

    11b Anthers dehiscing by short slits from base to middle, not reaching apex:

      12a Foliage leaf usually single, sometimes 2 (uppermost leaf ± entirely sheathing), blade ± linear, 1.0–4.5 mm wide; flowers usually 4–8 per spike; perianth tube ± 4.5 mm long . . . . 16. *I. simulans*

      12b Foliage leaves 3–7, blades lanceolate to falcate, mostly 4–20 mm wide, or closely undulate and then only 2–4 mm wide; perianth tube 2.5–4.0 mm long:

        13a Leaves 2–4(5) mm wide; corm tunics of fine, soft fibres; perianth tube 2–3 mm long:

          14a Leaf margins strongly undulate and crisped; anthers ± 1.5 mm long . . . . 12. *I. erubescens*

          14b Leaf margins plane; anthers 2–3 mm long:

            15a Corms producing slender stolons from base, terminating in a single, large cormlet; anthers ± 3 mm long . . . . 15. *I. flagellaris*

            15b Corms not producing stolons; anthers 2.0–2.5 mm long . . . . 14. *I. tenuis*

        13b Leaves 4–25 mm wide and margins plane or sometimes undulate(–crisped), corm tunics of firm, medium-textured to coarse fibres; perianth tube 2.5–5.0 mm long:

          16a Plants to 90 cm high, often branched, with branches held at right angles to main axis, becoming suberect distally; filaments expanded toward apex and becoming as wide as anthers; style branches widely flared at tips . . . . 17. *I. collina*

          16b Plants rarely more than 60 cm high, simple or branched, with branches ascending to suberect; filaments ± uniformly filiform, < 1/2 as wide as anthers; style branches not noticeably flared at tips . . . . 13. *I. scillaris*

fontein, seep in sandstone ground among *Kniphofia*, 900 m, (–CB), 5 Dec. 2005, *Pond 303* (NBG, holo.).

Plants 0.9–1.2 m high, stem surrounded at base by membranous cataphylls, simple or with 1–2 branches subtended by translucent, attenuate bracts and prophylls up to 8 mm long. *Corm* 15–18 mm diam., mature tunics not known, *Leaves* 4, lower 3 ± basal, reaching to ± middle of stem, blades linear to linear-sword-shaped, (3–)4–10 mm wide, firm-textured, margins and midrib and often a pair of secondary veins thickened and hyaline (at least when dry), uppermost leaf largely sheathing. *Spike* erect, elongate, weakly flexuose, 14–20-flowered, lateral spikes 6–14-flowered; bracts dry, membranous, translucent, flecked with brown in upper half, outer ± 7 mm long, usually with 3 prominent veins and 3-toothed, inner bracts ± as long as outer or slightly shorter, 2-veined and 2-toothed. *Flowers* rotate, mauve-pink; perianth tube narrowly funnel-shaped, ± 4 mm long, filiform and clasping style below, flared in distal ± 2 mm; tepals spreading, oblong-ovate, 12–13 × ± 4 mm. *Stamens* symmetrically disposed, erect; filaments filiform, inserted just above narrow part of tube, ± 4 mm long, ± white; anthers oblong, ± 3 mm long, erect, dehiscing fully by longitudinal slits, yellow. *Style* dividing opposite middle of anthers, branches ± 2 mm long, falcate, pale yellow, tubular, stigmatic apically, arching outward between or shortly overtopping anthers. *Flowering time*: Nov.–mid-Dec.

Distribution and habitat: Ixia altissima is known from two sites in the central and southern Cedarberg, from the Matjies River Nature Reserve and from Gonnafontein to the south (Figure 1). Plants grow along stream banks or in marshy seeps on sandstone derived soil, at the Gonnafontein site among *Kniphofia* plants. The habitat remains moist at least until December and possibly later into the summer. We suggest a conservation status for *I. altissima* of Rare (R), in light of its evident rarity, but we see no current threat to the species.

Diagnosis and relationships: Ixia altissima is a surprising plant, sometimes standing over 1 m in height and flowering late in the season, well into December. The anthers with vestigial connective and the involute style branches, stigmatic only at the tips, indicate its placement in sect. *Dichone* but the funnel-shaped

perianth tube with the filaments inserted in the middle of the upper part of the tube is unusual for the section, most species of which have the tube virtually filiform throughout with the filaments inserted at the apex of the tube. The perianth tube, only ± 4 mm long, and the pale floral bracts are reminiscent of two Roggeveld Escarpment species, *I. trifolia* and *I. rigida*, the latter also flowering relatively late in the season, often in November. These species share relatively unspecialized, longitudinally dehiscent anthers. *I. altissima* seems most like *I. rigida*, also a tall plant, up to 600 mm high and sometimes flowering as late as November. The latter has flowers with a similarly funnel-shaped perianth tube 2–3 mm long and silvery translucent floral bracts, contrasting with the slightly longer perianth tube, ± 4 mm long, and shorter anthers, just 3 mm long, in *I. altissima*, which is also unusual in the section in having the style dividing opposite the middle of the anthers.

Representative specimens

WESTERN CAPE.—**3219** (Wuppertal): Cedarberg, Matjies River Nature Reserve, stream bank, (–AD), 21 Nov. 1999, *Low 5838* (NBG); Cedarberg, Gonnafontein, seep in sandstone ground among *Kniphofia*, 900 m, (–CB), 3 Dec. 2000, *Pond 256* (NBG)

2. **Ixia rigida** Goldblatt & J.C.Manning, sp. nov.

TYPE.—Northern Cape, 3220 (Sutherland): ± 38 km north of Sutherland to Middelpos, Farm Geelhoek, rocky dolerite slope with red clay soil, 3 108 ft [950 m], (–AB), 7 Sept. 2006, *Goldblatt & Porter 12796* (MO, holo.; NBG, iso.).

Plants (150–)300–600 mm high, stem usually with 2–3 ± erect branches held close to axis, sheathed at base by firm, chestnut brown cataphylls, combining with old corm tunics and leaf bases to form collar around base. *Corm* 12–18 mm diam., tunics of firm, relatively hard fibres. *Leaves* (3–)5–7, lower (2–)3–6 with expanded blades, upper 1 or 2 leaves sheathing lower 1/3 to middle of stem, blades sword-shaped to sub-linear, (3–)5–8 mm wide, main vein, margins, and often secondary vein pairs moderately thickened, hyaline when dry. *Spike* weakly flexuose, nodding in bud, mostly 8–15-flowered, lateral spikes with fewer flowers; bracts membranous, silver-translucent, ± 5.5–7.0 mm long, outer with prominent brown or purple central vein and few secondary veins not reaching upper margin, acute or with 2 obscure lateral lobes, inner 2-veined and acutely 2-toothed. *Flowers* rotate, pale pink, darker at tepal bases and ± white in throat, sweetly rose-scented; perianth tube funnel-shaped, ± 2.5 mm long, filiform and clasping style in basal 1.5 mm or less, widely flared distally for up to 1 mm; tepals spreading, ovate-oblong, ± 12(–15) × 5(–7) mm, obtuse. *Stamens* symmetrically arranged, erect; filaments inserted at mouth of narrow part of tube, filiform, ± 1.5(2.5) mm long, ± white; anthers oblong to linear, curved back at base, 3–4(5) mm long, slightly recurved basally, thecae subacute at base, dehiscing fully by longitudinal slits, yellow. *Style* dividing at base of filaments, branches ± straight, 2.0–3.0(3.5) mm long, ± filiform-tubular, stigmatic apically, white. *Flowering time*: Sept. to mid-Nov. Figure 2.

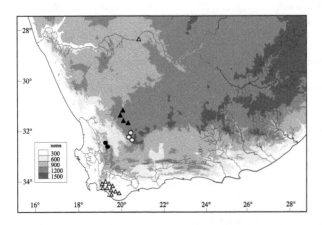

FIGURE 1.—Distribution of *Ixia altissima*, ●; *I. rigida*, ○; *I. curvata*, ▲; and *I. stricta*, △.

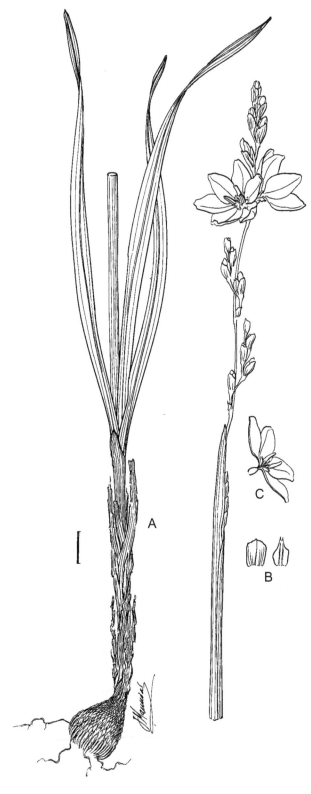

FIGURE 2.—*Ixia rigida*, *Goldblatt & Porter 12796* (NBG). A, whole plant; B, outer (left) and inner (right) bract; C, half flower. Scale bar: 10 mm. Artist: John Manning.

Distribution and habitat: *Ixia rigida* is restricted to the Roggeveld, where it is known from the Escarpment edge on the Farms Fransplaas, Quaggasfontein, Uitkyk, Houdenbek and a short distance inland northwest of Sutherland on the Farm Geelhoek (Figure 1). Plants are restricted to dolerite outcrops and grow in rocky ground in the heavy red clay characteristic of this habitat. The

first record of the species, made by Rudolf Marloth in 1920 and said to be from the 'Komsberg flats'[some distance south of Sutherland] may be erroneous, for *I. rigida* is otherwise unknown south of the town. The relatively few records show that the species is rare but plants often flower late, usually after mid-October on the escarpment edge, when relatively little collecting is done in the Roggeveld, so the present records may not fully reflect its range. There appears to be no imminent threat to its survival but at sites where we have seen the species, plants were heavily overgrazed and often do not set seed as a result. We tentatively suggest a conservation status of near-threatened, (NT).

Diagnosis and relationships: *Ixia rigida* is a distinctive species easily recognized by the moderately tall stem up to 600 mm high with 2–3 upright branches held close to the axis, and the silvery, translucent floral bracts subtending very short-tubed flowers, the tube up to 2.5 mm long with the narrow portion up to ±1.5 mm long. Plants have up to seven leathery leaves with prominently thickened margins and main and secondary veins. *I. rigida* was known to Lewis (1962) from only a single specimen, which she associated with *I. brevifolia* and De Vos (1999) likewise misunderstood the species, citing specimens of *I. rigida* under the name *I. brevifolia*.

Although fairly uniform across its range the populations of *Ixia rigida* inland of the Roggeveld escarpment and flowering three to four weeks earlier than those along the edge of the escarpment, thus in September, have slightly larger flowers and less fibrotic leaves. Too little is known about *I. rigida* to assess the significance of this variation. We note, however, that the escarpment populations mostly have 6 or 7 leaves, typically 5–7 mm wide, tepals ± 12 × 5 mm, anthers ± 3 mm long, and style branches ± 2 mm long. The inland populations have 4 or 5 leaves, mostly 7–8 mm wide, tepals ± 14–16 × 6–8 mm, slightly longer filaments, anthers 4(5) mm long, and style branches ± 3.5 mm long. The bracts of the inland plants also stand out in their darker purple veins.

Representative specimens

NORTHERN CAPE.—**3220** (Sutherland): Sutherland, Bo-Visrivier road, Farm Fransplaas, stony dolerite flats, (–AB), 19 Oct. 1995, *Goldblatt & Manning 10370* (MO); Roggeveld Escarpment, Farm Quaggasfontein, dolerite ridge near farm buildings, 4 900 ft [± 1 580 m], (–AB), 11 Oct. 2004, *Goldblatt & Porter 12667* (MO); Farm Uitkyk, Naaldegraskop, (–AD), 14 Nov. 1987, *Goldblatt & Manning 8645* (MO, NBG, PRE); Houdenhoek, Hoedenbek 4 × 4 trail, among dolerite boulders, 1 387 m, (–AD), 10 Oct. 2008, *Clark & Coombs 880* (NBG). Without precise locality and possibly an error: Komsberg flats, Oct. 1920, *Marloth 9782b* (PRE).

3. **Ixia trifolia** *G.J.Lewis* in Flowering Plants of South Africa 14: t. 543 (1934); Lewis: 171 (1962); De Vos: 69 (1999). Type: South Africa, [Western Cape], Tweedside, cultivated in Cape Town, Sept. 1932, *Lewis s.n.* (Nat. Bot. Gard. 2706/32 in BOL, lecto., here designated; BOL!, K, PRE!, SAM!, isolecto.) [Lewis: 171 (1962) cited this number at BOL as the holotype but there are two sheets, hence our designation of a lectotype].

Plants mostly 150–300 mm high, stem usually with 1–2(–3) short branches, or simple, sheathed at base by short, firm, green cataphylls. *Corm* mostly 10–18 mm diam., tunics of coarse fibres, vertical elements often becoming claw-like ridges. *Leaves* 3 or 4(5), lower 2 or 3 with expanded blades, lanceolate to falcate, 1/2–2/3 as long as stem, (2.5–)5–8(–12) mm wide, central vein and margins moderately thickened, upper 1(2) leaves sheathing lower 1/3–2/3 of stem, without expanded blade. *Spike* weakly flexuose, inclined, drooping in bud, mostly 3–7-flowered; bracts membranous, translucent, 5–6 mm long, outer with prominent central vein, sometimes with 2 prominent secondary veins, acute or trilobed, inner slightly shorter than outer, 2-veined and 2-toothed. *Flowers* rotate, pale to deep pink (rarely mauve), yellow or white edged with darker pink at base of tepals, unscented; perianth tube funnel-shaped, 3.5–4.0 mm long, filiform in basal 2.5–3.0 mm and clasping style, flared in distal 1 mm, tepals spreading, ovate-elliptic, obtuse to retuse, 12–15 × 5.0–6.5 mm. *Stamens* symmetrically arranged, erect, filaments inserted at mouth of narrow part of tube, 3–4 mm long, free or united basally for up to 0.7 mm, filiform, ± white, anthers oblong-linear, 3–4 mm long, yellow, thecae obtuse at base, dehiscing fully by longitudinal slits. *Style* dividing between base and lower 1/3 of filaments, style branches 4–5 mm long, projecting between filaments, gently arched outward, filiform-involute, stigmatic apically, pale pink. *Flowering time*: mid-Aug. to late Sept. Figure 4E, F.

Distribution and habitat: *Ixia trifolia* has a relatively wide range, extending along the Roggeveld Escarpment from the Farm Uitkyk southward through the Klein Roggeveld to the Laingsburg Karoo as far west as Tweedside (Figure 3). Plants grow on stony clay ground in low, karroid scrub or mountain renosterveld. The flowers offer no nectar and are pollinated by pollen-collecting female bees. The anthophorine *Amegilla spilostoma* (Apidae) has been captured visiting flowers near Sutherland (Goldblatt & Manning 2007) and we have also noted small hopliine beetles consistently visiting flowers at a population on the Farm Blesberg and clearly actively transferring pollen from flowers of one plant to those of another.

Diagnosis and relationships: *Ixia trifolia* is recognized by the combination of bright pink, rarely mauve flowers, white to yellow edged with dark pink in the centre, always held upright on inclined spikes, and two or three, relatively broad foliage leaves plus one (or two) more sheathing the stem. The specific epithet *trifolia* refers to the three foliage leaves, thus ignoring the presence of the upper, entirely sheathing leaf. Like other species of sect. *Dichone* from the Western Karoo, the style branches are slender and nearly straight and the anthers dehisce completely. Unique to this species in the section is that the filaments may be united basally for up to one quarter of their length, and the style is exserted up to 1.5 mm from the filiform part of the perianth tube. *Ixia trifolia* is perhaps most closely allied to *I. brevituba*, which has similarly oriented, upright flowers on an inclined spike but with narrower leaves, up to 1.5 mm wide and violet flowers darker in the centre. Both species stand out in the section in having the spikes drooping in bud.

Ixia trifolia is relatively common across its range but is poorly collected. The earliest collection of the species that we have found was made by Rudolf Marloth in 1920 at the top of Verlatekloof south of Sutherland

although the type population was discovered at Tweedside near Touws River sometime before 1932. The type specimens comprise plants cultivated at Kirstenbosch Botanic Gardens under G.J. Lewis's name.

Representative specimens

NORTHERN CAPE.—**3220** (Sutherland): Roggeveld Escarpment, entrance to Noudrif Farm, NW of Sutherland, (–AB), 23 Sept. 1981, *Goldblatt 6344* (MO); Farm Voëlfontein, banks of Bo-Visrivier, (–AB), 11 Oct. 2004, *Goldblatt & Porter 12665* (MO); Farm Koorlandskloof 70, N of Kruiskloof, (–AB), 26 Sept. 2009, *Helme 6202* (NBG); Farm Hottentotskloof, SW of Sutherland, (–AC), 2 Oct. 1999, *Goldblatt & Nänni 11193* (MO; PRE); top of Verlatekloof, ‚Farm Jakkalsvlei, (–AC), Oct. 1920, *Marloth 9646* (PRE); Klein Roggeveld, flats along stream at Farm Oranjefontein, (–DC), 30 Sept. 2004, *Goldblatt & Porter 12660* (MO).

WESTERN CAPE.—**3220** (Sutherland): Klein Roggeveld, Kruispad, at turnoff from Matjiesfontein–Sutherland road, (–DC), 19 Sept. 2003, *Goldblatt & Porter 12311* (MO). 3320 (Montagu): Memorial, flats W of the cemetery, (–AB), 9 Sept. 2006, *Goldblatt & Porter 12811* (MO).

4. **Ixia brevituba** G.J.Lewis in Journal of South African Botany 28: 170 (1962); De Vos: 70 (1999), in part. Type: South Africa, [Northern Cape], 'Quagas [sic] Pass [probably Ganagga Pass], road between Middelpos and Ceres', Sept. 1929, *Grant & Theiler 4931* (BOL, holo!).

I. amethystina J.C.Manning & Goldblatt: 139 (2006), syn. nov. Type: Northern Cape, 3220 (Sutherland): west of Farm Agterkop, near top of Gannaga Pass, 16 Sept. 1997, *Goldblatt & Manning 10745A* (NBG, holo.!, MO!, iso.).

Plants 150–300 mm high, stem either unbranched and then with scale-like leaves at aerial nodes, or with up to 2 short branches, erect below, inclined above, sheathed below by submembranous, red-brown cataphylls. *Corm* 12–20 mm diam., tunics of medium-textured, wiry, reticulate fibres, extending upward in papery collar 30–70 mm long. *Leaves* 3 or 4, basal, uppermost completely sheathing lower 2/3 of stem, blades suberect or lowermost slightly falcate, 1.5–2.0 mm wide, reaching to near top of stem, firm-textured, margins and midrib hyaline, lightly thickened. *Spike* inclined, nodding in bud, crowded, 5–7-flowered, branches fewer-flowered; bracts dry-membranous, translucent or flushed purple above, outer 5–7 mm long, acute or obscurely 3-dentate, inner ± as long as outer or slightly shorter, 2-veined

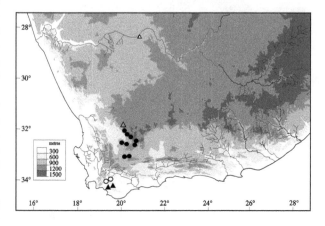

FIGURE 3.—Distribution of *Ixia bifolia*, ▲; *I. brevituba*, △; *I. trifolia*, ●; and *I. trinervata*, ○.

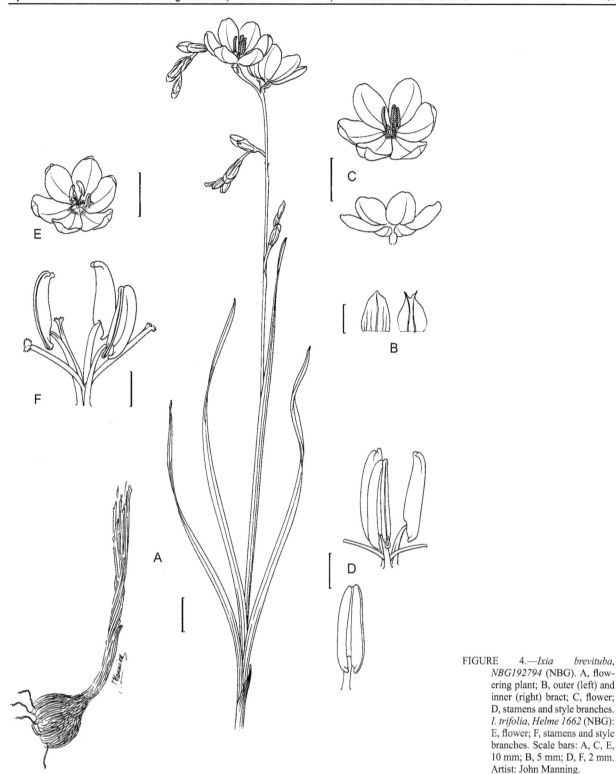

FIGURE 4.—*Ixia brevituba, NBG192794* (NBG). A, flowering plant; B, outer (left) and inner (right) bract; C, flower; D, stamens and style branches. *I. trifolia, Helme 1662* (NBG): E, flower; F, stamens and style branches. Scale bars: A, C, E, 10 mm; B, 5 mm; D, F, 2 mm. Artist: John Manning.

and 2-toothed, margins connate in lower 1.5 mm around ovary. *Flowers* rotate, light purple with small dark purple stain in centre, lightly scented; perianth tube ± funnel-shaped, ± 2.3 mm long, filiform below and clasping style, flaring in distal 1 mm; tepals spreading and slightly concave distally, obovate, ± narrowed below into a short claw, 12–13 × 7–8 mm. *Stamens* symmetrically arranged, erect, filaments inserted at apex of tube, diverging above, 2.5 mm long, blackish purple, anthers ± oblong, 4.5–5.0 mm long, blackish purple, dehiscing laterally by narrow slits along entire length. *Style* dividing at or slightly above base of filaments, ± 2 mm long,

branches filiform, involute, stigmatic at apex, purple, falcate, 2.0–2.5 mm long. *Flowering time*: late Sept. to early Oct. Figure 4A–D.

Distribution and habitat: Ixia brevituba is known from a small area on the edge of the Roggeveld Escarpment near Ganagga Pass southwest of Middelpos (Figure 3). The few known collections were made within a few kilometres of each other on Farms Zoekop and Agterkop (the type is not localized exactly but is very likely from the same area). Plants grow in stony, light clay in renosterveld. The flowers open widely around

08:00, even in cool, overcast conditions and remain fully open until late afternoon. Flowers of the closely related *I. trifolia*, like many other species of sect. *Dichone*, will not open fully or at all under cool conditions. The only pollinators recorded are small hopliine beetles, *Kabousia axillaris*. The form of the flowers is consistent with this pollination system. In view of its narrow range we concur with Raimondo *et al.'s* (2009) conservation status of vulnerable (VU), but we note that the entire range of *I. brevituba* falls within the Tanqua National Park and future disturbance seems unlikely.

Diagnosis and relationships: distinguished by its pale purple flowers with small, dark eye, *Ixia brevituba* has relatively broad, blackish anthers that dehisce laterally so that the light-coloured pollen forms a contrasting pale margin to each anther, and short style branches that are ± as long as the filaments, thus scarcely projecting from between the filaments. The lovely amethyst-coloured flowers are borne on inclined spikes so that they face directly upward in an elegant, arching spray. *I. brevituba* falls among a small group of species that are endemic or near-endemic to the Roggeveld Escarpment and Klein Roggeveld, including *I. curvata*, *I. rigida*, and *I. trifolia*, all of which share relatively unspecialized, longitudinally dehiscent anthers 3–5 mm long. Among these species *I. brevituba* appears to be most closely allied to *I. trifolia* on the basis of their unusual, inclined spikes, and lateral branches that are decurved and nodding when young. All other species have spikes that are erect from bud to fruit, a feature not explicitly mentioned by Lewis (1962) in the protologue but noted and illustrated by De Vos (1999). The two species also share a perianth tube 2–3 mm long. *I. brevituba* differs from *I. trifolia* in its blackish purple anthers and purple perianth tube, which gives the flowers a small, dark central eye, consistently narrower leaves 1.5–2.0 mm wide, and medium-textured corm tunics drawn into a well-developed neck. *I. trifolia*, like all other species in sect. *Dichone*, has yellow anthers, although the filaments may be pale mauve, and a pale perianth tube, giving the flowers a well-defined whitish central eye. The leaves of *I. trifolia* are broader than in *I. brevituba*, (2.5–)5.0–12.0 mm wide, and the tunics are more coarsely fibrous, with the lower fibres developed into woody claws but not drawn into a neck above. In addition, the style branches of *I. trifolia* are longer than the filaments, 4–5 mm long, and project conspicuously between them.

The ranges of *Ixia brevituba* and *I. trifolia* complement one another—only a short distance separates the two species and we infer that they share a common ancestor. The differences in their flowers appear to be related to their pollination biology. *I. brevituba* has flowers that conform to the hopliine beetle pollination syndrome (Goldblatt *et al.* 1998) and the beetle *Kabousia axillaris* has been captured while visiting the flowers. All individuals of this insect carried pure loads of *Ixia*-type pollen on their dorsal thorax and frons (Manning & Goldblatt 2006). In contrast, *Ixia trifolia* is pollinated by solitary female anthophorine bees (Goldblatt & Manning 2007) and possibly also by hopliine beetles.

When described by Lewis (1962), *Ixia brevifolia* included a collection of a second species, now *I. rigida*, and the protologue is a composite of the two species.

De Vos (1999) likewise misunderstood *I. brevituba* and later collections that she associated with the name are also *I. rigida*. Following De Vos's (1999) interpretation of *I. brevituba*, Manning & Goldblatt (2006) described *I. amethystina* for plants that we now realize conform exactly to the type of *I. brevituba*. That name now falls into synonymy.

Representative specimens

NORTHERN CAPE.—**3220** (Sutherland): Farm Zoekop, 3 km S of entrance along road to Gannaga Pass, (–AA), 26 Sept. 2002, *NBG 192794* (NBG); Farm Zoekop, past ruins near edge of escarpment, (–AA), 24 Sept. 2002, *Rösch 154* (NBG); W of farm Agterkop, near the top of Gannaga Pass, 16 Sept. 1997, *Goldblatt & Manning 10745A* (MO, NBG).

5. Ixia curvata G.J.Lewis in Journal of South African Botany 28: 171 (1962); De Vos: 70 (1999). Type: South Africa, [Northern Cape], Agterhantamsberg, Moordenaarspoort, 26 miles [± 39 km] NE of Calvinia, 25 Sept. 1952, *Lewis 3532* (SAM, holo.!; BOL [SAM 61919]!, PRE!, iso.).

Plants 120–350 mm high, stem simple or with 1 or 2 upright branches, sheathed at base by long, chestnut-brown cataphylls and a collar of fibres. *Corm* 12–20 mm diam., tunics of medium-textured fibres. *Leaves* usually 4 or 5, lower 3 or 4 with sword-shaped to falcate blades, ± 1/2 as long as stem and (2–)3–5 mm wide, uppermost leaf entirely sheathing, main veins and margins slightly to moderately thickened. *Spike* flexuose, erect, 2-ranked, mostly 6–12-flowered; bracts membranous, translucent, 4–5 mm long, outer with prominent central vein or secondary veins equally prominent, 3-lobed with central lobe largest or ± equally 3-lobed, inner slightly shorter than outer, 2-veined and 2-toothed. *Flowers* nodding with tepals held vertically, deep pink (sometimes called purple), white (sometimes edged with dark pink) at base of tepals, unscented; perianth tube filiform but expanded near mouth, 3–5 mm long; tepals spreading, oblong, (8–)10–14 × 4–8 mm. *Stamens* symmetrical in bud, becoming unilateral in open flower, straight, held horizontally at right angles to tepals; filaments inserted at mouth of tube, filiform, 2–3(–4) mm long, white; anthers oblong-linear, 2.5–4.0 mm long, yellow, dehiscing completely by longitudinal slits. *Style* dividing between base and lower 1/3 of filaments, style ± 2 mm long, branches tubular, strongly falcate, stigmatic apically, white. *Flowering time*: mid-Aug. to late Sept. Figure 5.

Distribution and habitat: *Ixia curvata* has a relatively modest range for sect. *Dichone*, extending in the western Karoo from the eastern end of the Hantamsberg at Moordenaarspoort northeast of Calvinia southward to Middelpos on the Roggeveld Plateau (Figure 1). The range remains poorly documented and there are no collections from Middelpos itself although we have seen the species on the commonage there. Our description is based largely on the Middelpos population and the illustration here is from this population. Plants favour the red clay soils derived from dolerite but are also found in pale clay ground derived from Karoo shales. Pollination of *I. curvata* has not been studied. Raimondo *et al.* (2009) regarded the conservation status as of Least Concern (LC) and we concur, noting the relatively modest range of the species.

FIGURE 5.—*Ixia curvata*, Middelpos, without voucher. Scale bar: 10 mm. Artist: John Manning.

Diagnosis and relationships: relatively low growing, *Ixia curvata* is a small plant, seldom exceeding 200 mm in height (although taller plants are known) and has nodding, pink flowers with tepals held vertically with stamens borne horizontally. Like other members of ser. *Euanthera*, the anthers split completely. The four or five leaves are typically falcate and leathery rather than fibrotic. Its immediate relationships remain uncertain. Other species of the series have upright flowers with tepals spreading horizontally. The plants are notable in the thick collar of fibres around the base and chestnut-brown cataphylls.

The species was first collected by Rudolf Marloth in October 1920, but the label data localizing the plants from 'plains near Tulbagh' is an obvious mistake. Marloth collected in the Roggeveld and nearby in October 1920 when he also made the first collections of *Ixia rigida* (initially assigned to *I. brevituba*) and *I. trifolia*, thereby establishing the presence of sect. *Dichone* in the Western Karoo.

Representative specimens

NORTHERN CAPE. **3119** (Calvinia): lower clay slopes of the Hantamsberg N of Downes Siding, (–BD), 12 Sept. 2004, *Goldblatt & Porter 12411* (MO); 4.5 miles [± 6.8 km] SSE of Calvinia, (–DB), 21 Sept. 1955, *Acocks 18504* (PRE). **3120** (Williston): 35 km SE of Calvinia to Middelpos on Blomfontein road, (–CA), 30 Sept. 1976, *Goldblatt 4281* (MO). Uncertain localities: 'Plains near Tulbagh,' Oct. 1920, *Marloth 9936* (PRE); Van Rhyns Pass, 15 Sept. 1953, *Taylor 3967* (NBG).

6. **Ixia trinervata** *(Baker)* G.J.Lewis in Journal of South African Botany 28: 169. (1962); De Vos: 75 (1999). *Tritonia trinervata* Baker: 191 (1892); Baker: 212 (1896). Type: South Africa, [Western Cape], Appelskraal, Riviersonderend, *Ecklon & Zeyher Irid. 242* (K, holo.!, G, Z, iso.).

Plants 250–450 mm high, stem usually unbranched or with a single short branch, with membranous cataphylls enclosing base. *Corm* 9–18 mm diam., tunics of firm, medium-textured, netted fibres. *Leaves* 2, lower 1 ± basal, upper 1 sheathing stem in lower 1/2, blades lanceolate, 4–10 mm wide, conspicuously 3-veined, main vein moderately thickened, secondary veins slightly less so, margins not or barely thickened. *Spike* flexuose, mostly 7–122-flowered; bracts membranous, translucent, sometimes turning rusty brown in upper 1/3, outer 5–6 mm long, with prominent central vein and up to six secondary veins, 1-toothed or ± fringed, inner ± as long as outer, 2-veined and 2-toothed. *Flowers* rotate, deep rose pink, unscented; perianth tube filiform, 4–5 mm long, clasping style; tepals spreading, ovate-elliptic, obtuse to retuse, 12–15 × 6–8 mm. *Stamens* symmetrically arranged, erect; filaments inserted at mouth of tube, filiform, 2–4 mm long, ± white; anthers erect, suborbicular, ± 2 mm long, yellow, thecae acute and recurved at base, dehiscing by narrow, longitudinal slits almost to apex. *Style* dividing at base of filaments, branches falcate, 2.5–3.0 mm long, tubular-filiform, stigmatic apically, pale pink.

Distribution and habitat: a narrow endemic of the Caledon District, *Ixia trinervata* extends from Elgin Station and the Farm Arieskraal in the west to Riviersonderend in the east, a linear distance of some 65 km (Figure 3). Plants favour clay soils and grow in renosterveld, often in stony ground. There is evidence that flowering is stimulated after fire and we have seen plants blooming well on the Farm Vleitjies after a summer burn. We agree with Raimondo *et al.*'s (2009) estimate of near threatened status (NT) for *I. trinervata*.

Diagnosis and relationships: first described by J.G. Baker (1892) as *Tritonia trinervata* and transferred to *Ixia* by G.J. Lewis (1962), *I. trinervata* is readily recognized by the presence of just one foliage leaf, the blade narrowly sword-shaped and with three ± equally conspicuous veins and unthickened margins. A second leaf is largely to entirely sheathing, although relatively broad. The flowers are an intense, bright mauve pink or red-magenta. Lewis (1962) considered the species most closely related to usually pale pink-flowered *I. stricta*, which blooms later in the season, at which time its 3–5 leaves are usually dry. Both species favour clay or clay loam soils. The new *I. bifolia*, described here, recalls *I. trinervata* in its single foliage leaf, but the leaf blade is much narrower than in *I. trinervata*, with a single main vein embedded in the leathery-succulent leaf tissue. When dry the leaf margins appear particularly thickened, unlike those of *I. trinervata*. The flowers of *I. bifolia* are slightly smaller than those of *I. trinervata* and have unusual, almost straight, deep pink style branches in marked contrast to the pale pink, falcate style branches of *I. trinervata*.

Representative specimens

WESTERN CAPE.—**3419** (Caledon): Diepklowe Nature Reserve, 14 km N of Botrivier, Farm Welgemoed, (–AA), 3 Sept. 2001, *Helme 2152* (NBG); Elgin, Die Hawe, near Post Office, clay ground, (–AA), 11 Oct. 2000, *Goldblatt 11639* (MO); Lebanon Forest Reserve, Houw Hoek, (–AA), 26 Sept. 1954, *Martin 1043* (NBG), *Lewis 4178* (PRE, SAM); slopes of Eseljag Pass, near Farm Vleitjies, clay soil in renosterveld, (–AB), 30 Sept. 2000, *Goldblatt & Nänni 11597* (MO, PRE); 18 Sept. 2011, *Goldblatt & Manning 13642* (MO, NBG).

7. **Ixia bifolia** Goldblatt & J.C.Manning, sp. nov.

TYPE.—Western Cape, 3419 (Caledon): northern slopes below Shaw's Pass, Farm Treintjiesrivier, (–AD), 13 Sept. 2009, *Goldblatt & Manning 13444* (NBG, holo.; MO, PRE, iso.).

Plants mostly 160–300 mm high, stem unbranched, with short scale ± 1 mm long, inserted 12–30 mm below base of spike, sheathed basally by dry, brown cataphylls. *Corm* 10–12 mm diam., tunics of fine, soft to firm fibres extending upward as collar, bearing cormlets at base. *Leaves* 2, lower 1with long sheath, ± lanceolate, blade 60–80 × (3–)5–8 mm, reaching up to middle of stem, leathery-succulent, main vein not raised when alive (visible as a pale line when held to light), slightly raised when dry, margins thickened, hyaline, transparent when alive, when dry raised above mesophyll and yellow, upper leaf largely sheathing, sometimes with short free blade similar to basal but 2–3 mm wide. *Spike* mostly 4–7-flowered, ± straight; bracts translucent pink with purple veins, 5–6 mm long, outer with 3 main veins terminating in acute teeth and 1 or 2 secondary veins along margins, inner 2-veined and acutely 2-toothed. *Flowers* rotate, upright, deep pink, darker toward tepal bases, rose-scented; perianth tube filiform, 5–6 mm long; tepals obovate, 12–14 × 5–7 mm. *Stamens* symmetri-

cally arranged; filaments filiform, inserted at mouth of narrow part of tube, ± 1.5 mm long, deep pink (drying purple); anthers suborbicular, 2.0–2.5 mm long, yellow, thecae obtuse basally, dehiscing completely or almost so by narrow slits. *Style* dividing at base of filaments, style branches 3.8–4.0 mm long, deep pink (drying purple), ± straight, extending outward below anthers, expanded and bilobed at tips. *Capsules* globose, ± 4 mm diam. *Flowering time*: mostly late Aug. to mid-Nov. but Dec. and early Jan. in the Caledon Swartberg. Figure 6.

Distribution and habitat: restricted to the Caledon District, *Ixia bifolia* is known from the northern slopes of Shaw's Mtn south of Caledon and from the lower slopes of the Swartberg east of Caledon (Figure 3). The soil at the Shaw's Mtn site is hard, stony clay and the habitat there is shared by the rare *Tritoniopsis flexuosa* (L.f.) G.J.Lewis (known from just two other localities). With its narrow range and just two known populations, one small and near a major road, we suggest a conservation status of (VU), Vulnerable.

Diagnosis and relationships: *Ixia bifolia* seems at first to belong in the *I. micrandra–confusa* complex, which also has very slender stems and small flowers but it stands out in the very short, leathery-succulent foliage leaf blade, up to 80 mm long and 5–7 mm wide. The leaf has a single, prominent vein and moderately thickened margins. The flowers seem most like those of *I. confusa* but the short, deep pink filaments ± 1.5 mm long and similarly deep pink, straight style branches ± 4 mm long (both drying purple) are quite different from *I. confusa*, which has white filaments 1.3–2.5 mm long and white, strongly falcate style branches 1–2 mm long. We suspect that *I. bifolia* may be more closely allied to *I. trinervata*, which has slightly larger flowers on longer spikes but also a single foliage leaf, this relatively broad and always with three prominent veins and margins not thickened. *I. bifolia* shares with other members of the *I. micrandra* complex the finely fibrous corm tunics that extend upward around the stem base. Stems are almost always unbranched and have an odd feature, a small, nearly microscopic, translucent scale and prophyll, ± 1 mm long, inserted 12–30 mm below the base of the spike. Stems of *I. confusa* may be branched and are normally taller than in *I. bifolia* but also have a translucent stem scale in populations from the Swartberg and sometimes elsewhere across its range. *I. confusa* has 3 or 4 linear leaves, usually ± 2 mm, but occasionally to 4 mm wide.

Representative specimens

WESTERN CAPE.—**3419** (Caledon): N slopes below Shaw's Pass E of main road, (–AD), 16 Sept. 2009, *Goldblatt & Manning 13640* (MO, NBG, K); lower slopes of Caledon Swartberg, grazed field E of Drayton Siding, (–BA), 16 Sept. 2011, *Goldblatt & Manning 13641* (MO, NBG).

8. **Ixia confusa** (G.J.Lewis) Goldblatt & J.C.Manning, stat. nov. *I. micrandra* var. *confusa* G.J.Lewis: 162 (1962); De Vos: 73 (1999). Type: South Africa, [Western Cape], Montagu, Donkerkloof, 26 Sept. 1946, *Compton 18481* (NBG, holo.!; BOL!, iso.).

Plants (150–)250–500 mm high, stem simple or 1-branched, with short scale ± 1 mm long, inserted

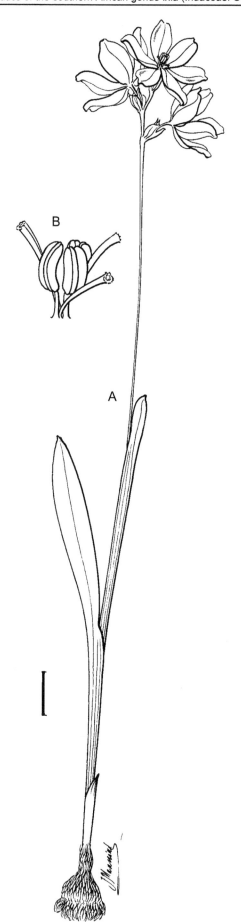

25–40 mm below base of spike, sheathed basally by green, leathery to firm, dry cataphylls. *Corm* 8–10 mm diam., tunics of fine, soft fibres sometimes extending upward in a weakly developed collar. *Leaves* (2)3(4), lower (1)2 or 3 with linear to narrowly sword-shaped blades, ± 1/3 as long to reaching top of stem, 1.5–4.0 mm wide, upper leaf largely sheathing, with short free blade. *Spike* slightly flexuose, (2–)5–9(–12)-flowered; bracts membranous, becoming dry from tips, 4–5 mm long, outer 1- or 3-veined, irregularly 2–3(–5)-lobed or shortly toothed, inner often slightly longer than outer, 2-veined and 2-toothed. *Flowers* rotate, upright, pink to light red-purple, rose-scented; perianth tube filiform, 3–4 mm long; tepals spreading, ± ovate, mostly 12–15 × 6–7 mm. *Stamens* symmetrically arranged, erect; filaments inserted at mouth of tube, filiform, 1.0–2.8 mm long; anthers oblong, 2.0–2.3 mm long, yellow, thecae acute and recurved at base, usually dehiscing longitudinally from base almost to apex. *Style* dividing at or up to middle of filaments, style branches 1.5–2.0 mm long, falcate, tubular with expanded ciliate, stigmatic tips, sometimes shortly forked apically. *Flowering time*: mostly Jul.–early Dec.

Distribution and habitat: *Ixia confusa* occurs in the mountains of southern Western Cape, from near Elim and the Riviersonderend Mtns in the west through the Langeberg to the Paardevlei Mtns near Farm Bonnievale at the western end of the Outeniqua Mtns (*Vlok 709* NBG, PRE), and inland in the Great Swartberg (Figure 7) where it has been recorded from Swartberg Pass, Meiringspoort, and Blesberg slightly further east. An outlying collection (*Marloth 2429* PRE) from Montagu Pass near George extends the range slightly south. We have encountered plants (*Goldblatt & Manning 12946*) on moist southwest-facing slopes in loamy clay in renosterveld–thicket transition vegetation. The flowers of our collection were strongly rose-scented. We suspect that plants flower particularly well after fire but our collection was made in veld that had not burned for over 10 years. Occurring mostly in montane habitats, there seems no current threat to the species and we agree with Raimondo *et al.*'s (2009) (LT, as *I. micrandra* var. *confusa*), Least Threat conservation status.

Diagnosis and relationships: known to G.J. Lewis (1962) as *Ixia micrandra* var. *confusa*, we raise the taxon to species rank after examining living plants and

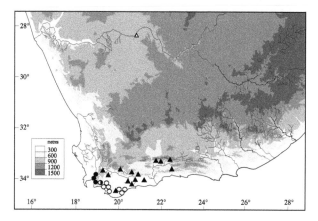

FIGURE 6.—*Ixia bifolia*, *Goldblatt & Manning 13444* (NBG). A, flowering plant; B, stamens and style branches. Scale bar: 10 mm. Artist: John Manning.

FIGURE 7.—Distribution of *Ixia confusa*, ▲; *I. micrandra*, ○; and *I. minor*, ●.

comparing them with *I. micrandra*. Both *I. micrandra* and closely related *I. minor* (= *I. micrandra* var. *minor* G.J.Lewis) have corms with finely fibrous tunics that extend upward in a collar around the base of the stem, terete to subterete or linear leaves mostly 1.0–1.5 mm at widest, and spikes of up to 6 flowers. *I. confusa* has corms with similar finely fibrous tunics, sometimes without a collar of fibres around the base, but plane leaves 1.5–4.0 mm wide, and spikes with as many as 12 flowers. *I. micrandra* has anthers 3–4 mm long, and slender, ± straight, ascending style branches 3–4 mm long, thus quite different from the shorter anthers, ± 1.0–2.5 mm long, and falcate style branches 1.5–2.5 mm long of *I. confusa*. The style branches of *I. confusa* match those of *I. minor* but the terete or subterete leaf blades, flattened filaments and consistently few-flowered spikes of the latter immediately separate them.

Ixia confusa may also be confused with the new *I. bifolia* but this species has a single, relatively broad, leathery-succulent foliage leaf and flowers with shorter filaments ± 1 mm and longer, straight style branches ± 4 mm long. Both the filaments and style branches of *I. bifolia* are dark pink and dry dark purple, unlike the pale filaments and style branches of *I. confusa*, which dry to a pale, almost colourless state.

Annotations by G.J. Lewis on collections of *Ixia confusa* from the Swartberg indicate that at one time she considered these populations a separate species, which she intended to call *I. aestivalis*, in part because of their late flowering, from late October to early December. We find that collections from the Swartberg also differ in having slightly shorter filaments, 1–2 mm long, than typical *I. confusa* (1.5–2.8 mm long), the style sometimes dividing above the mouth of the perianth tube (opposite the middle of the filaments in *Goldblatt 7962*), and anther dehiscence sometimes not reaching the middle of the thecae. The plants have a small, translucent scale and prophyll on the upper part of the stem, a feature occasionally present in southern populations of *I. confusa*. Additional study may show that the later-flowering Swartberg populations merit separate taxonomic status but we find no consistent character differences between them.

Representative specimens

[* = apparently no collar around base of stem]

WESTERN CAPE.—**3319** (Worcester): eastern foot of Stettynsberg, (–CD), 31 July 1949, *Esterhuysen 15590* (BOL, NBG, PRE); slopes of Boesmanskloof Pass, MacGregor, (–DC), 13 Sept. 1962, *Lewis 6067** (NBG, PRE); top of Boesmanskloof Pass, MacGregor, (–DC), 13 Sept. 1962, *Lewis 6080* (NBG). **3320** (Montagu): foot of Tradouw Pass, (–DC), 18 Sept. 1968, *Marsh 875* (PRE). **3321** (Ladismith): Paardevlei Mtns N of Farm Bonniedale, 1 500 ft [± 460 m], (–DD), 24 Sept. 1983, *Vlok 709** (NBG, PRE). **3322** (Oudtshoorn): summit, Swartberg Pass, (–AC), Dec. 1906, *H. Bolus 12336* (BOL); Swartberg Pass, (–AC), 5 Nov. 1958, *Werdermann & Oberdieck 843* (B, PRE), 22 Oct. 1986, *Goldblatt 7962* (MO, PRE); Swartberg, sandy flats SW of Blesberg, (–BC), 7 Jan. 1975, *Thompson 2274* (PRE); Meiringspoort, S slopes, (–BC), 16 Oct. 1955, *Esterhuysen 24880* (BOL, MO, PRE); Montagu Pass, George, (–CD), *Marloth 2429* (PRE). 3419 (Caledon): Uintjieskuil near Elim, (–DB), Sept. 1926, *Smith 3182** (PRE). **3420** (Bredasdorp): S of Swellendam on road to Bontebok Park, light stony clay, (–AB), 26 Aug. 2000, *Goldblatt 11430* (MO); coastal renosterveld hills at Goerreesoe, S slope, (–AD), 30 Aug. 1962, *Acocks 22672* (PRE); clay loam slopes between Heidelberg and Strawberry Hill, in renosterveld, (–BB), 4 Sept. 2007, *Gold-

blatt & Manning 12946 (MO, NBG, PRE). **3421** (Riversdale): near Oudebosch crossing, Korente River crossing, Riversdale, (–AA), 16 Sept. 1981, *Hugo 2773* (PRE).

9. Ixia micrandra *Baker* in Journal of Botany 1876: 237 (1876); Baker: 78 (1896); Lewis: 159 (1962); De Vos: 71 (1999). Type: South Africa, [Western Cape], Houw Hoek Mtns, July 1830, *Zeyher 4011* (K, lecto.! designated by Lewis: 161 (1962); PRE!, SAM!, S, Z, isolecto.).

Plants 250–350 mm high, stem unbranched, base sheathed by membranous to ± dry, brown cataphylls. *Corm* 9–18 mm diam., tunics of fine fibres extending upward as a collar. *Leaves* 2, lower 1 ± basal, upper 1 sheathing the stem in lower 1/2 to 2/3, free distally for up to 100 mm, blades linear, (0.5–)1.0–1.5(–2.0) mm wide, margins and midrib lightly thickened, hyaline when dry. *Spike* erect, ± straight, (1–)2–4(–6)-flowered; bracts membranous, translucent, sometimes suffused with purple distally, outer 5–7 mm long, prominently 3-veined and usually with secondary veins evident, acutely 3-cuspidate, sometimes central cusp smaller, inner ± as long as outer or slightly longer, 2-veined and 2-cuspidate. *Flowers* rotate, pale mauve-pink; perianth tube filiform, 3–4 mm long, clasping style; tepals spreading, ovate-elliptic, 12–15 × 4–6 mm. *Stamens* symmetrically arranged, erect; filaments inserted at mouth of tube, linear, straight, 0.6–1.8(–2.5) mm long, ± pale pink; anthers oblong, 2.0–2.6 mm long, erect, yellow, thecae slightly recurved basally and acute dehiscing completely from base by longitudinal slits. *Style* dividing at or shortly above base of filaments, branches ± straight and diverging to slightly arched, 2–5 mm long, usually longer than stamens, filiform-involute, stigmatic and sometimes forked at apex, pale pink. *Flowering time*: mostly late Jul.–early Sept., occasionally later.

Distribution and habitat: extending from Houw Hoek, Bot River and the Caledon Swartberg to Bredasdorp and De Hoop (Figure 7), *Ixia micrandra* typically occurs on sandstone slopes and flowers well only in seasons immediately after veld fires. A collection from De Hoop is from pockets of sandy soil in limestone pavement, an unusual habitat for the species. Largely montane, we see no current threat to the species and endorse Raimondo *et al.*'s (2009) (LT), least threat, status.

Diagnosis and relationships: treated by G.J. Lewis (1962) as including the two additional varieties, var. *confusa* and var. *minor*, we have no doubt that *I. micrandra* is allied to these two taxa, which are here raised to species rank. *I. micrandra* stands out in the group in its long, narrow, almost linear filaments and its style branches, which are unusually long at 2.0–3.5 mm, and straight or slightly arched, thus extending between the bases of the anthers. *I. confusa* has similarly slender but somewhat shorter filaments, slightly shorter anthers, and shorter, strongly arched style branches, 1.5–2.0 mm long. The leaves of *I. confusa* have plane, usually wider blades up to 4 mm wide (vs. 0.5–1.5(–2.0 mm) and without the prominently thickened main vein of *M. micrandra*. The style branches are also strongly arched in *I. minor* but that species has unusual flattened, lanceolate filaments, narrowed and kinked at the tips. The ranges of *I. micrandra* and *I. minor* are almost complementary, that of *I. micrandra* extending east of *I. minor*, but the two have been collected growing and flowering on the same day in the Houw Hoek Mtns, where their distributions overlap.

The only significant variant of *Ixia micrandra* we have seen is from near Bot River (*MacOwan in Herb. Norm. Austr. Afr. 262*). Plants of this population have unusual thick, fleshy leaf blades, but the flowers and other features accord with the species.

Representative specimens

WESTERN CAPE. **3419** (Caledon): Houw Hoek, (–AA), 17 Sept. 1987, [collected with *I. minor*, *Beyers 17*, on same day], *Beyers 22* (NBG); Bot River, Langhoogte, (–AA), Aug., *Ecklon & Zeyher Irid. 92* (MO); Caledon Baths, (–AB), July 1892, *F. Guthrie 2531* (BOL); Caledon Swartberg and vicinity of the baths, (–AB, BA), August 1830, *Zeyher 4009* (PRE); Highlands, low slopes between Kleinmond and Bot River (–AC), 27 July 1960, *Lewis 5733* (NBG); raised plain between the mountains and mouth of the Bot River, (–AC), July 1885, *MacOwan in Herb. Norm. Austr. Afr. 262* (BOL, PRE); Kleinmond commonage, (–AC), 15 Aug. 1982, *Burman 852* (BOL); Hermanus, behind Golf Course, deep sand, 60 m, (–AD), 1 Aug. 1984, *S. Williams 1042* (MO); Caledon Swartberg, well drained sandy soil, (–BA), 1 Aug. 2004, *Raimondo CR278* (NBG); Genadendal, mountains at ± 4 500 ft [1 400 m], (–BA), 27 Oct. 1899, *Galpin 4686* (BOL); Salmonsdam Reserve, Stanford, burnt slopes, 1 000 ft, (–BB), 31 Aug. *Goldblatt 421* (BOL); 2 miles [3 km] NW of Papiesvlei P.O., (–BC), 25 July 1962, *Acocks 22449* (PRE); Bredasdorp, hillside, (–BD), 27 July 1955, *Van Niekerk 430* (BOL). **3420** (Bredasdorp): De Hoop, Potberg Nature Reserve, upper slope above Boskloof, back sandy ground, (–BC), 6 Sept. 1978, *Burgers 1067* (PRE).

10. Ixia minor (*G.J.Lewis*) *Goldblatt & J.C.Manning*, stat. nov. *I. micrandra* var. *minor* G.J.Lewis: 162 (1962); De Vos: 73 (1999). Type: South Africa, [Western Cape], Hottentots Holland Mtns, 10 Sept. 1928, *J. Hutchinson 481* (BOL, holo.!; K, PRE!, iso.).

Plants 200–300 mm high, stem unbranched, sheathed by membranous to dry, brown cataphylls at base. *Corm* 8–10 mm diam., tunics of fine fibres extending upward as a collar, often in a dense mass. *Leaves* 2, lower 1 ± basal, upper 1 sheathing stem in lower 1/2 to 2/3, free distally for short distance, blades ± terete or oval in section, up to 1 mm diam., main vein evident only when dry. *Spike* erect, straight, mostly 2–4-flowered, bracts membranous, translucent, usually flushed purple, outer 5–7 mm long, usually prominently 3-veined, shallowly 3-lobed to acutely 3-toothed, inner bract ± as long as or slightly longer than outer, 2-veined and 2-lobed or 2-toothed. *Flowers* rotate, bright pink to mauve pink; perianth tube filiform, 3–5 mm long, clasping style; tepals spreading, ovate-elliptic, 12–14 × 5–6 mm. *Stamens* symmetrically arranged, erect, filaments flattened, ± lanceolate but narrowed and kinked at tips, inserted at mouth of tube, ± white, 1.0(–1.5) mm long, anthers ovoid, ± 1.0 mm long, erect, yellow, thecae acute but abruptly incurved and touching at base, dehiscing fully by longitudinal slits, often more widely at base, yellow. *Style* dividing opposite base to middle of filaments, branches up to 2 mm long, involute, tubular-falcate, ± compressed, narrowed at tips, stigmatic apically, white. *Flowering time*: late Jul.–mid-Sept., rarely later.

Distribution and habitat: a narrow endemic of the southwestern Cape, *Ixia minor* occurs from Franschhoek

Pass and the Wemmershoek Mtns south though the Hottentots Holland and Kogelberg Mtns to Houw Hoek (Figure 7). The habitat, stony sandstone slopes, is the same as that for *I. micrandra* and as in that species we see no current threat and suggest a least threat (LT) conservation status.

Diagnosis and relationships: *Ixia minor*, treated as var. *minor* of *I. micrandra* by Lewis (1962) but raised here to species rank, is evidently closely allied to the more widespread *I. micrandra*, with which it is easily confused unless floral details are examined carefully. Both share a slender, usually unbranched stem, small corms with fine, netted tunics that extend upward in a collar around the base, and a dry, often brown, upper cataphyll. The critical differences between the two species lie in their flowers: *I. minor* has flattened, ± wedge-shaped to oblanceolate filaments abruptly narrowed at the tips, shorter and broader anthers, often not dehiscing to the apex and kinked at the base with the bases of the thecae turned inwards and touching, and shorter, wider, strongly arched style branches slightly compressed and narrowed at the tips. To these microscopic details can be added a deep pink perianth. Flowers of *I. micrandra* have pale pink to purple-pink tepals, linear filaments, longer, oblong anthers dehiscing to the apex and longer, more slender style branches, ± straight to slightly arched.

The leaves of the two species also differ: those of *Ixia minor* are subterete to oval in section, thus when alive without margins or evident central vein, whereas *I. micrandra* has plane, linear leaf blades with somewhat thickened margins and, when dry, a raised main vein. Floral differences between *I. minor* and *I. micrandra* are illustrated in detail in the accounts by both Lewis (1962) and De Vos (1999). A color photograph of *I. minor* (Manning *et al.* 2002: 242) under the name *I. micrandra* clearly shows the short, recurved style branches of the species. Compelling evidence for their separation is provided by collections of both species made by the late J. Beyers in the Houw Hoek Mtns flowering on the same date and sufficiently different in appearance that Beyers gave them different collection numbers. Habitat differences, if any, were not recorded.

Representative specimens

WESTERN CAPE.—**3319** (Worcester): Wemmershoek, SE end, 3 000–4 000 ft [915–1 220 m], (–CC), 19 Oct. 1943, *Esterhuysen 9060* (BOL); Franschoek Pass, ± 1 km S of summit, (–CC), 9 Oct. 1973, *Boucher 2337* (PRE). **3418** (Simonstown): W foot of Sir Lowry's Pass, (–BB), Aug. 1985, *De Vos 2585* (NBG); Sir Lowry's Pass mountains, (–BB), July 1894, *H. Bolus 7850* (BOL); Steenbras River, 600 m, (–BB), 14 Oct. 1894, *Schlechter 5402* (PRE). **3419** (Caledon): Elgin, (–AA), 19 Sept. 1931, *Levyns 3341* (BOL); Hottentots Holland, Moordenaarskop and Sugar Loaf, (–AA), 30 Oct. 1943, *Esterhuysen 9123* (BOL); Houw Hoek, (–AA), 17 Sept. 1987 [collected with *I. micrandra*, *Beyers 22*, on same day], *Beyers 17* (NBG).

Ser. **Dichone**

Flowers upright or nodding, radially symmetric (*Ixia stricta*) or zygomorphic. *Stamens* symmetrically arranged (*Ixia stricta*) or unilateral and horizontal to pendent; anthers oblong, recurved and acute at base, dehiscing incompletely from base. *Style* branches always falcate.

7 spp., centred in the western half of Western Cape, but with isolated populations in northern Namaqualand and the Bokkeveld Mtns of Northern Cape.

11. **Ixia stricta** *(Eckl. ex Klatt)* G.J.Lewis in Journal of South African Botany 28: 167 (1962); De Vos: 73 (1999). *Tritonixia stricta* Eckl. ex Klatt: 357 (1882). *Tritonia scillaris* var. *stricta* (Eckl. ex Klatt) Baker: 191 (1892). Type: Western Cape, Caledon, Swartberg and vicinity of the Baths, December 1830, *Zeyher 4008* (SAM, lecto.!, designated by Lewis: 162 (1962); K, PRE!, isolecto.).

Agretta stricta Eckl.: 23 (1827), nom. nud.

Plants 200–450 mm high, stem simple or with 1 or 2 ± erect branches subtended by acute, translucent bracts and prophylls, sheathed at base by short, dry cataphylls. *Corm* mostly ± 10–15 mm diam., tunics of medium-textured, shaggy fibres. *Leaves* (3)4 or 5, often ± dry at flowering, basal, lower 2 or 3 with blades plane, 4–9 mm wide, upper 1 or 2 sheathing stem in lower 1/2 to 2/3, free distally for short distance, margins and main vein thickened (hyaline when dry). *Spike* erect, flexuose, 6–12(–18)-flowered, bracts membranous, translucent, outer 5–6 mm long, usually with 1 prominent vein, and 1-toothed, inner bract ± as long or slightly shorter or longer, 2-veined and 2-toothed. *Flowers* rotate, pink to purple, pale yellow edged with darker pink at tepal bases, sweetly scented (*Viviers 1150*), perianth tube filiform, 4–5 mm long, clasping style, tepals spreading, obovate, ± obtuse, ± 10 × 4–5 mm. *Stamens* symmetrically arranged, filaments 1.0–1.5 mm long, white, anthers oblong, ± 2 mm long, erect, yellow, thecae acute and slightly recurved at base, dehiscing near base by longitudinal slits. *Style* dividing opposite base of filaments, branches arching outward, ± 1.5 mm long, falcate, tubular, stigmatic apically, pale pink. *Flowering time*: Nov.–mid-Dec.

Distribution and habitat: restricted to the Caledon–Bredasdorp area of Western Cape, *Ixia stricta* extends from the Houw Hoek Mtns in the west to Bredasdorp in the southeast (Figure 1). Typically flowering late in the year, in November and December, *I. stricta* is poorly collected. Not surprisingly the foliage is often more-or-less dry at flowering. Plants occur mostly in montane habitats, but usually at fairly low elevations, in sandstone and loamy clay soils. Raimondo *et al.* (2009) suggest a conservation status of (LC), least concern, but we suggest that (NT), near threatened, is more appropriate.

Diagnosis and relationships: while not particularly distinctive in its radially symmetric flowers with short, erect stamens, *Ixia stricta* is identified in ser. *Dichone* by a combination of the rotate flowers and a basal fan of (3)4 or 5 sword-shaped leaves, the blades with a prominent main vein and thickened margins. The sweetly scented flowers have a filiform tube about half as long as the tepals and a particularly short style ± 1 mm long. Blooming late in the season, in November and December, the leaves are often dry or partly so, a feature unusual in the genus.

History: the plant illustrated by Jacquin (1792) and misidentified as *Ixia polystachya* L. is the earliest record of *Ixia stricta* and this illustration is also one of two sources cited by Ker Gawler (in Sims 1801) for his *I. scillaris* var. *angustifolia*. Both Lewis (1962) and De Vos (1999) associated Jacquin's illustration with *I. stricta* and the illustration shows clearly the relatively broad, prominently veined basal leaves and the sheathing upper leaves subtending straight, several-flowered branches. Ker Gawler's varietal name *angustifolia* is nomenclaturally illegitimate and superfluous because he included the earlier name *I. scillaris* var. *incarnata*, this now most likely *I. scillaris* subsp. *latifolia*.

Watsonia retusa (Salisb.) Klatt (1882), a combination based on indirect citation of the basionym *Ixia retusa* Salisb., was regarded by Lewis as a synonym of *I. micrandra* var. *confusa*. De Vos (1999), however, associated the species with *I. stricta* and we concur. Klatt evidently intended to provide a name for the plant called *I. polystachia* [sic] by Ker Gawler (1803) in *Curtis's botanical magazine* but as Ker Gawler specifically mentioned *I. retusa* Salisbury (1796) in his discussion of the plate in question, Klatt's name must be treated as a combination, but with a basionym that has no known type. Salisbury's brief protologue matches the plant depicted in *Curtis's botanical magazine* published seven years later in all respects, notably the style branches emerging between the filaments and the tepals twice as long as the perianth tube. But there is no connection between Salisbury's species and the painting.

Representative specimens

WESTERN CAPE.—**3419** (Caledon): Houw Hoek, 2 000 ft [610 m], (–AA), 25 Nov. 1896, *Schlechter 9411* (BM, BOL, K, PRE); Groenland Mtns, Klein Jakkalsrivier, red clay, 2 200 ft [± 700 m], (–AA), 2 Dec. 1982, *Viviers 1150* (NBG, PRE); Hermanus, Glenfruin, Onrus Mtn, fynbos on clay, recently burnt, (–AC), 16 Nov. 1991, *Barker 305* (MO); Vogelgat, Hermanus, at Fernkloof gate, on shale band, 450 m, (–AD), 27 Nov. 1983, *Williams 3523* (NBG, PRE); Drayton Siding, Caledon, (–BA), 16 Dec. 1968, *Goldblatt 393* (BOL); roadside between Napier and Elim, (–BD), 5 Dec. 1985, *De Vos 2642* (NBG); near Napier, (–BD), 30 Nov. 1933, *Salter 4151* (BOL, K); near Baardscheerdersbos, (–DA), 10 Nov. 1985, *De Vos 2636* (NBG); Boskloof, Bredasdorp, (–DB), 28 Dec. 1983, *Lavranos 21797* (MO, PRE); Bredasdorp, Farm Geelrug, (–DB), 17 Dec. 1975, *Barker 10940* (NBG, PRE). **3420** (Bredasdorp): Bredasdorp Mtns, (–CA), *Galpin s.n. PRE10463* (PRE).

12. **Ixia erubescens** *Goldblatt* in Journal of South African Botany 37: 233 (1971b), as a new name for *I. crispa* L.f.: 91 (1782), a superfluous name for *I. undulata* Burm.f. *Agretta crispa* Eckl.: 24 (1827), as a new name for *I. crispa* L.f. *Tritonia thunbergii* N.E.Br.: 137 (1929), as a new name for *Ixia crispa* in *Tritonia*, non *T. crispa* (L.f.) Ker Gawl. Type: South Africa without precise locality, *Thunberg s.n.* [LINN 58.25, lecto., designated by Lewis: 173 (1962)].

Plants 120–300 mm high, usually forming clumps, stem simple, rarely 1-branched, sheathed basally by membranous cataphylls. *Corm* mostly 5–8 mm diam., tunics of fine fibres, usually with large cormlets at base. *Leaves* 4–6, ± lanceolate, all basal, ± 1/3 to 1/4 as long as stem, 2–4(–6) mm wide, main vein well developed, margins or half blade crisped and deeply undulate. *Spike* erect, lax, secund, weakly flexuose, mostly 6–10-flowered, flowers mostly 15–25 mm apart; bracts mostly green, becoming translucent at tips, turning purple along distal margins, 3–4 mm long, outer with 3 acute-attenuate cusps, obscurely 3-veined, inner 2-veined and acutely 2-toothed. *Flowers* ± secund with tepals held ± vertically, zygomorphic with stamens unilateral, pink, yellow-green in throat edged with dark purple at tepal bases, unscented; perianth tube filiform, 2–3 mm long, tightly clasping style; tepals ovate, spreading ± at right angles to tube, 8–11 × 4–6 mm, outer tepals slightly narrower than inner. *Stamens* unilateral; filaments filiform, ± 2 mm long, greenish white, drying yellow; anthers oblong, ± 1.5 mm long, yellow, horizontal, becoming slightly pendent, thecae acute and recurved at base, dehiscing incompletely by a narrow slit at base, yellow. *Style* dividing at base of filaments, style branches 1.5–2.0 mm long, greenish white, falcate, recurved in distal third, tubular, stigmatic apically. *Flowering time*: mainly mid-Aug.–mid-Sept.

Distribution and habitat: predominantly a species of near southwestern Western Cape, *Ixia erubescens* extends from Tulbagh southwards to Stellenbosch and the Caledon District (Figure 8) where its southeasternmost station is at Franskraal, but is absent from the Cape Peninsula. Outlying populations occur on the lower slopes of the Piketberg. The species is most often encountered on heavy clay or granitic alluvium in seasonally damp places, where plants typically form tight clusters through the germination of daughter cormlets. The Piketberg plants are restricted to cooler south-facing slopes where they grow on well-drained clay slopes in the shade of shrubs and small trees, and grow singly rather than in clusters. We concur with the conservation status of *I. erubescens*, listed by Raimondo *et al.*(2009) as (LC), least concern, although we note that its range has been historically much reduced through agriculture, especially in the Swartland and Boland.

Diagnosis and relationships: a relatively small species, *Ixia erubescens* is immediately recognized by the fine corm tunics, the 4–6 narrow, deeply undulate and crisped leaves, usually unbranched stem, and the vertically oriented flower with the stamens unilateral and slightly pendent when fully open. Both Lewis (1962) and De Vos (1999) regarded the flowers as fundamentally actinomorphic with symmetrically arranged stamens and noted that only late in anthesis do the anthers become unilateral and pendulous, simply due to gravity. Our own observations counter this: the stamens are unilateral immediately the flower buds open. We

assume that *I. erubescens* is most closely allied to *I. scillaris*, which typically has medium to coarse corm tunic fibres, broader leaves, usually with plane margins, and somewhat to substantially larger flowers, also with unilateral stamens, but with a perianth tube up to 4.5 mm long and tepals up to 14 mm long (vs. tube 2–3 mm long and tepals 11–12 mm long). Occasional populations of *I. scillaris* with undulate or crisped leaves are readily distinguished from *I. erubescens* by their broader leaves, larger flowers, and coarser corm tunic fibres.

The taxonomic history of *Ixia erubescens* is somewhat confused. Linnaeus fil. (1782) described *I. crispa* based on C.P. Thunberg's collection but cited the earlier name *I. undulata* Burm.f. (Burman 1768) (now *Tritonia undulata*) in synonymy. According to current rules of nomenclature this rendered *I. crispa* superfluous and therefore illegitimate. The two were often confused by later authors and Baker (1877) used the name *Tritonia undulata* for the species, not realizing that its type is a different plant. Brown (1929) was the first to realize that *I. crispa* and *I. undulata* were different species and transferred *I. crispa* L.f. to *Tritonia* under a new epithet, *T. thunbergii*, because he believed that the name *T. crispa* (L.f.) Ker Gawl. (based on *Gladiolus crispus* L.f.) barred a combination based on Linnaeus fil's *I. crispa*. On transferring sect. *Dichone* to *Ixia*, Lewis (1962) used the name *I. crispa* for the species. Then Goldblatt (1971b), realizing that *I. crispa* was superfluous and thus illegitimate, established the new name *I. erubescens*. Transfer to *Ixia* of the legitimate *T. thunbergii* N.E.Br. is prevented by use of that epithet in *Ixia* by Roemer & Schultes (1817), their species being a synonym of *T. securigera* Ker Gawl.

Representative specimens

WESTERN CAPE.—**3218** (Clanwilliam): Piketberg, SE-facing slopes of Versveld Pass, damp clay and clay-loam slopes, (–AC), 6 Sept. 2009, *Goldblatt & Porter 13280* (MO, NBG, PRE), 2 Nov. 2011 (fr.), *Goldblatt & Porter 13712* (MO, NBG). **3318** (Cape Town): near Wellington, (–DB). 17 Aug. 1926, *Grant 2383* (BOL, PRE); stony hills near Paarl, (–DB), *Ecklon 10 'Agretta crispa'* (BM, G, M, K, PRE); Elsenberg, (–DD), *Grant 2516* (MO, PRE). **3319** (Worcester): Tulbagh, (–AC), Oct. 1916, *Marloth 7772* (PRE), Sept. 1915, *Marloth 7127* (PRE), 11 Sept. 1926, *Grant 2467* (PRE); Tulbaghskloof, etc., Sept. 1830, *Ecklon & Zeyher s.n. 'Agretta crispa'* (PRE); near Artois, Tulbagh, (–AC), Aug. 1885, *H. Bolus s.n. (BOL26898)*; Tulbagh, stony clay hillside, 12 Sept. 1950, *Esterhuysen 17474* (BOL). **3419** (Caledon): Houw Hoek, (–AA), Aug. 1912, *Rogers s.n.* (PRE 13553); 3 km W of Caledon, W of road to Badshoogte, (–AB), 12 Oct. 2001, *Helme 2294* (NBG); Farm Karwyderskraal, Bot River–Hermanus, (–AC), 15 Sept. 1988, *O'Callahan 1784* (NBG); Franskraal, N slope in damp clay, (–CB), 29 Sept. 2005, *Ebrahim & Raimondo CR1075* (NBG).

13. **Ixia scillaris** *L.*, Species plantarum ed. 2, 1: 52 (1762); Lewis: 173 (1962); De Vos: 78 (1999). *Tritonia scillaris* (L.) Baker: 163 (1877). *Tritonixia scillaris* (L.) Klatt: 357 (1882). Type: South Africa, without precise locality or collector [LINN 58.25—digital image!, lecto., designated by Lewis: 174 (1962)].

Plants mostly 150–500 mm high, stem often with 1–2, rarely up to 6 branches, sheathed basally by ± dry, papery cataphylls, often green distally, later turning brown. *Corm* 10–15 mm diam., tunics of medium-textured to coarse fibres; cormlets at base conspicuous in subsp. *toximontana*. *Leaves* 3–7, mostly in basal fan, upper 1 or 2 partly sheathing the stem, 1/3 to 1/2

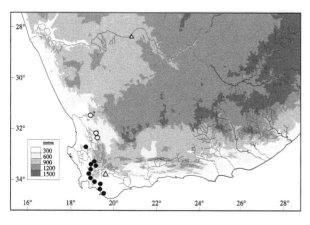

FIGURE 8.—Distribution of *Ixia collina*, △; *I. erubescens*, ●; and *I. flagellaris*, ○.

as long as stem, blade lanceolate or falcate, (3–)6–20 mm wide, main and sometimes additional veins raised but not thickened, margins markedly thickened in subsp. *toximontana*. *Spike* erect, fairly lax, flexuose to ± straight, secund in 2 ranks, mostly 10–20-flowered; bracts membranous, translucent, sometimes turning purple distally, 3–5 mm long, outer prominently 3-veined and 3-toothed, inner 2-veined and 2-toothed. *Flowers* ± nodding with tepals held vertically with lowermost slightly reflexed, zygomorphic with stamens unilateral, bright to pale pink, occasionally white, yellow or white edged in darker pink at tepal bases, lightly rose-scented or unscented; perianth tube filiform, 2.5–4.0 mm long, tightly clasping style; tepals ovate-elliptic, (8–)10–15 × 5–9 mm, inner 3 slightly larger than outer, lowermost tepal held apart and usually somewhat reflexed. *Stamens* unilateral; filaments inserted at mouth of tube, filiform, 1.6–3.5 mm long, arching upward near tips, white to pale yellow; anthers oblong-linear, 2.3–4.0 mm long, horizontal or slightly pendent, thecae recurved and acute at base, dehiscing incompletely from base by narrow slits, yellow. *Style* dividing ± opposite base of filaments, branches ± 1.5 mm long, pale yellow, falcate, tubular, stigmatic apically. *Capsules* ± obovoid–oblong, 4–5 mm long. *Flowering time*: mainly mid-Aug.–mid-Oct., occasionally later. Figure 9.

Distribution and habitat: predominantly a species of the west coast and near interior of Western Cape, *Ixia scillaris* extends from the Cape Peninsula north to the Gifberg and Kobee Mtns, along the coastal plain and mountain ranges parallel to the coast as far inland as the Cedarberg and the Pakhuis–Nardouw Mtns. Isolated populations occur in the Spektakel Mtns in northern Namaqualand, Northern Cape (Figure 10). Plants grow

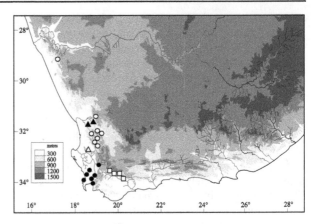

FIGURE 10.—Distribution of *Ixia scillaris* subsp. *latifolia*, ○.; subsp. *scillaris*, ●; subsp. *toximontana*, ▲; *I. simulans*, □; and *I. tenuis*, △.

on a variety of soils but most commonly on heavy clays derived from shale or granite. With so wide a range its conservation status (LC), Least Concern, seems appropriate. The range of subsp. *scillaris* is, however, much reduced and may be Vulnerable (VU).

Studies of the floral biology of *Ixia scillaris* (Goldblatt *et al.* 2000) show the flowers are adapted for buzz pollination by anthophorine bees. Pollen remains in the anthers until vibrated at high frequency by the wings of visiting bees. Unusual for flowers with this pollination strategy, the pollen falls back into the centre of the flower through narrow slits at the base of the anthers. Like all species of sect. *Dichone*, the flowers do not produce nectar.

Diagnosis and relationships: the most common and widespread species of sect. *Dichone*, *Ixia scillaris* is recognized primarily by the nodding flower with vertically oriented tepals with unilateral, slightly pendent and often relatively long anthers, combined with a fan of 3–7 basal leaves, (3)6–20 mm wide. Related *I. erubescens* has narrow, ± falcate leaves, 2–4(–6) mm wide, with closely crisped margins, and is a smaller plant with smaller flowers, the tepals 8–11 × 4–6 mm and anthers 2.0–2.5 mm long vs. tepals mostly 10–15 × 5–9 mm and anthers 3–4 mm long in *I. scillaris*. Apparently closely allied *Ixia simulans*, with half to fully nodding flowers and unilateral stamens with pendent anthers, has a single basal, ± linear foliage leaf and a second leaf sheathing the stem for most of its length. *Ixia tenuis* from the Piketberg is a particularly slender species with small corms with finely fibrous tunics, just three, narrow, soft-textured leaves and relatively small flowers with dark magenta tepals. *I. tenuis* grows in a habitat unique for species allied to *I. scillaris*, seeps on sandstone rocks, growing in moss in soils less than 10 mm deep.

Variation: Typical *Ixia scillaris* from Darling and Tulbagh to the Cape Peninsula has (4)5–7 relatively narrow leaves, mostly 5–8 mm wide, with 3–5 secondary veins conspicuous only when dry, mostly branched stems (rarely with up to 7 branches), and moderately dense spikes with flowers spaced 8–12 mm apart. The flowers are relatively small, with tepals 8–10 × 4–5 mm and ± symmetrically arranged, although the lowermost tepal is held slightly apart and recurved, and anthers

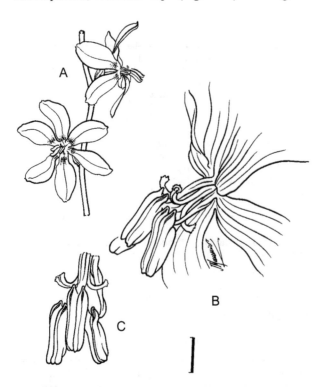

FIGURE 9.—*Ixia scillaris* subsp. *latifolia*, *Goldblatt & Manning 13482* (NBG). A, flowers; B, stamens and style branches, side view with base of tepals; C, stamens and style branches, dorsal view. Scale bar: 10 mm. Artist: John Manning.

2.0–2.8 mm long. In contrast, plants from the Olifants River Valley and low elevations in the Cedarberg ranging into the Bokkeveld and Spektakel Mtns are always taller, with 2–4, broader leaves mostly 15–20 mm wide, usually with at least two conspicuous secondary veins in addition to the main vein, stems simple or with 1(2) branches, and lax spikes with flowers distantly spaced, 15–25 mm apart. The flowers are larger, with tepals 14–16 × 5–9 mm, the upper tepals overlapping and separated from the three lower tepals, and anthers 3.2–4.0 mm long (Table 1). Although Lewis (1962) noted the variation in leaf number and venation between the southern and northern populations she preferred not to recognize the two forms taxonomically as she believed them to intergrade in leaf form and number to some extent. She did not, however, note that the distinction was accompanied by differences in size of the flowers, which we consider as significant as the leaf differences.

A third set of divergent populations occurs on the Gifberg–Matsikamma Mtn complex. These consist of unusually small plants, seldom exceeding 150 mm in height, with small flowers (tepals ± 8 × ± 4 mm, anthers ± 2 mm long), thus like those of the southern populations of *Ixia scillaris* but with just three leaves (rarely two in depauperate individuals), recalling the northern populations, at least in leaf number, although the leaves are shorter and narrower.

Our taxonomic solution to the variation in *Ixia scillaris* is to recognize the three sets of populations as separate subspecies, those of the larger-flowered, northern populations as subsp. *latifolia* and the Gifberg–Matsikamma populations as subsp. *toximontana*. We choose subspecies rank because of the small overlap between some populations of each subspecies and the comparatively small degree of difference between them relative to that between species of sect. *Dichone*.

In addition to these major variants, Lewis recognized populations with undulate leaves as var. *subundulata*. The type of the variety is from east of Pakhuis Pass (*Barker 4727*) and there are several more collections from this area. Plants with undulate leaves also occur locally near Tulbagh. Flowers of the Pakhuis–northern

Cedarberg have the relatively large flowers that correspond with subsp. *latifolia* whereas those from the south have smaller flowers of matching subsp. *scillaris*. We see no value in recognizing undulate-leaved plants taxonomically. Undulate leaf margins have evidently evolved independently in different parts of the range of *Ixia scillaris*.

History: the taxonomic history of *Ixia scillaris* begins with a brief polynomial entry in A. van Royen's (1740) *Florae leydensis prodromus*, an account of the plants grown in the botanical garden of Leiden University. Linnaeus (1753), in the first edition of the *Species plantarum*, adopted van Royen's species and, using his actual phrase name, provided the specific epithet *ramosus* to the description *Gladiolus caule ramoso, foliis linearibus*. The specimen is still at the Leiden Herbarium. Then, in the second edition of his *Species plantarum*, Linnaeus (1762) described *Ixia scillaris*, based on a specimen in his own herbarium. Evidently Linnaeus simply did not relate this specimen to the earlier name, although he may well have seen Van Royen's specimen or the living plants when he was in The Netherlands some 15 years earlier. *Gladiolus ramosus* was forgotten for many years but was listed by Ker Gawler (1827) as a synonym of *Melasphaerula graminea* (L.f.) Ker Gawl. and by Lewis *et al.* (1962) as a synonym of the same species under the name *M. ramosa* (Burm.f.) N.E.Br. [the basionym for *M. ramosa* is *Phalangium ramosum* Burm.f. (1768) and has no apparent connection with *G. ramosus* L.]. F.W. Klatt (1894) cited *G. ramosus* L. as the basionym for his combination *Melasphaerula ramosa* (L.) Klatt, an action that rendered Brown's later combination *M. ramosa* (Burm.f.) N.E.Br. a homonym. Transfer of *G. ramosus* to *Ixia* is prevented by *I. ramosa* Ker Gawl., a replacement name for the homonym *I. scillaris* Thunberg (1783), now *Geissorhiza scillaris* A.Dietr. (Goldblatt 1985). Thunberg was uncertain of the identity of *I. scillaris* L., which he cited as a possible synonym of *I. pentandra* L.f.

Key to subspecies

1a Leaves (4–)5–7, lower two up to 10 mm wide, usually lanceolate and lacking submarginal veins (margins rarely

TABLE 1.—Significant taxonomic features of the subspecies of *Ixia scillaris* and allied *I. tenuis*. Observations are taken from living plants or from well preserved specimens, bearing in mind that floral features may shrink up to 20% of original size, depending on the care with which specimens are prepared.

Character	Subsp. *scillaris*	Subsp. *latifolia*	Subsp. *toximontana*	*Ixia tenuis*
Corm tunic texture	medium	medium to coarse	medium to coarse	fine
Leaf number	(4–)5–7	3 or 4	(2)3(4)	3(2)
Leaf shape; width (mm)	lanceolate; 5–8	usually falcate; 8–25	usually falcate; 3–8	linear; 1.5–3.0
Prominent leaf veins	3–5, only main when alive	usually 2 or 3 plus main vein	main vein, secondary veins pair hyaline when alive, raised when dry	only main vein when dry
Margins	not raised	hardly raised	raised and thickened	not raised
Perianth tube length (mm)	2.5–4.0	3.5–4.0	± 3	3.0–3.5
Outer tepals (mm)	8–10 × 4–5	10–15 × 5–9	± 8 × 4	8–9 × 4–5
Stamen filaments (mm)	1.5–2.0	2.0–3.5	1.0–1.6	± 2
Anthers (mm)	mostly 2–3	3–4	1.6–2.5	2.0–2.5
Bracts (mm)	3.5–4.8	4–5	3–4	2–3
Flowers per main spike	mostly 10–15	mostly 15–20	mostly 10–15	mostly 4–10
Spike internodes (mm)	6–12	14–20	8–12	10–15

undulate); tepals 8–10 × 4–5 mm; anthers 2.0–3.5 mm long . . . . subsp. *scillaris*
1b Leaves 2–4, lower two 3–25 mm wide, usually ± falcate, sometimes with pair of prominent secondary veins, these often submarginal (margins sometimes undulate to crisped, then without prominent secondary veins); tepals 8–15 × 4–9 mm; anthers 2–4 mm long:
2a Leaves (9–)12–25 mm wide; tepals 10–15 × 5–9 mm; anthers mostly 3–4 mm long . . . . subsp. *latifolia*
2a Leaves 3–8 mm wide; tepals ± 8 × 4 mm; anthers 1.6–2.5 mm long . . . . subsp. *toximontana*

13a. subsp. **scillaris**

Gladiolus ramosus L.:37 (1753) [non *Ixia ramosa* Ker Gawl., (1802) = *Geissorhiza scillaris* A. Dietr.]. *Melasphaerula ramosa* (L.) Klatt: 203 (1894), epithet misapplied to *Melasphaerula*. Type: South Africa, without precise locality or collector, ex hort. Leiden, *A. van Royen s.n.* (L: Herb. Royen, holo.— digital image!).

Ixia pentandra L.f.: 92 (1782). *Agretta pentandra* (L.f.) Eckl. Type: South Africa, [Western Cape], near Groene Kloof, without date, *Thunberg s.n.* (UPS-THUNB, holo.—microfiche!).

Plants mostly 150–350 mm high, stem usually branched, sometimes with up to 6 branches. *Corm* mostly 12–14 mm diam. *Leaves* (4–)5–7, lanceolate to sword-shaped or subfalcate, margins sometimes undulate, 5–8 mm wide, upper leaves progressively smaller than lower. *Spike* mostly 10–15-flowered, flowers mostly 6–12 mm apart; bracts 3.5–4.8 mm long. *Flowers* pale to deep pink, darker at base of tepals, not pale in throat, perianth tube 2.5–4.0 mm long, tepals subequal, upper 5 tepals overlapping and evenly disposed with lower laterals extended horizontally, lowermost held apart, directed forward above base and then downward, 8–10 × 4–5 mm, inner narrower than outer. *Filaments* 1.5–2.0 mm long, anthers mostly 2–3 mm long. *Flowering time*: mainly mid-Sept.–late Oct. (rarely in late Aug.).

Distribution and habitat: restricted to the southwestern Western Cape, subsp. *scillaris* extends from Tulbagh and Darling to the Cape Peninsula and Somerset West (Figure 10). The only record outside this area, a collection from Hawston near Bot River, *Wilman s.n.* NBG, is the Clanwilliam form of *Ixia scillaris*, subsp. *latifolia*, and we assume the label is incorrect. Subsp. *scillaris* is common and fairly conspicuous where it occurs, mainly on granite-derived sandy gravels.

Diagnosis: subsp. *scillaris* can be distinguished by the several, (4)5–7, leaves mostly 5–8 mm wide, in which only the main is prominent when alive (Table 1). The flowers are slightly smaller than in subsp. *latifolia*, with tepals 8–10 × 4–5 mm, the lowermost slightly recurved and held separate from the upper five. The anthers are also slightly smaller than in subsp. *latifolia*, 2–3 mm vs. mostly 3–4 mm long.

Representative specimens

WESTERN CAPE.—**3318** (Cape Town): Groenekloof [Mamre] and hills nearby, (–AD), Oct., *Ecklon & Zeyher Irid 241* (MO); Mamre Hills (–AD), 26 Sept. 1941, *Compton 11768* (MO, NBG); Cape Town, above Camps Bay, (–CD), 26 Oct. 1944, *Barker 3214* (MO, NBG); Lions Head, (–CD), *H. Bolus 2826* (BM, BOL, K); Paarl Mtn, near Taal Monument, (–DA), 17 Oct. 1986, *Goldblatt 7919* (MO, PRE);

Langverwacht above Kuilsrivier, sandy west-facing slope, (–DC), 24 Aug. 1973, *Oliver 4362* (MO, NBG, PRE). **3319** (Worcester): fields near Tulbagh, leaves undulate, (–AC), Sept. 1916, *Marloth 7772* (PRE). **3418** (Simonstown): Helderberg, 240 m, (–BB), 17 Oct. 1948, *Parker 4360* (BOL, MO, NBG); Sir Lowry's Pass, (–BB), *Schlechter 5370* (BM, BOL, G, K); Vergelegen, Somerset West, (–BB), 13 Oct. 1982, *Viviers 709* (PRE).

13b. subsp. **latifolia** *Goldblatt & J.C.Manning*, subsp. nov.

TYPE.—Western Cape, 3218 (Clanwilliam): Cedarberg, Nieuwoudt Pass, burned slopes, (–BD), *Goldblatt & Porter 13475* (NBG, holo.; K, MO, PRE, iso.)

I. reflexa Andr.: t. 14 (1797). *I. rotata* [Ker Gawl.] in Andrews: 3 (1801), nom. illeg. superf. pro *I. reflexa* Andr. [Ker Gawl. is the author of this publication (Stafleu & Cowan 1979), but internal evidence for authorship is lacking]. Type: South Africa, without precise locality, cultivated in Great Britain by Messrs. Lee & Kennedy and sent to them by J. Pringle but possibly collected by F. Masson or W. Paterson (Gunn & Codd 1981), illustration in Andrews, *The botanist's repository* 1: t. 14 (1797)—no preserved specimen known.

I. polystachya var. *incarnata* Andr.: t. 128 (1800). Type: South Africa, without precise locality, cultivated in Great Britain by Messrs. Lee & Kennedy and sent to them by J. Pringle but possibly collected by Francis Masson or William Paterson (Gunn & Codd 1981), illustration in Andrews, *The botanist's repository* 2: t. 128 (1800)—no preserved specimen known.

I. scillaris var. *angustifolia* Ker Gawl. in Sims: sub t. 542 (1801), nom. illegit. superf., including type of *I. polystachya* var. *incarnata* Andr. Type: not designated, two illustrations cited.

I. scillaris var. *latifolia* Ker Gawl. in Sims: t. 542 (1801). Type: South Africa, without precise locality, cultivated in Great Britain, *Masson s.n.*, illustration in *Curtis's Botanical Magazine* 15: t. 542 (1801)—no preserved specimen known.

I. scillaris var. *subundulata* G.J.Lewis: 177 (1962). Type: Western Cape, Klipfonteinrand, 13 Sept. 1947, *Barker 4727* (NBG, holo.!, BOL!, iso.).

Plants mostly 300–500 mm high, stem simple or with 1 or 2(3) branches. *Corm* mostly 14–20 mm diam. *Leaves* 3 or 4, sometimes an additional leaf present at base of a branch, lower 2 or 3 broadly falcate, (9–)12–25 mm wide, usually with pair of conspicuous secondary veins and often well developed submarginal vein, upper leaf/leaves smaller, margins plane or undulate, then without well developed secondary veins. *Spike* mostly 15–20-flowered, flowers mostly 14–20 mm apart; bracts 4–5 mm long. *Flowers* pale to deep pink or white, then flushed pale pink near tepal bases; perianth tube 3.5–4.0 mm long; tepals subequal, upper 3 held close together, lowermost held slightly apart arched backward, 10–15 × 5–9 mm. *Filaments* 2.0–3.5 mm long, broader at base; anthers 3–4 mm long. *Style branches* ± 2.5 mm long, almost reaching anther bases. *Flowering time*: mainly mid-Aug.–mid-Sept., rarely to early Oct. at higher elevations. Figure 9.

Distribution and habitat: subsp. *latifolia* is centred in the Olifants River valley and lower elevations of the surrounding Cedarberg, Pakhuis–Bidouw Mts, extending from near Citrusdal to Bulshoek and locally in the Bokkeveld Mtns. Isolated populations occur on the Spektakel Mtns of northern Namaqualand, a notable disjunction (Figure 10). It is replaced on the Gifberg/Matsikamma Mtns by subsp. *toximontana*. Plants favour rocky situations, mostly in sandy ground. Subsp. *latifolia* is common throughout its range, but it flowers well only after fire or clearing of the veld. The spring following a fire plants are often seen in masses where before the fire none were evident.

Diagnosis and variation: subsp. *latifolia* is distinguished by the presence of 3 or 4 leaves and by the relatively larger flowers, the tepals 10–15 × 5–9 mm and anthers 3–4 mm long (vs. tepals 8–10 × 4–5 and anthers 2–3 m long). The upper three tepals are held more closely together than the lower three and the lowermost is conspicuously recurved.

Variation in leaf morphology in subsp. *latifolia* is notable. Plants from Clanwilliam and nearby have a conspicuous submarginal vein, but those from elsewhere have a prominent secondary vein pair distant from the margins, and plants from the Spektakel Mtns of Namaqualand show no prominent secondary veins and appear thinner in texture. No other feature seems to set the Spektakel population apart. Populations east of the Pakhuis–Nardouw Mtns have leaf margins slightly to markedly undulate and without conspicuous veins apart from the main vein.

Representative specimens

NORTHERN CAPE.—**2917** (Springbok): 14 miles [± 21 km] SW of Springbok, (–DA), *Acocks 19439* (NBG, PRE); Spektakel Pass, E-facing slope near the top, (–DA), 26 Sept. 1974, *Goldblatt 2804* (MO, PRE). Spektakel, (–DA), 25 Aug. 1941, *Esterhuysen 5853* (BOL). **3119** (Calvinia): between Nieuwoudtville and Oorlogskloof, (–?AC), Aug. 1941, *Leipoldt* (BOL, PRE).

WESTERN CAPE.—**3218** (Clanwilliam): Zeekoe Vlei, (–BA), 19 Aug. 1896, *Schlechter 8580 or 8500* (BM, G, K, MO, PRE); Clanwilliam, (–BB), 10 Sept. 1949, *Steyn 516* (MO, NBG); Farm Kransvlei, rocky sandstone ground, (–BB), 25 Aug. 2009, *Goldblatt & Manning 13482* (MO, NBG). **3219** (Wuppertal): Biedouw Valley on road to Wuppertal, (–AA), 7 Sept. 1992, *Goldblatt & Manning 9409* (MO, NBG, PRE); Brandewyn River, (–AA), 29 Aug. 1941, *Barker 1327* (BOL, NBG); Cedarberg, Langrug, (–AC), 12 Sept. 1982, *Viviers 579* (PRE); Clanwilliam Dam to Algeria, (–AC), 9 Sept. 1976, *Goldblatt 2567* (MO, PRE); banks of the Olifants River and at Villa Brakfontein, (–BA), without date, *Ecklon & Zeyher Irid. 243 or s.n.* (G, MO, PRE, SAM). Without precise locality: Grey's Pass to Clanwilliam, Sept. 1941, *Leipoldt 3832* (BOL, PRE)

13c. subsp. **toximontana** *Goldblatt & J.C.Manning*, subsp. nov.

TYPE.—Western Cape, 3118 (Vanrhynsdorp): Gifberg, summit plateau in sandy ground over rocky pavement, common after fire, (–DC), 1 Nov. 2011, *Goldblatt & Porter 13703* (NBG, holo.; K, MO, PRE, iso.).

Plants 120–250 mm high, simple or with 1 or 2, rarely 3 branches. *Corms* globose-conic, mostly 15–18 mm at widest diam., tunics of hard, coarse fibres; bearing conspicuous cormlets at base. *Leaves* (2)3(4), all ± basal, mostly 3–8 mm wide, main vein prominent, secondary pair of veins hyaline when alive, raised when dry, margins appearing thickened when dry. *Spike* mostly 10–15-flowered, flowers mostly 8–12 mm apart; bracts 3–4 mm long. *Flowers* opening pale pink turning deep pink next day, tepal bases yellow edged dark pink, throat pale yellow, perianth tube ± 3 mm long, tepals subequal, ± 8 × 4 mm lowermost held slightly apart and arched backward. *Filaments* 1.0–1.6 mm long, anthers 1.6–2.5 mm long. *Style branches* ± 1.5 mm long, almost reaching anther bases. *Flowering time*: late Sept.–mid-Nov.

Distribution: restricted to the Gifberg and Matsikamma Mtn complex of northern Western Cape, subsp. *toximontana* grows in shallow, sandy ground overlying sandstone pavement (Figure 10). The habitat is permanently wet in the rainy winter and early spring months but becomes dry by the end of September and remains so though the summer and autumn. Most of its range is virtually intact and pristine at present.

Diagnosis: subsp. *toximontana* is distinguished in *Ixia scillaris* by its low stature, seldom exceeding 150 mm. combined with relatively small flowers and only 2 or 3 leaves, these short and relatively narrow, 3–8 mm wide. The flowers have tepals ± 8 × 4 mm and anthers ± 2.3 mm long, smaller than in subsp. *latifolia* or subsp. *scillaris* (Table 1). In their small size they recall subsp. *scillaris* rather than subsp. *latifolia*, in whose range it is nested. The particularly small size of all parts may be an adaptation to the habitat, nutrient-poor, shallow, sandy ground.

Representative specimens

WESTERN CAPE.—**3118** (Vanrhynsdorp): Gifberg, (–DC), 17 Sept. 1961, *Barker 9578* (NBG, PRE); top of Gifberg, (–DC), 23 Sept. 1962, *Rycroft 2548* (NBG); summit of Gifberg Pass, open spaces and shallow rock pans, (–DC), 9 Oct. 1973, *Hall 4493* (NBG), 11 Oct. 1973, *Bayliss 6153* (MO); Matsikammaberg, on recently burnt sandstone, (–DB), 11 Nov. 1985, *Snijman 963* (NBG).

14. **Ixia tenuis** *Goldblatt & J.C.Manning*, sp. nov.

TYPE.—Western Cape, 3218 (Clanwilliam): Piketberg, Farm Noupoort, in wet mossy seeps on sandstone rocks, (–DC), 2 Nov. 2011, *Goldblatt & Porter 13710* (NBG, holo.; K, MO, PRE, iso.).

Plants slender, 180–350 mm high, stem simple or rarely 1-branched. *Corms* globose, 7–10 mm diam., tunics of fine fibres. *Leaves* mostly 3(2), occasionally 4 but then this very reduced in size, ± linear to narrowly sword-shaped, 1/4–1/2 as long as stem, mostly 1.5–3.0 mm wide, relatively soft-textured, pale grey-green, when alive main vein slightly raised (appearing thickened when dry), margins not raised (rarely undulate, then blade up to 5 mm wide—*Barker 4671, 5764*, see comments below). *Spike* mostly 4–10-flowered, flowers mostly 10–15 mm apart; bracts 2–3 mm long. *Flowers* deep magenta with tepal bases yellow edged with darker magenta (or pale pink or white—*Barker 4671, 5764*), throat pale yellow, perianth tube 3.0–3.5 mm long, expanded in upper 1 mm; tepals 8–9 × 4–5 mm, oriented vertically but with lowermost tepals recurved and ± horizontal. *Stamens* unilateral; filaments ± 2 mm long; anthers 2.0–2.5 mm long. *Style branches* ± 1 mm long, reaching to ± middle of filaments. *Flowering time*: mid-Oct.–mid-Nov.

Distribution and habitat: known from just a handful of sites, *Ixia tenuis* is restricted to rocky, sandstone habitats in the Piketberg Mtns, mostly at elevations of 650–800 m (Figure 10). We collected the type population on the Farm Noupoort on south-trending slopes of a rocky hill growing in wet moss on shallow soil on sandstone outcrops. The remaining collections lack detailed habitat notes. We note that the undulate-leaved variant (see below) is from the foot of the Piketberg Range and blooms earlier, in September. The species is rare, and at present merits a conservation status of Vulnerable (VU), but we see no imminent threat.

Diagnosis: *Ixia tenuis* is distinguished from the widespread *I. scillaris* by its very slender habit, the stems rarely branched, usually three, narrow leaves, mostly 1.5–3.0 mm wide, and relatively small flowers with tepals 8–9 × 4–5 mm and anthers 2.0–2.5 mm long (Table 1). The corms are also relatively small, mostly less than 10 mm in diameter and with relatively fine tunic fibres. Corms of *I. scillaris* are larger, typically 12–18 (rarely only 10) mm in diameter, with relatively coarse tunic fibres, and the species has broader leaves, mostly 8–25 mm wide. We provisionally include collections by W.F. Barker (*Barker 4671* and *5764*) from De Hoek and flowering in September here. These specimens have slightly broader but shorter leaves than typical *I. tenuis* and have undulate margins. This variant, which we have not seen alive, was included by Lewis (1962) as *I. scillaris* var. *subundulata*, the type of which we include in *I. scillaris* subsp. *latifolia*.

Representative specimens

WESTERN CAPE.—**3218** (Clanwilliam): Piketberg Mtns, S side of large boulders, 2 500 ft [800 m], (–DC), 3 Nov. 1951, *Barker 7584* (NBG); Piketberg, top of mountain on Versfeld Pass, 650 m, (–DC), 4 Nov. 1988, *Steiner 1863* (NBG); De Hoek, Piketberg, leaf margins undulate, (–DC), 28 Sept. 1943, *Barker 2560* (NBG), (flowers white), 12 Sept. 1947, *Barker 4671* (NBG), 10 Sept. 1949, *Barker 5764* (NBG).

15. Ixia flagellaris *Goldblatt & J.C. Manning*, sp. nov.

TYPE.—[Western Cape]. 3219 (Wuppertal): Heuningvlei, Groot Koupoort, common on wet sandy flats burnt last summer, 3 150 ft [± 960 m], (–AA), 11 Oct. 1975, *Kruger 1687* (NBG, holo.; PRE, iso.).

Plants 150–330 mm high, stem simple, rarely with 1 branch. *Corm* small, up to 10 mm diam., bearing stolons up to 50 mm long from base, with a large terminal cormlet up to 5 mm diam., tunics finely fibrous to moderately coarse. *Leaves* 4, uppermost sheathing for most of its length, foliage leaves with blades narrowly sword-shaped to falcate, ± 1/3 as long as stem, mostly 2–4 mm wide, main vein prominent, margins slightly thickened. *Spike* mostly 5–9-flowered; bracts translucent with purple veins, 3–4 mm long, outer 3-veined, acutely 3-toothed, inner slightly shorter than outer, 2-veined and 2-toothed. *Flowers* evidently zygomorphic, nodding with tepals vertical, pink, throat and tepal bases yellow edged with deep pink; perianth tube filiform, ± 3 mm long; tepals ovate, ± 9.0 × 4.5 mm. *Stamens* evidently unilateral; filaments slightly flattened, ± 2 mm long; anthers oblong-orbicular, ± 3 mm long, yellow, thecae recurved and acute at base, splitting incompletely from

base. *Style* dividing at base of filaments, branches ± 1.8 mm long, falcate, slender. *Flowering time*: Oct. and Nov.

Distribution and habitat: restricted to middle and upper elevations in the Cedarberg, *Ixia flagellaris* is currently known from just two collections from the Cedarberg, one from Heuning Vlei in the north and the other from Sneeuberg in the south (Figure 8). The type collection was made the season after a fire. The species evidently grows on damp flats, most likely on peaty sandstone ground, but this remains to be established. A collection from Kobee Pass (see below) is provisionally included here.

Diagnosis and relationships: the outstanding feature of *Ixia flagellaris* is the corm, which is remarkable in the *I. scillaris* complex in the finely fibrous corm tunics and in the production of slender stolons (Latin: flagellae) up to 50 mm long, each terminating in a relatively large cormlet. No other member of sect. *Dichone* produces stolons although they are known in a few species of sect. *Ixia* and *Morphixia* (Goldblatt & Manning 2011). *I. flagellaris* otherwise accords closely with *I. scillaris* subsp. *toximontana*, notably in its relatively low stature, 150–330 mm high, and small flowers with tepals ± 9.0 × 4.5 mm, but differs in lacking prominent, thickened leaf margins.. The late flowering and moist habitat, seeps and marshy sites, are unusual for the complex, although known in the Piketberg species, *I. tenuis*. Plants from Kobee Pass (*Hall 4517*) cited below may belong here as they have fine corm tunics and unthickened leaf margins but they evidently do not produce stolons. Their exact status remains uncertain.

Representative specimens

WESTERN CAPE.—**3219** (Wuppertal): Cedarberg, Sneeuberg, Duiwelsgat Kloof, damp places, (–AC), 24 Nov. 1982, *Viviers 789* (NBG). Provisionally included here: **3119** (Calvinia): summit of Kobee Pass, open spaces in shallow pans, (–CA), 14 Oct. 1973, *Hall 4517* (NBG).

16. Ixia simulans *Goldblatt & J.C. Manning*, sp. nov.

TYPE.—Western Cape, 3320 (Montagu): Langeberg Mtns, gorge near Montagu hot springs above Keisie River, on steep rocky sandstone slopes, (–CC), 1 Sept. 2007, *Goldblatt & Porter 12946* (NBG, holo., K, MO, PRE, iso.).

Plants 40–100 cm high, stems usually simple, rarely with a single branch, sheathed basally by membranous cataphylls. *Corm* ± 10 mm diam., tunics of fine, netted fibres. *Leaves* 2(3), lower 1(2) with narrowly sword-shaped to linear blade with single prominent vein, ± 1/2 as long as stem, 1.5–6.5 mm wide, margins thickened and slightly raised, hyaline when dry; upper leaf sheathing lower 1/2 of stem, free and unifacial distally. *Spike* erect, fairly lax, weakly flexuose or straight, 4–8-flowered; bracts membranous, translucent, purple at tips, ± 4 mm long, outer usually with 3 prominent veins and 3-toothed, inner 2-veined and forked at apex. *Flowers* ± nodding with tepals held vertically, zygomorphic with stamens unilateral, bright pink (rarely pale pink), white at tepal bases, unscented; perianth tube filiform, ± 4.5 mm long, tightly clasping style; tepals spreading, ovate-elliptic, (10–)12–14 × 5–6 mm, inner whorl slightly larger than outer. *Stamens* unilateral, filaments inserted at mouth of tube, filiform, ± 2.5 mm long, white, anthers oblong, 2.0–2.5 mm long, yellow, horizontal or slightly pendent, thecae acute and slightly recurved at base, dehiscing incompletely from base by narrow slits. *Style* dividing opposite lower 1/3 of filaments, branches ± 1.3 mm long, white, tubular, falcate, stigmatic apically. *Flowering time*: mid-Aug.–mid-Oct. Figure 11.

Distribution and habitat: *Ixia simulans* is currently known from just four collections, three from the western Langeberg, between Dassieshoek near Robertson and Swellendam, and one from Lemoenpoort in the northern foothills of the Riviersonderend Mtns (Figure 10). Records are all from stony ground, and at the Keisie River Gorge near Montagu hot springs plants grew on steep, rocky, south-facing slopes. Plants are rooted in shallow sandy soil, often growing in clumps of Restionaceae. All known localities for *I. simulans* are in virtually undisturbed montane vegetation and we see no threat to the species: a conservation status of Least Concern (LC) seems appropriate.

Diagnosis and relationships: *Ixia simulans* is a surprising novelty given that the western Langeberg has been botanized for over two centuries and intensely so in the past 50 years. The flowers closely resemble those of *I. scillaris* in general appearance and have unilateral, horizontal to slightly pendent anthers that dehisce incompletely by narrow slits near the base. *I. simulans* differs from *I. scillaris* most conspicuously in its single, or rarely two, long, basal leaves with ± linear blades 1.0–4.5 mm wide, spikes of only 4–8 flowers, relatively short floral bracts ± 4 mm long and the spreading, subequal tepals, those of the inner whorl being slightly shorter than the outer. The tepals are inclined and all held in the same plane whereas those of *I. scillaris* are always held vertically and the lowermost tepal is weakly to strongly recurved and held apart from the other five tepals.

The range of *Ixia simulans* and *I. scillaris* complement one another: *I. scillaris* extends from the Cape Peninsula and Somerset West in the south through the Olifants River Valley, Cedarberg–Pakhuis Mtn complex to the Bokkeveld Mtns, and locally well to the north in the Spektakel Mtns west of Springbok whereas *I. simulans* occurs well to the east of this line in the western Langeberg.

Representative specimens

WESTERN CAPE.—**3319** (Worcester): Lemoenpoort, rocky slopes, (–CD), 18 Aug. 1990, *Bruyns sub Perry 3766* (NBG); Dassieshoek near Robertson, (–DB), 27 Aug. 2006, *Goldblatt & Porter 12724* (MO, NBG, PRE). **3320** (Montagu): Langeberg Mtns, Marloth Reserve ± 100 m from Glenstroom, 212 m, (–CD), 9 Oct. 2007, *Turner et al. s.n.* (NBG).

17. **Ixia collina** *Goldblatt & Snijman* in South African Journal of Botany 51: 68 (1985). Type: South Africa, [Western Cape], Aan de Doorns, SW of Worcester, Farm Alfalfa, *Snijman 721* (NBG, holo!; MO!, PRE!, S!, iso.).

Plants 500–900 mm high, stem sheathed below by membranous cataphylls, usually 2–3-branched, branches held at right angles to main axis, ascending distally, subtended by short, acute bracts and prophylls. *Corm*

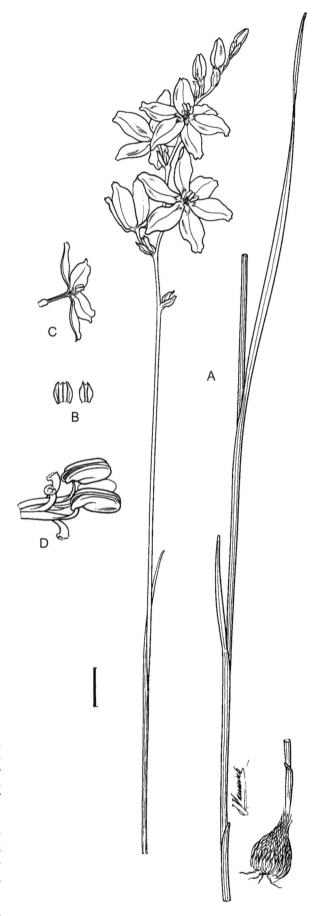

FIGURE 11.—*Ixia simulans*, *Goldblatt & Porter 12946* (NBG). A, flowering plant; B, outer (left) and inner (right) bract; C, half flower; D, stamens and style branches, side view. Scale bar: 10 mm. Artist: John Manning.

globose, 7–12 mm diam., tunics of medium-textured fibres, often with cormlets at base. *Leaves* 3 or 4, lower 2 or 3 with sublinear to narrowly sword-shaped blades, ± 1/2 as long as stem, 4–8 mm wide, main vein moderately thickened, uppermost leaf sheathing lower half of stem. *Spike* flexuose, 7–16-flowered, lateral spikes with fewer flowers; bracts membranous, translucent, 4.5–7.0 mm long, outer with 3 prominent, purple veins, 3- or 5-toothed, inner 2-veined and 2-toothed. *Flowers* zygomorphic, half nodding, pink, white to yellow edged deep pink at tepal bases, sweetly scented, perianth tube filiform, ± 5 mm long, expanded in upper 0.5 mm, tepals subequal, ovate, ± obtuse, 9–12 × ± 5 mm. *Stamens* unilateral, ± horizontal, inserted at mouth of filiform part of tube, filaments ± 3 mm long, filiform below, curved at right angles in upper 1 mm and becoming flattened and as wide as anthers, anthers ± 3 mm long, oblong, yellow, thecae acute and recurved near base, dehiscing incompletely from base. *Style* dividing at base of filaments, style branches ± 2.5 mm long, tubular, falcate, bifid at apex, stigmatic apically. *Flowering time*: late-Aug.–mid Sept.

Distribution and habitat: *Ixia collina* is a narrow endemic of the mid Breede River Valley, its only known locality the Farm Alfalfa near Aan de Doorns south of Worcester (Figure 8). Plants grow on the southern slopes of low clay hills above the Breede River flood plain and adjacent to cultivated land. With so narrow a range Raimondo *et al.* (2009) regard the species as endangered (EN) and we agree. We recommend efforts be made to locate additional populations but there is little doubt that *I. collina* is extremely rare.

Diagnosis and relationships: *Ixia collina* is easily distinguished from other members of ser. *Dichone* by the unusual branching pattern: the relatively short branches extend outward horizontally and curve upward distally as well as by an unusual floral feature. The filaments are unique in the section in being filiform below, curved at right angles in the distal 1 mm and broadened so that they are as wide as the anthers.

Additional specimen

WESTERN CAPE.—**3419** (Worcester): Aan de Doorns, SW of Worcester, Scherpenheuwel, on Farm Reiersrus, (–DC), *Bruwer sub Walters 750* (NBG).

EXCLUDED SPECIES

Ixia retusa Salisb., Prodromus stirpium in horto ad Chapel Allerton vigentium: 35 (1796). The identity of this species has remained a mystery as the description is inadequate to identify the plant (De Vos 1999) and no type is known. Lewis (1962) suggested *I. retusa* might be *I. stricta* but with a query. Ker Gawler (in Sims 1801; 1803) associated the name with what he called *Ixia polystachya* (actually *I. scillaris*, thus not the same as Linnaeus's species) [see extensive discussion under Taxonomic History]. Although we suspect that *I. retusa* is a member of sect. *Dichone*, it is impossible to tell which species and the name must be rejected. Likewise, Klatt's combination, *Watsonia retusa* (Salisb.) Klatt is rejected (see additional discussion under *Ixia stricta*).

ACKNOWLEDGEMENTS

Support for this study by grants 7316-02, 7799-05, and 8248-07 from the National Geographic Society is gratefully acknowledged. Collecting permits were provided by the Nature Conservation authorities of Northern Cape and Western Cape, South Africa. We thank Ingrid Nänni, Elizabeth Parker, and Lendon Porter for their assistance and companionship in the field; Roy Gereau for advice with nomenclatural queries; Clare Archer, South African National Herbarium, Pretoria for help with several questions; Mary Stiffler, Research Librarian, Missouri Botanical Garden, for providing copies of needed literature; Sharon Bodine for help with herbarium searches. We also thank the curators of the following herbaria for allowing us access to their collections or for the loan of material for extended study: BOL, K, MO, NBG, PRE, and SAM (acronyms following Holmgren *et al.* 1990)

REFERENCES

ANDREWS, H. 1797. *Ixia reflexa.* Reflex flowered Ixia. *The botanist's repository* 1: t. 14.

ANDREWS, H. 1800. *Ixia polystachya* var. *incarnata. The botanist's repository* 2: t. 128.

ANDREWS, H. 1801. *Recensiae plantarum.* London, J. White.

BAKER, J.G. 1876. New species of Ixieae. *Journal of Botany (London)* 5: 236–239.

BAKER, J.G. 1877 [as 1878]. Systema iridearum. *Journal of the Linnean Society, Botany* 16: 61–180.

BAKER, J.G. 1892. *Handbook of the Irideae.* Bell, London.

BAKER, J.G. 1896. Irideae. In W.T. Thiselton-Dyer (ed.), *Flora capensis* 6: 7–171. Reeve, London.

BROWN, N.E. 1929. The Iridaceae of Burman's *Florae capensis prodromus.* [Royal Botanic Gardens, Kew] *Bulletin of Miscellaneous Information* 1929: 129–137.

BURMAN, N.L. 1768. *Prodromus florae capensis.* Cornelius Haek, Amsterdam.

DE VOS, M.P. 1982. The African genus *Tritonia* Ker-Gawler 1. *Journal of South African Botany* 48: 105–163.

DE VOS, M.P. 1999. *Ixia.* In M.P. de Vos & P. Goldblatt, *Flora of southern Africa* 7, part 2, fascicle 1: 3–87.

ECKLON, C.F. 1827. *Topographisches Verzeichniss der Pflanzensammlung von C.F. Ecklon.* Esslingen.

GOLDBLATT, P. 1971a. Cytological and morphological studies in the southern African Iridaceae. *Journal of South African Botany* 37: 317–460.

GOLDBLATT, P. 1971b. A new species of *Gladiolus* and some nomenclatural changes in the Iridaceae. *Journal of South African Botany* 37: 229–236.

GOLDBLATT, P. 1985. Revision of the southern African genus *Geissorhiza* (Iridaceae: Ixioideae). *Annals of the Missouri Botanical Garden* 72: 277–447.

GOLDBLATT, P. & MANNING, J.C. 1999. New species of *Sparaxis* and *Ixia* (Iridaceae: Ixioideae) from Western Cape, South Africa, and taxonomic notes on *Ixia* and *Gladiolus. Bothalia* 29: 59–63.

GOLDBLATT, P. & MANNING, J.C. 2006. Notes on the systematics and nomenclature of *Tritonia* (Iridaceae: Crocoideae). *Bothalia* 36: 57–61.

GOLDBLATT, P. & MANNING, J.C. 2007. Pollination of *Romulea syringodeoflora* (Iridaceae: Crocoideae) by a long-proboscid fly, *Prosoeca* sp. (Diptera: Nemestrinidae). *South African Journal of Botany* 73: 56–59.

GOLDBLATT, P. & MANNING, J.C. 2011. Systematics of the southern African genus *Ixia* (Iridaceae): 3. Sections *Hyalis* and *Morphixia. Bothalia* 41: 83–134.

GOLDBLATT, P., BARI, A. & MANNING J.C. 1991. Sulcus and operculum structure in the pollen grains of Iridaceae subfamily Ixioideae. *Annals of the Missouri Botanical Garden* 78: 950–961.

GOLDBLATT, P., BERNHARDT, P. & MANNING, J.C. 1998. Pollination of petaloid geophytes by monkey beetles (Scarabaeidae: Ruteliinae: Hopliini) in southern Africa. *Annals of the Missouri Botanical Garden* 85: 215–230.

GOLDBLATT, P., BERNHARDT, P. & MANNING, J.C. 2000. Adaptive radiation of pollination mechanisms in *Ixia* (Iridaceae: Crocoideae). *Annals of the Missouri Botanical Garden* 87: 564–577.

GOLDBLATT, P., DAVIES, T.J., MANNING, J.C., VAN DER BANK, M. & SAVOLAINEN, V. 2006. Phylogeny of Iridaceae subfamily Crocoideae based on combined multigene plastid DNA analysis. *Aliso* 22: 399–411.

GOLDBLATT, P., RODRIGUEZ, A., POWELL, M.P., DAVIES, T.J., MANNING, J.C., VAN DER BANK, M. & SAVOLAINEN, V. 2008. Iridaceae 'Out of Australasia'? Phylogeny, biogeography, and divergence time based on plastid DNA sequences. *Systematic Botany* 33: 495–508.

GOLDBLATT, P. & SNIJMAN, D. 1985. New species and notes on the southern African genus *Ixia* (Iridaceae). *South African Journal of Botany* 51: 66–70.

GUNN, M. & CODD, L.E. 1981. *Botanical exploration of southern Africa*. Balkema, Cape Town.

HOLMGREN, P.K., HOLMGREN, N.H. & BARNETT, L.C. 1990. *Index Herbariorum. Part. 1: The Herbaria of the World*. New York Botanical Garden, New York.

JACQUIN, N.J. 1792. *Ixia polystachya. Icones plantarum rariorum* 2(9): t. 275. Wappler, Vienna.

KER GAWLER, J. 1802. *Ixia maculata*, var γ, *viridis*, green-stained ixia. *Curtis's Botanical Magazine* 16: t. 549.

KER GAWLER, J. 1803. *Ixia polystachia*. Lily of the valley-scented ixia. *Curtis's Botanical Magazine* 17: t. 629.

KER GAWLER, J. 1827. *Genera Iridearum*. De Mat, Brussels.

KLATT, F.W. 1882. Ergänzungen und Berichtigungen zu Baker's Systema Iridacearum. *Abhandlungen der naturforschenden Gesellschaft zu Halle* 15: 44–404.

KLATT, F.W. 1894. Iridaceae. Pp. 143–230 *in* T.A. Durand & H. Schinz, *Conspectus florae africae* 5. De Mat, Brussels.

LEWIS, G.J. 1934. *Ixia trifolia. Flowering Plants of South Africa* 14: t. 543.

LEWIS, G. J. 1954. Some aspects of the morphology, phylogeny and taxonomy of the South African Iridaceae. *Annals of the South African Museum* 40: 15–113.

LEWIS, J. 1962. South African Iridaceae: the genus *Ixia*. *Journal of South African Botany* 27: 45–195.

LINNAEUS, C. 1753. *Species plantarum*. Salvius, Stockholm.

LINNAEUS, C. 1762. *Species plantarum*, edn. 2, 1. Salvius, Stockholm.

LINNAEUS, C. (fil.) 1782 [as 1781]. *Supplementum plantarum*. Orphanotropheus, Brunswick.

MANNING, J.C. & GOLDBLATT, P. 2006. New species of Iridaceae from the Hantam-Roggeveld centre of endemism and the Bokkeveld, Northern Cape, South Africa. *Bothalia* 36: 139–145.

MANNING, J.C., GOLDBLATT, P. & SNIJMAN, D. 2002. *The color encyclopedia of Cape bulbs*. Timber Press, Portland, OR.

RAIMONDO, D., VON STADEN, L., FODEN, W., VICTOR, J.E., HELME, N.A., TURNER, R.C., KAMUNDI, D.A. & MANYAMA, P.A. (eds). 2009. *Red List of South African Plants. Strelitzia* 25. South African National Biodiversity Institute, Pretoria.

ROEMER, J.J. & SCHULTES, J.A. 1817. *Systema vegetabilium*. J. G. Cotta, Stuttgart.

SALISBURY, R. 1796. *Prodromus stirpium in horto ad Chapel Allerton vigentium*. London.

SALISBURY, R. 1812. On the cultivation of rare plants, etc. *Transactions of the Horticultural Society, London* 1: 261–366.

SIMS, G. 1801. *Ixia scillaris*, var. *latifolia* (a). Squill-flowered Ixia.— Broad-leaved variety. *Curtis's Botanical Magazine* 15: t. 542.

STAFLEU, F.A. & COWAN, R.S. 1979. Taxonomic Literature, vol. 2. W. Junk, The Hague.

THUNBERG, C.P. 1783. *Dissertatio de Ixia*. Edman, Uppsala.

VAN ROYEN, A. 1740. *Florae leydensis prodromus*. Luchtmans, Leiden.

8

A revision of Tecophilaeaceae subfam. Tecophilaeoideae in Africa

J.C. MANNING* and P. GOLDBLATT**

Keywords: Africa, *Cyanella* Royen ex L., *Eremiolirion* J.C.Manning & F.Forest, new species, systematics, Tecophilaeaceae, *Walleria* J.Kirk

ABSTRACT

Family Tecophilaeaceae subfam. Tecophilaeoideae is revised for the *Flora of southern Africa* region, with the inclusion of the tropical *Walleria mackenzii* J.Kirk for completeness. The genera *Cyanella* Royen ex L. (9 spp.), *Eremiolirion* J.C.Manning & F.Forest (1 sp.) and *Walleria* J.Kirk (3 spp.) are treated, with keys to the genera, species and subspecies; and full descriptions and distribution maps. A formal infrageneric classification is proposed for *Cyanella*, in which sect. *Trigella* (Salisb.) Pax & K.Hoffm. is revived for the species with a 3 + 3 arrangement of stamens. The new species, *C.* **marlothii** J.C.Manning & Goldblatt, is described from the Richtersveld; and *C. pentheri* Zahlbr. is resuscitated from the synonymy of *C. hyacinthoides* Royen ex L. Pink-flowered plants of normally yellow-flowered *C. lutea* have a separate geographical distribution and are recognized as subsp. **rosea** (Eckl. ex Baker) J.C.Manning & Goldblatt.

INTRODUCTION

Tecophilaeaceae is a small family of seven or eight genera and ± 25 species from California, Chile, and southern and tropical mainland Africa (Simpson & Rudall 1998). The reported occurrence of the family in Madagascar (Simpson & Rudall 1998) is based on *Walleria paniculata* Fritsch, a synonym of *Dianella ensifolia* (L.) DC. (Hemerocallidaceae). The family is best represented in Africa, where almost two thirds of the species are found. *Cyanastrum* Oliv. (3 spp.) and *Kabuyea* Brummitt (1 sp.) are strictly tropical, but *Walleria* J.Kirk (3 spp.), *Eremiolirion* J.C.Manning & F.Forest (1 sp.), and *Cyanella* Royen ex L. (9 spp.), are primarily distributed in subtropical and temperate southern Africa. Members of the family are perennial herbs with a cormous, usually tunicated rootstock, basal (rarely cauline) leaves, and long-lasting flowers, typically in racemose or paniculate, cymose inflorescences; but sometimes solitary and axillary. The flowers are actinomorphic or zygomorphic, with 3 + 3 petaloid tepals fused into a short tube adnate to the ovary, and six stamens, all fertile or some reduced to staminodes, with ± porose dehiscence. The ovary is inferior or semi-inferior and 3-carpellate, and matures into a loculicidal capsule (Simpson & Rudall 1998; Heywood *et al.* 2007).

The two tropical African genera, *Cynastrum* and *Kabuyea*, have been the subject of a detailed review (Brummitt *et al.* 1998), in which they were segregated as subfam. Cynastroideae, with the remaining genera of the family retained in subfam. Tecophilaeoideae. The taxonomy of the southern African species is relatively well understood, and both *Cyanella* and *Walleria* were revised fairly recently (Carter 1962; Scott 1991; Cowley & Brummitt 2001), including historical and morphological details. Since then, however, the genus *Eremiolirion* has been established to accommodate *Cyanella amboensis* Schinz, which was excluded from *Cyanella*

by Scott (1991), but unplaced. We have also published additional observations on the distribution and morphology of *Walleria gracilis* (Salisb.) S.Carter (Manning *et al.* 2001). It is now clear that there is more variation in some species of *Cyanella* than was recognized by Scott (1991), and three subspecies have since been described in *C. alba* L.f. (Manning *et al.* 2005). Field study and examination of herbarium material of *C. hyacinthoides* Royen ex L. suggest that this species is currently too broadly circumscribed, and that *C. pentheri* Zahlbr. should be resuscitated from synonymy. In addition, the clear geographical segregation between the typical yellow-flowered and the pink-flowered forms of *C. lutea* L.f. is appropriately reflected by the recognition of distinct subspecies for them. A collection from the Richtersveld, until now identified as *C. orchidiformis* Jacq., differs from that species and from all others in the genus in having all six filaments connate into a staminal tube. It evidently represents an unnamed species that we describe here.

Currently, therefore, there is no comprehensive treatment for the family in southern Africa and the available treatment of *Cyanella* is inadequate and incomplete in some respects. We provide here a complete review of the genera and species occurring in the *Flora of southern Africa* region, including also the tropical African *Walleria mackenzii* J.Kirk for completeness. We also propose a new infrageneric classification for *Cyanella* that associates morphologically similar species in two sections, with the larger of the two, sect. *Cyanella*, subdivided into two series.

MATERIALS AND METHODS

Type specimens or digital images of types from the relevant herbaria were examined for all names, as well as all available herbarium specimens in BOL, NBG, PRE, and SAM (herbarium acronyms after Holmgren *et al.* 1990). Particular use was made of high-resolution digital images on the Aluka website (www.aluka.org), and of the Herbarium of the Linnean Society of London (www.linnean-online.org).

---

* Compton Herbarium, South African National Biodiversity Institute, Private Bag X7, 7735 Claremont, Cape Town.
** B.A. Krukoff Curator of African Botany, Missouri Botanical Garden, P.O. Box 299, St. Louis, Missouri 63166, USA.

TAXONOMY

Key to genera

1a Corm not tunicated; leaves cauline; flowers solitary in leaf axils; seeds verrucose or papillate, with tufts of trichomes, brown .... *Walleria*
1b Corm with fibrous tunics; leaves basal; flowers in racemose or paniculate cymes; seeds rugose, glabrous, black:
  2a Foliage leaves 2; inflorescence a divaricate panicle; pedicels without a bracteole; flowers actinomorphic; stamens monomorphic, central, and symmetrical .... *Eremiolirion*
  2b Foliage leaves 3–12; inflorescence a raceme, usually branched, rarely condensed and flowers apparently solitary; pedicels bracteolate; flowers zygomorphic; stamens dimorphic, in two groups of 3 + 3 or 5 + 1 .... *Cyanella*

Walleria *J.Kirk* in Transactions of the Linnean Society 24: 497 (1864). Type species: *Walleria nutans* J.Kirk [lecto., designated by E.P.Phillips: 207 (1951)].

Androsyne Salisb: 61 (1866). Type species: *A. gracilis* Salisb. = *Walleria gracilis* (Salisb.) S.Carter.

Deciduous geophytes with deep-seated, non-tunicated corm; subterranean portion of stem developing paired adventitious roots at each node, aerial portion of stem erect or straggling, smooth, scabrid or armed with recurved prickles. *Cataphylls* numerous, scattered along subterranean portion of stem, small, tubular, membranous. *Foliage leaves* numerous, all cauline, alternate, sessile or amplexicaul, linear to ovate, acute or cirrhose and tendrilliferous, midrib sometimes armed with recurved prickles beneath. *Inflorescence* of solitary, axillary flowers, or rarely bracteole subtending a second flower; pedicels erect or cernuous, smooth or prickly, with solitary bracteole inserted ± midway. *Flowers* actinomorphic, erect or nodding, rotate, white to blue; tepals connate below into short tube, ± similar. *Stamens* 6, monomorphic, erect-symmetrical, inserted at mouth of tube; filaments short; anthers basifixed, erect, free, or connivent around style, narrowly lanceolate, dehiscing by apical pores, outer surface scabridulous in basal ± 1/2. *Ovary* ± superior, with several ovules per locule; style terete, erect, filiform. *Capsules* ovoid to subglobose. *Seeds* ovoid, brown, surface warty or produced into finger-like papillae, each with apical tuft of minute trichomes. *Basic chromosome number*: x = 12 (Goldblatt & Manning 1989).

3 spp., southern and southern tropical Africa.

Etymology: the genus is named for Horace Waller, who made the first collections of both tropical African species during an expedition to central Africa in 1863.

Ethnobotany: the corms comprise part of the traditional diet of the San, Tswana, and other indigenous tribes (*e.g.* Leffers 2008; also *Lugard 289, Maguire 2194, Snyman & Noailles 231, Story 6117*).

Key to species

1a Flowers erect, tepals 13–22 mm long; anthers free, not connivent, blue, purple, or black with yellow base and apex, 6–12 mm long; style 8–15 mm long .... 1. *W. mackenzii*
1b Flowers nodding, tepals 6–16 mm long; anthers connate and connivent, yellow at least in basal 1/2, 4–8 mm long; style 4.0–8.5 mm long:
  2a Plants erect or sprawling, free-standing, mostly unbranched; stems and pedicels smooth, scabrid or with hooked prickles; leaves not cirrhose; tepals plain white, pink, mauve, or blue .... 2. *W. nutans*
  2b Plants usually straggling or climbing, well branched; stems and pedicels always armed with hooked prickles; upper leaves cirrhose, with tendril-like apex; tepals white with basal purple blotch .... 3. *W. gracilis*

1. **Walleria mackenzii** *J.Kirk* in Transactions of the Linnean Society 24: 497, t. 52/2 (1864). Type: Nyasaland [Malawi], Manganja Hills, near Bishop Mackenzies Mission, 1863, *H. Waller sub J. Kirk s.n. K256015* (K, holo.!). Illustration: Cowley & Brummitt (2001).

W. angolensis Baker: 262 (1878). Type: Angola, Huilla, 18 Dec. 1859, *Welwitsch 1749* (BM, holo.!; K, iso.!).

Deciduous geophyte, 180–900 mm high. *Corm* subglobose or depressed-globose, 20–40 mm diam. *Stem* erect, mostly simple or with 1 or 2 branches, smooth or rarely scabrid or minutely prickly. *Leaves* ovate to narrowly lanceolate, 30–110 × (4–)5–20(–28) mm, upper narrower, base cuneate or weakly cordate but not amplexicaul, apex acute or rarely cirrhose, midrib smooth, sometimes scabrid or minutely prickly. *Flowers* solitary in axils in central portion of stem, erect, sometimes with additional flower developed in axil of bracteole; pedicels ascending and ± erect at flowering, straight or flexible, becoming deflexed or pendulous in fruit, 13–60 mm long, smooth or scabrid, with lanceolate bracteole 10–26 mm long inserted ± halfway, rarely lacking; tepals white, pink, or mauve to pale or bright blue, spreading, elliptic-lanceolate, 13–22 × 2.5–6.5 mm, inner slightly narrower than outer. *Stamens* erect, free and not connivent; filaments 1–3 mm long, awl-shaped; anthers 6–12 mm long, blue to purple or black with yellow base and apex, pores circular, apical. *Ovary* subglobose-pyramidal, 3-lobed above, ± 3 mm long; style 8–15 mm long. *Capsule* subglobose or ovoid, 10–20 mm long, maturing to dark yellow. *Seeds* ovoid, ± 5 mm long, dark mahogany-brown, papillate, papillae becoming longer and more finger-like in distal half, each with apical tuft of minute trichomes. *Flowering time*: mainly Nov.–Jan.(–Mar.), shortly after the onset of the rains.

Distribution and ecology: distributed across southern tropical Africa, from the higher-lying parts of central Angola, Zambia, and southern Democratic Republic of Congo, through Malawi into southern and western Tanzania [see Carter (1962) for map]. The species is largely restricted to higher rainfall areas, where it occurs in open woodland and savanna, often in rocky outcrops.

Diagnosis and relationships: distinguished from other species of *Walleria* by its generally more robust habit, erect, mostly larger flowers with tepals 13–22 mm long, and free anthers not cohering at the tips, predominantly blue to purple or black with only the base and tips yellow, and dehiscing through terminal, circular pores. *Walleria mackenzii* is likely to be confused only with *W. nutans*, which has nodding flowers with tepals 6–16 mm long and connivent anthers, connate at the tips, and dehiscing through short, subapical, introrse slits.

2. **Walleria nutans** *J.Kirk* in Transactions of the Linnean Society 24: 497, t. 52/1 (1864). *W. mackenzii* var. *nutans* (J.Kirk) Baker: 498 (1879). Type: Nyasaland [Malawi], Manganja Hills, near Bishop Mackenzies Mission, 1863, *H. Waller sub J. Kirk s.n. K256018* (K, holo.!). Illustration: Dyer: 1321 (1960).

W. muricata N.E.Br.: 145 (1909). Type: Bechuanaland [Botswana], near Palapye, Jan. 1898, *Lugard 289* (K, holo.!).

W. baumii Dammer: 361 (1912). Types: Angola, Kunene–Kubangoland, Kalolo, 22 Nov. 1899, *Baum 448* (BM, syn.); Angola, Habungo, 28 Nov. 1899, *Baum 448* (BM, syn.).

W. hockii De Wild.: 8 (1915). Type: Northern Rhodesia [Zambia], Kafue Valley, 1911, *A. Hock s.n. BR8642639* (BR, holo.!).

Deciduous geophyte (70–)100–300 mm high. *Corm* subglobose or depressed-globose, 20–30 mm diam. *Stem* erect or sprawling but never climbing, mostly simple or with 1 or 2 branches, rarely more, smooth or variously prickly with delicate, recurved prickles 0.5–1.5 mm long. *Leaves* linear to narrowly lanceolate, (30–)70–150 × (2–)5–7(–12) mm, upper narrower and attenuate, base cuneate or weakly cordate but not amplexicaul, midrib smooth or with recurved prickles beneath. *Flowers* solitary in axils in central portion of stem, nodding, sometimes with additional flower developed in axil of bracteole; pedicels suberect but sharply decurved distally, 20–50(–80) mm long, smooth or scabrid, with lanceolate bracteole 10–15 mm long inserted in upper third or quarter; tepals white, pink, or mauve to pale blue, recurved or reflexed, lanceolate, (6–)10–16 × 2–5 mm. *Stamens* connivent, connate at tips; filaments 0.5–1.0 mm long; anthers (4–)6–8 mm long, mostly yellow with narrow purple band across distal third and with grey tips, slits short, subapical, introrse. *Ovary* subglobose-pyramidal, 3-lobed above, ± 3 mm long; style 5.0–8.5 mm long. *Capsule* ovoid, shortly apiculate and 3-lobed above, 8–17 mm long, green, yellow or orange. *Seeds* ovoid, ± 5 mm long, dark mahogany-brown, papillate, papillae becoming longer and more finger-like in distal half, each with apical tuft of minute trichomes. *Chromosome number*: 2n = 12 (Goldblatt & Manning 1989). *Flowering time*: Nov.–Jan.(–Mar.). Figure 1A, B.

Distribution and ecology: widely distributed through subtropical Africa, from the higher-lying parts of central and northern Namibia and southern Angola through Zambia into eastern Botswana and the northern part of South Africa, where it has been recorded from the Soutpansberg into central Limpopo, adjacent Mpumalanga and North West Province, and southwest as far as Taung in Northern Cape (Figure 2). Plants occur in open savanna, mostly in sandy soils but also on limestone flats and dolomite rock sheets.

Diagnosis and relationships: closely resembling the southwestern Cape *W. gracilis*, with which it shares nodding flowers with apically connivent anthers dehiscing through introrse, subapical pores and sometimes prickly stems, pedicels, and abaxial leaf midribs. *Walleria nutans* is distinguished by its free-standing, mostly

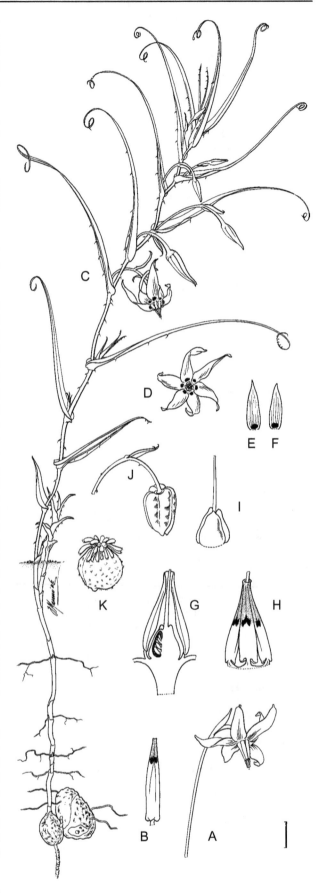

FIGURE 1.—A, B, *Walleria nutans*: A, flower; B, detached anther. C–K, *W. gracilis*: C, flowering plant; D, flower; E, outer tepal; F, inner tepal; G, half-flower; H, androecium with style; I, gynoecium; J, capsule; K, seed. Scale bar: A, C–F, J, 10 mm; B, G–I, 2 mm. Artist: John Manning.

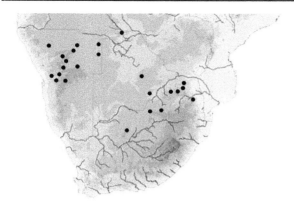

FIGURE 2.—Distribution of *Walleria nutans*.

unbranched stems, leaves without tendril-like tips, and unmarked, white, pink, or mauve to pale blue tepals. The stems, pedicels and underside of the leaf midribs may be smooth or variously armed with recurved prickles, but these are delicate, almost bristle-like, and mostly < 1 mm long, and the anthers are mostly yellow, with the purple and grey banding restricted to the apical third. The presentation of the flowers is subtly different in the two species: pedicels in *W. nutans* are essentially suberect up to the level of insertion of the bracteole in the upper third or quarter, at which point the pedicels are sharply decurved, whereas the bracteoles in *W. gracilis* are mostly inserted ± midway along the pedicels, which are therefore more arcuate.

Vernacular name: bush potato.

Representative specimens

NAMIBIA.—1723 (Singalamwe): Singalamwe, (–CB), 23 Nov. 1973, *Pienaar & Vahrmeijer 209* (PRE). 1820 (Tarikora): Gautscha Pan, E of Karakuwise, (–DD), 27 Dec. 1952, *Maguire 2194* (NBG); Cigarette, NE of Karakuwise, (–DD), 19 Jan. 1953 (fruiting), *Maguire 2275* (NBG). 1914 (Kamanjab): Ombutu, (–BC), 25 Feb. 1969, *Grobbelaar 85* (PRE). 1917 (Tsumeb): Tsumeb, (–BA), Dec. 1935, *Boss 35483* (PRE). 1920 (Tsumkwe): 157 miles [250 km] E of Grootfontein, Simkue, (–DA), 14 Jan. 1958, *Story 6117* (PRE). 2016 (Otjiwarongo): Farm Uitsig, 60 km E–NE of Otjiwarongo, (–BC), 5 Mar. 1984 (ex hort.), *Lavranos 21034* (NBG). 2017 (Waterberg): Waterberg, Farm Okamuru, (–CA), 5 Mar. 1974, *Merxmüller & Giess 30063* (PRE). 2118 (Steinhausen): 15 km along Kapps Farm road from Steinhausen to Windhoek, (–CC), 15 Mar. 1988 (fruiting), *Goldblatt & Manning 8802* (MO, PRE). 2215 (Trekkopje): Aukas, (–AA), 28 Nov. 1980, *Dinter 654* (SAM); Farm Neuschwaben, Undasbank, (–DB), 8 Mar. 1953 (fruiting), *Kinges 3061* (PRE). 2217 (Windhoek): Windhoek, Farm Lichtenstein, (–CD), 20 Jan. 1923, *Dinter 4310* (SAM).

BOTSWANA.—2225 (Mokatini): N of Lephephe, 100 km W of Serowe, (–BC), Feb. 1982 (fruiting), *Snyman & Noailles 231* (PRE). 2426 (Mochudi): Mochudi, (–AC), without date, *Rogers 6739* (BOL).

LIMPOPO.—2229 (Waterpoort): Soutpansberg, Wylies Poort, Ingwe Farm, (–DD), 18 Dec. 1960, *Hardy 407* (PRE). 2329 (Pietersburg) [Polokwane]: Buffelsberg near Munnik, (–DB), Dec. 1932, *Schweickerdt 1036* (PRE); Broederstroom, (–DD), 19 Nov. 1949, *Prosser 1361* (NBG). 2428 (Nylstroom): Vaalwater Poort on Nylstroom road, (–AC), 16 Dec. 1960, *Hardy & Bayliss 421* (PRE).

NORTH WEST.—2526 (Zeerust): Lichtenburg, Grasfontein, (–CC), Dec. 1929, *Sutton 338* (PRE). 2527 (Rustenburg): Broederstroom, (–DD), 19 Nov. 1949, *Prosser 1361* (PRE).

MPUMALANGA.—2430 (Pilgrim's Rest): Nooitgedacht mtn, near Branddraai, (–DA), 24 Nov. 1933, *Young A688* (BOL, PRE).

NORTHERN CAPE.—2724 (Taung): Barkly West, Madipelessa, (–CA), 26 Feb. 1937, *Acocks 1822* (PRE).

3. **Walleria gracilis** *(Salisb.) S.Carter* in Kew Bulletin 16: 189 (1962). *Androsyne gracilis* Salisb.: 61 (1866). Type: stated as from Nicobar Islands but probably from South Africa, Western Cape, comm. *William Marsden* [BM, holo.!; drawing in Salisbury mss. 8: 818 (BM)]. Illustration: Manning *et al.*:44–47 (2001).

W. armata Schltr. & K.Krause in Krause: 235 (1921). Type: South Africa, [Western Cape, near Klawer], [Farm] Windhoek, 8 July 1896, *R. Schlechter 8074* (B, holo. [not seen]; BM!, BR!, COI!, GRA!, K, MO!, PRE!, S!, iso.). [The collection was published as *Schlechter 2074* in the protologue, evidently a misprint].

Deciduous geophyte, 100–700 mm high. *Corm* subglobose or depressed-globose, 20–30 mm diam. *Stem* straggling or climbing, well branched, with recurved prickles 1.0–1.5 mm long in upper parts. *Leaves* lanceolate to narrowly lanceolate, (30–)70–120 × 5–10 mm, upper narrower and attenuate-cirrhose, apex coiling and tendril-like, amplexicaul, midrib with recurved prickles beneath. *Flowers* solitary in axils in central portion of stem, nodding, rose-scented; pedicels arcuate, 20–40 mm long, sparsely prickly, with lanceolate bracteole 6–10 mm long inserted ± midway; tepals white with purple blotch at base, recurved or reflexed, lanceolate, 10–16 × 2.5–3.5 mm. *Stamens* connivent, connate at tips; filaments 0.5–1.0 mm long; anthers 5–6 mm long, yellow in lower 1/2 and purple above with grey tips, slits short, subapical, introrse. *Ovary* subglobose-pyramidal, 3-lobed above, ± 2 mm long; style ± 4 mm long. *Capsule* ovoid, ± 15 mm long, shortly apiculate and 3-lobed above. *Seeds* ovoid, ± 5 mm long, dark mahogany-brown, with conspicuous apical cluster of finger-like papillae, each topped with tuft of minute trichomes, rest of seed ± smooth but covered with trichome-tufts. *Flowering time*: June and July. Figure 1C–K.

Distribution and ecology: the species has a limited, curiously scattered distribution along the west coast of South Africa. It is best known from the lower reaches of the Olifants River in Western Cape, where it has been recorded along the foot of the Gifberg east of Klawer and on Pakhuis Pass, some 50 km to the south (Figure 3). At these localities, the species occurs in deep sand among outcrops of Cape sandstone in arid fynbos vegetation. There is evidently a large disjunction in the distribution, based on a single enigmatic collection made by Rudolph Marloth in 1925 from near Kuboes in the Richtersveld. This locality, 60 km upstream from the mouth of the Orange River, is 350 km north of Klawer, and to date *W. gracilis* has not been re-collected there; nor from the intervening country. Although the identity of the Kuboes collection is not in doubt, it is unfortunately a plant that was cultivated to flowering in Cape Town five years later; and although the label is explicit in identifying the location at which the tuber was originally collected, the possibility that the locality has been confused must be considered until the species is rediscovered in the Richtersveld.

Plants may reach up to 600 mm in height when supported by small shrubs, but are much shorter in the open. The nodding, *Solanum*-like flowers are evidently adapted to buzz pollination, probably by solitary bees in the family Apidae: Anthophorinae (Manning *et al.* 2001).

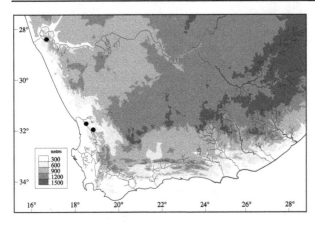

FIGURE 3.—Distribution of *Walleria gracilis*.

Diagnosis and relationships: the species closely resembles *W. nutans* from subtropical Africa and was treated as conspecific with it by Phillips (1951), but the two are quite distinct. *Walleria gracilis* is recognized by its straggling or climbing habit, well-branched stem, more robust prickles ± 1.0–1.5 mm long, upper leaves drawn into coiled, tendril-like tips, distinctive white flowers marked with a large purple blotch at the base of each tepal, and anthers that are yellow only in the lower half. The seeds of *W. gracilis* are also distinctive in being essentially smooth in the basal half (apart from the trichome-tufts) with a dense apical cluster of finger-like papillae. Flowering in *W. gracilis* takes place during the winter, whereas *W. nutans* blooms in summer.

Additional specimens seen

NORTHERN CAPE.—2817 (Vioolsdrif): Kubus [Kuboes] main kloof, 29 Aug. 1925 [fl. in cult. June 1930], *Marloth 12358* (PRE).

WESTERN CAPE.—3118 (Vanrhynsdorp): Klawer, Farm Windhoek, NW foothills of Gifberg, (–DA), mid-July 1998, *Manning 2180* (NBG), 25 June 2005, *Manning 2951B* (NBG), *Forest & Manning 542* (NBG). 3219 (Wuppertal): Clanwilliam, Cedarberg [Pakhuisberge], Farm Alpha, (–AA), 20 July 1941, *Bond 1053* (BOL, NBG).

Eremiolirion *J.C.Manning & F.Forest* in Bothalia 35: 117 (2005). Type species: *Eremiolirion amboense* (Schinz) J.C.Manning & C.A.Mannheimer.

Deciduous geophyte with deep-seated, tunicated corm, tunics decaying into firm-leathery, coarsely netted fibres extending into neck. *Cataphyll* 1, extending to ground level and enclosing leaf sheaths. *Foliage leaves* 2, basal, narrowly lanceolate-canaliculate, leathery. *Inflorescence* a divaricately branching, paniculate cyme with bracts subtending branches and pedicels only; pedicels ebracteolate, cernuous at tip, elongating slightly in fruit and straightening. *Flowers* actinomorphic, nodding, campanulate, white flushed pink or maroon abaxially; tepals connate below into short tube with minute, fringed corona present at mouth of tube, dimorphic, outer oblong, inner pandurate. *Stamens* 6, monomorphic, erect-symmetrical, inserted near mouth of tube; filaments short; anthers basifixed, erect and connivent around style, narrowly lanceolate, dehiscing by oblong apical pores. *Ovary* half inferior, with several ovules per locule; style terete, erect, filiform. *Capsules* ovoid to globose. *Seeds* ellipsoid-pyriform, blackish brown, testa surface rugose.

1 sp., central and northwest Namibia, southwest Angola.

Etymology: the name is a compound of the Greek *eremios* (desert or wilderness) and *lirion* (lily).

Ethnology: the corms are part of the traditional diet of the local tribes (*Giess, Volk & Bleissner 6039*).

Eremiolirion amboense *(Schinz) J.C.Manning & C.A.Mannheimer* in Bothalia 35: 117 (2005). *Cyanella amboensis* Schinz: 943 (1902). Type: South West Africa [Namibia], Amboland [Ovamboland], Ondonga, [Ondongwa], without date, *Rautanen 344* (Z, holo.!).

Plants (60–)100–250 mm high. *Corms* deep-seated, 30 mm diam; tunics decaying into firm-leathery, coarsely netted fibres extending into neck 10–60 mm long, pale whitish brown. *Leaves* 2, basal, suberect, narrowly lanceolate, (10–)15–25 × (8–)10–20 mm, attenuate, canaliculate with prominent midrib abaxially, leathery. *Inflorescence* a divaricately branching, paniculate cyme with (1–)3–7-branches, up to 30-flowered; pedicels cernuous at tip, 15–25 mm long, elongating slightly in fruit and straightening, ultimately 20–40 mm long. *Flowers* nodding, campanulate, white flushed pink or maroon abaxially at base of outer tepals, fragrant; perianth tube ± 4 mm long, with fringed corona 0.5–1.0 mm high at mouth of tube forming collar extending over ovary to surround base of style; outer tepals spreading from base, oblong, 15–20 × 5–7 mm, obtuse, margins revolute, inner tepals at first suberect but spreading in upper 1/2, pandurate and short-clawed, claw ± 2 mm long, blade ovate, 13–18 × 7–10 mm, apex slightly cucullate, margins crisped. *Stamens* monomorphic; filaments terete, ± 0.25 mm; anthers narrowly lanceolate, 9–10 mm long, yellow, dehiscing by oblong apical pores 1.5 mm long. *Ovary* half-inferior; ovules ± 6 per locule; style 10–12 mm long, extending shortly beyond anthers, white. *Capsules* ovoid to globose, 10–12 × 8–12 mm. *Seeds* ellipsoid-pyriform, 4.0–4.5 × 3.0–3.5 mm, blackish brown; testa surface rugose. *Flowering time*: (mid-Jan.–)Feb.–Mar.(–early Apr.). Figure 4.

Distribution and ecology: locally common through the higher-lying parts of west-central and northwestern Namibia, occurring along the better watered, western edge of the escarpment from west of Mariental in the south to Kaokoland in the north (Figure 5) and in southwestern Angola near Lake Arco. The species typically occurs in colonies, often numbering many individuals, in sandy loam or heavy clay soils, especially in stony or gravelly situations. Flowering is dependent on rainfall.

The flowers close at night ± 21:00, re-opening in the morning ± 09:00. They are fragrant during the day, with a jasmine-like fragrance at first but later smelling of stale urine, and are visited by bees and the occasional moth (*Ward, Ward & Ward 10518*).

Vernacular name: desert snowdrop.

Representative specimens

ANGOLA.—Namibe Prov., Lake Arco, Jan. 2009 (fl. ex cult. Mar. 2012), *Harrower 4061* (NBG).

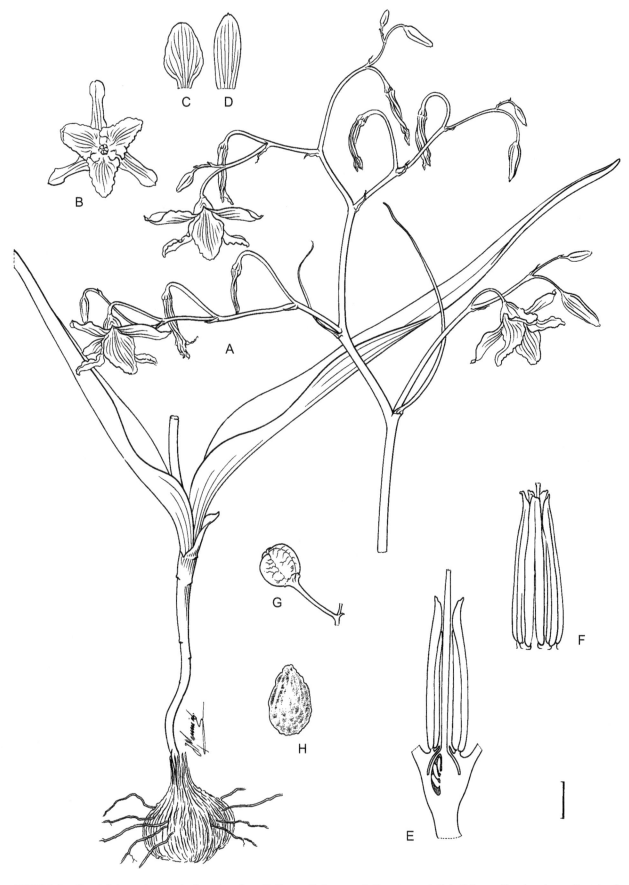

FIGURE 4.—*Eremiolirion amboense*: A, flowering plant; B, flower; C, inner tepal; D, outer tepal; E, half-flower; F, androecium with style; G, capsule; H, seed. Scale bar: A–D, G, 10 mm; E, F & H, 2 mm. Artist: John Manning.

NAMIBIA.—1713 (Swartbooisdrif): West of Ombazu, (–DD), 9 Apr. 1973, *Giess & Van der Walt 12658* (WIND). 1913 (Sesfontein): Kunene, Barab River, (–DB), 23 Mar. 2001, *Burke 1020* (WIND). 1914 (Kamanjab): Etendeka Mountain Camp, (–DD), 28 Feb. 2004, *Mannheimer 2510* (NBG, WIND). 1915 (Okaukuejo): Etosha, Adamax, (–BB), 16 Jan. 1974, *Le Roux 597* (PRE, WIND). 2014

(Khorixas): S side of watershed Ugab/Huab Rivers W of Brandberg, (–CA), 10 Apr. 1989, *Ward, Ward & Ward 10518* (PRE, WIND). 2114 (Uis): Omaruru, (–BA), 20 Mar. 1967, *Giess 9708* (PRE, WIND). 2315 (Rostock): Swakopmund, W of Kuiseb Canyon, (–BD), 10 Feb. 1966, *Giess 9131* (PRE, WIND); Farm Greylingshof SW 107, (–BD), 16 Feb. 1963, *Giess, Volk & Bleissner 5158* (PRE, WIND).

Cyanella *Royen ex L.*, Genera plantarum, edn. 5: 149 (1754). Type species: *Cyanella hyacinthoides* Royen ex L.

Pharetrella Salisb.: 47 (1866). *Cyanella* sect. *Pharetrella* (Salisb.) Pax & K.Hoffm.: 427 (1930). Type species: *P. alba* (L.f.) Salisb. = *Cyanella alba* L.f.

Trigella Salisb.: 46 (1866). *Cyanella* sect. *Trigella* (Salisb.) Pax & K.Hoffm.: 427 (1930). Type species: *T. orchidiformis* (Jacq.) Salisb. = *Cyanella orchidiformis* Jacq.

Note: Pax & Hoffmann (1930) inadvertently transposed the species and diagnoses of their sections *Pharetrella* and *Trigella*, assigning *Cyanella alba* to sect. *Trigella* and *C. orchidiformis* to sect. *Pharetrella*, thus precisely opposed to Salisbury's (1866) original placement. As Pax & Hoffmann were explicitly making combinations based on Salisbury's genera, however, the types are fixed according to Salisbury's designations, which are followed here.

Deciduous geophytes with deep-seated, tunicated corm, tunics decaying into fibrous or firm-leathery, coarsely netted fibres, sometimes extending into neck. *Cataphyll* 1, extending to ground level and enclosing leaf sheaths, entirely sheathing or with short leafy blade. *Stem* simple or branched, smooth or minutely and sparsely scabridulous. *Foliage leaves* 3–12, basal, lanceolate to linear-lanceolate and caniculate or filiform-terete, firm-textured or softer, margins plane, undulate or crispulate, smooth or scabridulous or ciliate, sur-face mostly glabrous, rarely puberulous. *Inflorescence* a raceme, usually branched, rarely highly condensed and flowers apparently solitary, with bracts subtending branches and pedicels; pedicels suberect or spreading, with solitary bracteole inserted ± midway. *Flowers* zygomorphic (perianth only weakly so through tepal orientation) or asymmetric (enantiomorphic) through stylar flexure, spreading-rotate, white, yellow, orange, pink, or mauve to blue, sometimes distinctly veined or patterned, scented; tepals free, spreading or reflexed, ± similar or weakly dimorphic with inner broader, ovate to oblanceolate, lower concave or ± cucullate. *Stamens* 6, dimorphic, either with 3 smaller posterior stamens plus 3 larger anterior stamens, or 5 smaller posterior stamens plus 1 larger anterior stamen and then lowermost either median or flexed laterally to left or right, suberect, upper stamens arcuate, lower stamen(s) declinate; filaments stout; anthers basifixed, upper sometimes adherent, narrowly lanceolate, dehiscing by apical pores or short, introrse slits. *Ovary* half-inferior, with several ovules per locule; style terete, declinate, filiform, median or flexed opposite lower stamen in enantiomorphic species. *Capsules* ovoid to globose. *Seeds* ovoid, black, or dark brown, testa surface rugose or scalariform. *Basic chromosome number*: $x = 12$ (Ornduff 1979).

9 spp., southern Namibia and southwestern South Africa, mainly winter rainfall parts.

Etymology: the name is a compound of the Greek *kyanus* (blue) and *-ella* (diminutive), alluding to the small blue flowers of *Cyanella hyacinthoides*, the first species to be described.

Ethnobotany: the corms comprise part of the traditional diet of the Nama tribes (*Archer 410*).

I. Section **Trigella** *(Salisb.) Pax & K.Hoffm.* in Die natürlichen Pflanzenfamilien 15a: 427 (1930). *Trigella*

Key to species

1a Stamens 3 + 3; flowers pink or mauve (sect. *Trigella*):
  2a Leaves linear, occasionally narrowly lanceolate, 2–8 mm wide; perianth not patterned; capsules subglobose-ovoid, 6–10 mm long; plants from southern Namibia and Richtersveld:
    3a Tepals 10–12 mm long; filaments connate < halfway into short tube ± 1 mm long; anthers yellow throughout; style 10–15 mm long, ± twice as long as lower stamens .... 1. *C. ramosissima*
    3b Tepals 13–20 mm long; filaments connate halfway or more into tube 1–2 mm long; anthers greyish or mauve distally; style ± 6 mm long, only slightly longer than lower stamens .... 2. *C. marlothii*
  2b Leaves lanceolate, 10–30 mm wide; perianth sometimes patterned; capsules ovoid-ellipsoid to oblong, 10–25 mm long; plants from Richtersveld to Western Cape:
    4a Tepals (8–)10–15(–20) mm long; posterior (upper) filaments arcuate or geniculate-sigmoid, ± evenly thick throughout, not flexuous distally; anterior (lower) anthers 5–6 mm long .... 3. *C. orchidiformis*
    4b Tepals 8–10 mm long; posterior (upper) filaments swollen basally, geniculate-sigmoid and filiform in distal half and strongly flexuous; anterior (lower) anthers 2.5–3.0 mm long .... 4. *C. cygnea*
1b Stamens 5 +1; flowers white, yellow, orange, pink, or mauve to blue (sect. *Cyanella*):
  5a Pedicels suberect; filaments connate at base only; style laterally deflexed to left or right opposite lower stamen and flowers enantiomorphic:
    6a Raceme not congested; pedicels 15–30 mm long .... 8. *C. lutea*
    6b Raceme congested, flowers apparently solitary among leaves; pedicels 80–120 mm long .... 9. *C. alba*
  5b Pedicels ± geniculate, spreading horizontally at first then sharply flexed upwards, rarely suberect or arcuate; filaments connate for half or more; style median and flowers not enantiomorphic:
    7a Raceme lax, lower flowers 1.5–3.0 × their length apart; bracteoles sub-basal; perianth orange .... 7. *C. aquatica*
    7b Raceme dense, lower flowers 0.5–0.6 × their length apart; bracteoles usually inserted in distal half of pedicel, rarely sub-basal; perianth white, pink, or mauve to blue:
      8a Upper cataphyll purple-reticulate; leaves linear, mostly 1–4 mm wide, margins conspicuously ciliate in basal half with long, shaggy cilia 2–3 mm long but ± smooth distally .... 6. *C. pentheri*
      8b Upper cataphyll usually pale, rarely purple-reticulate; leaves linear or lanceolate, mostly 4–15 mm wide, margins smooth or ciliolate along entire length with short hairs up to 1 mm long .... 5. *C. hyacinthoides*

Salisb.: 46 (1866). Type species: *Cyanella orchidiformis* Jacq.

Flowers never enantiostylous; perianth pink to mauve, sometimes patterned. *Stamens* 3 + 3, lower anthers tapering, upper anthers ± sagittate. *Ovary*: style median.

1. **Cyanella ramosissima** *(Engl. & Krause) Engl. & Krause* in Krause, Botanische Jahrbücher für Systematik 57: 239 (1921). *Iphigenia ramosissima* Engl. & Krause: 124 (1910). Type: Namibia, Aus, Kubub, Oct. 1906, *P. Range 139* (Z, holo.; SAM, iso.!).

C. krauseana Dinter & G.M.Schulze: 525 (1941). Type: Namibia, Klinghardtsgebirge, 23 Sept. 1922, *M.K. Dinter 3955* (B, holo.†; PRE!, SAM!, iso.).

Plants 80–200 mm high. *Corms* moderately or very deep-seated, 15–30 mm diam., tunics of coarsely netted, wiry fibres, extending shortly into a fibrous neck to 20 mm long, pale brown or grey. *Basal leaves* 4–6, spreading or suberect, linear to narrowly lanceolate, 50–150(–200) × 2–8 mm, acute to attenuate, plane, canaliculate or rarely involute, with prominent midrib and ribbed veins abaxially, firm-textured, glabrous, margins often ± undulate, usually ciliolate. *Inflorescence* a dense raceme up to 15(–20)-flowered, simple or 1- or 2-branched, lower flowers 0.2–0.5 × pedicel length apart; pedicels suberect but deflexed at bracteole, mostly 15–30 mm long; bracteoles mostly inserted in upper third or quarter. *Flowers* facing outwards, pale to deep pink or mauve with darker veins, fragrant; tepals spreading, outer elliptic, 13–20 × 3–4 mm, apiculate, inner oblanceolate, 13–20 × 4–7 mm, narrowed below. *Stamens* dimorphic, 3 + 3; filaments of posterior cluster sometimes almost geniculate, 2.5–3.0(–4.0) mm long, swollen basally and connate into short tube up to 1 mm long, yellow, anthers ± sagittate, outer smaller, ± 1.5 mm long, median ± 2 mm long, yellow; filaments of anterior cluster deflexed, 2.0–2.5 mm long, shortly connate for up to 1 mm, anthers 4–5 mm long, yellow. *Ovary* half-inferior, style medially deflexed, 10–15 mm long, almost twice as long as lower stamens. *Capsules* erect, subglobose-ovoid, 7–10 × 7 mm, 3-lobed. *Seeds* unknown. *Flowering time*: mainly Jul. and Aug.(–early Oct.).

Distribution and ecology: restricted to the winter rainfall part of southern Nambia, where it has been recorded on the higher ground, 350–1 050 m, from Aus and the Klinghardt Mtns along the Huib Hoch Plateau, extending into the central Richtersveld in South Africa as far south as Eksteenfontein (Figure 5). The species occurs on open stony flats, alluvial ridges, rocky terraces or sometimes on sandy or calcareous flats, in arid succulent karoo shrubland or sparse desert vegetation.

Diagnosis and relationships: readily recognized by the linear leaves, 2–8 mm wide, and dense raceme of large, pink to mauve flowers with 3 + 3 arrangement of stamens with plain yellow anthers, and a consistently long style, 10–15 mm long, thus almost twice as long as the lower stamens. *Cyanella ramosissima* may be confused with vegetatively similar *C. marlothii*, which has smaller flowers with the filaments of all six stamens connate for half or more of their length into a tube 1–2 mm long, bicoloured anthers, and a short style, ± 6 mm long.

The distinctive combination of narrow leaves and a long style separates *C. ramosissima* from the forms of *C. orchidiformis* with unpatterned tepals. The two species share smaller upper lateral anthers and otherwise resemble one another very closely although they are readily distinguished in fruit, as *C. ramosissima* has much smaller, subglobose or ovoid capsule, 7–10 mm long *vs.* the large, oblong or ellipsoidal capsules, 12–15 mm long of *C. orchidiformis*. Although *C. orchidiformis* mostly has the style shorter than the lower anthers, occasional collections (see below) have elongated styles like those of *C. ramosissima*. In the absence of fruits, such aberrant plants can be identified by their broader, soft-textured leaves and slightly larger anthers, tinged greyish distally. The two species are essentially allopatric, overlapping in their distribution only in the Richtersveld, where *C. ramosissima* is restricted to the mountainous central region whilst *C. orchidiformis* extends around the fringes.

Representative specimens

NAMIBIA.—2616 (Aus): Farm Klein Aus, (–CB), 11 Aug. 1959, *Giess & Van Vuuren 756* (BOL, PRE); 200 m N of T-junction, (–CB), 21 Oct. 1983, *Van Berkel 538* (NBG); Luderitz District, Farm Aub, (–CB), without date, *Lavranos & Pehlemann 21700* (MO); Aus Townlands, (–CB), Sept. 1983, *Lavranos & Pehlemann 21592* (MO). 2715 (Bogenfels): Klinghardtberge, (–BD), 17 Aug. 1986, *Van Berkel 571* (NBG, PRE); W Höckster Mtns, 2 km NW of Höckster, (–BD), 21 July 1986, *Van Berkel 558* (NBG). 2716 (Witpütz): Namuskluft, (–DD), 11 July 1988, *Bruyns 3191* (NBG); Farm Spitskop, Rosh Pinah, (–DD), Aug. 1981, *Lavranos 19935* (MO). 2817 (Vioolsdrif): Orange River just east of confluence with Fish River, (–AA), 1 July 1989, *Oliver 9177* (NBG).

NORTHERN CAPE.—2817 (Vioolsdrif): crest of ridge near Hottentotsparadys, (–AC), 9 Sept. 1996, *Bayer & Puttock SAF96157* (NBG); Stinkfontein Mtns, near foot of Cornellsberg, (–CA), 22 Aug. 1994, *Goldblatt & Manning 9952* (NBG); E of Eksteenfontein, (–CD), July 1989, *Williamson 4264* (NBG).

2. **Cyanella marlothii** *J.C.Manning & Goldblatt*, sp. nov.

TYPE.—Northern Cape, 2817 (Vioolsdrif): sandy flats between Jasper's werf and Doornpoort [Doringpoort Farm at W foot of Ploegberg], (–CA), 26 Aug. 1925, *R. Marloth 1211* (PRE, holo.).

Plants 200–350 mm high. *Corms* moderately deep-seated, 15–30 mm diam., tunics of coarsely netted, wiry

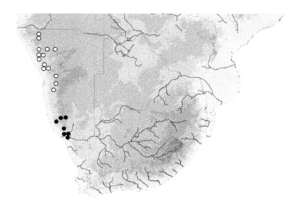

FIGURE 5.—Distribution of *Eremiolirion amboense*, ○; *Cyanella ramosissima*, ●.

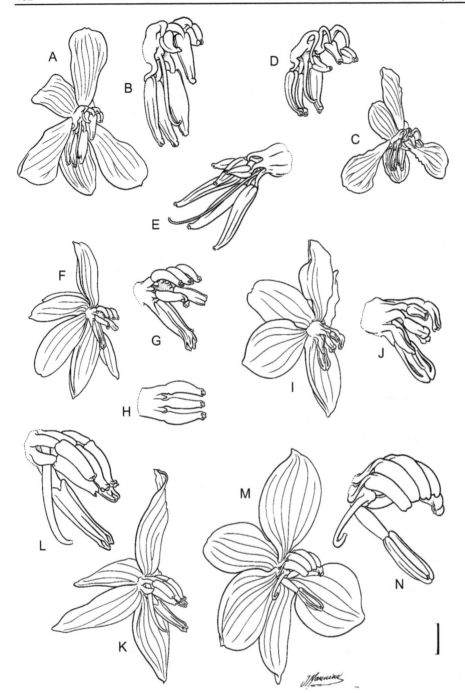

FIGURE 6.—*Cyanella* floral details. A, B, *C. orchidiformis*: A, flower; B, stamens and style. C, D, *C. cygnea*: C, flower; D, stamens and style. E, *C. marlothii*, stamens and style. F–H, *C. hyacinthoides*: F, flower; G, stamens and style; H, dorsal view of upper stamens. I, J, *C. aquatica*: I, flower; J, stamens and style. K, L, *C. lutea*: K, flower; L, stamens and style. M, N, *C. alba*: M, flower; N, stamens and style. Scale bar: A, C, F, I, K, M, 10 mm; B, D, E, G, H, J, L, N, 2 mm. Artist: John Manning.

fibres, extending shortly into a fibrous neck to 20 mm long, pale brown or grey. *Basal leaves* 4–6, suberect, linear or linear-lanceolate, 50–100 × 2–6 mm, acute to attenuate, canaliculate or involute, with prominent midrib and ribbed veins abaxially, firm-textured, glabrous, margins ± undulate, sparsely scabridulous-ciliolate. *Inflorescence* a moderately dense raceme up to 20-flowered, with up to 2 branches, lower flowers 0.5–0.6 × pedicel length apart; pedicels suberect, deflexed at bracteole, mostly 20–30 mm long; bracteoles inserted in upper third. *Flowers* facing outwards, pale mauve ('blue') with darker veins, presumably fragrant; tepals spreading, outer elliptic, 10–12 × 2–3 mm, apiculate, inner oblanceolate, 10–12 × 2–3 mm, narrowed below. *Stamens* dimorphic, 3 + 3; filaments erect but deflexed apically, 2–3 mm long, connate halfway or more into cylindrical tube 1–2 mm long; posterior anthers ± sag-

ittate, outer smaller, 1.5–2.0 mm long, median 2.0–2.5 mm long, yellow but greyish or mauve distally, anterior anthers 3–4 mm long, yellow basally but greyish or mauve in distal 2/3. *Ovary* half-inferior; style medially deflexed, ± 6 mm long, extending shortly beyond anthers. *Capsules* subglobose, 6–7 mm diam., 3-lobed. *Seeds* unknown. *Flowering time*: Aug.–Sept. Figure 6E.

Distribution and ecology: thus far known from a single collection from sandy flats near the Ploegberg, south of Kuboes in the Richtersveld (Figure 7).

Diagnosis and relationships: this distinctive species has the 3 + 3 arrangement of stamens that characterizes sect. *Trigella*, but is distinguished from other members in the section by having the filaments of all six stamens connate for half to two-thirds of their length into a cylindrical or conical tube that completely encloses

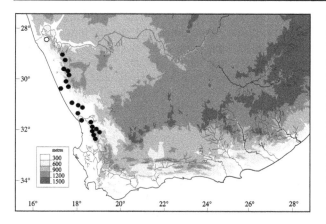

FIGURE 7.—Distribution of *Cyanella marlothii*, ○; *C. orchidiformis*, ●.

the ovary. Although the species is based on just a single collection, this comprises three essentially identical individuals. These plants were included in *C. orchidiformis* until now, despite their unique androecium. In this context it is significant that Marloth, who also collected true *C. orchidiformis* from Steinkopf on the same trip as *C. marlothii*, correctly identified the former but treated the latter as *C. capensis* (now *C. hyacinthoides*), a clear indication that he considered the Ploegberg collection to be distinct from *C. orchidiformis*, although he was misled by the connate filaments into misidentifying it as *C. hyacinthoides*. The latter does not occur in the Richtersveld, and is in any event immediately distinguished by its 5 + 1 arrangement of anthers and by the spreading-geniculate pedicels.

Among the members of sect. *Trigella*, *C. marlothii* resembles *C. ramosissima* in its narrow leaves, 2–6 mm wide, pale mauve or blue flowers with darker veins, and apparently ± globose capsules, but is separated from it by its smaller flowers with tepals 10–12 *vs.* 13–20 mm long, bicoloured *vs.* plain yellow anthers, and shorter style, ± 6 mm long and only slightly longer than the anthers *vs.* 10–15 mm long and ± twice as long as the anthers.

We have considered the possibility that the collection may be hybrid between a member of sect. *Trigella* (*C. cygnea* is recorded from the Kuboes area) and *C. hyacinthoides*, but discount this in view of the consistent appearance of the plants and the lack of other intermediate characters. This possibility did not suggest itself to Marloth, who did not record any potential parent species at the site. The absence of additional collections of the taxon is unfortunate but not unique—no further plants of *W. gracilis* have been recorded from the Richtersveld since Marloth's collection on 29 August 1925, just three days after his collection of *C. marlothii* (but see this species for further comment).

3. **Cyanella orchidiformis** *Jacq.,* Collectanea 4: 211 (1791). *Trigella orchidiformis* (Jacq.) Salisb.: 46 (1866). Type: South Africa, without locality or collector, illustration in Jacquin, Icones plantarum rariorum 2: t. 447 (1786–1793).

Plants 150–500 mm high. *Corms* moderately or very deep-seated, 15–30 mm diam., tunics of coarsely net-ted, woody fibres, sometimes connate below into flat claws, extending shortly into a fibrous neck up to 20 mm long, chestnut-brown. *Basal leaves* 4–6, suberect or spreading, lanceolate, 70–250 × 10–25(–30) mm, acute to attenuate, plane or canaliculate, with prominent midrib abaxially, soft-textured, glabrous, margins plane or undulate, smooth or ciliolate-scabridulous. *Inflorescence* a moderately dense raceme up to 35-flowered, with 1 or 2 branches, lower flowers 0.5–0.8 × pedicel length apart; pedicels suberect and deflexed at bracteole, mostly 15–30 mm long; bracteoles mostly inserted in upper third or quarter, sometimes in lower half or quarter. *Flowers* facing outwards, pink or mauve with darker veins, sometimes with darker centre, or with paler centre variously speckled with dark pink, the whole outlined with darker shading, fragrant; tepals spreading, outer elliptic, (8–)10–15(–20) × 4–5 mm, apiculate, inner oblanceolate, (8–)10–15(–20) × 5–6 mm, narrowed below. *Stamens* dimorphic, 3 + 3; filaments of posterior cluster arcuate to geniculate-sigmoid, 2–5 mm long, connate at extreme base only, ± evenly thick throughout, yellow with white base, anthers ± sagittate, outer smaller, 1–2 mm long, median 2–3 mm, yellow, but grey to purple distally; filaments of anterior cluster deflexed, 1.5–2.5 mm long, connate at extreme base, anthers 5–6 mm long, pale yellow at base, greyish or purple distally. *Ovary* half-inferior; style medially deflexed, (4–)5–13 mm long, ± as long as or extending well beyond anthers. *Capsules* erect, ovoid-ellipsoid to oblong, 14–25(–30) × 8–10 mm, pale with purplish reticulation. *Seeds* ovoid-ellipsoid, 3–4 × 1.5–2.0 mm, glossy black, rugose. *Chromosome number*: 2n = 24 (Ornduff 1979). *Flowering time*: (mid–)late Jul.–late Sept. Figure 6A, B.

Distribution and ecology: occurring along the western escarpment, from just north of Steinkopf in northern Namaqualand to Citrusdal in the Olifants River Valley (Figure 7). Collections from the Richtersveld cited under this species by Scott (1991) are referable to *C. cygnea*, evident from their filiform, sigmoid upper filaments and smaller anthers. Plants grow mostly in clay or loamy soils, often in rock crevices in granite or sandstone, where they benefit from extra moisture through runoff among rocks along the courses of seasonal streams, especially in Namaqualand.

Diagnosis and relationships: the most common and widespread of the three species of sect. *Trigella*, *C. orchidiformis*, is recognized by its lanceolate leaves, 10–25 mm wide, and racemes of pink to mauve flowers, mostly darker or patterned toward the centre, with the anthers partially or almost wholly greyish or purple, and large, ovoid-ellipsoid fruits, 14–25 mm long. The three species are essentially parapatric or allopatric, although both *C. cygnea* and *C. orchidiformis* have been collected near Steinkopf (*Marloth 6761, 6761A*). An exceptionally large-flowered variant with tepals 20 × 6–7 mm has been collected on the Gifberg Pass, growing in sandstone soil after fire (*Goldblatt & Porter 13190*), and may be polyploid.

Cyanella orchidiformis is closely allied to *C. cygnea*, with which it shares the distinctive large fruits, patterned perianth, and coloured anthers, but from which it is distinguished by its generally larger flowers, with tepals mostly 10–13 mm long *vs.* 8–10 mm long, and its unexceptional stamens. The upper filaments in *C. orchidi-formis* are arcuate or weakly geniculate, without a bulbous base and not evidently filiform in the distal half, and the lower anthers are relatively large, 5–6 mm long. The style is very variable in length, mostly 5–10 mm long, but occasionally up to 15 mm long. In contrast, *C. cygnea* has mostly smaller flowers, with tepals 8–10 mm long and very distinctive stamens, with the upper filaments geniculately sigmoid and sharply narrowed and filiform in the distal half, with much smaller lower anthers, 2.5–3.0 mm long, and a short style 3–4 mm long. The range of *C. orchidiformis* is largely to the south and east of *C. cygnea* but both species have been collected near Steinkopf.

The relatively broad leaves, 10–30 mm wide, and large capsules, readily distinguish *C. orchidiformis* from *C. marlothii* and *C. ramosissima*, which have narrow leaves 2–8 mm wide and smaller, subglobose-ovoid fruits 7–10 mm long.

Vernacular name: waterraap.

Representative specimens

NORTHERN CAPE.—2917 (Springbok): Steinkopf, (–BA), Aug. 1925, *Marloth 6761* (NBG); 6.5 km W of Steinkopf, (–BA), 29 Sept. 1986, *Perry & Snijman 3560* (NBG); between Springbok and Steinkopf beyond Bulletrap, (–BC), 29 Sept. 1986, *Perry & Snijman 3555* (NBG); Spektakel, (–DA), 25 Aug. 1941, *Compton 11398* (NBG); Eselsfontein, (–DA), 8 Sept. 1950, *Barker 519* (NBG). 3017 (Hondeklipbaai): Spoegivier, (–AD), 12 Sept. 1982, *Archer 295* (NBG). 3018 (Kamiesberg): 6 miles [9.6 km] north of Garies, (–CA), 3 Sept. 1945, *Leighton 1398* (PRE); Kamiesberg, 41.5 km from turn-off to Kliprand, (–DC), 15 Sept. 2006, *Goldblatt & Porter 12759A* (MO, NBG). 3117 (Lepelfontein): Towerberg Pass between Komkans and Kotzesrust, (–BB), 3 Sept. 1976, *Boucher 3160* (NBG). 3119 (Calvinia): Lokenburg, (–AC), 23 Aug. 1980, *Van Berkel 204* (MO).

WESTERN CAPE.—3118 (Vanrhynsdorp): Meerhofkasteel, (–AA), 8 Aug. 1984, *Snijman 805* (NBG); Farm Quaggaskop 125, (–AB), 11 Aug. 1977, *Le Roux 2282* (NBG); 15 miles [24 km] NW of Koekenaap, (–AD), 19 Aug. 1970, *Hall 3766* (NBG); between Trawal and Olifants River bridge, shale bank, (–DC), 27 Aug. 1991, *Goldblatt & Manning 9121* (MO); Gifberg Pass, Keurlandshoek, (–DD), 25 Sept. 2008, *Goldblatt & Porter 13190* (MO, NBG). 3218 (Clanwilliam): Clanwilliam, (–BB), 5 Aug. 1896, *Schlechter 8417* (MO, NBG); 29 July 1943, *Lewis NBG1814/32* (NBG); Olifants Dam, (–BB), 14 Sept. 1847, *Barker 4768* (NBG). 3219 (Wuppertal): Biedouw [Bidouw] Valley, (–AA), 23 Sept. 1952, *Barker 1748* (NBG); Cedarberg Forest Reserve, Langrug, (–AC), 21 Aug. 1983, *Viviers 496* (NBG); Rondegat River Valley 16 km NW of Algeria, (–BC), 8 Sept. 1976, *Thompson 2812* (NBG); near Citrusdal, (–CC), 6 Sept. 1949, *Steyn 390* (NBG).

Long-styled morphs

3017 (Hondeklipbaai): Grootvlei, (–BB), Sept. 1945, *Lewis 1380* (SAM); 7 Sept. 1945, *Barker 3716* (SAM). 3118 (Vanrhynsdorp): Holbak Farm, near Doornbaai [Doringbaai], (–CD), 5 Sept. 1964, *Hall 164* (NBG). 3218 (Clanwilliam): S of Clanwilliam, (–BB), 20 Sept. 1954, *De Vos 1719* (NBG); 10 miles [18 km] S of Clanwilliam, (–BB), July 1948, *Lewis 2999* (SAM).

4. Cyanella cygnea G.Scott in South African Journal of Botany 57: 50 (1991). Type: South Africa, [Northern Cape], 51.4 km from Springbok along road to Komaggas, 16 Sept. 1988 [cult. at Karoo Botanic Garden, Worcester from material collected ± 1978], *P.L. Perry 1119* (NBG, holo.!; K, MO, PRE!, iso.).

Plants (150–)200–500 mm high. *Corms* moderately or very deep-seated, 15–30 mm diam., tunics of coarsely netted, woody fibres, sometimes connate below into flat claws, extending shortly into a fibrous neck up to 20 mm long, chestnut-brown. *Basal leaves* 4–6, suberect, lanceolate, 80–200 × 10–20(–25) mm, acute to attenuate, plane or canaliculate, with prominent midrib and ribbed veins abaxially, soft-textured, glabrous, margins smooth or ciliolate-scabridulous. *Inflorescence* a dense or moderately dense raceme up to 35-flowered, with 1–4 branches, lower flowers 0.2–0.5 × pedicel length apart; pedicels suberect, deflexed at bracteole, mostly 15–30 mm long; bracteoles mostly inserted in upper third or quarter. *Flowers* facing outwards, pink with paler centre variously speckled with dark pink, the whole outlined with darker shading, fragrant; tepals spreading, outer elliptic, 8–10 × 4–5 mm, apiculate, inner obovate, 8–10 × 5–6 mm, narrowed and short-clawed below. *Stamens* dimorphic, 3 + 3; filaments of posterior cluster geniculate-sigmoid, 2–5 mm long, distally filiform and strongly flexuous, swollen basally, connate at extreme base only, yellow with white base, anthers ± sagittate, 1.5–2.0 mm long; filaments of anterior cluster deflexed, 1.0–1.5 mm long, connate at extreme base, anthers 2.5–3.0 mm long, pale yellow but greyish in distal half or third. *Ovary* half-inferior; style medially deflexed, 3–4 mm long, not extending beyond anthers. *Capsules* erect, ovoid-ellipsoid, (12–)15–20 × 8–10 mm. *Seeds* ovoid-ellipsoid, 3–4 × 1.5–2.0 mm, glossy black, rugose. *Flowering time*: late Aug.–early Oct.(–early Nov.). Figure 6C, D.

Distribution and ecology: restricted to the higher-lying parts of northern Namaqualand, where it has been collected in the Richtersveld along the Ploegberg and Stinkfontein Mtns, from Kuboes to Eksteenfontein, near Steinkopf, and along the edge of the escarpment around Komaggas, some 60 km to the south (Figure 8). Plants grow in rocky situations in open succulent karoo shrubland, typically where there is additional moisture such as along watercourses or in gorges.

Diagnosis and relationships: closely allied to *C. orchidiformis*, with which it shares characteristically mottled flowers and large, ovoid-ellipsoid capsules ± 15 mm long, and greyish or purple markings or speckling on the anthers. *Cyanella cygnea* typically has smaller flowers, with tepals 8–10 *vs.* (8–)10–15(–20) mm long, but is best identified by its stamens. The

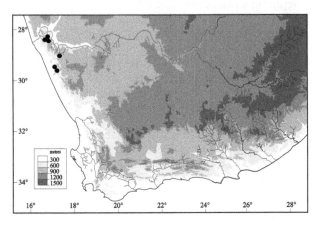

FIGURE 8.—Distribution of *Cyanella cygnea*.

strongly geniculate-sigmoid filaments of the posterior (upper) stamens are bulbous at the base and filiform in the distal half, giving them a characteristic flexuous form, the outer pair slightly longer than the median. All three anthers in the posterior cluster are subequal in size, 1.5–2.0 mm long, and the lower anthers are equally larger, 2.5–3.0 mm long. In contrast, the upper stamens in *C. orchidiformis* are ± uniformly thick except at the extreme apex and not evidently flexuous, the outer anthers are slightly smaller than the median, and the lower anthers are larger, 5–6 mm long, sometimes with the median larger than the laterals. The distributions of the two species are largely complementary, with *C. cygnea* occurring to the north and west of *C. orchidiformis*, but they overlap around Steinkopf.

Vernacular name: *wildebeet* (wild beet) (Scott 1991).

Representative specimens

NORTHERN CAPE.—2816 (Oranjemund): mtns SW of Kuboos [Khubus], (–BD), 11 Sept. 1973, *Lavranos 10834* (MO, PRE). 2817 (Vioolsdrif): Richtersveld, Kodaspiek, (–AA), 2 Sept. 1977, *Oliver, Tölken & Venter 492* (MO); Armmanshoek, (–AC), Aug. 1995, *G. & F. Williamson 5654* (NBG); Richtersveld, near Kubus [Khubus], (–CA), 13 Aug. 1983, *Archer 391* (NBG, PRE); Ploegwater at S portion of Ploegberg, (–CA), 7 Sept. 1991, *Germishuizen 5483* (PRE); Stinkfonteinberg SW of Vanzylsrus, (–CA), 4 Sept. 1977, *Oliver, Tölken & Venter 626* (NBG); stony flats 4 km N of Eksteenfontein, (–CD), 23 Aug. 2001, *Goldblatt & Porter 11751* (MO); 8 km N of Eksteenfontein, 22 Aug. 1994, *Goldblatt & Manning 9940* (MO). 2917 (Springbok): Steinkopf, (–BA), Aug. 1925, *Marloth 6761A* (NBG); Steinkopf, (–BC), 9 Aug. 1898, *M. Schlechter 119* (MO, PRE); Klipfontein, (–BA), Sept. 1929, *Grant 4840B* (MO); Komaggas, Van Reenen se Water, (–DC), 26 Aug. 1983, *Van Wyk 6501* (PRE).

II. Section **Cyanella**

Flowers sometimes enantiostylous; perianth white, yellow, orange, pink to mauve, or blue, never patterned. *Stamens* 5 + 1; anthers ± oblong. *Ovary*: style sometimes flexed to left or right.

Series *Hyacinthoides* J.C.Manning & Goldlbatt, ser. nov.

Flowers not enantiostylous; pedicels ± geniculate (horizontally spreading then flexed sharply upwards) or arcuate; perianth white, orange, pink, or mauve to blue. *Stamens*: filaments connate halfway or more. *Ovary*: style not flexed sideways. Type species: *Cyanella hyacinthoides* Royen. ex L.

5. Cyanella hyacinthoides *Royen ex L.*, Genera plantarum, edn 5: addendum [522] (1754). *C. capensis* L.: 985 (1759), nom. illegit. superfl. *C. pulchella* Salisb.: 249 (1796), nom. illegit. superfl. [*Note*: Scott's (1991) lectotypification of *C. pulchella* against Jacquin's (1776–1777) illustration of *C. capensis* L. is unwarranted and incorrect. There is no indication that Salisbury had any intention other than of replacing Linnaeus's name with his own]. Type: South Africa, without precise locality, date or collector, ex herb. Royen *Herb. Linn. 430.2* (LINN, holo.!).

Plants 150–400(–500) mm high. *Corms* deep-seated, 25–30 mm diam., tunics of coarsely netted, wiry or woody fibres, not or extending shortly into a fibrous neck to 20 mm long, pale brown or grey. *Basal leaves* 4–9(–12), suberect or spreading, linear to narrowly lan-

ceolate, 60–200(–250) × (2–)4–15(–25) mm, acute to attenuate, plane, canaliculate or rarely involute, midrib and veins prominent beneath (abaxially), firm-textured, usually glabrous but veins sometimes scabridulous or puberulous to villous abaxially with hairs up to 1 mm long, rarely both surfaces densely puberulous throughout, margins ± undulate or crispulate, usually ciliolate-scabridulous, sometimes flushed purple basally; upper cataphyll usually pale, rarely purple-reticulate or fenestrate. *Inflorescence* a moderately dense raceme up to 25-flowered, with 2–4 branches, rarely with second order branchlets and thus paniculate, lower flowers 0.3–0.6 × pedicel length apart; pedicels usually geniculate, horizontal in basal 1/2 or 2/3 then abruptly flexed upwards at ± right angles, rarely suberect or arcuate, mostly 20–30 mm long; bracteoles mostly inserted between lower and upper third, rarely sub-basal. *Flowers* facing outwards, pale to deep mauve or blue, rarely white or pink, fragrant; tepals spreading, ovate to obovate, 8–10 × 3–4 mm, apiculate. *Stamens* dimorphic, 5 + 1; filaments of posterior cluster 1.0–2.5 mm long, outer sometimes slightly longer than inner, connate ± halfway or almost completely into tube 1.0–2.0 mm long, yellow, sometimes with small intrastaminal lobules between bases of filaments, anthers 1.5–2.0(–2.8) mm long, yellow; anterior stamen with filament ± 1 mm long, connate to upper cluster for ± half length, anther 2.5–4.0 mm long, yellow. *Ovary* half-inferior; style medially deflexed, 3–4 mm long, not extending beyond anthers. *Capsules* erect on geniculate pedicels, subglobose, 5–6 mm diam., 3-lobed and retuse. *Seeds* ovoid, ± 2 mm diam., rugulose. *Chromosome numbers*: 2n = 24, 28 & 48 (Ornduff 1979). *Flowering time*: mainly mid-Sept.–mid-Dec. but mid-Aug.–mid Oct. in Namaqualand. Figure 6F–H.

Distribution and ecology: *Cyanella hyacinthoides* is widely distributed through the southern African winter-rainfall region, from just north of Steinkopf southwards through the higher-lying parts of Namaqualand into the southwestern Cape as far east as the Gouritz River (Figure 9), from near sea level to over 1 200 m. It has been recorded along the Roggeveld Escarpment south to Matjiesfontein but is absent from the arid Tanqua River basin and Little Karoo, apart from a single collection south of Oudtshoorn at the foot of the Outeniqua Mtns. The species has a wide edaphic amplitude and has

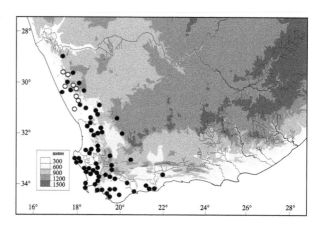

FIGURE 9.—Distribution of *Cyanella hyacinthoides* (pubescent forms, ○).

been collected on granite, sandstone, and limestone sub-strates, although it favours loamy or clay soils, where it is most often found as a component of renosterveld or succulent karooid communities. It is tolerant of distur-bance and thrives in old lands and along road verges.

Cyanella hyacinthoides is extremely variable in its foliage. Plants typically produce 4–6 lanceolate leaves but some forms may produce up to a dozen linear-invo-lute leaves. These narrow-leaved plants are scattered throughout the range of the species. A more circum-scribed ecotype occurs along the higher parts of cen-tral Namaqualand, between Kotzesrus and Springbok. Plants there tend to have the lower leaf surface vari-ously scabridulous or villous, with the hairs restricted to the leaf margins and the adaxial veins and midrib. In extreme forms, the hairs are shaggy and up to 1 mm long but there is a significant variation in the density and length of the vestiture, even within a single local-ity, from scarcely puberulous to densely villous leaves. A collection from north of Komaggas (*Barker 7412*) shows a second type of vestiture, with both leaf sur-faces closely and evenly puberulous. Populations from elsewhere in the range generally have the leaf surfaces glabrous, but some plants may have the lower surface sparsely and minutely scabridulous along the veins. There is no association between vestiture and other veg-etative features, such as leaf width or shape. The devel-opment of leaf pubescence in populations from this part of Namaqualand has also been recorded in species of *Trachyandra* (Asphodelaceae) (Manning & Goldblatt 2007) and *Haemanthus* (Amaryllidaceae) (Snijman 1984), and appears to represent a widespread ecological strategy.

Tetraploids have been detected among several wild populations of *Cyanella hyacinthoides* (Ornduff 1979), and it is thus possible that unusually robust specimens that have been remarked on by various collectors are polyploids.

Diagnosis and relationships: *Cyanella hyacinthoides* is distinguished by the moderately dense, branched racemes of mauve to blue (rarely white or pink) flow-ers with 5 + 1 arrangement of stamens with the filaments connate for ± half their length or more. The connate filaments and generally horizontally spreading pedi-cels serve to distinguish the species from pink-flowered forms of *C. lutea*, in which the stamens are ± free and the pedicels mostly suberect.

The species is closely allied to *C. pentheri*, with which it has been much confused, and the two were treated as conspecific by Scott (1991). They are essen-tially alike in their inflorescence, although the flowers in *C. pentheri* are typically paler, mostly white to pale mauve, but they differ strikingly in their foliage. The leaves of *C. pentheri* are linear-aristate and canalicu-late-involute with margins that are often crispulate and conspicuously ciliate only towards the base with shaggy hairs 2.0–3.0 mm long. Similar long cilia also fringe the upper cataphyll, which is funnel-shaped, and boldly pigmented with deep purple along the edges and veins, giving it a characteristic fenestrate appearance. Although *C. hyacinthoides* is highly variable in its foliage, the spe-cies only rarely produces similarly narrow, crispulate

leaves and in such cases they are either glabrous or are ciliolate-pubescent along their entire length, with much shorter hairs 0.2–1.0 mm long, and the upper cataphyll is usually unmarked, very rarely (*Goldblatt & Porter 11896*) purple-fenestrate. Although the two taxa have been recorded growing in close proximity in several localities (see discussion under *C. pentheri*), no interme-diates between them have been found.

Vernacular names: *raap, hotnotsraap, klipraap.*

Representative specimens

Typical form

NORTHERN CAPE.—2917 (Springbok): E of Kosies, (–BA), 15 Oct. 1988, *Williamson 3978* (NBG); Springbok, 15 miles [24 km] E of town, (–CB), 12 Oct. 1947, *Rodin 2193* (PRE); between Spektakel-berg and Komaggas, (–DA), 21 Aug. 1982, *Le Roux 2957* (NBG). 3017 (Hondeklipbaai): Spoegrivier, (–AD), 20 Sept. 1983, *Archer 410* (NBG); Kamieskroon, Skilpad Nature Reserve, (–BB), 18 Sept. 1995, *Cruz 92* (MO, NBG); sandy flats E of Kamieskroon at foot of pass, (–BB), 3 Nov. 1982, *Goldblatt 6651* (MO). 3118 (Vanrhynsdorp): 5 km S of Bitterfontein, (–AB), 9 Sept. 1985, *Duncan 184* (NBG); Vanrhynsdorp, Zandkraal Farm, (–DB), 7 Sept. 1949, *Barker 5662* (NBG); 13 km from Vanrhysdorp on road to Nieuwoudtville, (–DB), 31 Aug. 1986, *Fellingham 1116* (PRE). 3119 (Calvinia): Oorlogskloof Nature Reserve, Farm Driefontein, (–AC), 1 Nov. 1996, *Pretorius 398* (NBG); along Nieuwoudtville–Loeriesfontein road, (–AB), 11 Sept. 1986, *Steiner 1360* (NBG); Doringbos Valley, (–CC), 27 Sept. 1970, *Barker 10725* (NBG). 3217 (Vredenburg): Witteklip Rocks, (–DD), 19 Sept., *Perry 3197* (MO). 3218 (Clanwilliam): Lamberts Bay, Nortier Experimental Farm, (–AB), 6 Nov. 1974, *Boucher 2569* (NBG); irri-gation dam near Clanwilliam, (–BB), Sept. 1935, *Smuts PRE59124* (PRE); Clanwilliam, 6.2 km S of Ramskop, (–BB), 26 Sept. 1986, *Perry 3523* (NBG); Piketberg, approaching Moravian Mission at Goedverwag, (–DC), 3 Oct. 1984, *Perry 3214* (MO, NBG). 3219 (Wuppertal): Bidouw, Welbedacht Farm, (–AA), 22 Sept. 1952, *John-son 537* (NBG). Koue Bokkeveld, Ondertuin, (–CC), 28 Dec. 1978, *Hanekom 2519* (MO). 3220 (Sutherland): Roggeveld, Soekop Farm, (–AA), 11 Sept. 2006, *Rösch 660* (NBG). 3219 (Wuppertal): Citrus-dal, (–CC), 30 Sept. 1944, *Barker 3075* (NBG). 3318 (Cape Town): Langebaan, (–AA), 5 Oct. 1969, *Axelson 80* (NBG); Yzerfontein, De la Rey Farm, (–AC), 15 Oct. 1995, *Boucher 2557* (NBG); Groenekloof [Mamre], (–AC), 1850, *Zeyher 1718* (NBG); Buck Bay Farm, (–CA), 29 Nov. 1978, *Boucher 4156* (PRE); Robben Island, (–CD), 14 Nov. 1985, *Lloyd 574* (NBG); Cape Peninsula, Kamps [Camps] Bay, (–CD), Dec. 1897, *Thode s.n.* (NBG); Malmesbury, Burgers Post Farm, (–DA), 17 Oct. 1979, *Boucher & Shepherd 4839* (NBG); Langver-wacht above Kuils River, (–DC), 22 Nov. 1973, *Oliver 4806* (NBG); Paarl Mountains Nature Reserve, (–DD), 26 Oct. 1994, *Swanepoel 50* (NBG); Jonkershoek, (–DD), 27 Nov. 1973, *Smith 141* (NBG). 3319 (Worcester): Ceres, Lakenvlei Farm, (–BC), 19 Oct. 1941, *Barker 2004* (NBG); Rawsonville, (–CA), 18 Oct. 1980, *Walters 2322* (NBG); Worcester, (–CB), 17 Oct. 1980, *Walters 2310* (NBG); E approach to Franschhoek Pass, (–CC), 8 Nov. 1987, *Goldblatt & Manning 8583* (MO, PRE); Madeba Farm, W of Robertson, (–DD), 8 Oct. 1986, *Hilton-Taylor 1765* (NBG). 3320 (Montagu): Matjiesfontein, (–BA), 24 Oct. 1921, *Foley 120* (PRE). 3322 (Oudtshoorn): lower N slopes of Outeniqua Mtns, near Sebrafontein Farm, (–CC), 23 Oct. 1985, *Vlok 1216* (NBG). 3418 (Simonstown): Simonstown, Redhill Plateau, (–AB), 19 Nov. 1970, *Goldblatt 5168* (MO); Cape Peninsula, Noord Hoek, (–AB), 30 Nov. 1943, *Wasserfall 674* (NBG); Muizenberg, (–AB), Feb. 1907 (mostly in fruit), *Rogers TM25828* (PRE); Betty's Bay, 2 Dec. 1970, *Ebersohn s.n.* (NBG). 3419 (Caledon): Hermanus, Vogelgat, (–AD), 30 Oct. 1986, *Williams 3719* (MO, NBG); Genaden-dal, (–BA), 1854, *Roser PRE15439* (PRE); Gansbaai, Grootbos Nature Reserve, (–CB), 8 Dec. 2007, *Lutzeyer s.n.* (NBG). 3420 (Bredasdorp): Swellendam, Bontebok National Park, (–AB), Dec. 1962, *Liebenberg 6779* (NBG, PRE); De Hoop, Potberg Nature Reserve, (–AD), 28 Nov. 1978, *Burgers 1598* (NBG); Riversdale, Reisiesbaan Siding, (–AB), 31 Oct. 1979, *Bohnen 7043* (NBG). 3421 (Riversdale): near Still Bay on Rietvlei Road, (–AD), 13 Nov. 1982, *Bohnen 8152* (NBG); limestone hills S of Albertinia, (–AD), 4 Dec. 1985, *Goldblatt 7421* (MO); Farm Platbos, 2 km S of Aasvogelberg to Gouritz River, (–BC), 10 Dec. 1981, *Stirton 10261* (NBG).

Hairy forms

NORTHERN CAPE.—2917 (Springbok): Spektakel Pass, (–DA), 4 Sept. 1951, *Martin 835* (NBG), 11 Sept. 1993, *Goldblatt & Manning 9715* (MO); Ezelsfontein, (–DA), 8 Sept. 1950, *Barker 6656* (NBG); 5 miles [8 km] N of Komaggas, (–DB), 4 Sept. 1951, *Barker 7412* (NBG); between Brakwater and Komaggas, (–DB), 9 Sept. 1950, *Barker 6679* (NBG); 64.5 km W of Okiep towards Nababiep, (–DB), 26 Sept. 1986, *Perry 3550* (NBG). 3017 (Hondeklipbaai): 7 miles [11 km] NW of Kamieskroon, (–BB), 25 Sept. 1952, *Acocks 16477* (PRE); Kamieskroon, (–BB), 22 Aug. 1959, *Barker 9001* (NBG); Garies Hill, (–BD), 2 Sept. 1951, *Barker 7403* (NBG); 19 km S of Kotzesrus, (–DD), 16 Sept. 2001, *Goldblatt & Porter 11896* (MO, NBG). 3018 (Kamiesberg): 26 km S of Garies on road to Bitterfontein, Farm Mostertsvlei, (–CA), 30 Sept. 1987, *Reid 1310* (PRE).

6. **Cyanella pentheri** *Zahlbr.* in Annalen des kaiserlichen naturhistorischen Museums 15: 26 (1900). Type: South Africa, [Western Cape], Olifantrivier [Olifants River], Aug. [without year], *Penther 400* (W, holo.†). Neotype: South Africa, [Western Cape], Clanwillam, Biedouw [Bidouw], Welbedacht Farm, 22 Sept. 1952, *A.J. Middelmost 1741* (NBG, neo., designated here; SAM, iso.).

Note: The type of *Cyanella pentheri* is presumed lost (Scott 1991) but Zahlbruckner's (1900) description is quite clear and we designated an extant specimen to serve as a neotype.

Plants 100–400 mm high. *Corms* deep-seated, 25–30 mm diam., tunics of coarsely netted, wiry or woody fibres, extending in a short or very long fibrous or papery neck to 100 mm long, pale brown. *Basal leaves* (5–)9–17, suberect, often ± twisted or coiled apically, linear, 60–150 × 1–4(–5) mm, attenuate, canaliculate-involute, with prominent midrib and ribbed veins abaxially, firm-textured, glabrous or veins puberulous abaxially, margins straight or ± undulate or crispulate, conspicuously ciliate in basal parts only with shaggy hairs 2.0–3.0 mm long but glabrous distally; upper cataphyll prominent, with crispulate margins villous as in leaves, strongly flushed purple towards edge and along veins, thus fenestrate, sometimes also villous on veins. *Inflorescence* a moderately dense raceme up to 25-flowered, simple or up to 4-branched, lower flowers 0.3–0.6 × pedicel length apart; pedicels geniculate, horizontal in basal half or 2/3 then abruptly flexed upwards at ± right angles, mostly 20–30 mm long; bracteoles mostly inserted between lower and upper third, rarely sub-basal. *Flowers* facing outwards, white to pale mauve or blue, fragrant; tepals spreading, ovate, 7–11 × 3–4 mm, apiculate. *Stamens* dimorphic, 5 + 1; filaments of posterior cluster 2.0–3.5 mm long, connate ± 1/3 to 2/3 into tube 1.0–1.5 mm long, yellow, anthers 1.5–2.5 mm long, yellow; anterior stamen with filament ± 1 mm long, connate to upper cluster for ± half length, anther 2.5–3.5 mm long, yellow. *Ovary* half-inferior; style medially deflexed, 3–4 mm long, not extending beyond anthers. *Capsules* erect on geniculate pedicels, subglobose, 5–6 mm diam., 3-lobed and retuse. *Seeds* unknown. *Flowering time*: late Aug.–early Oct.

Distribution and ecology: *Cyanella pentheri* has a restricted distribution through the middle reaches of the Olifants River Valley from north of Citrusdal to Klawer, extending along the foot of the Gifberg onto the Bokkeveld Escarpment, and inland to the Bidouw and Doring River Valleys (Figure 10). Plants favour rocky

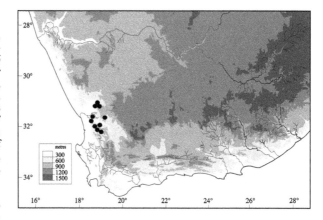

FIGURE 10.—Distribution of *Cyanella pentheri*.

places, often sandstone, mainly in arid fynbos.

Diagnosis and relationships: *Cyanella pentheri* has mostly been treated as conspecific with *C. hyacinthoides*, essentially because of the confusion between true *C. pentheri* and what we regard as pubescent forms of *C. hyacinthoides*. Florally, the two species are certainly alike in their moderately dense, branched racemes of spreading, white or mauve to blue flowers with 5 + 1 arrangement of stamens with the filaments connate for ± half their length or more, but they differ significantly in their foliage. The leaves of *C. pentheri* are consistently linear and canaliculate-involute, mostly 1–4 mm wide, with margins that are conspicuously ciliate only towards the base with long, shaggy hairs 2.0–3.0 mm long. Similar, long cilia also fringe the upper cataphyll, which is funnel-shaped, and strikingly pigmented with deep purple along the edges and veins, giving it a characteristic fenestrate appearance. The leaves of *C. hyacinthoides*, in contrast, are mostly lanceolate and 4–15 mm wide, rarely narrower, with margins either smooth or ciliolate-pubescent along their entire length, with much shorter hairs 0.2–1.0 mm long, and the upper cataphyll is usually unmarked. Pubescent forms of *C. hyacinthoides* from central Namaqualand have leaves that are variously puberulous to villous, but never with the long cilia characteristic of *C. pentheri*.

The variation in vestiture in *C. hyacinthoides* is not correlated with leaf shape, unlike the situation in *C. pentheri*. This is compelling evidence that *C. pentheri* represents a distinct genotype, which is further corroborated by the fact that the vegetative differences between the two species are maintained wherever the two have been collected together, notably north of Klawer at Zandkraal Farm (*Barker 5648* vs *Barker 5662*), Welbedacht Farm in the Bidouw Valley (*Middelmost 1741* vs *Johnson 537*) and Clanwillam (*Perry 3526* vs *Barker 4771*). We have examined both taxa growing together just outside Clanwillam ourselves and at none of these localities have we found intermediates between them.

Vernacular name: *klipraap*.

Representative specimens

NORTHERN CAPE.—3119 (Calvinia): Nieuwoudtville, Willems River Farm, (–AC), Sept. [without year], *Leipoldt 789* (NBG); Nieuwoudtville, hills near Groenrivier, (–AC), Sept. [without year], *Leipoldt 790* (NBG).

WESTERN CAPE.—3118 (Vanrhynsdorp): Zandkraal, (–DA), 7 Sept. 1949, *Barker 5648* (NBG). 3119 (Calvinia): foot of Van Rhyn's Pass, (–AC), 22 Aug. 1950, *Barker 6447* (NBG, SAM). 3218 (Clanwilliam): intersection of Citrusdal road with Klawer–Clanwilliam road, (–BB), 14 Sept. 1985, *Scott 25* (NBG); Olifants Dam, (–BB), 14 Sept. 1947, *Barker 4771* (NBG, SAM); Clanwilliam, near dam, (–BB), Sept. 1947, *Lewis 2400* (SAM); Botterkloof Pass SE of Kameelberg, (–CD), 9 Sept. 1983, *Oliver 8052* (NBG); Kanolvlei, (–DD), 6 Sept. 1951, *Barker 7448* (NBG). 3219 (Wuppertal): Diamond Drift, Biedouw River between Pakhuis and Wuppertal, (–AA), Aug. 1939, *Leipoldt 3114* (PRE); Biedouw Valley, 2 km along road to Doorn River, (–AA), 22 Aug. 1993, *Goldblatt & Manning 9632* (MO); road to Algeria, (–AC), 6 Sept. 1980, *Le Roux 2813* (NBG). *Without precise locality*: Olifantsrivier, Dec. [without year], *Zeyher s.n. SAM20551* (SAM).

7. **Cyanella aquatica** *Oberm. ex G.Scott* in South African Journal of Botany 57: 40 (1991). Type: South Africa, [Northern Cape], Nieuwoudtville, Klipkoppies, 21 Sept. 1986, *G. Scott 66* (NBG, holo.!; PRE, iso.!).

Plants up to 500 mm high. *Corms* shallow or moderately deep-seated, 20 mm diam., tunics of papery or leathery layers, not extending into neck, pale whitish brown. *Basal leaves* ± 5 or 6, suberect, linear-lanceolate or narrowly lanceolate, 200–350 × 10–15 mm, attenuate, canaliculate with prominent midrib abaxially, soft-textured, bright green, glabrous. *Inflorescence* a lax raceme, up to 15-flowered, simple or with 1–3 branches from near base, lower flowers 1.5–3.0 × pedicel length apart; pedicels geniculate, horizontal in basal 1/2 to 2/3 then abruptly flexed upwards at ± right angles, mostly 15–20 mm long but lowermost up to 30 mm long; bracteoles basal or sub-basal. *Flowers* facing outwards, bright orange, veined green on reverse, fragrant; tepals spreading, outer ovate, 9–12 × 3–4 mm, recurved-apiculate, inner short-clawed, claw ± 1 mm long, blade ovate, 9–11 × 4–5 mm. *Stamens* dimorphic, 5 + 1; filaments of posterior cluster 2.0–2.5 mm long, outer slightly longer than inner, connate ± halfway or more into tube 1.5–2.0 mm long, yellow, anthers 1.5–2.0 mm long, yellow; anterior stamen with filament ± 1 mm long, connate to upper cluster for most of length, anther ± 3 mm long, yellow. *Ovary* half-inferior; style medially deflexed, ± 3 mm long, not extending beyond anthers. *Capsules* erect on geniculate pedicels, subglobose, ± 8 × 6 mm, 3-lobed and retuse. *Seeds* ovoid, 3.0 × 2.5 mm, rugulose. *Chromosome number*: 2n = 24 (Ornduff 1979: as 'Klipkoppies' population of *C. hyacinthoides*). *Flowering time*: mid-Sept.–early Nov. Figure 6I, J.

Distribution and ecology: known originally only from the rocky outcrops immediately east of Nieuwoudtville, inland of the edge of the Bokkeveld Escarpment, *C. aquatica* has recently been collected significantly further inland just south of Calvinia, but is still the most local one of species in the genus (Figure 11). Plants are restricted to dolerite dykes, along watercourses or drainage lines where the soil becomes seasonally waterlogged during the winter months.

Diagnosis and relationships: distinguished by the lax, sparsely branched racemes, ± basal bracteoles on sharply sigmoid pedicels, and bright orange flowers. *Cyanella aquatica* is superficially similar to *C. hyacinthoides*, which also has a 5 + 1 arrangement of stamens with the filaments connate for ± half their length or more, but which differs in its fibrous corm tunics and dense racemes of white or pink to blue flowers with

the bracteoles usually inserted near the middle of the pedicels or above, only rarely near the base. The two taxa are ecologically separated, with *C. hyacinthoides* favouring better drained, sandy or gritty soils. In perianth colour, *C. aquatica* might be confused with yellow-flowered *C. lutea*, but that species has suberect pedicels with the bracteoles inserted ± midway along, filaments that are ± free to the base, and a laterally deflexed style. *Cyanella lutea* is also ecologically separated, favouring fine-grained clay soils in renosterveld or drier karroid vegetation.

Representative specimens

NORTHERN CAPE.—3119 (Calvinia): Niewoudtville, Klipkoppies, (–AC), 15 Sept. 1961, *Barker 9531* (BOL, NBG, PRE); 5 Nov. 1962, *Barker 9764* (NBG); trek path E of Nieuwoudtville near Calvinia road, (–AC), 29 Oct. 1996, *Goldblatt & Manning 10581A* (MO); Farm Driefontein, SW of Calvinia, SW slopes of Driefontein-se-Berg, in watercourse among dolerite rocks, (–DA), 23 Sept. 2009, *Goldblatt & Manning 13419* (NBG, MO).

Series *Luteae* J.C.Manning & Goldblatt, ser. nov.

Flowers ± enantiomorphic; pedicels suberect; perianth white, yellow, or pink. *Stamens*: filaments free, anthers sometimes spotted or maculate. *Ovary*: style and lower anther weakly or strongly flexed sideways in opposite directions. Type species: *Cyanella lutea* L.f.

8. **Cyanella lutea** *L.f.*, Supplementum plantarum: 201 (1782). Type: South Africa, without precise locality or date, *Sparrman s.n. Herb. Linn. 430.1* (LINN, holo.!).

Plants (120–)150–350 mm high. *Corms* moderately to deep-seated, 20–25 mm diam., tunics of coarsely netted, fibrous, leathery or woody fibres, sometimes connate below into claws, extending shortly in a neck to 30 mm long, rarely into a fibrous neck up to 100 mm long, brown. *Basal leaves* 4–15 mm, suberect or spreading, linear-hemiterete to lanceolate, 30–200(–250) × 2–15(–20) mm, acute to attenuate, leathery, plane or canaliculate, glabrous, margin smooth or ciliolate-scabridulous. *Inflorescence* a moderate or dense raceme up to 15-flowered, with 1–3 branches congested near base, thus emerging from among leaves, rarely with accessory branchlets and thus paniculate, lower pedicels 0.2–0.8 × their length apart; pedicels suberect, rarely arcuate or almost geniculate, 15–30(–50) mm long; bracteoles mostly inserted between ± halfway and upper third, sometimes in basal third or sub-basal. *Flowers* ±

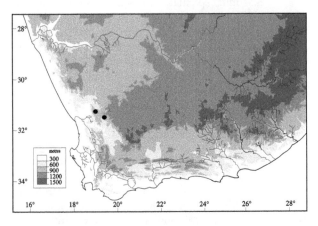

FIGURE 11.—Distribution of *Cyanella aquatica*.

enantiomorphic, facing outwards, yellow or pink to purple, usually flushed darker on reverse, with dark veins, fragrant; tepals spreading, outer oblong-elliptic, 10–15(–18) × 2–4 mm, apiculate, inner elliptic-ovate, 10–15(–18) × 3–7 mm, acute, narrowed basally or very short-clawed. *Stamens* dimorphic, 5 + 1; filaments of posterior cluster 2.5–4.0 mm long, connate only at extreme base, ± linear, yellow, anthers 2–4 mm long, yellow, usually finely spotted black or maroon; anterior stamen with filament deflexed ± laterally, 4–5 mm long, linear, connate to upper cluster at extreme base only, anther 4–7 mm long, thus ± twice as large as upper, yellow, brown, or mauve. *Ovary* half inferior; style ± laterally deflexed to left or right opposite lower stamen, 6–10 mm long, not

extending beyond lower anther. *Capsules* erect, sub-globose-retuse, 6–8 mm diam., 3-lobed. *Seeds* ovoid, ± 2 mm diam., rugulose. *Chromosome number: 2n* = 24 (subsp. *lutea*: Ornduff 1979). *Flowering time*: mainly Aug.–Nov. Figures 6K, L; 12.

Distribution and ecology: the most widely distributed species in the genus, *C. lutea*, extends through the winter rainfall region of southern Namibia and South Africa and around the interior margin of the central plateau but is absent from the central and Great Karoo (Figure 13).

Pink-flowered plants, often with narrower leaves, have been distinguished taxonomically several times, but differ consistently from the typical yellow-flowered form only in perianth colour. Baker (1871) initially recognized var. *rosea* from the Eastern Cape but subsequently (Baker 1880) changed his mind. This decision was followed by Scott (1991). However, the two colour morphs are geographically segregated: pink-flowered plants are recorded from the edges of the winter rainfall region into interior southern Africa, typically in sandy soils; and yellow-flowered plants are restricted to the southwestern Cape and nearby, on clay soils. We accordingly treat them here as distinct subspecies.

Diagnosis and relationships: distinguished from other members of sect. *Cyanella* by its racemes of pink or yellow, ± enantiostylous flowers with almost free filaments, connate only at the extreme base, and the lower anther ± twice as large as the upper anthers. Yellow-flowered plants are readily recognized by their colour but pink-flowered plants could be confused with *C. hyacinthoides* around Springbok in Namaqualand, where both occur. *Cyanella hyacinthoides* is recognized by its partially connate upper filament cluster, with the lower anther mostly less than twice as long as the upper, and by its spreading-geniculate pedicels. Subspecies *rosea* has also been confused with *C. ramosissima* (sect. *Trigella*), but the arrangement of the stamens is quite different in the two species.

Key to subspecies

1a Leaves mostly lanceolate, (2–)5–15(–20) mm wide; perianth pale to golden yellow, rarely orange, often flushed reddish on reverse; plants from southwestern Cape, from Nieuwoudtville to Uitenhage .... 8a. subsp. *lutea*
1b Leaves linear to linear-lanceolate, 2–10(–12) mm wide; peri-

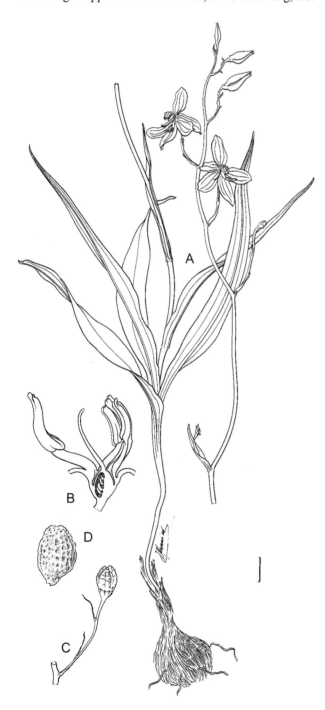

FIGURE 12.—*Cyanella lutea*: A, flowering plant; B, half-flower; C, capsule; D, seed. Scale bar: A–C, 10 mm; D, 2 mm. Artist: John Manning.

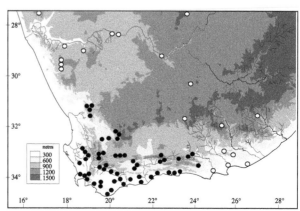

FIGURE 13.—Distribution of *Cyanella lutea* subsp. *lutea*, ●; *C. lutea* subsp. *rosea*, ○.

anth pink to purple; plants from southern Namibia and northern Namaqualand across interior of South Africa into Eastern Cape as far as Humansdorp . . . . 8b. subsp. *rosea*

8a. subsp. **lutea**

C. racemosa Schinz: 394 (1895). Type: South Africa, [Western Cape], in arenosis [sandy] Camp Ground propre [proper], Cape Town, 12 June 1892, *Schlechter 839* (Z, holo.; PRE, iso.!).

C. lutea forma *angustior* Zahlbr.: 27 (1900). Type: South Africa, [Western Cape], Caledon, Oct. [without year], *Penther 494* (W, holo.†).

Leaves 4–10, mostly lanceolate, rarely linear, (2–)5–15(–20) mm wide. *Flowers* pale to golden yellow, often flushed reddish on reverse or tinged orange.

Distribution: endemic to winter rainfall South Africa, where it has been recorded from the Bokkeveld Escarpment and southern Roggeveld to the Cape Flats and Bredasdorp in the south and eastwards through the Little Karoo to Uitenhage. The subspecies is essentially restricted to renosterveld shrubland on fine-grained clay or laterite soils, rarely on stony limestone flats.

Diagnosis: recognized by the yellow perianth, often flushed reddish on the reverse and thus with an orange tinge, and the typically lanceolate leaves, mostly 5–15(–20) mm wide, rarely narrower and grass-like. *Flowering time*: mainly Sept.–Oct. but to Nov. in the southern Cape.

Representative specimens

NORTHERN CAPE.—3119 (Calvinia): Nieuwoudtville Reserve, (–AC), 12 Oct. 1983, *Perry & Snijman 2372* (NBG, PRE); Oorlogskloof Nature Reserve, 15 km SW of Nieuwoudtville, (–AC), 14 Oct. 1996, *Pretorius 388* (NBG); Lokenburg, (–AC), 23 Aug. 1980, *Van Berkel 207* (MO). 3220 (Sutherland): Roggeveld Escarpment, Ouberg Pass, (–AD), 6 Sept. 2006, *Rösch HR538* (NBG); Sutherland, Houthoek, (–CA), 13 Sept. 1971, *Hanekom 1575* (PRE); Koedoesberg, (–CC), 1 Sept. 1973, *Oliver 4378* (NBG).

WESTERN CAPE.-3218 (Clanwilliam): Farm Nurust, about 8 miles [13 km] N of Porterville, (–DD), 22 Sept. 1966, *Loubser 2107* (NBG). 3318 (Cape Town): Bobbejaanberg above Groene Kloof [near Mamre], (–AD), Oct., *Ecklon & Zeyher 269* (MO); N of Tigerberg [Tygerberg], (–CC), 20 Sept. 1947, *Barker 4808* (NBG); Stellenbosch, Elsenburg, (DD), 5 Oct. 1938, *Penfold 153* (NBG). 3319 (Worcester): Saron, (–AA), Oct. 1896, *Schlechter 10633* (MO); Tulbagh, (–AC), Oct. 1920, *Marloth 9939* (NBG); 5 miles [8 km] from Ceres at bottom of Theron's Pass, (–AD), 11 Nov. 1974, *Snijman 9* (NBG); Karoopoort, (–BA), 26 Sept. 1944, *Compton 16054* (NBG); Tanqua Karoo, near Bloukop, (–BD), 22 Sept. 1975, *Thompson 2549* (NBG); Karoo Garden, Worcester, (–CB), 11 Sept. 1969, *Tarr s.n.* (NBG); Worcester, Langerug Koppie, (–CB), 23 Sept. 1974, *Walters 1207* (NBG); Rooihoogte Pass, (–DB), 28 Oct. 1980, *Mauve, Reid & Wikner 197* (NBG). 3320 (Montagu): Laingsburg, Cabidu, (–AB), 28 Sept. 1951, *Compton 22890* (NBG); Whitehill, (–BA), 20 Sept. 1943, *Compton 14874* (NBG); S of Ashton, (–CC), 21 Sept. 1941, *Barker 2032* (NBG); 14 km E of Montagu, Kliepheuwel Farm, (–CC), 16 Oct. 1998, *Manning 2195* (NBG). 3321 (Ladismith): Vleiland, N of Klein Swartberge, (–AC), 10 Oct. 1976, *Thompson 3183* (NBG); road to Waterkloof NW of Ladismith, (–AD), 23 Oct. 1980, *Mauve, Reid & Wikner 105* (NBG); S side of Rooiberg, (–CB), 22 Nov. 1983, *Mauve, Van Wyk & Pare 40* (NBG); Van Wyksdorp, (–DA), 12 Sept. 1983, *Bohnen 8297* (NBG). 3322 (Oudtshoorn): Prince Albert route 407 to Klaarstroom, Farm Welgelegen, (–AC), 1 Sept. 2006, *Roux 4199* (NBG); George Forest, (–CD), 25 Nov. 1950, *Martin 638* (NBG); De Rust, Ostekloof Farm, (–DA), 28 Sept. 1971, *Dahlstrand 2088* (MO, PRE); Knysna, Barrington, (–DD), 14 Nov. 1949, *Barker 6068* (NBG). 3419 (Caledon): Kogelberg State Forest, Remhoogte, (–AA), 25 Oct. 1984, *Brits 23* (NBG); Greyton, (–BA), 21 Oct. 1967, *Bayliss 4019*

(MO, NBG); 5 miles [8 km] NW of Riviersonderend, (–BB), 17 Sept. 1949, *Heginbotham 83* (NBG); Swellendam to Stormsvlei, (–BB), 3 Oct. 1974, *Goldblatt 2924* (MO); slopes of Kleinberg, ± 3 km NW of Napier, (–BD), 19 Oct. 1976, *Thompson 3206* (NBG, PRE); ± 15 km NW of Napier, Fairfield Farm, (–BD), 3 Oct. 1994, *Kemper IPC644* (NBG); Bredasdorp, Bosheuwel, (–BD), 6 Oct. 1982, *Cowling 1882* (NBG). 3420 (Bredasdorp): Kathoek Farm, 30 km E of Bredasdorp, (–AD), 11 Oct. 1981, *Mauve & Hugo 140* (NBG); De Hoop, Potberg Nature Reserve, (–AD), 12 Oct. 1978, *Burgers 1276* (NBG); Swellendam, Bontebok Park, (–AB), 20 Sept. 1965, *Grobler 490* (NBG); Struisbaai, ± 5 km on Bredasdorp road, (–CC), 26 Oct. 1987, *Fellingham 1366* (NBG). 3421 (Riversdale): Blombos Road, 8–10 km S of Riversdale, (–AA), 11 Oct. 1993, *Goldblatt & Manning 9792* (NBG); Reisiesbaan siding, (–AB), 31 Oct. 1979, *Bohnen 7051* (NBG); Still Bay, (–AD), 16 Oct. 1978, *Bohnen 4463* (NBG). 3422 (Mossel Bay): Great Brak, (–AA), 21 Sept. 1959, *Lewis 5601* (NBG). 3423 (Knysna): Plettenberg Bay, (–AB), 21 Nov. 1953, *Taylor 4320* (NBG).

EASTERN CAPE.—3323 (Willowmore): flats between Hotsprings and Toorwater, (–AC), 5 Oct. 1971, *Oliver 3646* (NBG, PRE); Vledermuis area between Fullerton & Heuningklip, (–BA), 14 Sept. 1973, *Oliver 4582* (NBG); Baviaanskloof, Adamskraal, (–BC), 22 Oct. 1999, *Desmet 2095* (NBG); Bellvue, ± 4 km from Avontuur, (–CC), 11 Nov. 1978, *Botha 2188* (PRE); Suuranysberge, Vöelkraal Farm, (–CC), 1 Oct. 1984, *Stirton 10903* (NBG). 3324 (Steytlerville): Kruisrivier–Hankey Dist., (–CB), [without date], *Manson 297* (NBG); poort between Patensie and Cambria, (–DA), 11 Sept. 1973, *Thompson 1885* (NBG).

8b. subsp. **rosea** *(Eckl. ex Baker) J.C.Manning & Goldblatt*, stat. nov. *Cyanella lutea* var. *rosea* Eckl. ex Baker: t. 259 (1871). Type: South Africa, [Eastern Cape], Queenstown, 1860, *T. Cooper 270* (K, holo.!).

Note: Scott (1991) was of the opinion that no material of Cooper's collection had been preserved and thus lectotypified the name against the illustration in *Refugium Botanicum*, which was drawn from plants collected and cultivated by Thomas Cooper. There exists, however, a specimen at Kew, collected by Cooper in 1860 at Queenstown in the Eastern Cape where this form has since been re-collected, and labelled with the name *Cyanella rosea*. There seems no reason to doubt that it represents the original collection from which the cultivated plants were derived. This material, as the holotype, takes precedence over the illustration (McNeil *et al.* 2006: Art. 9.10 & 9.17). Baker's (1871) citation of the Ecklon manuscript name, *Cyanella rosea* Eckl., which appeared as a printed label on some herbarium collections, including *Ecklon 255* (NBG), is a clear indication that the correct author citation for the epithet is Eckl. ex Baker.

C. lineata Burch.: 589 (1812). Type: South Africa, Bechuanaland [Northern Cape], near Moshowa [Moshaweng] River, without exact date [1811–1812], *Burchell 2256-2* (K, holo.!).

C. odoratissima Ker Gawl.: t. 1111 (1827). Type: South Africa, Cape of Good Hope, without precise locality, date or collector, cultivated in Tate's nursery, London, apparently not preserved, illustration in Ker Gawl., The Botanical Register 13: t. 1111 (1827). [*Note*: Scott's (1991) attribution of the name to Lindley is incorrect, as John Bellenden Ker [-Gawler] wrote the text for the first 14 volumes (Stafleu & Cowan 1976), and John Lindley only assumed authorship from vol. 15].

C. lutea var. *angustifolia* Schinz: 48 (1896). Type: Namibia, Oas [Huib-Hoch Plateau], Oct. 1891, *Fleck 232* (Z, holo.!).

Leaves 6–12, linear-hemiterete to linear-lanceolate,

2–10(–12) mm wide. *Flowers* pale to deep pink or purple. *Flowering time*: mainly Aug.–Sept. in Namaqualand and Bushmanland; Oct.–Dec. in the interior and Eastern Cape.

Distribution: recorded from central Namaqualand around Springbok and the Huib-Hoch Plateau in southern Namibia, inland through Bushmanland along the Orange and Vaal Rivers as far as Kuruman in Northern Cape and Smithfield in the southern Free State, thence southwards through the eastern Upper Karoo to Humansdorp (Figure 13). Plants have been recorded mainly from sandy, sometimes calcareous, flats in Nama-Karoo shrubland or drier grassland, in the Kuruman area typically beneath small bushes. The subspecies is relatively poorly documented for such a large range.

Diagnosis: distinguished by its generally narrower, often grass-like leaves 2–12 mm wide, and its pink perianth. Plants from Namaqualand-Bushmanland and southern Namibia are especially distinctive in their very small stature, numerous, semi-terete leaves, and ± congested inflorescence branching near the base, giving them a characteristic caespitose appearance.

Representative specimens

FREE STATE.—3026 (Aliwal North): Smithfield, (–BA), Oct. [without year or collector], *STE12787* (NBG).

NORTHERN CAPE.—2623 (Morokweng): Vryburg, (–DB), Sept. 1924, *Henrici 160* (PRE). 2723 (Kuruman): 36 miles [57.6 km] E-NE of Van Zylsrus, 2 miles [3 km] N of Kuruman River on Tsabong road, (–AD), 17 Oct. 1961, *Leistner 2886* (PRE). 2818 (Warmbad): 2 miles [3 km] S of Goodhouse, (–DD), 27 July 1950, *Lewis 3003* (SAM), *63739* (PRE); Goodhouse, (–BD), 27 July 1950, *Barker 6262* (NBG). 2819 (Ariamsvlei): Augrabies, (–DB), 21 Aug. 1954, *Compton 24474* (NBG); Augrabies Falls National Park, (–DB), 22 Aug. 2005, *Steyn 759* (NBG, PRE). 2820 (Kakamas): 12 miles [19 km] E of Kakamas, (–DB), 28 Aug. 1963, *Hardy & Rauh 1560* (PRE). 2823 (Griekwastad): Brakfontein, (–CD), 20 Sept. 1988, *Saaiman 227* (PRE). 2824 (Kimberley): Kuruman River 16 miles [25.6 km] W of Kuruman-Gordonia boundary, (–BA), 18 Oct. 1961, *Leistner 2893* (PRE). 2917 (Springbok): along Goodhouse road, (–BD), 20 Sept. 1980, *Van Berkel 260* (NBG); near Springbok, (–DB), Sept. 1939, *Lewis 750* (SAM); Droëdap [SE of Springbok], (–DD), 27 Aug. 1941, *Barker 2029* (NBG). 2918 (Gamoep): Aggenys, (–BD), 13 Oct. 1971, *Wisura 2264* (NBG). 2922 (Prieska): Prieska, (–DA), [without date], *Bryant s.n. PRE38351* (PRE). 3017 (Hondeklipbaai): Theunis se Dam, 36 km S of Little Rock Caravan Park on Droëdap road, (–BB), 25 Aug. 1977, *Thompson & le Roux 37* (NBG); Droëdap, (–BB), 27 Aug. 1941, *Esterhuysen 5894* (PRE). 3023 (Britstown): De Aar, (–DB), 30 Aug. 1895, *Solly s.n. PRE38315* (PRE). 3024 (De Aar), Rolfontein Nature Reserve, Springbok Flats, (–BB), 9 Sept. 1982, *Coetzee s.n. PRE61030* (PRE).

WESTERN CAPE.— 3223 (Rietbron): 20 km from Farm Rietbron on road to Murraysburg, (–BA), 13 Oct. 1983, *Retief & Reid 521* (PRE).

EASTERN CAPE.—3126 (Queenstown): lower slopes, (–DD), 1893, *Galpin 1568* (PRE). 3127 (Lady Frere): Little Bushy near Cala, (–DA), Dec. 1910, *Royffe s.n. TM25721* (PRE). 3225 (Somerset East): Mountain Zebra National Park, (–AD), 4 Oct. 1979, *Du Toit 155* (PRE); Addo National Park, (–BC), Nov. 1962, *Liebenberg 6620* (PRE). 3226 (Fort Beaufort): Bushman's River Mouth, (–DB), 2 Dec. 1941, *Barker 2034* (NBG). 3227 (Stutterheim): Queenstown, Bram Neck, (–AA), 28 Oct. 1946, *Thorns s.n.* (NBG); between Fish River and Governor's Kop, (–BD), 16 Oct. 1961, *Batten 1-Pl.83* (NBG). 3325 (Port Elizabeth): Kommadagga, (–BB), 27 Nov. 1973, *Bayliss 6199* (MO); Vanstadensberg, (–CC), Dec. [without year], *MacOwan 1086* (SAM); near Zwartkop River, (–DC), Nov. [without year], *Ecklon 255* (NBG, SAM). 3424 (Humansdorp): Humansdorp, (–BB), 14 Oct. 1928, *Gillett 2397* (NBG).

9. **Cyanella alba** *L.f.*, Supplementum plantarum: 201

(1782). *Pharetrella alba* (L.f.) Salisb.: 47 (1866). Type: South Africa, without precise locality or date, *Thunberg s.n. Herb. Linn. 430.4* [LINN, lecto.!, designated by Scott: 46 (1991)].

Plants 80–200 mm high. *Corms* deep-seated, 15–25 mm diam., tunics of coarsely netted fibres, extending into neck up to 50 mm long, pale brown. *Basal leaves* ± 10–20, erect, filiform to linear, (40–)50–100 × 0.5–3.0 mm, attenuate, leathery, bright green, glabrous. *Inflorescence* a highly congested, simple raceme such that flowers apparently solitary among leaves; pedicels suberect, (80–)100–200 mm long; bracteoles either subbasal or inserted in upper half. *Flowers* enantiomorphic, facing outwards, white or pale pink or pale yellow, fragrant; tepals spreading, cucculate, outer elliptic, 12–20 × 5–7 mm, recurved-apiculate, inner ovate, 12–20 × 7–12(–15) mm, acute, narrowed basally or short-clawed, claw up to 1 mm long. *Stamens* weakly dimorphic or submonomorphic, 5 + 1; filaments of posterior cluster 3–5 mm long, connate only at extreme base, awl-shaped, white, anthers 3.5–5.5 mm long, yellow, sometimes marked with black spot on upper surface near base, sometimes cohering; anterior stamen with filament deflexed laterally, 3–4 mm long, awl-shaped, connate to upper cluster at extreme base, anther 4–6 mm long, yellow. *Ovary* half-inferior; style laterally deflexed opposite lower stamen, 7–9 mm long, not extending beyond lower anther. *Capsules* erect, ellipsoid, 13–15 × 7–8 mm, 3-lobed. *Seeds* ovoid, ± 2 mm diam., rugulose. *Chromosome number*: $2n = 24$ (subsp. *flavescens*: Ornduff 1979). *Flowering time*: (late Aug.–)mid-Sept.–mid-Oct.(Nov.). Figures 6M, N; 14.

Distribution and ecology: the species has a scattered distribution along the western mountains in Western Cape, where it is known from the Bokkeveld Escarpment, the Cedarberg and Olifants River Mtns, and the base of the Swartruggens (Figure 15). These three areas of occurrence correspond to the distribution of the three subspecies that we recognize. *Cyanella alba* is restricted to clay soils in renosterveld shrubland.

Diagnosis and relationships: one of the easiest species to identify on account of its highly congested inflorescence axis with extremely elongate pedicels, the flowers thus apparently borne on 1-flowered peduncles rather than in a raceme. The raceme is never branched, and up to a maximum of nine flowers are produced, thus very much fewer than in other species. The flowers are strongly enantiostylous, and either white to pale pink with uniformly yellow anthers, or pale yellow with maculate anthers. These colour morphs, which are geographically segregated, correlate with the position of the bracteole on the pedicels, and we recognize them as three subspecies. The large, ellipsoid capsule, 13–15 mm long, is unique in sect. *Cyanella*, resembling those of *C. cygnea* and *C. orchidiformis* in sect. *Trigella*.

Key to subspecies

1a Leaves filiform, 0.5–1.5 mm diam.; flowers white; bracteoles subbasal, not readily visible among leaves and thus apparently absent . . . . 9c. subsp. *minor*
1b Leaves linear-filiform, 1–3 mm wide; flowers white or yellow; bracteoles inserted in distal half of pedicel, thus

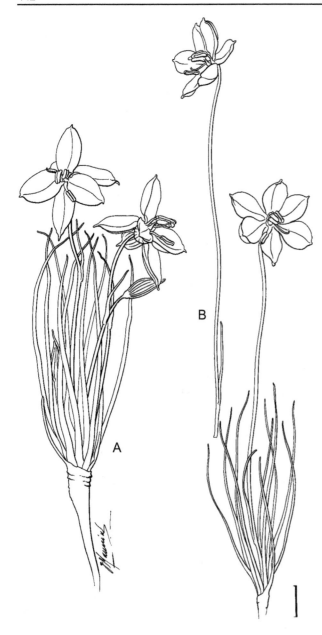

FIGURE 14.—*Cyanella alba*: A, subsp. *alba*, showing distal bracteole; B, subsp. *minor*, showing subbasal bracteole. Scale bar: 10 mm. Artist: John Manning.

clearly present:

2a Flowers 3–9 per plant, white or pale pink; anthers uniformly yellow .... 9a. subsp. *alba*
2b Flowers 1–4 per plant, pale yellow or outer tepals white; upper anthers marked with black blotch adaxially near base .... 9b. subsp. *flavescens*

9a. subsp. **alba**

Plants (80–)100–200 mm high. *Leaves* linear, 1–3 mm wide. *Inflorescence* 3–9-flowered; pedicels with bracteole in distal half. *Flowers* white to pale pink. *Stamens*: anthers uniformly yellow. Figure 14A.

Distribution: endemic to the Bokkeveld Escarpment, from just north of Nieuwoudtville southward to Menzieskraal near Botterkloof (Figure 15).

Diagnosis: characterized by the long pedicels, (80–)100–200 mm long, with the bracteole inserted between one third and three-quarters along, and white or pale

pink flowers flushed darker pink on the reverse. The anthers are uniformly yellow, with the upper cluster free or coherent. Plants are often well grown, producing 3–9 flowers. The position of the bracteoles in the distal half of the pedicels distinguishes subsp. *alba* from subsp. *minor* from the Tanqua Basin to the south, which has similar flowers but subbasal bracteoles.

Representative specimens

NORTHERN CAPE.—3119 (Calvinia): N of Nieuwoudtville, Grasberg Farm, (–AC), 16 Sept. 1961, *Barker 9457* (NBG); Nieuwoudtville Reserve, (–AC), 8 Sept. 1983, *Perry & Snijman 2351* (NBG); ± 15 km S of Nieuwoudtville, Matjiesfontein Farm, (–AC), 13 Sept. 1976, *Thompson 2902* (NBG); Lokenberg Farm, (–CA), 26 Sept. 1933, *Acocks 17263* (PRE); 4 Sept. 1985, *Snijman 905* (NBG); Menzieskraal Farm, (–CA), 29 Sept. 1933, *Markotter s.n.* (NBG).

9b. subsp. **flavescens** *J.C.Manning* in Manning *et al.* in Bothalia 35: 119 (2005). Type: South Africa, Western Cape, Biedouwberg, 26 Aug. 1896, *Schlechter 8686* (SAM, holo.!, BOL!, PRE!, iso.).

Plants 120–200 mm high. *Leaves* linear-filiform, 1–2 mm wide. *Inflorescence* 1–4-flowered; pedicels with bracteole in distal 1/2. *Flowers* pale yellow or outer tepals white. *Stamens*: anthers yellow, upper five coherent and maculate with dark blotch on upper side near base.

Distribution: restricted to the northern Cedarberg and Olifants River Valley, between Clanwilliam and Wuppertal, and especially common in the Biedouw River Valley (Figure 15).

Diagnosis: a very distinctive taxon recognized by its pale yellow flowers (sometimes the outer tepals white) with the upper anthers coherent and marked on the upper side with a black blotch near the base. Up to four flowers are produced per plant.

Representative specimens

WESTERN CAPE.—3218 (Clanwilliam): Clanwilliam, (–BB), 4 Aug. 1896, *Schlechter 8405* (BOL, PRE); 10 km S of Clanwilliam, (–BB), 12 Sept. 1997, *Goldblatt & Manning 10741* (MO, NBG). 3219 (Wuppertal): Biedouw Mtn, (–AA), 20 Sept. 1937, *Lewis s.n.* (NBG); bottom of hill to Biedouw Valley, (–AA), 9 Aug. 1984, *Perry 3145*

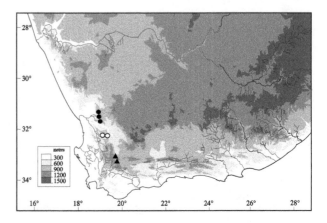

FIGURE 15.—Distribution of *Cyanella alba* subsp. *alba*, ●; subsp. *flavescens*, ○; subsp. *minor*, ▲.

(NBG); Farm Welbedacht, (–AA), 20 Sept. 1937, *Barker 283* (NBG); Koudeberg near Wuppertal, (–AA), 4 Oct. 1897, *Bolus 9095* (NBG); Citadel Kop, (–AA), 7 Sept. 1953, *Compton 24237* (NBG); near Wuppertal, (–AA), 28 Aug. 1951, *Martin 811* (NBG).

9c. subsp. **minor** *J.C.Manning* in Manning *et al.* in Bothalia 35: 119 (2005). Type: South Africa, Western Cape, Karoopoort, 27 Sept. 1944, *Barker 3024* (NBG, holo.!).

Plants 80–150 mm high. *Leaves* filiform, 0.5–1.5 mm wide. *Inflorescence* 1–3-flowered; pedicels with bracteole subbasal. *Flowers* white to pale pink with darker pink on reverse. Stamens: anthers uniformly yellow. Figure 14B.

Distribution: highly localized and known only from just north of Karoopoort in the southern Tanqua Karoo basin (Figure 15).

Diagnosis: distinguished from the typical subspecies, which has similar white or pale pink flowers and uniformly yellow anthers, by the shorter pedicels, mostly < 100 mm (rarely up to 150 mm long) with the bracteoles sub-basal and thus difficult to distinguish from the leaves. This led Manning *et al.* (2005) to conclude that bracteoles were absent, and we were only able to establish the true situation after having the opportunity of dissecting live plants. The plants are typically small in stature, with only 1–3 flowers per plant.

Representative specimens

WESTERN CAPE.—3319 (Worcester): Karoopoort, (–BA), 19 Sept. 1938, *Levyns 6236* (BOL); Tanqua Karoo N of Karoopoort, (–BA), 9 Sept. 2007, *Goldblatt & Porter 12970* (NBG); 13 Sept. 2009, *Goldblatt, Manning & Porter 12970* (MO, NBG).

EXCLUDED SPECIES

Walleria paniculata Fritsch: 493 (1896). Type: Madagascar, Ins. St Marie, without date, *Paulay s.n.* (GZU, holo.) = *Dianella ensifolia* (L.) DC. (Hemerocallidaceae) (Perrier de la Bathie 1938).

ACKNOWLEDGEMENTS

Our sincere thanks to Anne-Lise Fourie, Head Librarian at the South African National Biodiversity Institute, and Mary Stiffler, Research Librarian at the Missouri Botanical Garden, for their assistance in locating necessary references. This is the last of our contributions edited by Beverley Momberg before her retirement, and we are deeply grateful to her for her years of sterling service.

REFERENCES

BAKER, J.G. 1871 '1870'. *Cyanella lutea* var. *rosea. Refugium botanicum* 4: t. 259.
BAKER, J.G. 1878 '1880'. Report on the Liliaceae, Iridaceae, Hypoxidaceae, and Haemodoraceae of Welwitsch's Angolan Herbarium. *Transactions of the Linnean Society, Botany* 1: 245–273.
BAKER, J.G. 1879. A synopsis of Colchicaceae and the aberrant tribes of Liliaceae. *Journal of the Linnean Society, Botany* 17: 405–510.
BROWN, N.E. 1909. List of plants collected in Ngamiland and the northern part of the Kalahari Desert. *Bulletin of Miscellaneous Information, Kew* 1909: 89–146.
BRUMMITT, R.K., BANKS, H., JOHNSON, M.A.T., DOCHERTY, K.A., JONES, K., CHASE, M.W. & RIDALL, P.J. 1998. Taxonomy of Cynastroideae (Tecophilaeaceae): a multidisciplinary approach. *Kew Bulletin* 53: 769–803.
BURCHELL, W.J. 1812. *Travels in the interior of southern Africa.* Batchworth, London.
CARTER, S. 1962. Revision of *Walleria* and *Cyanastrum* (Tecophilaeaceae). *Kew Bulletin* 16: 185–195.
COWLEY, E.J. & BRUMMITT, R.K. 2001. Tecophilaeaceae. In G.V. Pope, *Flora zambesiaca* 12, 3: 18–25.
DAMMER, C.L.U. 1912. Liliaceae africanae IV. *Botanische Jahrbücher für Systematik* 48: 360–366.
DE WILDEMAN, É. 1915. Decades novarum specierum florae congolensis XVI. *Bulletin du Jardin de l'État à Bruxelles* 5: 3–8.
DINTER, M.K. & SCHULTZE, G.M. 1941. Neue Amaryllidaceen aus Deutsch-Südwest-Afrika. *Botanische Jahrbücher für Systematik* 71: 520–525.
DYER, R.A. 1960. *Walleria nutans. Flowering Plants of Africa* 34: t. 1321.
ENGLER, A. & KRAUSE, K. 1910. Lilaceae africanae 2. *Botanische Jahrbücher für Systematik* 45: 123–155.
FRITSCH, K. 1896. Zur Flora von Madagascar. *Annalen des K.K. Naturhistorischen Hofmuseums, Wien* 5: 492–494.
GOLDBLATT, P. & MANNING, J.C. 1989. Chromosome number in *Walleria* (Tecophilaeaceae). *Annals of the Missouri Botanical Garden* 76: 925, 926.
HEYWOOD, V.H., BRUMMITT, R.K., CULHAM, A. & SEBARG, O. 2007. *Flowering plant families of the world.* Royal Botanic Gardens, Kew.
HOLMGREN, P.K., HOLMGREN, N.H. & BARNETT, L.C. 1990. *Index Herbariorum, par. 1: the herbaria of the World.* New York Botanical Garden, New York.
JACQUIN, N.J. 1776–1777. *Hortus botanicus vindobonensis* 3. Wappler, Vienna.
JACQUIN, N.J. 1786–1793. *Icones plantarum rariorum* 2. Wappler, Vienna.
JACQUIN, N.J. 1791 ['1790']. *Collectanea* 4. Wappler, Vienna.
KRAUSE, K. 1921 Liliaceae africanae 6. *Botanische Jahrbücher für Systematik* 57: 235–239.
KER GAWLER, J.B. 1827. *Cyanella odoratissima. The Botanical Register* 13: t. 1111. Ridgway, London.
KIRK, J. 1864. On a new genus of Liliaceae from East tropical Africa. *Transactions of the Linnean Society* 24: 497–499.
LEFFERS, A. 2008. Gemsbok Bean and Kalahari Truffle. Macmillan Education Namibia, Windhoek.
LINNAEUS, C. 1754. *Genera plantarum*, edn 5. Salvius, Stockholm.
LINNAEUS, C. 1759. *Systema naturae*, edn 10. Salvius, Stockholm.
LINNAEUS, C. (fil.). 1782 ['1781']. *Supplementum plantarum.* Braunschweig, Uppsala.
MANNING, J.C. & GOLDBLATT, P. 2007. *Trachyandra arenicola* and *T. montana* (Asphodelaceae), two new species from South Africa. *Bothalia* 37: 26–31.
MANNING, J.C., GOLDBLATT, P. & BATTEN, A. 2001. *Walleria gracilis. Flowering Plants of Africa* 57: 44–47.
MANNING, J.C., FOREST, F. & MANNHEIMER, C.A. 2005. *Eremiolirion*, a new genus of southern African Tecophilaeaceae, and taxonomic notes on *Cyanella alba. Bothalia* 35: 115–120.
McNEILL, J., BARRIE, F.R., BURDET, H.M., DEMOULIN, V., HAWKSWORTH, D.L., MARHOLD, K., NICOLSON, D.H., PRADO, J., SILVA, P.C., SKOG, J.E., WIERSMA, J.H. & TURLAND, N.J. 2006. International Code of Botanical Nomenclature (Vienna Code) adopted by the seventeenth International Botanical Congress, Vienna, Austria, July 2005. Gantner, Liechtenstein. [*Regnum Vegetabile* 146].
ORNDUFF, R. 1979. Chromosome numbers in *Cyanella* (Tecophilaeaceae). *Annals of the Missouri Botanical Garden* 66: 581–583.
PAX, F. & HOFFMANN, K. 1930. Amaryllidaceae. In A. Engler & K. Prantl, *Die natürlichen Pflanzenfamilien*, edn 2, 15a: 391–430. Engelman, Leipzig.
PERRIER DE LA BATHIE, H. 1938. Liliacées. In H. Humbert, *Flore de Madagascar.* Gouvernement Général de Madagascar, Tananarive.
PHILLIPS, E.P. 1951. *The genera of South African flowering plants.* Memoirs of the Botanical Survey of South Africa No. 25. Department of Agriculture, Pretoria.
SALISBURY, R.A. 1796. *Prodromus stirpium in horto ad Chapel Allerton vigentium.* Hooker, London.
SALISBURY, R.A. 1866. *The genera of plants.* Van Voorst, London.

SCHINZ, H. 1895. Beiträge zur Kenntnis der afrikanischen Flora III. Amaryllidaceae. *Bulletin de l'Herbier Boissier* 3, sér. 1: 394, 395.

SCHINZ, H. 1896. Die Pflanzenwelt Deutsch-Südwest-Afrikas. *Bulletin de l'Herbier Boissier* 4, sér. 1, App. III: 1–57.

SCHINZ, H. 1902. Beiträge zur Kenntnis der afrikanischen Flora XIV. Haemodoraceae. *Bulletin de l'Herbier Boissier* 2, sér. 2: 943, 944.

SCOTT, G. 1991. A revision of *Cyanella* (Tecophilaeaceae) excluding *C. amboensis*. *South African Journal of Botany* 57: 34–54.

SIMPSON, M.G. & RUDALL, P.J. 1998. Tecophilaeaceae. In K. Kubitzki, *The families and genera of vascular plants III. Flowering plants—monocotyledons*: 429–436.

SNIJMAN, D. 1984. A revision of the genus *Haemanthus* L. (Amaryllidaceae). *Journal of South African Botany*, Suppl. vol. 12. National Botanical Gardens, Cape Town.

STAFLEU, F.A. & COWAN, R.S. 1976. *Taxonomic literature* 1: A–G. Bonn, Scheltema & Holkema, Utrecht.

ZAHLBRUCKNER, A. 1902. Plantae pentherianae. *Annalen des kaiserlichen naturhistorischen Museums* 15: 1–73, t. I–IV.

9

Nomenclature and typification of southern African species of *Euphorbia*

P.V. BRUYNS*

Keywords: Euphorbia, Euphorbiaceae, lectotypes, neotypes, typification

ABSTRACT

Types have been located for most of the 185 species of *Euphorbia* L. that are known to occur naturally in southern Africa and also for most of their synonyms. Lectotypes or neotypes are selected where possible for those names for which a holotype cannot be found. The synonymy largely follows previous accounts and reasons are given where new synonymy is proposed. *Euphorbia huttonae* N.E.Br. is reinstated at the level of species and *E. franksiae* var. *zuluensis* A.C.White *et al.* is raised to the level of species as *E.* **gerstneriana** Bruyns, nom. nov. A new name, *E.* **radyeri** Bruyns, is provided for the rhizomatous plants previously referred to as *E. caerulescens* Haw., which is synonymous with *E. ledienii* A.Berger.

INTRODUCTION

The large, cosmopolitan genus *Euphorbia* L. is represented in southern Africa (taken here to include the countries Botswana, Lesotho, Namibia, South Africa, and Swaziland) by 185 species. For these there are 368 validly described names at the level of species. In this listing only naturally occurring species are included. These are distributed among the four subgenera established by Bruyns *et al.* (2006) and based on molecular and morphological data as shown in Table 1, where additional information on growth habit is also supplied. Many of the succulent species such as *E. globosa* Sims, *E. meloformis* Aiton and *E. obesa* Hook.f. are popular subjects in specialist collections around the world, but the other species have little economical use, though their medicinal use is probably underestimated. The last revisions of the genus for this region were those of N.E. Brown (1911–1912, 1915), with the succulent species receiving further attention by White *et al.* (1941). The taxonomy of *Euphorbia* in southern Africa remains disorganised, with many names applied in different herbaria in South Africa to quite different species. In an attempt to bring order to the situation, types have been located and, where these are missing, lectotypes or neotypes are selected as applicable.

MATERIALS AND METHODS

For each validly published name for a southern African species of *Euphorbia*, the protologue was consulted in the relevant literature. Type specimens have been searched for among material in the herbaria B, BM, BOL, G, GRA, K, KMG, M, NBG, NH, NU, NY, OXF, P, PRE, S, SAM, SBT, W, WIND, WRSL, WU, Z (herbarium acronyms according to Holmgren *et al.* (1990)). A specimen is taken as the holotype if it was indicated as such by the author, or if it is clear from where it is located relative to where the author worked that it must be the holotype. In many cases it has proved to be impossible to be sure which specimen is the holo-

type. In such instances, a lectotype is generally selected from among the duplicates of the 'type number' or from among the 'syntypes' mentioned by the author in the protologue. A particular specimen is chosen over others according to its quality or, in the case of syntypes, according also to how widely duplicates (if any) are represented. In cases where no appropriate material for use as a lectotype was located, a neotype was selected. All material cited has been seen unless it is expressly stated otherwise. The JSTOR Plant Science website (http://plants.jstor.org/) has been consulted in all applicable cases and the Kew Herbarium Catalogue (http://apps.kew.org/herbcat) was consulted for many species and names as well. Data on localities is given as on the specimens, with the present-day country where the specimen was collected added.

RESULTS

The species are arranged alphabetically within the four subgenera of *Euphorbia* that were established in Bruyns *et al.* (2006). The synonymy is as in Bruyns *et al.* (2006), except where otherwise mentioned and discussed.

NOMENCLATURAL ACCOUNT

1. **Euphorbia** subg. **Chamaesyce** *Raf.*

1a. Sect. **Anisophyllum** *Roeper*

E. austro-occidentalis *Thell.*, Vierteljahrsschrift der Naturforschenden Gesellschaft in Zürich 61: 431 (1916). Type: Namibia, Okahandja, sandy bushveld, cultivated land, 1 300 m, Oct., *Dinter 105* (Z, lecto., designated here; BOL, GRA, SAM-3 sheets, isolecto.). [Thellung cited also *Dinter 222* (Z), *222a* (Z) and *822* (Z). A lectotype is selected.]

E. chamaesycoides *B.Nord.*, Dinteria 11: 20 (1974). *Chamaesyce chamaesycoides* (B.Nord.) Koutnik: 263 (1984). Type: Namibia, Brandberg, Upper Tsisab Valley, ± 1 600 m, 6 May 1963, *Nordenstam 2567* (S, holo.; M, iso.).

* Bolus Herbarium, University of Cape Town, 7701

Table 1.—Numbers of species, available (i.e. validly published) names and showing the numbers of species exhibiting different growth forms (annuals, herbs, succulents, and geophytes) in the subgenera of Euphorbia.

Subgenus	Species	Available names	Annuals	Perennial, non-succulent herbs	Succulents	Geophytes
Chamaesyce	37	84	15	0	21	1
Esula	15	33	0	12	3	0
Euphorbia	50	66	0	0	50	0
Rhizanthium	83	183	0	5	73	5

E. eylesii *Rendle*, Journal of Botany 43: 52 (1905). *Chamaesyce eylesii* (Rendle) Koutnik: 263 (1984). Type: Zimbabwe, Deka siding along Bulawayo-Victoria Falls railway line, May 1904, *Eyles 130* (BM, holo.; SRGH, iso.).

E. leshumensis N.E.Br., Flora of Tropical Africa 6(1): 513 (1911). Type: Botswana, Leshumo Forest, received May 1883, *Holub* (K, lecto., designated here). [Brown (1911) also cited *Macaulay 423* (K), from Zambia.]

E. glanduligera *Pax*, Botanische Jahrbücher für Systematik 19: 142 (1894). *Chamaesyce glanduligera* (Pax) Koutnik: 263 (1984). Type: Namibia, bei Nawas am Swakop, bei Salem, 12 Dec. 1888, *Gürich 3* (missing; sketch of type at K). Neotype (designated here): Namibia, Naukluft Mtns, between Ababes and Homnus, *Pearson 9106* (BOL).

E. pfeilii Pax.: 534 (1897). Type: Namibia, Stolzenfels, Rietfontein, 1890/1891, *Pfeil 91* (missing).

E. glaucella Pax.: 737 (1898). Type: Namibia, Okahandja, Mar. 1883, *Köpfner 68* (Z, lecto., designated here). [Pax (1898) also cited *Fleck 454a* (Z), so a lectotype is selected.]

E. anomala Pax: 636 (1908), *nom. illegit.*, non Boissier (1862).

E. kwebensis N.E.Br.: 137 (1909). Type: Botswana, Kwebe Hills, 3300', 7 Jan. 1897, *Lugard 143* (K, lecto., designated here). [Brown (1909) also cited *Lugard 81* (K), from the same locality and marked both specimens as 'Type'.]

E. gueinzii *Boiss.* in A.P. de Candolle, Prodromus 15(2): 71 (1862). Type: South Africa, at Natal Bay, *Gueinzius* (G, lecto., designated here; W, isolecto.). [Boissier (1862) also cited an unnumbered collection of *Sanderson* (in 'h. Kew', missing).]

E. gueinzii var. *albovillosa* (Pax) N.E.Br.: 252 (1915). *E. albovillosa* Pax: 373 (1904). Type: South Africa, Natal, Inchanga, 1 180 m, 16 Sept. 1893, *Schlechter 3245* (BOL, lecto., designated here; GRA, K, PRE, isolecto.). [A lectotype is designated as no specimens seen by Pax have been located.]

E. inaequilatera *Sond.*, Linnaea 23: 105 (1850). *Anisophyllum inaequilaterum* (Sond.) Klotzsch & Garcke: 22 (1860). *Chamaesyce inaequilatera* (Sond.) Soják: 169 (1972). Type: South Africa, Port Natal, *Gueinzius 167* (MEL [501275], holo.; F, MEL, iso.).

Anisophyllum mundii Klotzsch & Garcke: 25 (1860). Type: South Africa, Cape, Gamka R., Prince Albert div.,

Jan. 1820, *Mund & Maire 15* (K 000253186, lecto., designated here). [There are two specimens under this number at K and it is not certain that either was seen by Klotzsch & Garcke. This is the larger specimen, the other is a 'branch from the type', according to N.E. Brown.]

A. setigerum E.Mey. ex Klotzsch & Garcke: 29 (1860). Type: South Africa, Cape, *Drège* (missing).

E. parvifolia E.Mey. ex Boiss.: 34 (1862). Type: South Africa, Cape, Jan. 1820, *Mund & Maire 15* (K 000253186, lecto., designated here). [Boissier cited: near Gariep, *Drège*; Beaufort distr., *Lund*; 'Anis. Mundtii Kl. et Gke l.c. p. 25'. From the latter, a lectotype is selected.]

E. parvifolia var. *laxa* Boiss.: 34 (1862). Type: none located. [Boissier cited: *Drège 8191; 8198* and 'Sieb. Cap. n. 154, which have not been located.]

E. sanguinea var. *setigera* E.Mey. ex Boiss.: 35 (1862). Type: South Africa, near Kei and Bashee Rivers, *Drège* (missing).

E. sanguinea var. *natalensis* Boiss.: 35 (1862). Type: South Africa, Port Natal, *Gueinzius 167* (F, lecto., designated here; MEL-2 sheets, isolecto.). [Boissier (1862) did not say in which herbarium he had seen this specimen so a lectotype is selected.]

E. nelsii Pax: 737 (1898). Type: Namibia, Hereroland, 1886, *L. Nels 91* (Z, holo.; K, iso.). [N.E. Brown annotated the piece of *Nels 91* at K as 'fragment from type' and mentioned also that the type was at Z.]

E. inaequilatera var. *perennis* N.E.Br.: 246 (1915). Type: South Africa, Natal, near Tugela, 4 Jan. 1886, *Wood 3552* (K, lecto., designated here). [Brown also cited many other syntypes for this variety.]

E. livida *E.Mey. ex Boiss.*, in A.P. de Candolle, Prodromus 15(2): 14 (1862). *Chamaesyce livida* (E.Mey. ex Boiss.) Koutnik: 263 (1984). Neotype (designated here): South Africa, Natal, without precise locality, *Gerrard 1171* (K). [Boissier (1862) cited 'ad Natal Bay, *Drège*' and *Gueinzius 177*, which are both missing. The Drège specimens found do not have this locality on them so a neotype is selected.]

E. mossambicensis *(Klotzsch & Garcke) Boiss.* in A.P. de Candolle, Prodromus 15(2): 36 (1862). *Anisophyllum mossambicense* Klotzsch & Garcke: 30 (1860). *Chamaesyce mossambicensis* (Klotzsch & Garcke) Koutnik: 263 (1984). Type: Moçambique, Rios de Sena, *Peters 33* (K, lecto., designated here). [No type has been found at B. This is a fragment of the type so is designated as lectotype.]

E. neopolycnemoides *Pax & K.Hoffm.*, Botanische Jahrbücher für Systematik 45: 240 (1910). *Chamaesyce neopolycnemoides* (Pax & K.Hoffm.) Koutnik: 263 (1984). Type: South Africa, Transvaal, between Nylstroom and Naboomspruit, sandy places along Machalaquana R., 24 Jan. 1894, *Schlechter 4278* (missing). Neotype (designated here): South Africa, Waterberg, Boschpoort near Warmbaths, 3650', Jan 1906, *Bolus 12280* (K; duplicate at BOL). [The type is missing and the specimen at K of *Bolus 12280* was matched by N.E. Brown against *Schlechter 4278* so it is selected as neotype.]

E. arabica var. *latiappendiculata* Pax: 85 (1909). Type: South Africa, Waterberg, Boschpoort near Warmbaths, Jan 1906, *Bolus 12280* (K, lecto., designated here; BOL, isolecto.). [The specimen at K is 'part of the type' and the other was not seen by Pax so a lectotype is selected.]

E. pergracilis *P.G.Mey.*, Mitteilungen aus der Botanischen Staatssammlung München 6: 247 (1966). *Chamaesyce pergracilis* (P.G. Mey.) Koutnik: 263 (1984). Type: Namibia, 7 miles east of Purros towards Sesfontein, 23 June 1963, *Giess 3211* (M, holo.; MO, PRE, WAG, iso.).

E. phylloclada *Boiss.* in A.P. de Candolle, Prodromus 15(2): 66 (1862). Type: South Africa, between Verleptpram and mouth of Gariep, Sept., *Drège 238* (S, lecto., designated here). [Boissier (1862) cited *Drège* 'in h. Bunge' (missing), from the same locality.]

E. hereroensis Pax: 35 (1889). Type: Namibia, Hereroland, Hykamkab, 300 m, May 1886, *Marloth 1190* (missing).

E. rubriflora *N.E.Br.*, Flora of Tropical Africa 6(1): 509 (1911). Type: Zimbabwe, Victoria Falls, Jan. 1906, *Allen 264* (K, lecto., designated here; SRGH, isolecto.). [Brown (1911) also cited: Zambia, Livingstone, *Rogers 7132* (K, BOL).]

E. schlechteri *Pax*, Botanische Jahrbücher für Systematik 28: 26 (1900). *Chamaesyce schlechteri* (Pax) Koutnik: 263 (1984). Type: Moçambique, Ressano Garcia, 1 000', 24 Dec. 1897, *Schlechter 11915* (PRE, lecto., designated here; BOL, BR, COI, G, GRA, HBG, K, WAG, isolecto.). [No material definitely seen by Pax in known and so a lectotype is selected.]

E. spissiflora *S.Carter*, Kew Bulletin. 45: 331 (1990). Type: Zimbabwe, Nhongo, 8 km north of Gokwe, 6 Mar. 1964, *Bingham 1158* (K, holo.; SRGH, iso.).

E. tettensis *Klotzsch* in W.C.H. Peters, Naturwissenschaftliche Reise nach Mossambike 1: 94 (1861). *Chamaesyce tettensis* (Klotzsch) Koutnik: 263 (1984). Type: Moçambique, Tete, *Peters* (missing). [The name *Anisophyllum tettense* Klotzsch & Garcke: 34 (1860) was not validly published since it only cited the above, which had not yet appeared.]

E. zambesiana *Benth.*, Hooker's Icones Plantarum 14: t. 1305 (1880). *Chamaesyce zambesiaca* (Benth.)

Koutnik: 263 (1984). Type: Malawi, Zomba Mtn, 1861, *Meller, Livingstone's Expedition* (K, lecto., designated here). [Bentham (1880) also cited: Malawi, Shire Highlands, Blantyre, *Buchanan 10* (K). A lectotype is selected.]

1b. Sect. **Articulofruticosae** *Bruyns*

E. angrae *N.E.Br.*, Flora capensis 5(2): 279 (1915). Type: Namibia, Lüderitz (Angra Pequeña), 18 Jan. 1907, *Galpin & Pearson 7549* (K, lecto., designated here; drawing; PRE, SAM, isolecto.). [Both the specimen at SAM and that at K were annotated as 'Type' by N.E. Brown, so a lectotype is selected.]

E. einensis G.Will.: 57 (2004). Type: Namibia, southern Schakalberg, 70 km NE of Oranjemund, *Williamson 5143* (BOL-2 sheets, holo.).

E. einensis var. *anemoarenicola* G.Will.: 62 (2004). Type: South Africa, Kortdoorn, *Williamson 5985* (BOL-2 sheets, holo.).

E. burmannii *E.Mey. ex Boiss.* in A.P. de Candolle, Prodromus 15(2): 75 (1862). Type: South Africa, Cape, towards Blauwberg, *Drège 2920* (P, lecto., designated here). [Boissier (1862) cited *Drège 2920* (P); Tygerberg, *Bergius; Krauss*. Neither of the latter two specimens has been located.]

E. biglandulosa Willd.: 27 (1814), *nom. illegit. non* Desf. (1808). [Willdenow (1814) gave a description but cited no specimens.]

Arthrothamnus burmannii E.Mey. ex Klotzsch & Garcke: 62 (1860). Type: South Africa, Cape, *Drège* (missing). Neotype (designated here): South Africa, Cape, *Drège 2920* (P). [No specimen as cited by Klotzsch & Garcke (1860) has been found, so a neotype is selected.]

Arthrothamnus bergii Klotzsch & Garcke: 63 (1860). Type: South Africa, Cape, *Bergius* (missing).

E. phymatoclada Boiss.: 24 (1860). Type: South Africa, rocky hills at Ebenezer, *Drège 2943* (GRA, lecto., designated here). [Boissier (1860) cited a specimen 'in h. Bunge', which has not been located so a lectotype with the same number is selected. This is from a plant of *E. burmannii*, although this name is usually placed as a synonym of *E. mauritanica* (e.g. White *et al.* 1941).]

E. hydnorae E.Mey. ex Boiss.: 95 (1862). Type: South Africa, between Kaus and Doornpoort, *Drège 2943* (GRA, lecto., designated here). [Apart from *Drège 2943*, Boissier (1862) also cited 'in montibus Niueweweld alt. 3 000–4 000 ped.', apparently another collection of Drège and both were 'in h. Bunge'. The lectotype selected here is a specimen of *E. burmannii*, although this name is also usually placed as a synonym of *E. mauritanica* (e.g. White *et al.* 1941).]

E. corymbosa N.E.Br.: 279 (1915). Type: South Africa, Cape, near Albertinia, 16 Nov. 1910, *Muir* (K, holo.).

E. karroensis (Boiss.) N.E.Br.: 290 (1915). *E. bur-mannii* var. *karroensis* Boiss. in DC.: 75 (1862). Type: South Africa, Cape, Karoo between Hol River and Mierenkasteel, 500–1 000', 5 Aug. 1830, *Drège 2947* (P, lecto., designated here; K, isolecto.). [Boissier (1862) did not say in which herbarium the specimen was located so a lectotype is selected.]

E. macella N.E.Br.: 288 (1915). Type: South Africa, Cape, near Little Brak River, 10 Oct. 1814, *Burchell 6197/2* (K, holo.).

E. ephedroides *E.Mey. ex Boiss.* in A.P. de Candolle, Prodromus 15(2): 75 (1862). Type: South Africa, Cape, Karoo at Goedemanskraal, 2 500', 8 Sept. 1830, *Drège 2949* (P, lecto., designated here; K, MO, S, isolecto.). [Boissier (1862) cited also *Burchell 1424* ('in h. DC.'), which has not been located.]

E. ephedroides var. *imminuta* L.C.Leach & G.Will.: 72 (1990). Type: South Africa, Cape, Alexander Bay, *Williamson 3652* (NBG, holo.; K, PRE, iso.).

E. ephedroides var. *debilis* L.C.Leach: 73 (1990). Type: Namibia, north of Rosh Pinah, *Leach & Brunton 15893* (NBG, holo.; K, MO, PRE, iso.).

E. exilis *L.C.Leach*, South African Journal of Botany 56: 76 (1990). Type: South Africa, Cape, Aties, May 1984, *Leach & Bayer 17129* (NBG-2 sheets, holo.; K, iso.).

E. glandularis L.C.Leach & G.Will.: 75 (1990). Type: South Africa, Cape, near Steinkopf, *Leach & Hilton-Taylor 17019* (NBG, holo.; K, PRE, iso.).

E. gentilis *N.E.Br.*, Flora capensis 5(2): 289 (1915). Type: South Africa, Cape, Vanrhynsdorp Div., hills near Zout River, 500', 14 Jul. 1896, *Schlechter 8136* (BOL, lecto., designated here; GRA, HBG, K, PRE, S, iso-lecto.). [Brown (1915) also cited Graafwater (Grauwa-ter), 15 Dec. 1908, *Pearson 3271* (BOL, K, SAM); near Bitterfontein, *Zeyher 1531* (G, K, S, W).]

E. vaalputsiana L.C.Leach: 534 (1988a). Type: South Africa, Cape, Vaalputs, near Gamoep, *Leach & Perry 17232* (NBG, holo.; K, MO, PRE, iso.).

E. gentilis subsp. *tanquana* L.C.Leach: 538 (1988a). Type: South Africa, Cape, near turnoff to Skitterykloof, *Leach & Perry 17247a* (NBG, holo.; K, M, MO, PRE, iso.).

E. giessii *L.C.Leach*, Dinteria 16: 27 (1982). Type: Namibia, 18 km east of Henties Bay, Dec. 1976, *Giess 14809 (sub Leach 15940)* (PRE, holo.; M, WIND, iso.).

E. herrei *A.C.White* et al., The Succulent Euphor-bieae 2: 962 (1941). Type: South Africa, Cape, near Swartwater, 1930, *Herre sub PRE 46025* (PRE, holo.). [Although White *et al.* (1941) did not mention the number PRE 46025, it is assumed that this is the same specimen as the one they cited.]

E. juttae *Dinter*, Neue und wenig bekannte Pflanzen Deutsch-SWA's: 30 (1914). Type: Namibia, Garub, 900 m, 9 Jan. 1910, *J.Dinter 1047* (SAM, lecto., designated by Leach (1988a); NY, isolecto.).

E. siliciicola Dinter: 31 (1914). Type: Namibia, Büll-sport, 5 Apr. 1911, *Dinter 2132* (SAM, lecto., designated by Leach (1988a)).

E. aequoris N.E.Br.: 279 (1915). Type: South Africa, Cape, Middelburg div., Schoombie, Feb. 1897, *Trol-lip (sub SAM 20091)* (SAM, lecto., designated here; K, isolecto.). [Brown (1915) cited also: Rosmead Junc-tion, 4 000', 22 Mar. 1900, *Sim sub Galpin 5626* (PRE); between Colesburg & Hanover, 1871, *Bolus 2201* (K).]

Leach (1988a) discussed *E. juttae* in detail but con-sidered that *E. aequoris*, although closely related, was 'sufficiently distinct in vegetative characters and habit alone for it to be disregarded' in those discussions. He did not say in what way it was so distinct. It appears that this distinctiveness lay in the much more robust plants formed by *E. aequoris*, with longer and more slender stems and branches, with more widely spaced and less prominent tubercles and a longer rootstock, as well as the lack of the peculiar habit that the branches have in *E. juttae* of bending over to the north or west. Nevertheless, among the material that he cited under *E. juttae* were two specimens from near Olifantshoek and near Kenhardt respectively that are rather more typical of *E. aequoris* than of *E. juttae*. While many specimens of *E. aequoris* are unmistakeable (especially those from the Great Karoo and drier parts of the Eastern Cape), those from the Northern Cape and calcareous pans on the southern edge of the Kalahari are not clearly refer-able to either species. Some of these (especially plants from exposed spots) may even exhibit a similar, almost prostrate habit to *E. juttae*, and have shorter stems and branches with more prominent tubercles while more pro-tected plants are erect, slender, and more typical of *E. aequoris*. I have found no clear distinctions between the two species and have placed *E. aequoris* in synonymy.

E. lavrani *L.C.Leach*, The Journal of South African Botany 49: 807 (1983). Type: Namibia, Namuskluft, 1 200 m, *Lavranos & Newton 16872* (PRE holo.; NBG, SRGH, iso.).

E. muricata *Thunb.*, Prodromus plantarum capen-sium 2: 86 (1800). Type: South Africa, Cape, *Thunberg* (UPS-THUNB 11499, holo.; drawing and fragment at K, iso.).

E. spicata E.Mey. ex Boiss.: 97 (1862). Type: South Africa, 31 Aug. 1830, *Drège 2946* (K, lecto., designated here; S, isolecto.). [Boissier (1862) cited also Cape, near Bitterfontein, *Zeyher 1531* (G, K, S, W), which is *E. gentilis*.]

E. aspericaulis Pax: 26 (1899). Type: South Africa, Cape, Hantam Mtns, *ex Dr Meyer* (holo. missing; draw-ing and fragment at K, iso.). [According to Carter (2002) this specimen is at B, but it cannot be located.]

E. rhombifolia *Boiss.*, Centuria Euphorbiarum: 19 (1860). Type: South Africa, Cape, arid places on south-ern Karoo, *Drège 8217* (G, lecto., designated here; K, S, W, isolecto.). [Boissier (1860) cited also *Ecklon & Zey-her, Euphorb. 23, 83* (G, W).]

Arthrothamnus densiflorus Klotzsch & Garcke: 62 (1860). Type: South Africa, Cape, Karoo near Olifants

River, Oudtshoorn distr., Jan. 1820, *Mund & Maire* (K, lecto., designated here). [Since there is no evidence that Klotzsch & Garcke saw this specimen it is designated as lectotype.]

E. brachiata (E.Mey. ex Klotzsch & Garcke) Boiss.: 74 (1862). *Arthrothamnus brachiatus* E.Mey. ex Klotzsch & Garcke: 62 (1860). Type: South Africa, Cape, near Ebenezer, *Drège 2948* (K, lecto., designated here; S, isolecto.). [Although Boissier (1862) included *E. muricata* Thunb. in the synonymy of *E. brachiata*, this is considered here as a separate species. Since it is unlikely that Klotzsch & Garcke saw any of the sheets listed, a lectotype is selected. The specimen at K of *Drège 2948* was taken from 'the type' in 'Drège's herbarium' by N.E. Brown.]

E. decussata E.Mey. ex Boiss.: 74 (1862), *nom illegit.*, non Salisb. (1796). [Boissier (1862) cited *Drège 3926* (missing), *Drège 8218* (K, MO, S, W) and 'Olifants River, *Mund & Maire*' (K). However, since it is an illegitimate name, a lectotype is not selected here.]

E. amarifontana N.E.Br.: 275 (1915). Type: South Africa, Cape, near Springbokkuil River, Bitterfontein, *Zeyher 1534* (K, lecto., designated here; BOL, SAM '1534b'-2 sheets, isolecto.). [Brown (1915) cited also *Pearson 5532* (BOL).]

E. chersina N.E.Br.: 274 (1915). Type: Namibia, Lüderitz (Angra Pequeña), 18 Jan. 1907, *Galpin & Pearson 7584* (K, lecto., designated here; PRE, isolecto.). [Brown (1915) also cited *Marloth 4638* (K) and marked both specimens as 'Type', so a lectotype is selected.]

E. caterviflora N.E.Br.: 286 (1915). Type: South Africa, Cape, Nieweveld, Beaufort West, *Drège 8218* (K, lecto., designated here; G, MO, W, isolecto.). [Brown (1915) also cited and wrote 'Type' on *Tyson 167* (K, SAM), so a lectotype is selected.]

E. hastisquama N.E.Br.: 288 (1915). Type: South Africa, Cape, fields by the Swartkops River, *Zeyher 1099* (BOL, lecto., designated here; K, isolecto.). [Brown (1915) cited also *Zeyher 3854* (SAM) and *Ecklon & Zeyher, Euphorb. 25* (K, SAM).]

E. mundii N.E.Br.: 287 (1915). Type: South Africa, Cape, Montagu, 1 Jan. 1903, *Marloth 2805* (K, lecto., designated here; PRE, isolecto.). [Brown (1915) also cited *Marloth 3904* (PRE) and *Marloth 4878* (K) and several others as syntypes. Brown gave this as a new name for *Arthrothamnus densiflorus* Klotzsch & Garcke, which could not be transferred to *Euphorbia*. He regarded these syntypes as identical to the Mund & Maire specimen that typified *A. densiflorus* and therefore named this plant after Mund.]

E. perpera N.E.Br.: 277 (1915). Type: South Africa, Cape, along Orange River, between Verleptpram and its mouth, *Drège* (K, holo.).

E. rudolfii N.E.Br.: 276 (1915). Type: South Africa, Cape, Vanrhynsdorp div., Bitterfontein, Sept. 1897, *Schlechter 11047* (K 000252612, lecto., designated here; BR, GRA, K. L-2 sheets, PRE, S, isolecto.). [Brown

cited also: between Bitterfontein and Stinkfontein, 5 Dec. 1910, *Pearson 5533* (BOL, K). He only wrote 'Type' on the specimens at K of *Pearson 5533* and on one of the specimens at K of *Schlechter 11047* (K 000252612). Therefore a lectotype is selected here.]

E. bayeri L.C.Leach: 539 (1988b). Type: South Africa, Cape, 2 km west of Mossel Bay, 11 Sept. 1985, *Bayer 4875* (NBG, holo.; K, MO, PRE, iso.).

E. spartaria *N.E.Br.* Flora of Tropical Africa 6(1): 558 (1911). Type: Namibia, Hoffnung, Feb. 1907, *Galpin & Pearson 7560* (K, holo.; PRE, SAM, iso.). [Brown annotated the sheet of *Galpin & Pearson 7560* (K) as 'Type' and that at SAM as 'Part of the type'. He did not do this for any of the specimens of *Dinter 983* (K, SAM-2 sheets) designated by Leach & Williamson (1990) as lectotype and also not on *Dinter 255* (K). Consequently their lectotype is set aside here in favour of Brown's preferred 'type'.]

E. racemosa E.Mey. ex Boiss.: 75 (1862), *nom. illegit.*, *non* Tausch ex Rchb. (1832). [Boissier (1862) cited: South Africa, Cape, near Hamerkuil, *Drège* (MO, S '8204'); distr. Beaufort, *Ecklon in h. Petrop* (missing). As this is an illegitimate name a lectotype is not selected.]

E. indecora N.E.Br.: 274 (1915). Type: South Africa, Cape, between Dabenoris and Houms Drift, 11 Jan. 1909, *Pearson 3387* (K 000252597, lecto., designated here; K, isolecto.). [There are two specimens of this collection at K, both annotated by N.E. Brown as 'Type', so a lectotype is selected. The size of these plants (2–3' tall, according to the specimens) suggests that they are *E. spartaria* rather than *E. rhombifolia*.]

E. rhombifolia var. *laxa* N.E.Br.: 285 (1915). Type: South Africa, Cape, among rocks along Chichaba River between Komgha and Kei Mouth, Aug. 1891, 1 000', *Flanagan 838* (GRA, lecto., designated here; PRE, SAM, isolecto.). [In this case (unlike for *E. spartaria*) Brown annotated the sheets of the different collections *MacOwan 1612* (GRA) and *Flanagan 838* (GRA, PRE, SAM) as 'type' so a lectotype is designated. Another syntype is *Sutherland* (K).]

E. rhombifolia var. *triceps* N.E.Br.: 285 (1915). Type: South Africa, Cape, Queenstown distr., mountains near Imbasa River, 1860, *Cooper 318* (K, lecto., designated here; BOL, W, isolecto.). [Brown (1915) cited several specimens as representing var. *triceps*, so a lectotype is selected.]

E. cibdela N.E.Br.: 275 (1915). Type: Namibia, on hills at Schakalskuppe, 4 900–5 600', 18 Jan. 1909, *Pearson 4428* (K, holo.; BOL, LD, SAM, iso.).

E. rectirama N.E.Br.: 283 (1915). Type: South Africa, Cape, Klipfontein, Griqualand West, 29 Dec. 1812, *Burchell 2633* (K, lecto., designated here). [Brown (1915) cited several specimens, of which the above is selected as lectotype.]

E. spinea *N.E.Br.*, Flora capensis 5(2): 272 (1915). Type: Namibia, among rocks near Dabegabis, *Pearson 4380* (K, lecto., designated here; BOL, isolecto.).

[Brown (1915) also cited *Pearson 3296* (BOL, K, SAM), which is *E. rhombifolia* and *Pearson 4585* (K) and wrote 'Type' on all three specimens, so a lectotype is designated.]

E. stapelioides *Boiss.*, Centuria Euphorbiarum: 26 (1860). Type: South Africa, Cape, at the mouth of the Gariep (Orange), 4 Oct. 1830, *Drège 8199* (P, holo.; S, W, iso.).

E. lumbricalis L.C.Leach: 369 (1986b). Type: South Africa, Cape, north of Koekenaap, 10 May 1984, *Leach & Bayer 17123* (NBG, holo.; K, MO, PRE-2 sheets, iso.).

E. suffulta *Bruyns*, South African Journal of Botany 56: 129 (1990). Type: South Africa, Cape, Tierberg, Prince Albert distr., 6 Dec. 1987, *Bruyns 2902* (BOL, holo.; K, PRE, iso.).

E. tenax *Burch.* Travels in the Interior of southern Africa, 1: 219 (1822). Type: South Africa, Cape, Hang-klip, near Ongeluks River, Ceres div., 17 July 1811, *Burchell 1219* (K, holo.).

E. arceuthobioides Boiss.: 20 (1860). Type: South Africa, Cape, 70.10, *Ecklon & Zeyher, Euphorb. 76,* (*Ecklon 1312*) (G, holo.; W, iso.).

Arthrothamnus ecklonii Klotzsch & Garcke: 63 (1860). Type: South Africa, Cape, *Ecklon & Zeyher, Euphorb. 24,* (*Ecklon 1871*) (W, lecto., designated here). [Klotzsch & Garcke (1860) cited 'Ecklon n. 23. 25 & 24 ex parte'. The only specimens from the Ecklon and Zeyher collections with similar numbering are those labelled 'Euphorb. 23', 'Euphorb. 24' and 'Euphorb. 25' and so it must be to these that Klotzsch & Garcke referred.]

Arthrothamnus scopiformis Klotzsch & Garcke: 63 (1860). Type: South Africa, Cape, *Bergius* (missing).

E. rhombifolia var. *cymosa* (Klotzsch & Garcke) N.E.Br.: 285 (1915). *Arthrothamnus cymosus* Klotzsch & Garcke: 63 (1860). Type: South Africa, Cape, *Ecklon & Zeyher, Euphorb. 24* (W, lecto., designated here). [Klotzsch & Garcke (1860) cited 'Ecklon n. 24 ex parte'. This is assumed to be the same as '*Ecklon & Zeyher, Euphorb. 24*', of which there is a piece in W. This piece belongs to *E. tenax*. However, there is no evidence that they saw this specimen and so it is selected as a lectotype.]

E. serpiformis Boiss.: 75 (1862). Type: South Africa, Cape, Berg River Valley, *Zeyher 1535* (BOL, lecto., designated here; K, S, SAM, W, WU, Z, isolecto.). [Boissier (1862) cited also 'Eckl. & Zeyh. 24' (i.e. *Ecklon & Zeyher, Euphorb. 24* (W)) and 'Riesvallei (*Bergius h. Berol.*)' (missing).]

E. mixta N.E.Br.: 585 (1925). *E. arrecta* N.E.Br.: 283 (1915), *nom. illegit., non* N.E.Br. (1914). Type: South Africa, Cape, Berg River Valley, *Zeyher 1535* (K, holo.; BOL, S, SAM, W, WU, Z, iso.).

In Bruyns *et al.* (2006), *E. tenax* was treated as a synonym of *E. arceuthobioides*. The respective types make it clear that they are the same species. However, *E. tenax* was published first and so this treatment was wrong.

E. verruculosa *N.E.Br.*, Flora capensis 5(2): 585 (1925). Type: Namibia, Lüderitz (Angra Pequeña), 10 miles from coast, Nov. 1908, *Marloth 4639* (PRE, holo.; K, iso.). [Brown annotated the specimen at PRE as 'Type' and that at K as 'half of the Type sheet, presented to Kew by Dr Marloth'. So the sheet at PRE is taken as the holotype.]

1c. Sect. **Espinosae** *Pax & K.Hoffm.*

E. guerichiana *Pax*, Botanische Jahrbücher für Systematik 19: 143 (1894). Type: Namibia, rocks south of Khorixas, 14 Nov. 1888, *Gürich 73* (missing). Neotype (designated here): Namibia, Ababes, banks of Tsondap River, 30 Dec. 1915, *Pearson 9119* (BOL).

E. commiphoroides Dinter: 90 (1909). Neotype (designated here): Namibia, Tsumeb distr., Auros, 10 Feb. 1925, *Dinter 5596* (BOL; duplicate at SAM). [Dinter (1909) cited no specimens and only mentioned 'Häufig in Hererolande: Salem, Modderfontein, Omburo, Tsaobis, Omatako'. No specimens from any of these localities have been found. A neotype is therefore selected.]

E. frutescens N.E.Br.: 270 (1915). Type: Namibia, lower mountain slopes of Aus, 3 000', Jan. 1909, *Pearson 4714* (K, holo.; BOL, SAM, iso.). [Although several of these sheets are labelled 'Type', only that at K was annotated by N.E. Brown himself and so this specimen is taken as the holotype.]

E. espinosa *Pax*, Botanische Jahrbücher für Systematik 19: 120 (1894). Type: Tanzania, without precise locality, *Fischer 285* (K, lecto., designated here). [No material definitely seen by Pax in known and so a lectotype is selected.]

E. gynophora Pax: 374 (1904). Type: Tanzania, Pare Mountains, betweem Kisuani and Madji-ya-juu, 700 m, 13 Oct. 1902, *Engler 1579* (K, drawing, lecto., designated here). [Pax (1904b) cited *Engler 1579* and *Engler 1586*. No material definitely seen by Pax is known, but drawings of both these specimens are at K. One of these is selected as lectotype.]

1d. Sect. **Frondosae** *Bruyns*

E. leistneri *R.Archer*, South African Journal of Botany 64: 258 (1998). Type: Namibia, east of Epupa Falls, Jul. 1976, *Leistner et al. 264* (PRE, holo.; B, K, WIND, iso.).

E. transvaalensis *Schltr.*, Journal of Botany 34: 394 (1896). Type: South Africa, Transvaal, near Edwin Bray Battery, shady kloofs in Kap River Valley, Barberton, 2 000', fl. Nov. 1890, *Galpin 1198* (GRA, lecto., designated here; K, NH, SAM, Z, isolecto.). [Since there is no sign that Schlechter saw any of the sheets listed, a lectotype is selected. Brown compared the specimen at K with the type, but did not state where the latter was.]

E. galpinii Pax: 742 (1898). Type: South Africa, Transvaal, near Edwin Bray Battery, Barberton, 2 000', fl. Nov. 1890, *Galpin 1198* (SAM, lecto., designated here; GRA, K, NH, Z, isolecto.). [No material definitely seen by Pax in known and so a lectotype is selected.]

E. ciliolata Pax: 743 (1898). Type: Angola, Sierra Chella and Gambos, 900–1 100 m, *Antunes & Dekindt 781* (BR, lecto., designated here; LISC, Z, isolecto.). [No material definitely seen by Pax in known and so a lectotype is selected. Specimens labelled '*Dekindt 781*'are at Z and BR while at LISC there is a specimen labelled '*Antunes 781*'. These are all assumed to be the same collection, namely *Antunes & Dekindt 781*.]

2. **Euphorbia** subg. **Esula** Pers.

E. albanica *N.E.Br.*, Flora capensis 5(2): 258 (1915). Type: South Africa, Albany div., Brookhuisens Poort, near Grahamstown, *MacOwan 657* (GRA, holo.; K, iso.)

E. berotica *N.E.Br.*, Flora of Tropical Africa 6(1): 600 (1912). Type: Angola, Moçamedes distr., foot of Sierra Negros, behind the mouth of the Bero River, July 1859, *Welwitsch 633* (BM, holo.; LISU, iso.).

E. epicyparissias *E.Mey. ex Boiss.* in A.P. de Candolle, Prodromus 15(2): 168 (1862). Type: South Africa, Transvaal, near Vaal River, *Burke* (K, lecto., designated here). [Boissier (1862) cited: Cape, near Zwangerberg, *Drège; Mund & Maire* in h. Berol; near Vaal R, in h. Kew, *Burke* (K). The lattermost is selected as lectotype.]

Tithymalus epicyparissias E.Mey. ex Klotzsch & Garcke: 88 (1860). Type: South Africa, Cape, *Drège* (HBG, holo.; MO, W-3 sheets, iso.).

Tithymalus involucratus E.Mey. ex Klotzsch & Garcke: 91 (1860). Type: South Africa, *Drège* (HBG, lecto., designated here; MO, isolecto.). [Klotzsch & Garcke (1860) cited: *Drège* (MO, HBG), *Ecklon & Zeyher n. 6*; *Ecklon & Zeyher n. 8* (HBG, S, SAM); *Krebs* (K).]

E. involucrata E.Mey. ex Boiss.: 168 (1862). Type: South Africa, near Phillipstown, *Ecklon & Zeyher n. 8* (HBG, lecto., designated here; S, SAM, isolecto.). [Boissier (1862) cited: near George, *Drège* (BM, MO); between Langekloof and 'Zoëga' R., *Krauss*; near Phillipstown, *Ecklon & Zeyher n. 6*; *Ecklon & Zeyher n. 8* (HBG, S, SAM).]

E. bachmannii: Pax: 535 (1897). Type: South Africa, Pondoland, end Oct. 1888, *Bachmann 755* (missing).

E. involucrata var *megastegia* Boiss.: 168 (1862). Type: South Africa, Cape, near Katberg, *Drège; Krebs*.

E. epicyparissias var. *puberula* N.E.Br.: 267 (1915). Type: South Africa, Kentani, 1 200', 8 Oct. 1910, *Pegler 460* (K, holo.; SAM, iso.).

E. epicyparissias var. *wahlbergii* (Boiss.) N.E.Br.: 267 (1915). *E. wahlbergii* Boiss.: 169 (1862). Type: South Africa, 1842, *Wahlberg* (S, lecto., designated here). [Boissier (1862) cited: 'South Africa, between Umtata and Omgaziana, *Drège*; *Wahlberg*, h. Bunge & Holm'. The latter is in S.]

E. ericoides *Lam.*, Encyclopédie méthodique 2(2): 430 (1788). Type: South Africa, Cape of Good Hope, *Sonnerat* (P-LAM P00381881, holo.).

E. erythrina *Link*, Enumeratio plantarum horti regii berolinensis altera 2: 12 (1822). *Tithymalus erythri-nus* (Link) Klotzsch & Garcke: 91 (1860). Type: South Africa, Cape of Good Hope, *Bergius* (missing). Neotype (designated here): South Africa. Cape, Paarl Mountain, *Drège 2197* (K 000253220; duplicate at K).

E. erythrina var. *meyeri* N.E.Br.: 262 (1915). *E. meyeri* Boiss.: 35 (1860), nom. illegit., non Steud. (1840). Type: South Africa. Cape, Paarl Mountain, *Drège 2197* (K 000253220, lecto., designated here; K, isolecto.). [Since *E. meyeri* Boiss. was illegitimate, I treat var. *meyeri* as described by Brown. Brown (1915) cited several specimens: without locality, *Mund & Maire*; Malmesbury, *Schlechter 5348*; Paarl Mountain, *Drège 2197* (K); mountains near Cape Town, *Ecklon & Zeyher Euphorb. 14* (LE).]

Tithymalus apiculatus Klotzsch & Garcke: 94 (1860). Type: South Africa. Cape, *Mund & Maire* (K, lecto., designated here). [Cited were: South Africa. Cape, *Ecklon & Zeyher 14* (LE); *Mund & Maire* (K) so a lectotype is selected. The latter is annotated by N.E. Brown as 'from the type' from the Berlin Herbarium.]

Tithymalus confertus Klotzsch & Garcke: 94 (1860). Type: South Africa. Cape, *Mund & Maire* (K, lecto., designated here). [Cited were: South Africa. Cape, *Ecklon & Zeyher 5* (SAM); *Mund & Maire* (K) so a lectotype is selected. The specimen *Ecklon & Zeyher 5* (SAM) is of *E. ericoides* rather than *E. erythrina* (though the label on it gives '*Euphorbia striata* Thunb.').]

E. erythrina var. *burchellii* Boiss.: 169 (1862). Type: South Africa, *Burchell 458* (missing). [This specimen was said to be 'in herb. DC.']

E. foliosa *(Klotzsch & Garcke) N.E.Br.* Flora capensis 5(2): 262 (1915). *Tithymalus foliosus* Klotzsch & Garcke: 67 (1860). Type: South Africa, Cape Flats, near Cape Town, *Ecklon & Zeyher 12* (K 000253222, lecto., designated here; K, SAM, isolecto.). [The type of Klotzsch & Garcke has not been located but Brown kept part of it at K (comment on 000253222).]

E. dumosa E.Mey. ex Boiss.: 168 (1862), nom. illegit., non A.Rich. (1850). Types: South Africa, Pondoland, near the Umsikaba River, *Drège 4619* (K, 2 sheets, MO); '*Eckl. & Zeyh 86*' (missing). [Since this is an illegitimate name, a lectotype is not selected here.]

E. artifolia N.E.Br.: 263 (1915). Type: South Africa, Milkwoodfontein, Riversdale div., ± 600', 7 Oct. 1897, *Galpin 4562* (K, holo.; PRE, iso.). [The specimen at K was annotated as 'Type Specimen' by N.E. Brown while that at PRE was not annotated by him. Consequently the one at K is the holotype.]

E. genistoides *P.J.Bergius*, Descriptiones Plantarum ex Capite Bonae Spei: 146 (1767). *Tithymalus genistoides* (P.J.Bergius) Klotzsch & Garcke: 97 (1860). *Galarhoeus genistoides* (P.J.Bergius) Haw.: 144 (1812). Type: South Africa, Cape of Good Hope, *Auge (Grubb)* (SBT 3.1.6.13, holo.).

Bergius only cited 'Herm. Afr. 23', which refers to page 23 in J. Burman's *Catalogi duo plantarum africanorum* of 1736 that was in turn part of his *Thesaurus zeylanicus*. No illustration or specimen is listed, only a phrase which corresponds to the same phrase on page 23 in Burman's *Catalogi*. However, there is a specimen at SBT annotated by Bergius as 'Euphorbia mihi genistoides' and 'e. Cap. b. sp. Grubb'. It is known that a consignment of specimens collected at the Cape by J.A. Auge was bought from Auge by Michael Grubb during a brief visit to the Cape in 1764 and presented to Bergius, and that these formed the basis of Bergius' *'Descriptiones'* (Gunn & Codd 1981). Consequently, this specimen is taken as the type. Haworth (1812) did not refer to Bergius' publication directly, but to 'Willd., Sp. Pl. 2: 908' where references 'Mant. 564' and 'Berg. cap. 146' were given, the latter clearly the same as above.

Tithymalus revolutus Klotzsch & Garcke: 99 (1860). Type: South Africa. Cape of Good Hope, *Ecklon & Zeyher 2* (missing).

E. genistoides var. *puberula* N.E.Br.: 264 (1915). Type: South Africa, Cape, Lion Mountain, *Wolley-Dod 3104* (K, lecto., designated here; BOL, isolecto.). [Brown (1915) cited: without locality, *Thunberg*; *Mund*; *Harvey 444* (K); near Hopefield, *Bachmann 85*; New Kloof, *Drège*; Lion Mountain, *Drège 8192* (HBG); *Schlechter 1381*; *Wolley-Dod 3104* (BOL, K); near Cape Town, *Prior* (K); Simon's Bay, *Wright 447*.]

E. genistoides var. *corifolia* (Lam.) N.E.Br.: 264 (1915). *E. corifolia* Lam.: 431 (1788). Type: South Africa, Cape of Good Hope, *Sonnerat* (P-LAM P00381882, holo.; K, iso.).

E. kraussiana *Bernh. ex C.Krauss*, Flora 28: 87 (1845). Type: South Africa, Natal, forest margins near Pietermaritzburg, Sept. 1839, 2 000–2 500', *Krauss 256* (MO, holo.; BM, K-2 sheets, iso.). [Bernhardi's herbarium was bought by MO (Gunn & Codd 1981) and, since Bernhardi drew up the description and Krauss published it, the holotype is taken as the specimen at MO.]

Tithymalus truncatus Klotzsch & Garcke: 75 (1860). Type: South Africa, Cape, *Krebs* (missing).

Tithymalus meyeri Klotzsch & Garcke: 75 (1860). Type: South Africa, Cape, *Ecklon & Zeyher Euphorb. 13* (Z, lecto., designated here; SAM, isolecto.). [Klotzsch & Garcke (1860) also cited '*Drège*' and '*Krebs*', which have not been located.]

E. kraussiana var. *erubescens* (E.Mey. ex Boiss.) N.E.Br.: 268 (1915). *E. erubescens* E.Mey. ex Boiss.: 116 (1862). Type: South Africa, Natal, between Umzimkulu & Umkomaas, Apr., *Drège* (S, lecto., designated here; BM, isolecto.). [Boissier (1862) cited 'Zuurbergen ('2347' K); near Grahamstown, *Drège* (K); near Vanstadensriver, *Krauss*; between Umzimkulu & Umkomaas, *Drège* (BM, S); 'Winterberg, *Ecklon & Zeyher*'. Only that at S is annotated by Boissier.]

E. mauritanica *L.*, Species Plantarum 1: 452 (1753). *Tithymalus mauritanicus* (L.) Haw.: 139 (1812). Type: Illustration in Dillen., Hort. Eltham. 2: 384, t. 289, f. 373 (1732) (lecto., designated by Croizat 1945).

Tithymalus zeyheri Klotzsch & Garcke: 71 (1860). Type: South Africa, Cape, *Ecklon & Zeyher, Euphorb. 26* (missing).

T. brachypus Klotzsch & Garcke: 74 (1860). Type: South Africa, Cape, *Bergius* (missing).

E. melanosticta E.Mey. ex Boiss.: 95 (1862). Type: South Africa, Kaus Mountain, towards Goedemanskraal, 2 500', *Drège 2945* (K, lecto., designated here; MO, isolecto.). [Boissier (1862) cited a specimen at 'h. Bunge' that has not been located, so a lectotype is selected.]

E. mauritanica var. *namaquensis* N.E.Br.: 292 (1915). Type: South Africa, Pofadder distr., Groot Rosynbos, 9 Jan. 1909, *Pearson 3845* (K, lecto., designated here; BOL, NBG, Z, isolecto.). [Brown (1915) cited (among others): Namibia, koppie about 20 km south of Warmbad, 27 Jan. 1909, *Pearson 4432* (BOL, K); South Africa, between Groot Rosynbos and Wortel, 10 Jan. 1909, *Pearson 3628* (BOL, K).]

E. sarcostemmatoides Dinter: 304 (1921b). Type: Namibia, (Tsamkubis ?) Klein Aub, 7 Apr. 1911, *Dinter 2149* (SAM, lecto., designated here). [Dinter (1921b) cited 2 collections: *Dinter 2149* (SAM) and *2532a* (missing).]

E. paxiana Dinter: 265 (1921a). Type: Namibia, Klein Aub, am schwarzem Kam Rivier im Bastardland, *Dinter 2652* (SAM, holo.).

E. mauritanica var. *foetens* Dinter ex A.C.White *et al.*: 961 (1941). Type: Namibia, 8 km east of Pomona, 14 June 1929, *Dinter 6418* (PRE, holo.; BOL, HBG-2 sheets, K, M, NBG, S, SAM, iso.).

E. mauritanica var. *minor* A.C.White *et al.*: 961 (1941). Type: South Africa, Cape, 30 miles north of Laingsburg, Aug. 1939, *Dyer 4105* (PRE, holo.; K, iso.).

E. mauritanica var. *lignosa* A.C.White *et al.*: 961 (1941). Type: Namibia, Namib near Lüderitzbucht, Nov. 1908, *Marloth 4638* (PRE 0248633-0, holo.; PRE, iso.).

E. mauritanica var. *corallothamnus* Dinter ex A.C.White *et al.*: 961 (1941). Type: Namibia, dunes near Buchuberge, 1 July 1929, *Dinter 6467* (PRE, holo.; BOL, HBG-3 sheets, K, LD, M, NBG, S, SAM, iso.).

E. muraltioides *N.E.Br.*, Flora capensis 5(2): 264 (1915). Type: South Africa, Albany div., Brookhuisens Valley, *MacOwan 642* (K, lecto., designated here; GRA, isolecto.). [Brown (1915) also cited *MacOwan 329* (GRA, K) and *Glass 665* (K, SAM) and wrote 'Type' on all of them.]

E. natalensis *Bernh. ex Krauss* Beiträge zur Flora des Cap- und Natallandes: 150 (1846). Type: South Africa, Natal, base of Tafelberg, Aug. 1839, *Krauss 434* (MO, holo.; BM, FI, K, M, iso.). [Krauss (1845) mentioned the number '*434*', though this did not appear in Krauss (1846). As for *E. kraussiana*, the holotype is at MO.]

Tithymalus capensis Klotzsch & Garcke: 98 (1860). Type: South Africa, Cape of Good Hope, *Ecklon & Zeyher* (missing), *Drège* (missing).

E. ruscifolia *(Boiss.) N.E.Br.*, Flora capensis 5(2): 259 (1915). *E. sclerophylla* var. *ruscifolia* Boiss.: 169 (1862). Type: South Africa, between Kei and Gekau, *Drège 4621* (missing). Neotype (designated here): South

Africa, Cape, Krielis Country, *Bowker* (K). [Boissier (1862) cited a specimen at 'h. Bunge' that has not been located, so a neotype is selected. This was compared by N.E. Brown with *Drège 4621* in Lübeck.]

E. sclerophylla *Boiss*., Centuria Euphorbiarum: 37 (1860). Type: South Africa, Cape, ad Grahamstown, Jul. 1829, *Ecklon et Zeyher n° 11* (G, lecto., designated here; LE, MO (only piece on right hand side), SAM, W, isolecto.). [Boissier (1860) cited: Ad. Prom. B. spei, *Krebs pl. exs. n° 296* (G-DC, LE); ad Grahamstown, *Ecklon & Zeyher n° 11* (G, LE, SAM, W).]

Tithymalus multicaulis Klotzsch & Garcke: 98 (1860). Type: South Africa, Cape of Good Hope, *Krebs* (missing).

E. ovata (E.Mey. ex Klotzsch & Garcke) Boiss.: 167 (1862). *Tithymalus ovatus* E.Mey. ex Klotzsch & Garcke: 97 (1860). Type: South Africa, Cape of Good Hope, *Drège* (LD, lecto., designated here; MO, NY, isolecto.). [A lectotype is designated as it cannot be ascertained whether Klotzsch & Garcke saw any of these sheets.]

E. sclerophylla var. *myrtifolia* E.Mey. ex Boiss.: 169 (1862). Type: South Africa, near Assegaaibosch, *Drège 3563* (P, holo.; K-2 sheets, iso.). [Sheets at HBG and MO do not have the number '3563' on them and are not included here.]

E. striata var. *brachyphylla* Boiss.: 170 (1862). Type: South Africa, Sterkstroom div., plains on top of Katberg, *Drège* (K 000253210, lecto., designated here). [Boissier (1862) cited 'South Africa, Sterkstroom div., plains on top of Katberg, *Drège* (K 000253210); Los Tafelberg, 5 000–6 000', *Drège*' (missing). The first specimen is of *E. ovata* = *E. sclerophylla* and so selecting it as the type means that this name becomes a synonym of *E. sclerophylla* rather than of *E. striata*.]

E. sclerophylla var. *puberula* N.E.Br.: 260 (1915). Type: South Africa, Bathurst div., Rietfontein, between Kariega River and Port Alfred, *Burchell 3961* (K, holo.).

E. stolonifera *Marloth ex A.C.White* et al., The Succulent Euphorbieae 2: 961 (1941). Type: South Africa, Cape, near Matjiesfontein and 'Dwars in die Weg', 900 m, Oct. 1920, *Marloth 9836* (PRE 0838532-0, holo.; PRE, iso.).

E. striata *Thunb*., Prodromus plantarum capensium 2: 86 (1800). *Tithymalus striatus* (Thunb.) Klotzsch & Garcke, Abh. Königl. Akad. Wiss. Berlin 1859: 98 (1860). Type: South Africa, *Thunberg* (UPS-THUNB 11560, holo.).

E. striata var. *cuspidata* Boiss.: 170 (1862). *E. cuspidata* Bernh. ex Krauss: 150 (1846), *nom. illegit. non.* Bertol. (1843). Type: South Africa, Natal, summit of Tafelberg, 2 000–3 000', Sept. 1839, *Krauss 441* (MO, holo.; BM, BOL, M, TCD, iso.). [Krauss (1845) mentioned the number '441', though this did not appear in Krauss (1846). Boissier (1862) did not cite any specimens and only cited Krauss' illegitimate name. Consequently the type of Boissier's name is the same as that of Krauss'. As for *E. kraussiana*, the holotype is at MO.]

3. **Euphorbia** subg. **Euphorbia**

3a. Sect. **Euphorbia**

E. aeruginosa *Schweickerdt*, Bulletin of Miscellaneous Information 1935: 205 (1935). Type: South Africa, Transvaal, Soutpan, Soutpansberg, 12 Apr. 1934, *Schweickerdt & Verdoorn 688* (K, lecto., designated here; PRE, isolecto.). [Schweickerdt (1935) cited also 'Soutpan, 23 Nov. 1932, *Obermeyer, Schweickerdt & Verdoorn 151*' (PRE) and indicated that both were 'syntypes'.]

E. avasmontana *Dinter*, Sukkulentenforschung in Südwestafrika, II. Teil: 96 (1928). Type: Namibia, near Windhoek, Auas Mtns, *Dinter* (PRE, lecto., designated here). [Although Carter (2002) cited a specimen at B, this does not exist. A specimen at PRE was annotated by Dinter himself as '*Euph. avasmontana Dtr msc*'. This is selected as lectotype.]

E. volkmanniae Dinter: 124 (1928). Type: Namibia, near Otavi, Auros, 1924, *Dinter* (B, photo).

E. hottentota Marloth: 336 (1930). Type: South Africa, Cape, Richtersveld, Kubus Kloof, 300 m, 29 Aug. 1925, *Marloth 12520* (PRE, lecto., designated here). [Marloth (1930) also cited *Marloth 13357* (missing).]

E. kalaharica Marloth: 338 (1930). Type: South Africa, Cape, Neusberg, near Kakamas, 700 m, 15 Aug. 1928, *Marloth 14039* (PRE, lecto., designated here). [Marloth (1930) also cited *Marloth 13555* (missing).]

E. sagittaria Marloth: 337 (1930). *E. avasmontana* var. *sagittaria* (Marloth) A.C.White et al.: 817 (1941). Type: South Africa, Cape, 12 miles south of Upington towards Prieska, Aug. 1929, *Marloth 14035* (PRE, lecto., designated here). [Marloth (1930) also cited *Marloth 13385* (missing).]

E. venenata Marloth: 337 (1930). Type: Namibia, Tsarris Mtns, west of Maltahöhe, *Marloth 4687* (K, holo.). [Although Carter (2002) cited a specimen at PRE, this does not exist. Marloth's description of *E. venenata* is vague about such things as the size of the cyathia and the number of glands in each cyathium. Nevertheless, the fairly weak spines of the photograph that he included (figure 7) and the type specimen show that this is not *E. virosa* but *E. avasmontana*.]

The name *E. hottentota* was maintained as distinct from *E. avasmontana* in Bruyns et al. (2006). Marloth (1930: 335) separated *E. avasmontana* and *E. hottentota* by the number of angles on the branches (7-angled in *E. avasmontana*; 5–6-angled in *E. hottentota*) but White et al. (1941: 824) pointed out that 'some of Marloth's herbarium specimens do not agree entirely with the typical form' so that the identity of this 'species' is less clear than Marloth thought. Over the large area where it occurs branches are frequently 4-angled and may have up to eight angles and no clear separation into 5–6-angled and 7-angled plants is possible. No differences in the floral structures have been detected on which they could be separated.

E. barnardii *A.C.White* et al., The Succulent Euphorbieae 2: 965 (1941). Type: South Africa, Transvaal, Sekukuniland, farm Driekop, east of Lulu Mountain, 3 000', 6 Jan. 1937, *Barnard 449* (PRE, holo.; MO, iso.).

E. caerulescens *Haw.*, The Philosophical Magazine, or Annals of Chemistry, Mathematics, Astronomy, Natural History and General Science, Ser. 2, 1: 276 (1827). *E. virosa* var. *caerulescens* (Haw.) A.Berger: 81 (1906a). Type: South Africa, Cape of Good Hope, *Bowie*, cultivated plant at Kew Gardens, pressed Nov. 1876 by N.E. Brown (K, lecto., designated here).

Possible types for *E. caerulescens* include (1) a specimen 'Cape of Good Hope, *Bowie* (K)', which was made by N.E. Brown in November 1876 from 'the type plant (still in cultivation at Kew) dried by myself' (Brown 1915: 365) and (2) a drawing by Bond (423/292) of the apex of a branch and annotated 'drawn from the plant from which Haworth described' and 'Received in 1823 from the Cape of Good Hope by Mr Bowie'. I propose that we accept that the plant in cultivation was among those (if there were more than one) from which Haworth drew up his description so that I have designated the specimen made by N.E. Brown as the lectotype.

E. canariensis Thunb.: 86 (1800), *nom. illegit.*, *non* L. (1753). Type: South Africa, *Thunberg* (UPS-THUNB 11416, holo.).

E. ledienii A.Berger: 80 (1906a). Type: South Africa, fl. & fr. Aug. 1906, received from collection of *F. Ledien* (NY, holo.).

E. ledienii var. *drègei* N.E.Br.: 366 (1915). Type: South Africa, near Port Elizabeth, received 9 Sept 1912, *I.L.Drège* (K, lecto., designated here). [For *E. ledienii* var. *drègei*, Brown (1915) cited two collections: Humansdorp div., near Zeekoe River, *Thunberg*; near Port Elizabeth, received 9 Sept 1912, *I.L.Drège* (K). He annotated both the specimen UPS-THUNB 11416 and that of Drège as 'var. *dregei*' so one is designated as lectotype.]

Brown (1915) mentioned that he had not seen any flowers of *E. caerulescens*, nor any dried specimens that he could definitely refer to it, other than the 'type'. He distinguished *E. caerulescens* and *E. ledienii* by the glaucous or bluish-green stems, with spines 6–12 mm long in the former; green, not glaucous stems, with spines 2–6 mm long in the latter (Brown 1915: 244). Dyer (1931) and White *et al.* (1941) found that these distinctions were not useful and they maintained that the only difference between *E. caerulescens* and *E. ledienii* was the rhizomatous habit of the former. This character was neither mentioned by Haworth nor is it visible in either the type specimen or the drawing by Bond. It was also not mentioned by N.E. Brown, who knew the type specimen in cultivation. Therefore the association by Dyer (1931) and White *et al.*(1941) of a rhizomatous habit with *E. caerulescens* and a non-rhizomatous habit with *E. ledienii* is erroneous and the name *E. caerulescens* must refer to the same non-rhizomatous plants as *E. ledienii*. Consequently, *E. ledienii* is a synonym of *E. caerulescens*. This confusion was not recognised in Bruyns *et al.* (2006), where *E. ledienii* was treated as a

separate species from *E. caerulescens*. The rhizomatous plants are here treated as a separate species, *E. radyeri* Bruyns and the differences between them are discussed under that species.

E. clavigera *N.E.Br.*, Flora capensis 5(2): 362 (1915). Type: Swaziland, Bremersdorp (Manzini), 1 800', 5 Jan. 1905, *Burtt-Davy 3010* (K 000253371, holo.; K, PRE, iso.). [The sheet at K has two specimens of the same number mounted on it, of which the lower one is annotated as 'type'. This is therefore designated as holotype. The specimen at PRE is 'part of type'.]

E. persistens R.A.Dyer: t. 713 (1938). Type: Moçambique, east of Ressano Garcia, July 1936, *F.Z.van der Merwe E14 sub PRE 23395* (PRE, holo.; K, PRE, iso.).

E. clivicola *R.A.Dyer*, Bothalia 6: 221 (1951). Type: South Africa, Transvaal, Lunsklip, 20 miles north of Potgietersrust, 13 Sept. 1946, *Plowes sub PRE 28386* (PRE, holo.; K, iso.).

E. complexa *R.A.Dyer*, The Flowering Plants of South Africa 17: t. 643 (1937). Type: South Africa, Transvaal, road from Louw's Creek to Kaapmuiden, June 1936, *Van der Merwe 100 sub PRE 21373* (PRE, holo.; K-2 sheets, W, iso.).

E. confinalis *R.A.Dyer*, Bothalia 6: 222 (1951). Type: South Africa, Transvaal, Kruger Nat. Park, 2 miles east of 'The Gorge Camp', 900', 20 May 1949, *Codd & De Winter 5580* (PRE, holo.; K, NH, iso.).

E. cooperi *N.E.Br. ex A.Berger*, Sukkulente Euphorbien: 83 (1906). Type: South Africa, Natal, Umgeni Valley, 1862, *Cooper*, cultivated plant at Kew Gardens, pressed Sept. 1899 by N.E. Brown (K 00025338, lecto., designated here; K, isolecto.). [Brown made two specimens in September 1899 from the plant introduced to Kew by Cooper in 1862. He labelled both of these 'Type specimen'. Leach (1970) selected one of these specimens (though it is not specified which of them) as a neotype for *E. cooperi*. However, although Berger (1906a) described it from material at La Mortola in Italy, he was familiar with the plants at Kew and so one of Brown's specimens is taken as the lectotype.]

E. eduardoi *L.C.Leach*, Boletim da sociedade broteriana 42: 161 (1968). Type: Angola, Namibe distr., Dois Irmaos, 550 m, 5 May 1960, *Mendes 3959* (LISC 011538, holo.; BM, LISC, LUAI, PRE, iso.).

E. enormis *N.E.Br.*, Flora capensis 5(2): 362 (1915). Type: South Africa, Pietersburg, Sept. 1905, *Marloth 5144* (PRE, holo.; K, iso.).

E. excelsa *A.C.White* et al., The Succulent Euphorbieae 2: 966 (1941). Type: South Africa, Transvaal, Lydenburg distr., hills near Olifants River, Apr. 1938, *Van der Merwe 1677a* (PRE, holo.).

E. grandialata *R.A.Dyer*, The Flowering Plants of South Africa 17: t. 641 (1937). Type: South Africa, Transvaal, Penge mine, *Van der Merwe 1002 sub PRE 21372* (PRE, holo.; K, W, iso.).

E. grandicornis *A.Blanc*, Catalogue and Hints on Cacti, ed. 2: 68 (1888). Type: Illustration on left hand side of figure on page 68 of A. Blanc, Catalogue & Hints on Cacti, ed. 2 (1888) (lecto., designated here).

E. grandicornis Goebel: 42, fig. 15 (1889), *nom. illegit., non* A. Blanc (1888).

E. grandicornis J.E.Weiss: 291 (1893), *nom. illegit., non* A. Blanc (1888).

The authorship of this species is usually given as 'Goebel' (e.g. Brown (1915); White *et al.* (1941)) or 'Goebel ex N.E.Br.' (e.g. Carter (2002)). However, while N.E. Brown (1897) published the first detailed description of *E. grandicornis*, the name was in use for a long time before this and there are several earlier brief descriptions that validated the name. The first known published appearance of the name *E. grandicornis* is Oudemans (1865), but the name was not validly described there. The earliest validation of the name is that by A. Blanc (1888), in which it is said that '*Euphorbia grandicornis* is still more remarkable on account of its tremendous spines and queer, contorted form'. According to White *et al.* (1941), a figure of *E. grandicornis* appeared in an earlier catalogue of A. Blanc of 1887, but I have not been able to trace this. The next one that has been detected is that of Goebel (1889), in which the diagnosis is similarly rudimentary but still constitutes valid publication. In J.E. Weiss' account of 1893 a more detailed diagnosis of *E. grandicornis* appeared. Since both Weiss' and Goebel's names are illegitimate, lectotypes are not selected for either of them.

E. grandidens *Haw.*, Philosophical magazine and journal 66: 33 (1825). Type: Illustration number 807/323 by T. Duncanson at K of specimen received 1822 from Cape of Good Hope collected by Bowie (lecto., designated here).

E. evansii Pax: 86 (1909). Type: South Africa, Transvaal, Lowveld, near Barberton, *Pole Evans* (missing). [Carter (2002) cited the type specimen at PRE, but this does not exist, nor is there any material known elsewhere that could have been seen by Pax.]

Euphorbia evansii was said to differ (White *et al.* 1941) from *E. grandidens* in being shorter (reaching 10 m as opposed to 16 m), with 3- to 4-angled secondary branches with gently sinuate margins (as opposed to 3-angled or rarely 2- to 4-angled in *E. grandidens* with more prominently toothed margins), spines lacking the pairs of prickles at their bases, these often present in *E. grandidens*. None of these differences are clear-cut and I have found it impossible to separate the known collections into two distinct species. Consequently, the name *E. evansii* is placed in synonymy, although it was kept separate in Bruyns *et al.* (2006).

E. griseola *Pax*, Botanische Jahrbücher für Systematik 34: 375 (1904). Type: Botswana, Lobatsi, *Marloth 3413* (missing). Neotype (Leach 1967): Botswana, 2 miles north of Lobatsi, 16 Jan 1960, *Leach & Noel 121* (SRGH, duplicates at BR, G, K, LISC, PRE). [The type has not been located.]

E. groenewaldii *R.A.Dyer*, The Flowering Plants of South Africa 18: t. 714 (1938). Type: South Africa, Transvaal, 10 miles northeast of Pietersburg towards Mokeetsi, Nov. 1936, *B.H.Groenewald sub Van der Merwe 1186* (*sub PRE 23397*) (PRE 253379, holo.; K, PRE, iso.).

E. ingens *E.Mey. ex Boiss.* in A.P. de Candolle, Prodromus 15(2): 87 (1862). Type: South Africa, Natal, in woods near Durban, *Drège 4614* (S, holo.; K, iso.). [Boissier (1862) cited a specimen at 'h. Bunge' and that at S was annotated by him, so is taken as the holotype. That at K is a 'fragment from type'.]

E. similis A.Berger, Sukk. Euph.: 69 (1906a). Type: South Africa, Natal ? (missing).

N.E. Brown pressed two specimens from plants in cultivation at Kew that were reputed to be *E. similis* and mentioned that he had sent a branch to Berger who had confirmed that this was what he named *E. similis*. However, many of the pressed branches on the two specimens at K bear foliage-leaves 15–80 mm long and consequently they cannot represent either *E. ingens* or *E. similis* in which the leaves were 'minute' according to Berger and where such foliage-leaves are only present on the young stem. P.R.O. Bally determined one of these specimens at K as *E. obovalifolia* A.Rich. (= *E. ampliphylla* Pax) and this is more likely to be the correct identity of this plant, which Brown (1915) used for his description of *E. similis*, but which is not the same as that which Berger (1906a) described.

E. kaokoensis (*A.C.White* et al.) *L.C.Leach*, Dinteria 12: 33 (1976). *E. subsalsa* var. *kaokoensis* A.C.White *et al.*: 965 (1941). Type: Namibia, Kaokoveld, Kauas Okawe, 28 Nov. 1939, *C.J.Hahn sub Otzen 3* (PRE, holo.).

E. keithii *R.A.Dyer*, Bothalia 6: 223 (1951). Type: Swaziland, western edge of Lebombo Mtns, near Stegi, fl. 1949, *Keith sub PRE 28423* (PRE, holo.; GRA, K, NH, S, SRGH, iso.).

E. knobelii *Letty*, The Flowering Plants of South Africa 14: t. 521 (1934). Type: South Africa, Transvaal, Enselsberg near Zeerust, Sept. 1933, *Knobel sub PRE 15854* (K, holo.). [Although Carter (2002) cited the type from PRE, the specimen is not present there. It is assumed that this was sent to K on this occasion. This specimen was collected from the same plant from which the figure was painted.]

E. knuthii *Pax*, Botanische Jahrbücher für Systematik 34: 83 (1904). Type: Moçambique, Ressano Garcia, 1 000', 27 Dec. 1897, *Schlechter 11949* (K, lecto., designated here; BM, BOL, BR, G-2 sheets, GRA, HBG, PRE, WAG, isolecto.). [The sheet at K was annotated by Pax ('Knuthii Pax !') and here he also scratched out Schlechter's proposed name for the plant. Nevertheless, N.E. Brown annotated it as 'part of type'. This sheet is then taken as the lectotype. Carter & Leach (2001) informally selected the specimen at K as lectotype, but this is invalid and so it is formally designated here.]

E. limpopoana *L.C.Leach ex S.Carter*, Kew Bulletin 54: 960 (2000). Type: Zimbabwe, Fulton's Drift, 25.5 km NNW of Beitbridge, Sept. 1963, *Leach 11582a* (SRGH, holo.).

E. malevola subsp. *bechuanica* L.C.Leach: 6 (1964). Type: Botswana, halfway between Palapye and Francistown, Jul. 1937, fl. 1942, *Obermeyer* (PRE 0645765-0, holo.; K, PRE, iso.).

E. louwii *L.C.Leach*, The Journal of South African Botany 46: 207 (1980). Type: South Africa, Transvaal, c. 14 km east of Marken, 900 m, 1 Nov. 1975, *Leach et al.15555* (PRE 0548997-0, holo.; K, PRE, SRGH, iso.)

E. lydenburgensis *Schweickerdt & Letty*, The Flowering Plants of South Africa 13: t. 486 (1933). Type: South Africa, Transvaal, Steelpoort Valley, 30 miles north of Lydenburg, 7 July 1932, *Van Balen & De Wyn sub PRE 14398* (PRE, lecto., designated here; K, isolecto.). [Schweickerdt & Letty (1933) cited two specimens: *Van Balen & De Wyn sub PRE 12465* (PRE) and *Van Balen & De Wyn sub PRE 14398* (PRE, K). The latter is selected as lectotype.]

E. otjingandu *Swanepoel*, S. African J. Bot. 75: 497 (2009). Type: Namibia, Kunene Region, along Van Zyl's Pass 1 km west of Otjihende, 1 305 m, 1 May 2007, *Swanepoel 268* (WIND, holo.; PRU, iso.).

E. otjipembana *L.C.Leach*, Dinteria 12: 29 (1976). Type: Namibia, north of Otjipemba, *Leach & Cannell 15044* (PRE, holo.; BM, K, LISC, M, MO, SRGH, WIND, iso.).

E. perangusta *R.A.Dyer*, The Flowering Plants of South Africa 18: t. 716 (1938). Type: South Africa, Transvaal, Koedoesrant, north of Zeerust, Jan. 1936, *Louw 99* (*sub PRE 23399*)(PRE, holo.; BOL, GRA, K-2 sheets, MO, P, SRGH, iso.).

E. pseudocactus *A.Berger*, Sukkulente Euphorbien: 78 (1906). Type: Country unknown, but probably India, branch from the type plant, received from A. Berger Oct. 1910 (K, lecto., designated here).

Euphorbia radyeri *Bruyns*, sp. nov., *a E. caerulescente caulibus crassioribus, plus profunde articulatis, exterioribus rhizomatosis differt.* Type: South Africa, Cape, 20 miles from Kendrew towards Jansenville, Jan. 1930, *Dyer 2357* (GRA, holo.; PRE, iso.).

Bisexual spiny glabrous succulent shrub 1–2 m tall, 1–3 m broad, branching extensively mainly from base of similar main stem with woody and fibrous roots, with many peripheral branches spreading underground from plant for up to 0.5 m by rhizomes and then rising erect from soil. *Branches* 30–70 mm thick, strongly constricted into many ± spherical segments, smooth, grey-green; *tubercles* fused into 3–7 wing-like often sinuate angles, laterally flattened and rounded and projecting 3–10 mm from angles, spine-shields around apex and united into continuous horny and later somewhat corky brown to grey or black margin, 4–6 mm broad in upper part tapering to 2–3 mm below, bearing 2 spreading and widely diverging brown to grey spines (2–)6–15 mm long; *leaf-rudiments* on tips of new tubercles towards apex of branches and main stem, 1–4 mm long, 2–4 mm broad, spreading, fleeting, broadly ovate, obtuse, sessile, with green-brown obtuse ± pyramidal stipule on either side at base. *Inflorescences* in large numbers per branch towards apex, each a group of 1–3 cymes in axil of tubercle, on peduncle 2–4(6) mm long, 2–3 mm thick, each cyme with 3 vertically disposed cyathia, central

male, outer 2 female only (or bisexual) and developing later, with 2 ovate bracts 1.0–1.5 mm long and 1.5–2.0 mm broad subtending cyathia; *cyathia* cupular-conical, glabrous, 3.5–6.0 mm broad (2–3 mm long below insertion of glands), with 5 lobes with deeply incised margins, bright yellow; *glands* (3–)5, transversely oblong to kidney-shaped or rectangular, 2–3 mm broad, bright yellow, ascending-spreading, slightly convex to concave above, outer margins entire and slightly raised; stamens entirely glabrous, bracteoles palmate and enveloping groups of stamens, deeply and finely divided, glabrous; *ovary* globose, glabrous, included to slightly exserted on erect pedicel 1.5–2.0 mm long and soon becoming slightly exserted, calyx slightly extended around base; styles 2–4 mm long, branched in upper third. *Capsule* 6–7 mm diam., obtusely 3-angled, glabrous, erect and exserted on short pedicel 2–4 mm long.

Although *E. caerulescens* and *E. radyeri* are similar, they are easily separated. Branches around the perimeter of most plants of *E. radyeri* are usually rhizomatous and this phenomenon is unknown in *E. caerulescens*. The branches tend to have a more bluish green colour in *E. radyeri* than in *E. caerulescens*, though the colour varies greatly in the latter, with greener branches on plants from more sheltered habitats. The branches of *E. radyeri* are thicker, deeply articulated into almost spherical segments, while those of *E. caerulescens* are generally more slender and only indistinctly articulated into considerably longer, cylindrical segments. In *E. radyeri* the tubercles are often much longer and broader and the leaf-rudiments are somewhat larger than in *E. caerulescens*. Florally *E. caerulescens* and *E. radyeri* are very similar. In *E. caerulescens* the cyathia are often slightly narrower, becoming more abruptly narrow beneath the glands, while the female florets are borne on a slightly longer pedicel and are without the elongated calyx of *E. radyeri*.

E. restricta *R.A.Dyer*, Bothalia 6: 224 (1951). Type: South Africa, Transvaal, The Downs, 4 500', 14 Oct. 1947, *Codd & De Winter 3092* (PRE 0248764-0, holo.; GRA, K-2 sheets, NH, PRE-2 sheets, SRGH, iso.).

E. rowlandii *R.A.Dyer*, Bothalia 7: 28 (1958). Type: South Africa, Transvaal, Kruger Nat. Park, 8 miles north of Punda Maria, 1 600', 25 July 1951, *Rowland Jones 48* (PRE 0248767-0, holo.; K-2 sheets, PRE, SRGH-2 sheets, iso.).

E. schinzii *Pax*, Bulletin Herbier Boissier 6: 739 (1898). Type: South Africa, Transvaal, Berea Ridge, Barberton, 3 100', 13 Feb. 1891, *Galpin 1297* (BOL, lecto., designated here; K, isolecto.). Pax (1898) also cited 'South Africa, Transvaal, Pretoria, *Rehmann 4347'* (missing).

E. sekukuniensis *R.A.Dyer*, The Flowering Plants of South Africa 20: t. 775 (1940). Type: South Africa, Transvaal, Steelpoort River, north of Roossenekal, Aug. 1938, *Van der Merwe 1765* (*sub PRE 25475*) (PRE 0248772-1, holo.; GRA, PRE, SRGH, iso.).

E. stellata *Willd.*, Species Plantarum 2: 886 (1799). Type: Illustration in F. le Vaillant, Reise Itin. Ed. Germ. Francof. 4: 245, t. 11 (1797) (lecto., designated here).

E. procumbens Meerburgh: t. 55 (1789), *nom. illegit., non* Mill. (1786).

E. radiata Thunb.: 86 (1800). Type: South Africa, Cape, *Thunberg* (UPS-THUNB 11547, holo.).

E. uncinata DC.: 151 (1805). Type: Illustration in DC (1805) by Redouté opposite p. 151 (lecto., designated here). [De Candolle (1805) did not cite any specimens and none annotated as *E. uncinata* by him have been found.]

E. squarrosa Haw.: 276 (1827). Type: Illustration number 295/423 by G. Bond at K of specimen from Cape of Good Hope (lecto., designated here). [No type was designated by Haworth (1827) nor, in this case, did he refer to a collection of Bowie. There is a specimen at Kew made by N.E. Brown soon after he arrived at Kew in 1873. This was from a very old plant which was 'believed to have been introduced by Bowie and so may have been one of the original plants from which Haworth described the species'. Since there is some uncertainty surrounding whether Haworth saw this specimen, the drawing number 295/423 by G. Bond is selected as lectotype.]

E. micracantha Boiss.: 25 (1860). Type: South Africa, Cape, between Zuurberg and Klein Bruintjieshoogte, 2 000–2 500', Oct. *Drège 8206a* (K, lecto., designated here; MO, S, isolecto.). [Boissier (1860) cited 'inter Zuurebergen et Klein Bruintjeshoogte et inter Vischrivier et Fort Beaufort (*Drège n° 8206*)'. The collection from 'between Fish R. & Fort Beaufort' is now labelled *Drège 8206c* (K) and the other as *Drège 8206a* (K, MO, S).]

E. gilbertii A.Berger: 39 (1906a). Type: South Africa, Cape, *Cooper* (missing).

E. lombardensis Nel: 194 (1933b). Type: South Africa, Cape, Mortimer, 1 200–1 300 m, Dec. 1933, *M.Lombard sub SUG 1564* (NBG).

White *et al.* (1941) recognised three species: *E. micracantha* (plants with mainly 4-angled, erect branches, low tubercles less than 4 mm long and relatively long spines), *E. squarrosa* (plants with mainly 3-angled, often spreading branches, particularly prominent tubercles 4–8 mm long and relatively short spines) and *E. stellata* (plants with mainly 2-angled, spreading branches usually pressed to the ground, relatively low tubercles less than 4 mm long and relatively short spines). However, they illustrated many plants which were intermediate between these three and expressed doubt that three species could be distinguished: 'And in the event that distinct species are involved, their limits can hardly be defined accurately' (p. 730). This arrangement of three species was followed in Bruyns *et al.* (2006). However, it is quite often impossible to place a plant with certainty under one of these three names and so a broader view is taken here and a single species is recognised.

E. subsalsa subsp. **fluvialis** *L.C.Leach*, Dinteria 12: 29 (1976). Type: Angola, Ruacana Falls, *Leach & Cannell 14509* (LISC, holo.; BM, K, LUAI, M, MO, PRE, SRGH, iso.).

E. tetragona *Haw.*, The Philosophical Magazine, or Annals of Chemistry, Mathematics, Astronomy, Natural History and General Science Ser. 2,1: 276 (1827). Type: Illustration number 291/1060 by G. Bond at K of specimen received in 1823 from Cape of Good Hope collected by Bowie (lecto., designated here). [There are two paintings of *E. tetragona* by Bond and this one, where details of the cyathia are shown, is selected as the lectotype.]

E. tortirama *R.A.Dyer*, The Flowering Plants of South Africa 17: t. 644 (1937). Type: South Africa, Transvaal, Bandolierskop, *Soll & S.W.Smith sub PRE 21371* (PRE 0258980-1, holo.; K, PRE, W, iso.).

E. triangularis *Desf. ex A.Berger*, Sukkulente Euphorbien: 57 (1906). Type: South Africa, Cape, cultivated plant at Kew Gardens, pressed 30 Oct. 1913 by N.E. Brown (K, lecto., designated by Dyer 1974b).

E. umfoloziensis *Peckover*, Aloe 28: 37 (1991). Type: South Africa, Natal, near Dingaanstat, 10 Apr. 1981, *Peckover* (PRE, holo.).

E. vandermerwei *R.A.Dyer*, The Flowering Plants of South Africa 17: t. 660 (1937). Type: South Africa, Transvaal, White River, Sept. 1936, *Van der Merwe sub PRE 22436* (PRE, holo.; K-2 sheets, P, SRGH, iso.). [The specimens at P, SRGH and one at K lack the PRE number but are 'from Type Specimen' so are taken as isotypes as well.]

E. venteri *L.C.Leach ex R.Archer & S.Carter*, The Flowering Plants of Africa 57: 86 (2001). Type: Botswana, near Tsessebe, c. 45 km north of Francistown, 12 Dec. 1991, *Venter et al. 174* (PRE, holo.; K, UNIN, iso.).

E. virosa *Willd.*, Species Plantarum 2: 882 (1799). Type: Illustration in Paterson, Reisen: 60, t. 9, 10 (1790) (lecto., designated here). [These two figures were cited by Willdenow (1799) and are considered here to constitute a single plate, suitable as a lectotype. This figure was cited by Leach (1971) and Carter (2002), but in neither case was it formally designated as lectotype.]

E. bellica Hiern: 945 (1900). Type: Angola, Moçamedes distr., frequent in sandy coastal hills from Giraul up to Cape Negro, Jul. 1859, *Welwitsch 643* (BM, holo.).

E. dinteri A.Berger: 109 (1906b). Type: Namibia, Khan River, received 1904, *Dinter* (NY, holo.). [The specimen in the Alwyn Berger Herbarium consists of seeds only. These are annotated by Berger as follows: '11069, von C. Dinter als E. virosa eingeführt. 1904'. They are therefore the seeds which Berger (1906b) mentioned, that had been sent to him by Dinter. Their large size makes it clear that they came from plants of *E. virosa*.]

E. virosa f. *caespitosa* H.Jacobsen: 81 (1955). Type: none cited.

E. virosa f. *striata* H.Jacobsen: 81 (1955). Type: none cited.

E. waterbergensis *R.A.Dyer*, The Flowering Plants of Africa 28: t. 1095 (1951). Type: South Africa, Transvaal, 2.5 miles north of Elmerston P.O. towards Ellisras, 3 300', Apr. 1948, *Codd & Erens 4018* (PRE 0248809-0, lecto., designated here; BOL, K, PRE, SRGH, isolecto.). [There are two sheets of this at PRE, neither annotated as 'Type' and so the present one is selected as lectotype.]

E. zoutpansbergensis *R.A.Dyer*, The Flowering Plants of South Africa 18: t. 715 (1938). Type: South Africa, Transvaal, Wylliespoort, Sept. 1937, *Dyer 3873 sub PRE 23393* (PRE 0248810-0, holo.; E, K, MO, PRE-2 sheets, US, iso.).

3b. Sect. **Monadenium** *(Pax) Bruyns*

E. lugardiae *(N.E.Br.) Bruyns*, Taxon 55: 413 (2006). *Monadenium lugardiae* N.E.Br.: 138 (1909). Type: Botswana, foot of Kwebe Peak, Kwebe Hills, 3 500', fl. Aug. 1897 & leaves Feb. 1898, *Mrs Lugard 22* (K, holo.).

3c. Sect. **Tirucalli** *Boiss.*

E. gummifera *Boiss.*, Centuria Euphorbiarum: 26 (1860). Type: South Africa, Cape, low-lying areas between Verleptpram and the mouth of the Orange River, Sept. 1830, *Drège 2944* (P, holo.; S, iso.).

E. gregaria *Marloth*, Transactions of the Royal Society of South Africa 2: 36 (1910). Type: Namibia, Kuibis, *Marloth 4683* (PRE, holo.; K, iso.).

E. congestiflora *L.C.Leach*, Boletim da sociedade broteriana, sér. 2, 44: 197 (1970). Type: Angola, Namibe distr., between Cumilunga & Curoca Rivers, 11 Jan. 1956, *Mendes 1265* (LISC, holo.; BM, LUA, M, SRGH, iso.).

E. damarana *L.C.Leach*, Bothalia 11: 500 (1975). Type: Namibia, Damaraland, c. 64 km west of Khorixas, 27 July 1973, *Leach & Cannell 15064a* (LISC, holo.; K, M, PRE, SRGH, WIND, iso.). [Although Leach (1975b) stated that the holotype is at PRE, it is at LISC.]

4. **Euphorbia** subg. **Rhizanthium** *(Boiss.) Wheeler*

E. albipollinifera *L.C.Leach*, South African Journal of Botany 51: 281 (1985). Type: South Africa, Cape, Springbokvlakte, Dec. 1978, *Bruyns 1826* (NBG, holo.; K, PRE, iso.).

E. arida *N.E.Br.*, Flora capensis 5(2): 319 (1915). Type: South Africa, Cape, Britstown div., near De Aar, *Schonland* (K, holo.).

E. benthamii *Hiern*, Catalogue of the African plants collected by Dr. Friedrich Welwitsch in 1853–61, 1: 943 (1900). Type: Angola, between Lopollo and Ivantala, Feb. 1860, *Welwitsch 283* (BM, holo.; K, LISU, iso.).

E. brakdamensis *N.E.Br.*, Flora capensis 5(2): 324 (1915). Type: South Africa, Cape, Brakdam, 1 600', 7 Sept. 1897, *Schlechter 11123* (K, holo.; BOL, BR, GRA, HBG, L-2 sheets, PRE, S, WAG, iso.).

In Bruyns *et al.* (2006) *E. brakdamensis* was included under *E. filiflora*. Careful examination of Schlechter's many pressings of the type collection of *E. brakdamensis*, shows, however, that this is not correct. In *E. filiflora* the stem and branches are very similar in shape and thickness, with the stem usually slightly longer than the branches, if it can be detected at all. In *E. brakdamensis*, on the other hand, the branches are very much more slender than the stem, which is largely buried in the ground and is greatly exceeded in height by the branches. *E. filiflora* has unusually long cyathia (often around 8 mm long), with especially long styles (7–9 mm long) and long male pedicels. The cyathia in *E. brakdamensis* do not exceed 5 mm long and the styles are not longer than 6 mm. The marginal processes on the glands in *E. brakdamensis* are much more brightly coloured than those of *E. filiflora* where, however, they are longer, more slender and considerably more numerous.

E. braunsii *N.E.Br.*, Flora capensis 5(2): 326 (1915). Type: South Africa, Cape, Aberdeen distr., without precise locality, *Brauns* (K, holo.). [Although Brown (1915) cited two specimens, he mentioned, in addition, that the species was described from the collection of Brauns and so the Brauns collection at K is taken as the holotype.]

E. rudis N.E.Br.: 322 (1915). Type: Namibia, sandy plains northeast of Narudas Süd, 28 Dec. 1912, *Pearson 8141* (BOL, lecto., designated here; SAM, isolecto.). [Of the collections cited by Brown (1915) only *Pearson 4310* (BOL, K) and *Pearson 8141* (BOL, SAM) have duplicates and so *Pearson 8141* is selected as the lectotype.]

E. marientalii Dinter: 31 (1914). Type: Namibia, Mariental, *Dinter 3164* (SAM, holo.).

E. rangeana Dinter: 31 (1914). Type: none cited. [*Euphorbia rangeana* was very similar to *E. marientalii* and was distinguished by '*E. rangeana* ist grünbraun und graubraun' (Dinter 1914: 31), which does make it validly published. However, no specimens were cited here.]

Euphorbia rudis was maintained as a distinct 'species' in Bruyns *et al.* (2006). However, for *E. rudis* and *E. braunsii* White *et al.* (1941: 474) mentioned that 'there is really no sharp line of distinction between the two plants, but rather a gradation. The typical forms of the two are fairly clearly distinguishable, while many of the intermediate forms are very confusing indeed and difficult to classify satisfactorily.' The distinctions between the two included: the smaller 'average size of the main stem', the 'more slender' branches with the tubercles 'somewhat more recurved at the apex' and 'rather smaller' cyathia and 'more completely united styles' *in E. rudis*. These are all subject to considerable variation so that the name *E. rudis* has been abandoned here.

E. brevirama *N.E.Br.*, Flora capensis 5(2): 317 (1915). Type: South Africa, Cape, Jansenville div., near Klipplaat, *Schonland 1716* (K, holo.). [Carter (2002) listed a specimen at GRA but this does not exist.]

E. bruynsii *L.C.Leach*, The Journal of South African Botany 47: 103 (1981). Type: South Africa, Cape, Steytlerville, *Bruyns 1814* (PRE, holo.; SRGH, iso.).

E. bubalina *Boiss.*, Centuria Euphorbiarum: 26 (1860). Type: South Africa, Cape, among thorn-bushes near Buffelsrivier, *Drège 4615* (P, holo.). [Boissier (1860) cited a specimen in 'h. Bunge', so this sheet is taken as the holotype.]

E. laxiflora Kuntze: 286 (1898). Type: South Africa, East London, 5 Mar. 1894, *Kuntze* (NY, holo.; K, iso.).

E. bupleurifolia *Jacq.*, Plantarum rariorum horti caesari schoenbrunnensis descriptiones et icones 1: 55, t. 106 (1797). *Tithymalus bupleurifolius* (Jacq.) Haw.: 138 (1812). Type: Illustration in Jacq., Pl. Hort. Schönbr. 1: t. 106 (1797) (lecto., designated here).

E. proteifolia Boiss.: 92 (1862). Type: South Africa, near Umtata, *Drège 8196* (missing). [Boissier (1862) cited a specimen in 'h. Bunge', but this has not been located.]

E. caperonioides *R.A.Dyer & P.G.Mey.*, Mitteilungen aus der Botanischen Staatssammlung München 6: 245 (1966). Type: Namibia, Kaokoland, 3 miles west of Etanga, 7 Apr. 1957, *De Winter & Leistner 5420* (PRE, holo.).

E. caput-medusae *L.*, Species Plantarum 1: 452 (1753). Type: J. Burm., Rar. Afric. Pl.: t. 8 (1738) (lecto., designated by Wijnands 1983).

E. fructus-pini Mill.: Euphorbia no. 10 (1768). *Medusea fructus-pini* (Mill.) Haw.: 134 (1812). Neotype (designated here): J. Burm., Rar. Afric. Pl.: t. 8 (1738).

When Miller (1768) 'described' *E. fructus-pini*, he referred to Linnaeus (1737) and Boerhaave (1720: 258). He also referred to it as 'Euphorbium Afrum facie fructus pini' and then added 'African Euphorbium with the appearance of Pine fruit, commonly called Little Medusa's Head'. In the longer discussion after the literature citations, he added 'The tenth sort hath a thick short stalk, which seldom rises more than eight or ten inches high, from which come out a great number of trailing branches which are slender, and grow about a foot in length; these intermix with each other like those of the seventh sort, but they are much smaller, and do not grow near so long, but have the same appearance, from whence it is called Little Medusa's Head: the ends of these branches are beset with narrow leaves, between which the flowers come out, which are white, and shaped like those of the other species.'

Linnaeus (1737) referred to Boerhaave (1720: 258) and 'Breyne, Prodr. 2: 100'. In Boerhaave (1720: 258) one finds '8...in capitis Medusae' and '9. Euphorbium; Afrum; facie fructus pini'....Tithymalus, Africanus, arborescens, squamato caule, spinosis MH 3:344'. 'MH 3' refers to the third volume of 'Planta Historia universalis' (Morison 1699). On page 344 of this work, Morison referred to 'Pluk. Phyt. t. 230'. In Plukenet (1692), the phrase 'Tithymalus, Africanus, arborescens, squamato caule, spinosis' appears under t. 230, fig. 5 as well as 'pini fructu facie'. This figure is of *E. loricata*.

Euphorbia loricata does not produce trailing branches from a 'thick short stalk', nor is it 'without spines, having tubercles furnished with very narrow leaves'. Therefore Miller's information makes it clear that his name cannot be applied to *E. loricata*, even though some of the references he gave refer to that species. The reference to 'very narrow leaves' makes it more likely that this name refers to *E. caput-medusae* than *E. inermis*, among the species with a 'thick short stalk' and 'trailing branches'. At present no preserved material of Miller's Euphorbia no 10 is known and so a neotype is selected.

E. caput-medusae var. *geminata* Aiton: 136 (1789). Type: Illustration in J. Burm., Rar. Afric. Pl.: t. 9, fig. 1 (1738) (lecto., designated here).

E. caput-medusae var. *major* Aiton: 135 (1789). Type: Illustration in Commelijn, Praeludia Bot.: t. 7 (1703) (lecto., designated here).

E. caput-medusae var. *minor* Aiton: 135 (1789). Type: Illustration in Breyne, Prodr. rar. pl. sec.: t. 19 (1739) (lecto., designated here).

E. tuberculata Jacq.: 43, t. 208 (1797). *Dactylanthes tuberculata* (Jacq.) Haw.: 133 (1812). *Medusea tuberculata* (Jacq.) Klotzsch & Garcke: 61 (1860). Type: Illustration in Jacq., Pl. Hort. Schönbr. 2: t. 208 (1797) (lecto., designated here).

E. medusae Thunb.: 86 (1800). Type: South Africa, Cape, *Thunberg* (UPS-THUNB 11494, lecto., designated here). [Thunberg (1800) placed two of his collections under *E. medusae*, namely UPS-THUNB 11494 and 11495. The latter is a piece of *E. hamata*.]

Medusea major Haw.: 134 (1812). Type: Illustration in Commelijn, Praeludia Bot.: t. 7 (1703) (lecto., designated here).

Medusea tessellata Haw.: 135 (1812). *E. tessellata* (Haw.) Sweet: 107 (1818). Type: none cited.

E. commelinii DC.: 110 (1813). Type: Illustration in Commelijn, Praeludia Bot.: t. 7 (1703) (lecto., designated by Wijnands 1983).

E. fructus-pini var. *geminata* Sweet: 356 (1826). Type: Illustration in J.Burm., Rar. Afric. Pl.: t. 9, fig. 1 (1738) (lecto., designated here).

E. bolusii N.E.Br.: 333 (1915). Type: South Africa, Transvaal, near Middelburg?, Sept. 1886, *H.Bolus 9767* (BOL, holo.; K, iso.). [The locality given is considered to be an error (White *et al.* 1941: 372).]

E. ramiglans N.E.Br.: 306 (1915). Type: South Africa, Namaqualand, 1883, *H. Bolus sub BOL 9448* (BOL, holo.; K, iso.).

E. marlothiana N.E.Br.: 331 (1915). Type: South Africa, Cape, near Neu Eisleben, fl. Oct.-Nov. 1914, *Marloth 5733* (PRE, holo.; BOL, NBG, K, iso.).

E. muirii N.E.Br.: 331 (1915). Type: South Africa, Cape, Platbos, Still Bay, *Muir 174* (BOL, lecto., designated here; PRE, SAM, isolecto.). [Brown (1915) also cited the following: Albertinia, *Muir* (K), *Pearson sub SAM 2261* (K, SAM).]

E. tuberculatoides N.E.Br.: 332 (1915). Type: South Africa, Cape, Theefontein, Malmesbury div., *Bachmann 1042* (K, lecto., designated here). [Brown (1915) also cited the following: *Grey* (K), *Bolus 4359* (BOL).]

E. macowanii N.E.Br.: 334 (1915). *E. tuberculata* var. *macowanii* (N.E.Br.) A.C.White *et al.*: 372 (1941). Type: South Africa, Clanwilliam [wrongly labelled as Cannon Hill, Uitenhage], *MacOwan 3286* (K, lecto., designated here; SAM, WU, isolecto.). [Brown (1915) also cited the following: *Schlechter 8419* (GRA, K, PRE).]

E. confluens Nel: 193 (1933b). Type: South Africa, Cape, open flats, Kliphoogte, Sept. 1929, *Herre sub SUG 5549* (missing). Type: Illustration in Kakteenkunde: 194 (1933) (lecto., designated here). [Although Carter (2002) cited a specimen at STE (now incorporated into NBG), this does not exist.]

E. celata *R.A.Dyer*, Bothalia 11: 278 (1974). Type: South Africa, Vanrhynsdorp distr., Moedverloor, 100 m, 12 May 1973, *Hall 4272* (PRE, holo.).

E. miscella L.C.Leach: 341 (1984a). Type: South Africa, Cape, near Lekkersing, *Leach et al.16545* (NBG, holo.; PRE, iso.).

E. clandestina *Jacq.*, Plantarum rariorum horti caesari schoenbrunnensis descriptiones et icones 4: 43, t. 484 (1804). Type: Illustration in Jacq., Pl. Hort. Schönbr. 4: t. 484 (1804) (lecto., designated here).

E. clava *Jacq.*, Icones plantarum rariorum 1 (4): 9, t. 85 (1784). *Treisia clava* (Jacq.) Haw.: 131 (1812). Type: Jacq., Icon. 1: t. 85 (1781) (lecto., designated by Wijnands 1983).

E. canaliculata Lam.: 417 (1788). Type: South Africa, *collector unknown* (P-LAM P00381883, holo.). [A specimen of this 'species' is present in the Lamarck herbarium at P and is taken as the holotype and Wijnands' lectotype (Wijnands 1983: 99) is set aside.]

E. coronata Thunb.: 86 (1800). Type: South Africa, Cape, *Thunberg* (UPS-THUNB 11434, holo.). *Treisia tuberculata* Haw.: 65 (1819). Type: Introduced by D. Young to Epsom, 1815, fl. Chelsea 1818 (missing).

E. pubiglans N.E.Br.: 338 (1915). Type: South Africa, Cape, near Port Elizabeth, Sept. 1912, *I.L.Drège* (K, holo.).

E. clavarioides *Boiss.*, Centuria Euphorbiarum: 25 (1860). Type: South Africa, Cape, Sneeuberge at Poortjie, *Drège 8200* (P; duplicates at K, S, W) (lecto., designated here). [Boissier (1860) did not state which herbarium he saw this collection in. This suggests that there was a specimen at G, but this has not been located. A lectotype is selected.]

E. clavarioides var. *truncata* (N.E.Br.) A.C.White *et al.*: 309 (1941). *E. truncata* N.E.Br.: 309 (1915). Type: South Africa, Standerton, *Burtt-Davy 1953* (K, lecto., designated here). [Brown (1915) also cited the following: Transvaal, 23 Nov. 1905, *Leendertz 670* (K); *Leendertz 1873* (K) and *Wilms 1339* (missing); *Kolbe* (BOL).]

E. basutica Marloth: 408 (1910a). Type: Lesotho, Leribe, *Dieterlin* (cult. Phillips, fl. Cape Town in Mar. 1909) *sub Marloth 4671* (K, holo.; NH, PRE, SAM, iso.). [Though Marloth (1910a) cited no number, it is assumed that this is the same specimen as his type.]

E. colliculina *A.C.White* et al., The Succulent Euphorbieae 2: 962 (1941). Type: South Africa, Cape, 2.5 miles north of Oudtshoorn, Aug. 1939, *Dyer 4053* (PRE 0247438-2, lecto., designated here; BOL, K, PRE, isolecto.). [According to Dyer's collecting book the number should be 4053 not 4052, as given in White *et al.* (1941); the latter has no entry next to it while the former is '*E. colliculina* WDS sp. nov. type'. White *et al.* (1941) designated *Dyer 4052* the 'type' and *Marloth 10577* (K, PRE) the 'type of capsule' so a lectotype is designated here.]

In Bruyns *et al.* (2006), *E. colliculina* was included under *E. esculenta*. However, while they bear a close resemblance to one another, there are many differences and two distinct species are involved. Mature specimens of *E. colliculina* are altogether more delicate than those of *E. esculenta* and neither the main stem nor the branches reach the thickness that are normal for *E. esculenta*. *E. esculenta* also produces several swollen roots from the base of the tap-root and this phenomenon is unknown in *E. colliculina*, where the thick taproot tapers off quite abruptly into slender, fibrous roots. Florally *E. colliculina* is also easily separated from *E. esculenta* in that the cyathial lobes and the bracteoles within the cyathium lack the densely bushy hairiness at their apices that make the cyathium of *E. esculenta* distinctly furry or woolly. The cyathial glands are also much larger than those of *E. esculenta*.

E. crassipes *Marloth*, Transactions of the Royal Society of South Africa 1: 318 (1909). Type: South Africa, Cape, Biesiespoort, *Marloth (4399)4397* (PRE, holo.; K, iso.).

E. fusca Marloth: 38 (1910b). Type: South Africa, Cape, Britstown, Sept. 1909, *Marloth 4682* (PRE, holo.; K, iso.). [In the cases of both *E. crassipes* and *E. fusca*, Brown annotated the specimens at K as parts from Marloth's type specimens and so the holotype is the specimen at PRE in each case, with isotypes at K.]

E. baliola N.E.Br.: 327 (1915). Type: Namibia, Great Karas Mountains, between 1st & 2nd outspan between Kraikluft and Narudas Süd, 5400', 26 Dec. 1912, *Pearson 8095* (K, holo.; BOL-2 sheets, GRA, SAM, iso.). [Brown annotated the specimen at K himself as 'Type' but those at BOL and SAM were not annotated by him. Therefore that at K is taken as the holotype.]

E. inornata N.E.Br.: 586 (1925). *E. inelegans* N.E.Br.: 322 (1915), *nom. illegit., non* N.E.Br. (1911). Type: South Africa, Cape, near Kimberley, Sept. 1912, *Moran (sub Schonland 1718)* (K 000253322, holo.; GRA, K, iso.). [Brown (1915) mentioned that *E. inornata* was described from a living plant sent by Schonland in 1912, grown at Kew and pressed by Brown himself in June 1913. This is the specimen 'near Kimberley, Moran, living plant sent to Kew by Schonland' (K000253322). Mounted on the same sheet is another specimen, namely '*Moran sub Schönland 1718*' (K000253323) and both had 'Type' written on them by Brown. The former is taken as the holotype.]

E. eendoornensis Dinter: 196 (1932). Type: Namibia, between Wittsand and Eendorn, 26 Mar. 1924, *Dinter* (missing). Neotype (designated here): Namibia, Vrede, *Bruyns 11362* (NBG).

E. hopetownensis Nel: 192 (1933b). Type: South Africa, Cape, Hopetown, 1930, *E.Markoetter sub SUG 5529* (missing). Type: Illustration in Kakteenkunde: 192 (1933) (lecto., designated here). [Although Carter (2002) cited a specimen at STE (now incorporated into NBG), this does not exist.]

Marloth (1910b) said that *Euphorbia fusca* differed from *E. crassipes* by the non-persistent peduncles (some peduncles being persistent in *E. crassipes*). In most populations of *E. crassipes* one finds plants with persistent peduncles and others without them so this character cannot be used to distinguish between them and the type of *E. crassipes* at PRE is a typical specimen of what is usually referred to as '*E. fusca*'. White *et al.* (1941) maintained that the main differences between *E. crassipes* and *E. fusca* were the slightly more cylindrical stem, thicker branches, the deeper involucres and the green glands. However, in the description Marloth did not mention the glands at all and they were only represented in a small black and white drawing so that their colour was unknown. None of these other differences are significant in this widely distributed and quite variable species.

Although the glands of *E. inornata* were given as olive-green on their upper surface, which is unusually pale for *E. crassipes*, the shape of the plant, the relative thickness of the branches and the shape of the cyathia and glands all fit *E. crassipes*, under which it is included here.

Euphorbia hopetownensis was described from a small plant (only 5 cm broad) with ascending, relatively stout branches which bore unusually short peduncles at 5–7 mm long and 'pink-purple' glands with five teeth. The small figure in the text and these few details are strongly suggestive of *E. crassipes*, under which this name is subsumed here.

Euphorbia baliola was not listed in Bruyns *et al.* (2006) but is included here under *E. crassipes*. Brown (1915) believed it to differ from *E. crassipes* by the different manner in which the tubercles on the stem are formed (from the persistent bases of the branches), the presence of branches right to the centre of the stem and the longer pedicels of the male florets with longer hairs. However, collections made near where the type was collected are typical of *E. crassipes* except for somewhat more slender branches and it seems improbable that two distinct species are involved here.

E. crotonoides Boiss. in A.P. de Candolle, Prodromus 15(2): 98 (1862). Type: Sudan, Kordofan, near El Obeid in shade of *Adansonia*, *Kotschy 419* (S, holo., K, iso.). [Boissier (1862) cited a specimen in 'h. Vindob.' and since the sheet at S is from 'Herb. Musei Palat. Vindob.', this is taken as the holotype.]

E. cumulata *R.A.Dyer*, Records of the Albany Museum 4: 92 (1931). Type: South Africa, Cape, Botha Ridge, 10 miles from Grahamstown on Queen's Road, *Dyer 669* (GRA, holo.; K, iso.).

E. cylindrica *Marloth ex A.C.White* et al., The Succulent Euphorbieae 2: 962 (1941). Type: South Africa, Cape, Kubiskow Mtn, 7 Sept. 1926, *Marloth 12860* (PRE, holo.).

E. davyi *Pax ex N.E.Br.*, Flora capensis 5(2): 305 (1915). Type: South Africa, Transvaal, near Pretoria, 19 Nov. 1901, *J.W.C. Kirk 48* (K, lecto., designated here; PRE, isolecto.). [Brown (1915) listed three specimens: *Kirk 48* (K, PRE), *Burtt-Davy 2196* (K) and *Burtt-Davy 5562* (K) and annotated the first and last as 'Type', so a lectotype is selected. At K there is a letter written by Pax from Breslau in Feb. 1906 to Burtt-Davy that requested his permission to name this species after him.]

E. pseudohypogaea Dinter: 265 (1921a). Type: Namibia, am Wege von Oas nach Gobabis, *Dinter 3144* (missing).

E. bergii A.C.White *et al.*: 963 (1941). Type: South Africa, Orange Free State, Koffiefontein, *Scholtz* (missing).

E. pseudoduseimata A.C.White *et al.*: 963 (1941). Type: Namibia, Hohenhorst, 45 miles SW of Windhoek, Nov. 1940, *Otzen* (PRE, holo.; K, iso.). [White *et al.* (1941) cited a specimen from '45 miles SW of Windhoek, *Otzen 37*' as the type, and mentioned the farm-name 'Hohenhorst' (p. 414) as well. At PRE there is a specimen 'Hohenhorst, SW of Windhoek, Nov. 1940, *Otzen PRE 45881*'. The material at Kew lacks the number '37', but is from the same locality and was annotated by Dyer as 'Part of type specimen'. Therefore the specimen at PRE is taken as the holotype.]

E. decepta N.E.Br., Flora capensis 5(2): 320 (1915). Type: South Africa, Cape, near Willowmore, *Brauns 1712* (K, holo.).

E. albertensis N.E.Br.: 323 (1915). Type: South Africa, Cape, near Prince Albert, between railway and village near Prince Albert, May 1907, *Marloth 4397* (K, holo.; PRE, iso.).

E. astrophora Marx: 311 (1996). Type: South Africa, Cape, north of Klipplaat, *Marx 204* (GRA, holo.).

E. gamkensis Marx: 38 (1999a). Type: South Africa, Cape, south of Calitzdorp, *Marx 225* (GRA, holo.).

E. suppressa Marx: 33 (1999a). Type: South Africa, Cape, near Seekoeigat, *Marx 227* (GRA, holo.).

According to N.E. Brown (1915), Marloth considered his number 4397 from between Prince Albert and Prince Albert Road ('the railway') to belong to *E. crassipes*. However, Brown believed that the absence of a 'flat top to the stem' was significant and that it represented a distinct species, which he named *E. albertensis*. The plants pressed (K, PRE) have a relatively slender stem (far too slender to belong to *E. crassipes*) with numerous short branches towards their apex (which are also much more slender than in *E. crassipes*) with many long, slender, spine-like persistent, sterile peduncles. Vegetatively these plants are extremely similar to *E. decepta* and, although Brown (1915) was unable to supply much detail about the floral parts of *E. albertensis*, this name is included here under *E. decepta*.

Both *E. gamkensis* and *E. suppressa* were treated as distinct species in Bruyns *et al.* (2006). but are here relegated to synonymy.

Euphorbia suppressa was compared extensively with *E. albertensis* and *E. arida* (Marx 1999a). The basis for comparison with *E. albertensis* was mainly Figure 445 of White *et al.* (1941). However, it is uncertain whether this figure is of the 'species' described by N.E. Brown as *E. albertensis*. Apart from the fact that Dyer (GRA records) had tentatively attributed two specimens from the area between Prince Albert and Klaarstroom to *E. arida*, it remains unclear what this new species has to do with *E. arida* (a species of the north-eastern Great Karoo and southern Free State) and why it was not compared with *E. decepta*, which is fairly well-known on the southern portion of the Great Karoo between Beaufort West and Willowmore. Florally *E. arida* and *E. decepta* are not easily separated except by the somewhat shallower cyathium (and slightly shorter styles) with fewer, often obsolete teeth on the outer margins of the cyathial glands in *E. decepta* (deeper cyathium, longer style and more prominent and more numerous marginal teeth in *E. arida*). However, although the plant appears to be very similar in both species, beneath the soil plants of *E. arida* have a system of swollen tuberous roots which develop from and extend the tap-root. These structures are entirely absent in *E. decepta*. In all these respects *E. suppressa* is identical to *E. decepta* and so this name is included here under *E. decepta*.

Euphorbia gamkensis was compared extensively with *E. crassipes* (and its synonym *E. fusca*). However, it differs from *E. crassipes* by its much smaller stature (main stem at most 90 mm thick) by the considerably deeper cyathium whose glands are more-or-less without marginal processes (these are particularly prominent in *E. crassipes* and are usually strongly deflexed). Again, it ought to have been considered how it differs from *E. decepta*. Vegetatively the two are difficult to separate and I have been unable to find any reliable differences. The cyathia differ in that the styles are shorter and more deeply divided in *E. gamkensis*, but no other significant differences have been detected. As I consider this to be insufficient on which to base a separate and otherwise so similar species, I have included *E. gamkensis* under *E. decepta*.

While *E. astrophora* was compared with many species, including *E. decepta* (Marx 1996), it was said that it 'very closely resembles' *E. decepta*, differing by the slightly shorter branches and the convex glands. The glands may be concave in *E. decepta* as well, and plants of *E. decepta* are very variable in size so that there are no substantial differences between them. There are therefore no grounds for separating *E. astrophora* from *E. decepta*.

E. dregeana *E.Mey. ex Boiss.* in A.P. de Candolle, Prodromus 15(2): 95 (1862). Type: South Africa, Cape,

between Koussie and Silverfontein in Kaus Mtn, 2 000', 29 Aug. 1830, *Drège 2942* (P, holo.; G, K-2 sheets, S, iso.). [Boissier (1862) cited a specimen in 'h. Bunge', so the specimen in P is taken as the holotype. A specimen at MO is excluded as it is unnumbered and has been annotated 'must be 2942', for which no grounds are known.]

E. elastica Marloth: 37 (1910b), *nom. illegit.*, *non* Poisson & Pax (1902). Type: South Africa, Cape, near Anenous, Nov. 1908, *Carstens sub Marloth 4684* (PRE, holo.).

E. duseimata *R.A.Dyer*: t. 530 (1934). Type: Botswana, ± 100 miles northwest of Molepolole, flowered in cultivation in Pretoria in Nov. 1931, *G.J.de Wyn sub PRE 12426* (PRE, holo.).

E. ecklonii *(Klotzsch & Garcke) Baill.*, Adansonia 3: 144 (1863). *Tithymalus ecklonii* Klotzsch & Garcke: 68 (1860). Type: South Africa, Cape of Good Hope, Swellendam district, Breede River at Swellendam (70.10), hills under 1000', Aug., *Ecklon & Zeyher, Euphorb. 16* (W, holo.; P, S, iso.).

E. pistiifolia Boiss.: 93 (1862). Type: South Africa, Cape of Good Hope, Swellendam district, Breede River at Swellendam (70.10), hills under 1 000', Aug., *Ecklon & Zeyher, Euphorb. 16* (S, lecto., designated here; P, W, isolecto.). [Boissier (1862) also cited *Drège 8195* (S, W), which is from the same locality.]

E. esculenta *Marloth*, Transactions of the Royal Society of South Africa 1: 319 (1909). Type: South Africa, Cape, Klipplaat (Graaff-Reinet), received living Sept. 1907, *Marloth 4162* (PRE, holo.; BOL, K, SAM, iso.).

E. inermis var. *laniglans* N.E.Br.: 328 (1915). Type: South Africa, Cape, near Klipplaat, received Oct. 1912, *Marloth 5270* (K, holo.; PRE, iso.).

E. fasciculata *Thunb.*, Prodromus plantarum capensium 2: 86 (1800). Type: South Africa, Cape, *Thunberg* (UPS-THUNB 11456, holo.).

E. ferox *Marloth*, Transactions of the Royal Society of South Africa 3: 122 (1913). Type: South Africa, Cape, Klipplaat, 1905, *Marloth 5147* (PRE, holo.; BOL, iso.).

E. alternicolor N.E.Br.: 344 (1915). *E. aggregata* var. *alternicolor* (N.E.Br.) A.C.White *et al.*: 616 (1941). Type: South Africa, *N.S.Pillans* (K, holo.).

E. captiosa N.E.Br.: 345 (1915). Type: South Africa, Cape, near Aberdeen, flow. Sept. 1904, *Schonland 1661* (GRA, holo.).

E. filiflora *Marloth*, Transactions of the Royal Society of South Africa 3: 123 (1913). Type: South Africa, Cape, near Concordia, Apr. 1912, *Krapohl sub Marloth 5119* (NBG, lecto., designated here; K, PRE, isolecto.). [Marloth (1913) mentioned two collections, one made by himself at Chamis in Great Namaqualand (i.e. southern Namibia) in October 1910 and another sent to him from Concordia in Namaqualand (i.e. in north-western South Africa) by Krapohl in March 1912. He appears to have recorded both of these under his number 5119 but

I have not been able to locate any material of the collection from Chamis. He wrote 'Type' on the specimen at NBG and not on any of the other pieces distributed under this number.]

E. filiflora var. *nana* G.Will.: 49 (2003). Type: South Africa, Cape, T'Gabies Plateau, northwest of Kosies, Oct. 1999, *Williamson 5933* (BOL, holo.).

E. nelii A.C.White *et al.*: 484 (1941). *E. meyeri* Nel: 134 (1933a), *nom. illegit. non* Steud. (1840) nec Boiss. (1860). Type: South Africa, Cape, Klipfontein, c. 1 000 m, Sept. 1929, *Herre sub SUG 5545* (missing). Type: Illustration in Kakteenkunde: 134 (1933) (lecto., designated here). [Since the type of *E. meyeri* Nel and consequently of *E. nelii* White *et al.* is missing, a lectotype is selected.]

E. versicolores G.Will.: 284 (1995). Type: South Africa, Cape, near Eksteenfontein, *Williamson 4453* (NBG, holo.).

E. flanaganii *N.E.Br.*, Flora capensis 5(2): 314 (1915). Type: South Africa, Cape, grassy slopes near Kei Mouth, 100', June 1893, *Flanagan 1800* (PRE 0254449-0, holo.; K, PRE, iso.). [The specimen at K is annotated 'branches from the Type Specimen (in Cape Town Herb.)'. There is no type specimen of this species in any Herbarium in Cape Town but there are two sheets at PRE, one of which is annotated by Brown as 'Type'. This is taken as the holotype.]

E. ernestii N.E.Br.: 307 (1915). Type: South Africa, Cape, Hospital Hill, near Queenstown, 3 600', 17 Sept. 1911, *Galpin 8066* (K 000253285, holo.; K, PRE, iso.).

E. gatbergensis N.E.Br.: 310 (1915). Type: South Africa, Cape, near Gatberg (south of Elliott), 3 000–3 500', *Baur 251* (K, holo.).

E. franksiae N.E.Br.: 315 (1915). Type: South Africa, Natal, Camperdown, 2 000', 19 Oct. 1910, *Franks sub Medley-Wood 11727* (K, holo.; NH, PRE, iso.). [The specimens at NH and PRE were not seen by Brown, though that at NH is annotated as 'part of Type Spec.'. The sheet at K contains two specimens, one collected by Franks on 19 October 1910, pressed by Wood and sent to K (this being the other 'part of Type Spec.') and another made from two plants sent in Apr. 1913 to, and cultivated at, Kew. Only the former specimen is annotated by Brown as 'type' and is taken as the holotype.]

E. woodii N.E.Br.: 315 (1915). Type: South Africa, Natal, Clairmont Flats, *Wood 4090* (K, lecto., designated here; NH, isolecto). [Brown (1915) cited also: Clairmont Flats, *Wood 11803* (K) and *Wood 12612* (K).]

E. passa N.E.Br.: 313 (1915). Type: South Africa, Natal, *Cooper*, cult. J.Corduroy, 6 July 1905 (K 000253311, lecto., designated here, K, isolecto.). [Brown (1915) cited also: Scottsburg, *Pole Evans* (missing); Umzumbi, *Wood* (K).]

E. discreta N.E.Br.: 316 (1915). Type: South Africa, Natal, banks of Umzimkulu River near shore, 25 Feb. 1837, *Bachmann 757* (K, holo.).

Brown (1915) recognised a host of 'species' here, including *E. discreta*, *E. ernestii*, *E. flanaganii*, *E. franksiae*, *E. gatbergensis*, *E. passa*, and *E. woodii*. White *et al.* (1941) reduced the number slightly by placing *E. discreta* and *E. passa* in synonymy under *E. woodii* and recognising *E. ernestii*, *E. flanaganii*, *E. franksiae*, *E. gatbergensis*, and *E. woodii* as distinct species. Brown (1915: 314) commented on the remarkable extent to which these plants can vary in size; in particular, how one of them increased in size in cultivation from 30–40 branches at 3–8 inches long to 140 branches that were 9–14.5 inches long and this underlines the vegetative variability that one may observe here. Nevertheless, he distinguished *E. flanaganii* from *E. woodii* by the 'much shorter branches' (Brown (1915): 314) and *E. discreta* from *E. woodii* by the fact that the 'body of the plant is much smaller' (Brown (1915): 316). As commented on extensively by White *et al.* (1941), this makes no sense in view of such strong variation in the size of individuals. Plants producing more than one rosette of branches are not unusual and are found in many populations. Although this feature was not mentioned by Brown (1915) in his descriptions, this was supposed to separate *E. gatbergensis* from *E. ernestii* (White *et al.*, 1941: 75), but they recognised that plants of both 'species' could produce several rosettes. *E. flanaganii* and *E. woodii* were separated by 'Ovary puberulous = *E. flanaganii*'; Ovary glabrous to thinly pubescent with long hairs = *E. woodii*' (White *et al.* (1941): 75, adapted from Brown (1915): 239). In practise some populations have plants with pubescent ovaries and others with glabrous ovaries and to distinguish two species on the basis of the length and density of this pubescence is untenable. Consequently all these names are reduced here to synonymy under a single species.

E. fortuita *A.C.White* et al., The Succulent Euphorbieae 2: 962 (1941). Type: South Africa, Cape, 27 miles from Ladismith towards Barrydale, Aug. 1939, *Dyer 4074* (PRE, Sheet I, holo.; K, PRE-2 sheets, iso.).

Euphorbia fortuita was included under *E. esculenta* in Bruyns *et al.* (2006). However, although in both species the cyathial glands are mostly dark and the centre of the cyathium is densely filled with white hairs, there are significant differences between them that warrant their recognition as distinct species. In *E. fortuita* the glands are much broader and the cyathium is more conical, having a rather rounded, almost spherical shape in *E. esculenta*. Furthermore, the pedicels of the male florets in *E. esculenta* are glabrous (densely pubescent in *E. fortuita*) but in *E. fortuita* the bracteoles are uniformly pubescent in their upper half, while in *E. esculenta* they are densely pubescent only at their apices. The ovary is entirely glabrous in *E. esculenta* and densely pubescent above in *E. fortuita*.

E. friedrichiae *Dinter*, Neue und wenig bekannte Pflanzen Deutsch-SWA's: 29 (1914). Type: Namibia, Warmbad, comm. Sept. 1913, M. *Friedrich sub Dinter 3253* (SAM, holo.).

E. gariepina *Boiss.*, Centuria Euphorbiarum: 28 (1860). Type: South Africa, Cape, Verleptpram, interior at Orange River, *Drège 8214* (G, holo.; K, S, W, iso.).

E. gariepina subsp. **balsamea** *(Welw. ex Hiern)* L.C.Leach, Excelsa Taxonomic Series 2: 78 (1980). *E. balsamea* Welw. ex Hiern: 951 (1900). Type: Angola, *Welwitsch 634* (K, holo.; G, P, iso.).

E. bergeriana Dinter: 28 (1914). Type: Namibia, Okawayo near Karibib, *Dinter 1385* (SAM, holo.).

E. schaeferi Dinter: 304 (1921b). Type: Namibia, Klein Karas, *Schäfer sub Dinter 1233* (SAM, lecto., designated here). [Dinter (1921b) cited: Holoog, *Dinter 1233*; Klein Karas, *Schäfer*. Since Holoog is close to Klein Karas, it is assumed that the specimen cited here is one of these, although the details do not quite correspond.]

E. gerstneriana *Bruyns*, nom. nov.

E. franksiae var. *zuluensis* A.C.White *et al.*: 962 (1941). Type: South Africa, Natal, near Mahlabatini, 18 Oct. 1935, *Gerstner 687* (PRE, holo.).

E. gerstneriana is closely allied to *E. flanaganii*. In *E. flanaganii*, the branches form a dense, usually strongly spreading crown around the apex, which is itself devoid of branches. This bare apex of the stem is green with prominent tubercles and is somewhat depressed towards the centre. In *E. gerstneriana* the branches are produced right to the apex of the stem so that the apex of the stem is not visible at all. The branches in *E. flanaganii* are usually distinctly swollen towards their bases while in *E. gerstneriana* the branches are uniformly thick to their bases. They are also much less densely clustered around the apex of the stem and form an ascending, usually lax rosette. The cyathia differ in that they are pale green and distinctly red-veined on the lobes and in the subtending bracts in *E. gerstneriana*, with deep brownish purple, comparatively small glands that are widely spaced around the cyathium. In *E. flanaganii* the cyathia and their subtending bracts are yellow-green, the glands are usually bright yellow and are far broader, usually almost contiguous around the cyathium. The styles of *E. gerstneriana* are particularly broad (more than twice the breadth of those in *E. flanaganii*) and form an almost mushroom-like top to the female floret.

E. globosa *(Haw.) Sims*, Curtis' Botanical Magazine 53: t. 2624 (1826). *Dactylanthes globosa* Haw.: 382 (1823). *Medusea globosa* (Haw.) Klotzsch & Garcke: 61 (1860). Type: Illustration number 808/15 by T. Duncanson at K of specimen received 1821 from Cape of Good Hope collected by Bowie (lecto., designated here).

E. glomerata A.Berger: 104 (1906a). Type: South Africa, Cape (missing).

E. hallii *R.A.Dyer*, The Journal of South African Botany 19: 135 (1953). Type: South Africa, Cape, Botterkloof, May 1953, *Hall sub PRE 28532* (PRE, holo.; GRA, K, iso.).

E. hamata *(Haw.) Sweet*, Hortus suburbanus Londinensis: 107 (1818). *Medusea hamata* (Haw.) Klotzsch & Garcke: 251 (1859). *Dactylanthes hamata* Haw.: 133 (1812). Type: Illustration in J.Burm., Rar. Afric. Pl.: t. 6, figure 3 (1738) (lecto., designated here). [This figure was cited by Haworth (1812). It was cited by Carter

(2002) as 'T: icono' but this does not constitute valid lectotypification.]

E. cervicornis Boiss.: 27 (1860). Type: South Africa, Cape, Heerenlogement, *Zeyher 1530* (G, lecto., designated here; BOL, SAM, isolecto.). [Boissier (1860) also cited *Drège 2950* (missing).]

E. peltigera E.Mey. ex Boiss.: 91 (1862). Type: South Africa, Cape, on rocks at Orange River near Verleptpram, 19 Sept. 1830, *Drège 2951* (S, lecto., designated here; K, isolecto.). [Boissier (1862) cited a specimen 'in h. Bunge'. The specimen in S may be that formerly in Bunge's herbarium and could be the holotype but this is not certain and so it is chosen as lectotype. A sheet at MO, 'assumed to be 2951' is excluded.]

E. heptagona *L.*, Species Plantarum 1: 450 (1753). Type: Illustration in Boerh., Ind. Alter. Hort. Lugd.-Bat. 1: figure opposite p. 258 (1720) (lecto., designated here). [This figure was cited by Linnaeus (1753). It was also cited by Carter (2002) as 'T: icono' but this does not constitute valid lectotypification. Jarvis (2007) stated that it remained untypified.]

Anthacantha desmetiana Lem.: 64 (1858). Type: South Africa, Cape, cult. L. Desmet (missing).

E. enopla Boiss.: 27 (1860). Type: South Africa, Cape, Witpoortsberg, 2 000–3 000', Aug., *Drège 8207* (S, holo.; BM-2 sheets, K, MO, P, W-2 sheets, iso.). [Boissier (1860) did not cite a herbarium here and so a lectotype is chosen The specimen at MO is a mixed sheet of which only the left hand and middle pieces are this species.]

E. heptagona var. *fulvispina* A.Berger: 109 (1902b). Type: none cited.

E. morinii A.Berger: 98 (1906a). Type: South Africa, Cape, cultivated material sold by Co. Haage & Schmidt-Erfurt (missing).

E. atrispina N.E.Br.: 342 (1915). Type: South Africa, Cape, near Prince Albert, received 1912, *Pearson* (K, holo.).

E. heptagona var. *dentata* (A.Berger) N.E.Br.: 351 (1915). *E. enopla* var. *dentata* A.Berger: 95 (1906a). Type: South Africa, Cape, Witpoortsberge, *Drège* (P, lecto., designated here). [Berger (1906a) did not state where the specimen was that he saw, so a lectotype is designated.]

E. heptagona var. *ramosa* A.C.White *et al.*: 964 (1941). Type: South Africa, Cape, 17 miles north of Oudtshoorn, Aug. 1939, *Dyer 4049* (PRE, holo.; GRA, iso.).

E. heptagona var. *subsessilis* A.C.White *et al.*: 964 (1941). Type: South Africa, Cape, 17 miles east of Ladismith (15 miles west of Calitzdorp), Aug. 1939, *Dyer 4067* (PRE, holo.).

E. heptagona var. *viridis* A.C.White *et al.*: 964 (1941). Type: South Africa, Cape, 11 miles west of Calitzdorp in Huis River Pass, Aug. 1939, *Dyer 4065* (PRE, holo.).

E. enopla var. *viridis* A.C.White *et al.*: 964 (1941). Type: South Africa, Cape, 17 miles north of Jansenville towards Graaff-Reinet, Aug. 1939, *Dyer 4008* (PRE, holo.).

E. atrispina var. *viridis* A.C.White *et al.*: 964 (1941). Type: South Africa, Cape, 12–15 miles from Montagu near Ouberg Pass, Aug. 1939, *Dyer 4094* (PRE, holo.).

E. huttonae *N.E.Br.*, Flora capensis 5(2): 316 (1915). *E. inermis* var. *huttonae* (N.E.Br.) A.C.White *et al.*: 395 (1941). Type: South Africa, Cape, Carlisle Bridge, on the Fish River, fl. Nov. 1903, *H. Hutton* (K, holo.; GRA, iso.). [N.E. Brown based his description on a small dried specimen sent to Kew by Schonland in June 1913. Brown kept two branches and one 'flower' at Kew and sent one branch back to GRA. Thus, although he annotated each as 'Half of the type specimen', that at Kew is actually two thirds of the specimen and is taken as the holotype.]

E. superans Nel ex Herre: 15 (1950). Type: South Africa, Eastern Cape, July 1948, Rosenbrock *sub SUG 7215* (missing). Neotype (designated here): South Africa, Carlisle Bridge, Nov. 1903, *H. Hutton* (GRA, holo.; duplicate at K). [The specimen cited here by Herre is missing (though Carter (2002) cited it as being at STE, now incorporated into NBG). The name is usually cited as '*E. superans* Nel' but the article in which it was published was written by H. Herre. No photograph was included with the protologue and so a neotype is selected.]

Euphorbia huttonae is re-instated at the level of species for various reasons. Vegetatively it differs from *E. inermis* in that the rootstock does not develop a series of swollen, fusiform roots below the stem, but tapers rapidly off into fine roots. There are several clear differences in the cyathia. In *E. huttonae* the whole of the upper surface of the gland is bright yellow. Each gland may be divided deeply down the middle into two broad, convex, yellow structures which remain pressed together towards their bases or it may be an entire, solid wedge-shaped structure that is convex above. The outer edges of the glands are irregularly toothed and notched and may be slightly paler in some populations. In *E. inermis* each gland possesses a dark green part towards the base above which it is divided deeply and finely into antler-like, white processes. Other floral differences are the spreading, white cyathial lobes in *E. inermis* (rather than the pale yellowish green inwardly pressed lobes of *E. huttonae*) and the longer styles in *E. inermis* which are only divided near their apex (divided much more deeply to near their middle in *E. huttonae*).

Some confusion exists over the identity of *Euphorbia superans*, which was maintained as a distinct species in Bruyns *et al.* (2006). A figure appeared in the Euphorbia Journal (Vol. 2: 138, as 'supernans') which was cited by Carter (2002) as *E. superans*, but the slender, bright green branches and finely toothed, broad cyathial glands make it clear that this figure is of *E. flanaganii*. Herre (1950) compared *E. superans* with *E. inermis* and mentioned that the glands were 'yellow...shortly bifid with two processes denticulate at the apex, divided [to] about a third with two diverging processes...slightly revolute'.

This is very similar to the structure of the glands in *E. huttonae* but is not similar at all to that in *E. flanaganii*. The length of the styles and the length to which they are divided also correspond closely to *E. huttonae* under which *E. superans* is now included.

E. hypogaea *Marloth*, Transactions of the Royal Society of South Africa 2: 37 (1910). Type: South Africa, Cape, on the Nieuweveld near Beaufort West, 1 300 m, Nov. 1908, *Marloth 4692* (PRE, holo.; K, iso.).

E. inermis *Mill.*, The Gardener's Dictionary, ed. 8: Euphorbia no. 13 (1768). Neotype (designated here): South Africa, Cape, near Swartkops R. and on hills near Addo, *Zeyher 1098* (K; duplicate at SAM,). [Miller (1768) cited no material and none is known to exist from this date. Therefore a neotype has been selected.]

E. insarmentosa *P.G.Mey.*, Mitteilungen aus der Botanischen Staatssammlung München 6: 246 (1966). Type: Namibia, Outjo distr., Welwitschia, 19 Mar. 1967, *Giess, Volk & Bleissner 6128* (M, holo.).

E. jansenvillensis *Nel*, Jahrbuch der Deutschen Kakteen-Gesellschaft 1: 32 (1935). Type: South Africa, Cape, near Jansenville, Apr. 1932, *Le Roux sub SUG 6550* (missing). Neoype (designated here): South Africa, Cape, 1.5 miles east of Jansenville, *Dyer 4012* (PRE). [Although Carter (2002) cited a specimen at STE (now incorportated into NBG), this does not exist.]

E. tubiglans Marloth ex R.A.Dyer: 268 (1935). Type: South Africa, Cape, near Steytlerville, Aug. 1929, *Herre 1596* (K, holo.; PRE, iso.).

E. lignosa *Marloth*, Transactions of the Royal Society of South Africa 1: 316 (1909). Type: Namibia, near Tschaukaib, 400 m, Nov. 1908, *Marloth 4637* (PRE, holo.; BOL, K, iso.). [Brown wrote 'Part of Type' on a specimen of *Marloth 5070* at K, but this is incorrect.]

E. engleriana Dinter: 263 (1921a). Type: Namibia, zwischen Ababis und Habis, Apr. 1913, *Dinter 2815* (SAM, holo.).

E. curocana L.C.Leach: 111 (1975a). Type: Angola, ± 18 km southeast of Cumilunga, *Mendes 1260* (LISC, holo.; BM, COI, LUAI, iso.).

E. loricata *Lam.*, Encyclopédie méthodique 2(2): 416 (1788). Type: Illustration in Pluk., Phytographia 3: t. 230, figure 5 (1692) (lecto., designated here). [Lamarck (1788) also cited 'Petiver Gaz., t. 86, fig. 519' and 'Buc'hoz, Dec. 9, t. 3', but of these three, Plukenet's figure appeared first. These figures all appear to be copies (sometimes modified by the author) of the figure, assumed to be by Heinrich Claudius, that is among the collection of paintings made during the expedition of Simon van der Stel to the Copper Mountains of Namaqualand in 1685-6 and known as the Codex Witsenii (Wilson *et al.* 2002). The original figure of Claudius appears to have been unknown to Lamarck. There is no specimen of this species in Lamarck's herbarium at P.]

E. hystrix Jacq.: 43, t. 207 (1797). *Treisia hystrix* (Jacq.) Haw.: 131 (1812). Type: Illustration in Jacq., Pl. Hort. Schönbr. 2: t. 207 (1797) (lecto., designated here).

E. armata Thunb.: 86 (1800). Type: South Africa, Cape, *Thunberg* (UPS-THUNB 11412, holo.).

E. eustacei N.E.Br.: 122 (1913). Type: South Africa, Cape, near Matjiesfontein, Oct. 1912, *C.E. Pillans* (K 000253356, holo.; K, PRE, iso.). [From the material sent by Pillans and cultivated at Kew, N.E. Brown made and annotated three specimens on two sheets at K and also sent 'part of the type' to PRE. Brown annotated only one of them (K 000253356) as 'Type Specimen' (others as 'Type, branches from type plant' and 'Type Plant') and so this is taken as the holotype and the others as isotypes.]

In Bruyns *et al.* (2006) *E. eustacei* was maintained as distinct from *E. loricata*. This does not reflect the position correctly. Dense, low-growing and mound-forming plants with slightly broader, more obovate leaves and spines drying out white have always been taken as typical of *E. eustacei* and were assumed to be restricted to the Matjiesfontein area (White *et al.* 1941), while the more diffuse, taller plants with narrower leaves and spines drying out brown that are characteristic of the valley of the Olifants River between Citrusdal and Clanwiliam are typical of *E. loricata*. Nevertheless, White *et al.* (1941) hinted at a wider distribution for *E. loricata* and included some more densely branched plants (e.g. figure 264) in their concept of this species. Now that the respective distributions have become better known it has been found that there is a gradation from the one into the other as one progresses eastwards from the valley of the Olifants River to the Great Escarpment (rather than two disjunct and distinct species each confined to particular areas) so that *E. eustacei* and *E. loricata* are ecotypes of one considerably more widespread species.

E. maleolens *E.Phillips*, The Flowering Plants of South Africa 12: t. 459 (1932). Type: South Africa, near Bandolierskop, Dec. 1925, *C.A.Smith sub PRE 8465* (PRE, holo.).

E. mammillaris *L.*, Species Plantarum 1: 451 (1753). Type: Commelijn, Praeludia Bot.: t. 9 (1703) (lecto., designated by Wijnands 1983).

E. fimbriata Scopoli: 8 (1788). Type: Illustration in Delic. Fl. Faun. Insubr. 3: 8, t. 4 (1788) (lecto., designated here).

E. enneagona Haw.: 184 (1803). Type: none located.

E. erosa Willd.: 27 (1814). [Willdenow (1814) gave a description but cited no specimens.]

E. scopoliana Steud.: 615 (1840), *nom. superfl.* [Steudel (1840) believed that *E. fimbriata* Scopoli was illegitimate, with the name used earlier by Raeuschel, but no such name has been traced.]

E. mammillaris var. *spinosior* A.Berger: 109 (1902b). Type: South Africa, Cape, probably ex hort. F. Ledien (missing).

E. mammillaris var. *submammillaris* A.Berger: 125 (1902c). Type: South Africa, Cape, cultivated plant from Berlin Botanic Garden (missing).

E. latimammillaris Croizat: 331 (1933). Type: none cited.

E. platymammillaris Croizat: 333 (1933). Type: none cited.

E. matabelensis *Pax*, Annalen des K. Naturhistorischen Hofmuseums 15: 51 (1900). Type: Zimbabwe (Matabeleland), *Penther 944* (W, holo.; BM, iso.).

E. currorii N.E.Br.: 545 (1911). Type: Angola, Elephant's Bay, *Curror 29* (K, holo.).

E. ohiva Swanepoel: 249 (2009). Type: Namibia, Kaokoveld, Hartmann Valley above Cunene River, 470 m, 12 Jan. 2006, *Swanepoel 250* (WIND, holo.; PRE, iso.).

E. melanohydrata *Nel*, Jahrbuch der Deutschen Kakteen-Gesellschaft 1: 31 (1935). Type: South Africa, Cape, flats at Swartwater, Oct. 1930, *Herre sub SUG 6533* (missing). Neotype (designated here): South Africa, 4 km east of Beesbank, March 1985, *Williamson 3401* (BOL). [Although Carter (2002) cited the type at STE, now incorporated into NBG, no material of the type has been located and a neotype is selected of material from the same area where the type originated.]

E. meloformis *Aiton*, Hortus Kewensis, ed. 1, 2: 135 (1789). Type: Illustration by F. Masson at BM of specimen introduced 1774 from Cape of Good Hope collected by Masson (lecto., designated here).

E. pomiformis Thunb.: 86 (1800). *E. meloformis* var. *pomiformis* (Thunb.) Marloth: 45 (1928). Type: South Africa, Zwartkops, *Thunberg* (missing).

E. falsa N.E.Br.: 586 (1925). *E. meloformis* subsp. *meloformis* f. *falsa* (N.E.Br.) Marx: 32 (1999b). *E. infausta* N.E.Br.: 358 (1915), *nom. illegit., non* N.E.Br. (1912). Type: South Africa, Cape, sheet 332, specimen annotated 'dead plant-split-1810' by Haworth (OXF) (lecto., designated here). [Brown (1915) cited two specimens: South Africa, without locality, *N.S.Pillans sub BOL* 10684 (BOL) and 'Herb. Haworth'. The latter is designated as lectotype.]

E. pyriformis N.E.Br.: 359 (1915). Type: cultivated plant at Kew of unknown origin, pressed by N.E. Brown 14 Jan. 1913 (K, holo.).

E. valida N.E.Br.: 356 (1915). *E. meloformis* subsp. *valida* (N.E.Br.) G.D.Rowley: 97 (1998). Type: South Africa, Cape, Jansenville div., near Waterford, received 26 Aug. 1912, *I.L. Drège* (K, holo.).

E. meloformis var. *prolifera* Frick: 74 (1934). Type: Cultivated material from seed imported from South Africa, *A.C.S. 5-112-006* (missing).

E. meloformis subsp. *meloformis* f. *magna* R.A.Dyer ex Marx: 13 (1999b). Type: South Africa, Cape, Kwa Ncwane, Peddie (3327AA), 18 Mar. 1999, *Marx 550* (GRA, holo.).

E. monteiroi *Hook.f.*, Bot. Mag. 91: t. 5534 (1865). Type: Angola, *Monteiro* (K, holo.).

E. marlothii Pax: 36 (1889). Type: Namibia, Karibib, 1 000 m, May 1886, *Marloth 1425* (PRE, lecto., designated here). [The number of the type collection was given as 4425 by Pax (1889), but this is assumed to be an error. There is no evidence that Pax saw this specimen so it is designated as a lectotype.]

E. longibracteata Pax: 742 (1898). Type: Namibia, Rehoboth, 1892, *Fleck 447a* (Z, holo.).

E. baumii Pax: 636 (1908). Type: Angola, left bank of Cubango River above Kui marva, 1 100 m, 23 Nov. 1899, *Baum 458* (Z, holo.).

E. monteiroi subsp. **ramosa** *L.C.Leach*, Kirkia 6: 138 (1968a). Type: South Africa, Transvaal, 10 miles south of Mica, *Leach 11999* (PRE, holo.; BM, BOL, COI, G, K, LISC, M, SRGH, WIND, Z, ZSS, iso.).

E. monteiroi subsp. **brandbergensis** *Nordenstam*, Dinteria 11: 23 (1974). Type: Namibia, Brandberg, between Tsisab and Königstein, c. 1 750 m, 29 May 1963, *Nordenstam 2786* (S, holo.).

E. multiceps *A.Berger*, Monatsschrift für Kakteenkunde 15: 182 (1905). Type: South Africa, Cape, Karoo near Matjiesfontein, 950 m, *Marloth 3450* (missing). Type: South Africa (NY, lecto., designated here). [The specimen in the Herbarium of Alwyn Berger at NY consists of several small stems but has no information apart from the name on it. It was undoubtedly seen by Berger and may well be the Marloth specimen, but is designated here as lectotype.]

E. multifolia *A.C.White* et al., The Succulent Euphorbieae 2: 962 (1941). Type: South Africa, Cape, 30 miles from Laingsburg towards Ladismith, Aug. 1939, *Herre* (PRE, lecto., designated here). [White *et al.* (1941) listed two specimens from the same locality, collected by Smith and Herre respectively and designated that by Smith the type. This being missing, the specimen of Herre is designated as lectotype.]

E. namaquensis *N.E.Br.*, Flora capensis 5(2): 325 (1915). Type: South Africa, between Aggeneys and Pella, *Pearson 2992* (BOL, lecto., designated by Williamson 2007; K, SAM, isolecto.).

E. multiramosa Nel: 29 (1935). Type: South Africa, Cape, Little Bushmanland, flats between Jakkalswater and Vioolsdrift, Oct. 1930, *Herre sub SUG 5890* (missing). Neotype (designated here): South Africa, Cape, between Jakkalswater and Vioolsdrift, 600 m, Sept. 2006, *Williamson 6048* (BOL, duplicate at E). [The specimen *SUG 5890* cited by Carter (2002) at STE, now incorporated into NBG, does not exist. In designating a neotype, Williamson (2007) cited '*Williamson 6048* (BOL, E)'. Since two specimens are cited, this neotypification was invalid and this is rectified here.]

Euphorbia namaquensis was included under *E. friedrichiae* in Bruyns *et al.* (2006). These two species are very similar and (among various 'medusoid' species occurring in the arid south of Namibia and north-western South Africa), they share the feature of particularly slender branches which become thicker towards their bases. The two differ in that the cyathia-bearing peduncles arise in *E. friedrichiae* at or near the tips of the branches (the tip of the branch elongating into a peduncle in some cases) around the apex of the plant while in *E. namaquensis* the cyathia-bearing peduncles are shorter (and more densely tuberculate) and arise lower on the branches mainly in the lower half of the plant. The cyathia in both are of a similar size but the glands have longer and more slender processes in *E. friedrichiae*, while the ovary is densely pubescent with short styles (glabrous to pubescent with often much longer styles in *E. namaquensis*). In *E. friedrichiae* in Namibia, the capsules often have an unusual array of warts and slightly raised wing-like ridges along the three angles while they are quite without these in *E. namaquensis*. However, these excrescences are usually (though not always) absent in plants in South Africa of *E. friedrichiae* (from east of Onseepkans), where the capsules are also often larger than in Namibian plants.

White *et al.* (1941) expressed doubt as to whether the two names *E. multiramosa* and *E. namaquensis* represented distinct species. *E. multiramosa* was also included under *E. friedrichiae* in Bruyns *et al.* (2006). Williamson (2007) made extensive notes on *E. multiramosa* and *E. namaquensis*. He concluded that they represented distinct species, since 'the general appearance of both plants is quite different....Cymes in *E. multiramosa* are only produced on the leeward aspect mostly from half to the lower third of the plant....the cymes are solitary with very short peduncles and the involucral glands smaller, sessile, horizontally curving outwards with 4–8 marginal processes. The capsules are glabrous and ± 8 mm in diameter. *Euphorbia namaquensis* has a single or up to two pairs of cyathia with elongated peduncles at branch apices and with involucral glands larger, shortly stipitate, suberect to erect and with 3–6 marginal processes and capsules densely pubescent ± 10–12 mm in diameter.' In practice, the 'general appearance' of plants from north of Steinkopf (taken to be typical of *E. multiramosa*) and those from west of Gamoep (taken to represent *E. namaquensis*) is identical; all the other features mentioned are actually very variable within populations. Consequently *E. multiramosa* and *E. namaquensis* differ only in the glabrous vs. pubescent capsules, though even this feature has been found to be variable in *E. multiramosa*.

E. namibensis *Marloth*, Transactions of the Royal Society of South Africa 1: 318 (1909). Type: Namibia, near Tschaukaib about 31 miles from Angra Pequeña, 800 m, Nov. 1908, *Marloth 4635* (PRE, holo., K, SAM, iso.). [The specimen under this number at BOL is from a different locality (in desert near Lüderitzbucht, 50 m, Aug. 1909) and so is not part of the same collection, although it bears the same number.]

E. argillicola Dinter: 27 (1914). Type: Namibia, flats around Jakkalskuppe, Jan. 1910, *Dinter 3145* (SAM, holo.).

E. namuskluftensis *L.C.Leach*, The Journal of South African Botany 49: 189 (1983). Type: Namibia, Namuskluft, ± 1 200 m, Oct. 1978, *Lavranos & Pehlemann 20796* (PRE holo.; WIND, iso.).

E. nesemannii *R.A.Dyer*, Bulletin of Miscellaneous Information 1934: 267 (1935). Type: South Africa,

Cape, koppie west of Robertson, 300', Jul. 1930, *Nesemann sub Dyer 2441* (GRA, lecto., designated here, K, isolecto.). [Dyer (1935) cited *Dyer 2440* (GRA, K) and *Dyer 2441* (GRA, K), so one has been selected as lectotype.]

E. oatesii *Rolfe* in Oates, Matabeleland Victoria Falls, ed. 2, appendix V: 408 (1889). Type: Zimbabwe, Matabeleland, Apr. 1878, *F. Oates* (K, lecto., designated here). [Rolfe also cited: Zambia, *Rogers 8466* (K); Zimbabwe, 160 km northeast of Bulawayo, *Rand 218* (missing).]

E. obesa *Hook.f.*, Curtis' Botanical Magazine 129: t. 7888 (1903). Type: South Africa, Kendrew, near Graaff-Reinet, 2 000', Mar. 1897, *MacOwan 3153* (K, holo.).

E. symmetrica A.C.White *et al.*: 964 (1941). *E. obesa* subsp. *symmetrica* (A.C.White *et al.*) G.D.Rowley: 97 (1998). Type: South Africa, Cape, 19 miles northwest of Willowmore on road to Rietbron, Aug. 1939, *Dyer 4038* (PRE, lecto., designated here; K, isolecto.). [Dyer did not specify which of these specimens is the holotype and so a lectotype is selected.]

E. oxystegia *Boiss.*, Centuria Euphorbiarum: 27 (1860). Type: South Africa, Cape, between Goedemanskraal and Kaus, *Drège* (S; lecto., designated here; K, P, W, isolecto.). [Boissier (1860) did not cite a herbarium here and so a lectotype is selected.]

E. patula *Mill.*, Dict., ed. 8: Euphorbia no. 11 (1768). *Dactylanthes patula* (Mill.) Haw., Syn. Pl. Succ.: 132 (1812). *Medusea patula* (Mill.) Klotzsch & Garcke, Monatsber. Königl. Akad. Wiss. Berlin 1859: 251 (1859). Neotype (designated here): South Africa, Cape, sheet 328, specimen (one of two) labelled 'Grimwood's St' by Haworth (OXF). [There are two specimens on this sheet in Haworth's Herbarium at Oxford. The one selected here is fertile, while the other, labelled 'My own' is sterile.]

E. ornithopus Jacq.: 76, t. 120, fig. 2 (1809). Type: Jacq., Fragm. Bot., 6: t. 120, figure 2 (1809) (lecto., designated here).

The name *Euphorbia patula* Mill. has been a source of considerable confusion. N.E. Brown (1915: 293) suggested that it was a weak form of *E. mauritanica* and this was taken up by White *et al.* (1941: 120), while Carter (2002) referred it to *E. tridentata*. Both Brown (1915) and White *et al.* (1941) considered, wrongly, that *Dactylanthes patula* was published Haworth (1812), while it was merely a new combination for Miller's name *E. patula*. White *et al.* (1941) also believed that Robert Sweet (1818) described a new species '*Euphorbia patula*'. However, there he referred to 'H.S.', which meant 'Haworth on Succulent Plants', i.e. Haworth (1812). Since this provided a clear reference to Haworth's book and hence back to Miller (1768), it did not constitute publication of a new, and then illegitimate name *Euphorbia patula* Sweet, as was assumed in White *et al.* (1941) and Carter (2002) but merely referred to Miller's *E. patula*. White *et al.*(1941) also considered that Klotzsch & Garcke (1859) published a new name *Medusea patula* but this, too, is wrong and this was also a new combination for *E. patula* Mill. Consequently,

they missed the fact that Miller's name *E. patula* was the earliest valid name for *E. ornithopus*.

E. pedemontana *L.C.Leach*, South African Journal of Botany 54: 501 (1988). Type: South Africa, Cape, foot of Matsikamma, Vanrhynsdorp distr., *Lavranos & Bleck 20828* (NBG, holo.).

E. pentagona *Haw.*, The Philosophical Magazine, or Annals of Chemistry, Mathematics, Astronomy, Natural History and General Science, Ser. 2, 3: 187 (1828). Type: South Africa, Cape of Good Hope, received 1823, *Bowie* (missing). Neotype (designated here): South Africa, Cape, Kei River Mouth, *Flanagan 2344* (BOL; duplicates at GRA, PRE). [The painting number 296/926 at K by G. Bond represents a very weak and imperfectly developed branch without spines (as noted by N.E. Brown on the painting) and it is doubtful whether this is a reasonable lectotype as it could belong to one of several species. Brown (1915) also doubted whether it was made from the plant from which Haworth described the species, since Haworth (1828) mentioned spines and these are absent from this painting. Rather than use this painting of somewhat doubtful identity as lectotype, a neotype has been selected.]

E. pentops *Marloth ex A.C.White* et al., The Succulent Euphorbieae 2: 963 (1941). Type: South Africa, Cape, near Komaggas, 10 June 1930, *Herre 5562* (PRE, holo.).

E. pillansii *N.E.Br.*, Bulletin of Miscellaneous Information 1913: 122 (1913). Type: South Africa, Cape, near Doornkloof River, between Muiskraal and Ladismith, Aug. 1907, *N.S.Pillans sub BOL 12543* (BOL, holo.; K, iso.). [N.E. Brown wrote '1 piece kept for Kew' on the specimen at BOL, which is a much larger specimen, so this is taken as the holotype.]

E. pillansii var. *albovirens* A.C.White *et al.*: 965 (1941). Type: South Africa, Cape, Paardekop near Spes Bona, 650 m, 3 Oct. 1925, *Marloth 12543* (PRE 0258928-1, lecto., designated here; PRE, isolecto.). [There are two specimens at PRE, neither was selected by the authors as type so a lectotype is selected here.]

E. pillansii var. *ramosissima* A.C.White *et al.*: 965 (1941). Type: South Africa, Cape, between Montagu and Touws River, Aug. 1939, *Dyer 4100* (missing).

E. polycephala *Marloth*, South African Gardening & Country Life 21: 133 (1931). Type: South Africa, Cape, near Mortimer, Aug. 1913, *Shoesmith sub Marloth 5295* (PRE, lecto., designated here). [Marloth (1931) cited also *Marloth 12644* but this is missing.]

E. polygona *Haw.*, Miscellanea naturalia, sive dissertationes variae ad historiam naturalem spectantes: 184 (1803). Neotype (designated here): South Africa, Cape, Witpoortsberg, 2 000–3 000', Aug., *Drège 8212* (S 2583; duplicates at BM, HBG-2 sheets, K, MO, P, S, W-3 sheets). [Haworth (1803) mentioned that *E. polygona* was described from material introduced before 1790, but nothing was preserved. A neotype has therefore been selected.]

E. horrida Boiss.: 27 (1860). Type: South Africa, Cape, Witpoortsberg, 2 000–3 000', Aug., *Drège 8212* (S 2583, lecto., designated here; BM, HBG-2 sheets, K, MO, P, S, W-3 sheets, isolecto.). [Boissier (1860) did not cite a herbarium and so a lectotype is chosen.]

E. horrida var. *striata* A.C.White *et al.*: 964 (1941). Type: South Africa, Cape, 15 miles north of Steytlerville, *Lückhoff 123* (missing).

E. horrida var. *noorsveldensis* A.C.White *et al.*: 965 (1941). Type: South Africa, Cape, 1.3 miles north of Jansenville, Aug. 1939, *Dyer 4010* (PRE, holo.).

E. horrida var. *major* A.C.White *et al.*: 965 (1941). Type: South Africa, Cape, Kruidfontein, 19 miles from Willowmore towards Rietbron, Aug. 1939, *Dyer 4041* (missing).

E. polygona var. *nivea* Schnabel: 25 (2011). Type: South Africa, Long Kloof, Kleinrivier, 508 m, 15 Nov. 2010, *Schnabel 1* (GRA, holo.).

E. polygona var. *exilis* Schnabel: 20 (2012). Type: South Africa, Eastern Cape, foothills of Kouga Mountains, 65 m, 16 Nov. 2011, *Schnabel 4* (GRA, holo.).

E. procumbens *Mill.*, The Gardener's Dictionary, ed. 8: Euphorbia no. 12 (1768). *Medusea procumbens* (Mill.) Haw.: 134 (1812). Neotype (designated here): J.Burm., Rar. Afric. Pl.: t. 10, fig. 1 (1738). [Miller cited neither material nor figures. The figure designated here as a neotype was cited by Haworth (1812) under his 'account' of *Medusea procumbens* and so gives an indication of what was then understood by Miller's name. Haworth's association of this figure with Miller's name is unlikely to have been co-incidental. Burmann (1738: t. 10) referred to the plant in this figure as '*Euphorbium humile, procumbens,...*' so that it is likely that Miller adopted Burman's adjective '*procumbens*', as White *et al.* (1941) suggested.]

E. pugniformis Boiss.: 92 (1862). Type: J.Burm., Rar. Afric. Pl.: t. 10, fig. 1 (1738) (lecto., designated by Wijnands 1983).

E. gorgonis A.Berger: 230 (1910). Type: South Africa, Cape, neither collector nor locality (missing). [Carter (2002) cited a specimen of Burtt-Davy at PRE as the type, but this does not exist, nor is there any evidence that it could possibly be the type of Berger's name. In his discussion of *E. gorgonis*, Berger (1910) mentioned having obtained plants of the recently described *E. davyi* from Burtt-Davy but not that Burtt-Davy had supplied him with *E. gorgonis*. This appears to have been mis-interpreted by Carter (2002).]

The name *E. procumbens* was not used in Bruyns *et al.* (2006). This followed White *et al.* (1941), who did not adopt *E. procumbens* Mill. as the name for these plants, even though it antedated *E. pugniformis* (based on the same figure) by nearly 100 years, apparently because Miller's 'description is too incomplete to permit of any certainty' (p. 337) in its identity and 'that name cannot be maintained at all' (p. 338). However, its identity is clear from Haworth's references which lead to the present neotypification and the replacement of *E. pug-*

niformis by this name. White *et al.* (1941) assumed that Sweet's (1818) use of '*Euphorbia procumbens*' was a new name but, since Sweet (1818) referred to Haworth (1812) and thus indirectly to Miller, they were not correct.

E. pseudoglobosa *Marloth*, South African Gardening & Country Life 19: 191 (1929). Type: South Africa, Cape, near Krombeks River, Riversdale distr., Sept. 1933, *Muir 4089* (PRE, holo.).

E. frickiana N.E.Br.: 491 (1931). Type: South Africa, Riversdale div., Ferguson comm. *Frick* (K, holo.).

E. juglans Compton: 126 (1935). Type: South Africa, Cape, about 20 miles west of Ladismith, Feb. 1932, *Compton 3951* (BOL, holo.).

E. pseudotuberosa *Pax*, Bulletin de L'Herbier Boissier, sér. 2, 8: 637 (1908). Type: South Africa, Transvaal, Pretoria, 1892, *Fehr 43* (Z, holo.).

E. pulvinata *Marloth*, Transactions of the Royal Society of South Africa 1: 315 (1909). Type: South Africa, Cape, Queenstown, *Marloth 4372* (missing). Neotype (designated here): South Africa, Cape, Queenstown, Nov. 1898, *Galpin 2527* (PRE). [The type has not been located. A neotype from the same locality is selected here.]

E. quadrata *Nel*, Jahrbuch der Deutschen Kakteen-Gesellschaft 1: 42 (1935). Type: South Africa, Cape, near summit of Stinkfonteinberg, Oct. 1930, *Herre sub SUG 6519* (BOL, holo.). [Carter (2002) cited this specimen at STE (now incorporated into NBG) and BOL, but the former does not exist.]

E. stegmatica Nel: 43 (1935). Type: South Africa, Cape, Stinkfonteinberg, Oct. 1930, *Herre sub SUG 6518* (BOL, holo.). [Although the illustration in Nel (1935) is clearly of *E. oxystegia*, the type is a specimen of *E. quadrata*.]

E. francescae L.C.Leach: 563 (1984b). Type: South Africa, Cape, Cornellsberg, Sept. 1984, *Williamson 3248* (NBG, holo.).

E. restituta *N.E.Br.*, Flora capensis 5(2): 339 (1915). *E. radiata* E.Mey. ex Boiss.: 90 (1862), *nom. illegit.*, *non* Thunb. (1800). Type: South Africa, Cape, between Stinkfontein and Garies, *Pillans 5579* (BOL, lecto., designated here; K, isolecto.). [Brown (1915) also cited: *Schlechter 11098* (BOL); between Zwartdoorn R. and Groen R., Aug, *Drege 2941* (missing). The latter is probably the same as the specimen which Boissier (1862) cited: between Zwartdoorn R. and Groen R., Aug., *Drège* (S, W).]

E. graveolens N.E.Br.: 253 (1915). Type: South Africa, Cape, between Stinkfontein and Garies, Dec. 1910, *Pillans 5579* (BOL 137769, lecto., designated here; BOL, K, isolecto.). [Brown (1915) also cited: Bakhuis, *Pillans 5486* (K).]

E. schoenlandii *Pax*, Jahresbericht der Schlesischen Gesellschaft für vaterländische Cultur 82: 24 (1905). Type: South Africa, Cape, 'Clanwilliam (Woodifield)', fl. May 1904, *Schonland* (GRA, lecto., designated here). [This specimen was annotated as 'Co-type' by Schonland and is unlikely to have been seen by Pax. It is presumed that the other part was sent to Pax and this part remains missing.]

E. silenifolia *(Haw.) Sweet*, Hortus Britannicus, ed. 1, 2: 356 (1826). *Tithymalus silenifolius* Haw.: 61 (1821). Type: Illustration number 810/147 by T. Duncanson at K of specimen received 1823 from Cape of Good Hope collected by Bowie (lecto., designated here).

E. elliptica Thunb.: 86 (1800), *nom. illegit.*, *non* Lam (1786). *Tithymalus ellipticus* (Thunb.) Klotzsch & Garcke: 69 (1860). Type: South Africa, Cape, *Thunberg* (UPS-THUNB 11446, holo.).

Tithymalus bergii Klotzsch & Garcke: 68 (1860). Type: South Africa, Cape, *Bergius* (missing).

Tithymalus longipetiolatus Klotzsch & Garcke: 68 (1860). Type: South Africa, Cape, *Bergius* (missing).

Tithymalus attenuatus Klotzsch & Garcke: 69 (1860). Type: South Africa, Cape, *Bergius* (missing).

E. elliptica var. *undulata* Boiss.: 93 (1862). Neotype (designated here): Type: Illustration number 810/147 by T. Duncanson at K of specimen received 1823 from Cape of Good Hope collected by Bowie. [Boissier (1862) cited '*Tithymalus silenifolius* & *Tith. crispus* Haw., revis. pl. Succul. p. 61 (*ex descriptione*)', so he took these two names as applying to the same species and combined them under this variety. This view is not supported here. By selecting a neotype as above, this name becomes a synonym of *E. silenifolia*.]

E. mira L.C.Leach: 10 (1986a). Type: South Africa, Cape, near Tulbagh, *Bayer sub Leach 17175* (NBG, holo.; K, PRE, iso.).

Although the name *E. mira* L.C.Leach was maintained as a distinct species in Bruyns *et al.* (2006), observations of populations of *E. silenifolia* have made it clear how this species may begin its growth extremely early (in February, well before winter) and how narrow the leaves may be in some populations, often mixed up with plants with considerably broader leaves. Thus, while Leach (1986a: 11) believed he had found three, possibly even four geophytic species of *Euphorbia* growing together at the type locality of *E. mira*, it is clear from the photograph (Leach 1986a: figure 2) and the specimens made, that he found *E. tuberosa* and various forms of *E. silenifolia* at this locality.

E. stellispina *Haw.*, The Philosophical Magazine, or Annals of Chemistry, Mathematics, Astronomy, Natural History and General Science, Ser. 2, 1: 275 (1827). Type: Illustration number 803/324 by T. Duncanson at K of specimen received 1822 from Cape of Good Hope collected by Bowie (lecto., designated here). [The painting selected as lectotype was made from the plants seen by Haworth (of which no material was preserved). There are two figures of *E. stellispina* by Duncanson and this one is selected as the other exhibits very odd growth and is not representative of the species.]

E. stellispina var. *astrispina* (N.E.Br.) A.C.White *et al.*: 716 (1941). *E. astrispina* N.E.Br.: 355 (1915). Type:

South Africa, Beaufort West distr., Willowmore side, *Brauns 1711* (K, holo.).

E. susannae *Marloth*, South African Gardening & Country Life 19: 191 (1929). Type: South Africa, Cape, Phisantefontein, Oct. 1923, *Muir 2762* (BOL 137790, lecto., designated here; BOL, PRE, isolecto.). [Marloth (1929) cited also: *Marloth 12155* (NBG, PRE).]

E. systyloides *Pax* subsp. **porcaticapsa** *S.Carter*, Kew Bulletin 45: 336 (1990). Type: Zimbabwe, Hurungwe distr., Zambesi Valley, Rifa R., 520 m, 24 Feb. 1953, *Wild 4085* (K, holo.; EA, SRGH, iso.).

E. trichadenia *Pax*, Botanische Jahrbücher für Systematik 19: 125 (1894). Type: Angola, Lunda, between Kimbundo and the Quango, Sept. 1876, *Pogge 116* (missing). Neotype (designated here): Angola, Huilla, near Lopollo towards Nene, Oct.-Nov. 1859, *Welwitsch 282* (BM; duplicates at G, K). [Although Carter (2002) cited this specimen as being at B, there is no such material there.]

E. benguelensis Pax: 741 (1898). Type: Angola, Huilla, source of Luala, *Antunes 362* (missing).

E. subfalcata Hiern: 948 (1900). Type: Angola, Huilla, near Lopollo towards Nene, Oct.–Nov. 1859, *Welwitsch 282* (BM, holo.; G, K, iso.).

E. gossweileri Pax: 88 (1909). Type: Angola, Malandsche, *Gossweiler 994* (K, lecto., designated here). [There is no sign that Pax saw this specimen although N.E. Brown wrote 'Type' on it. Consequently it is selected as lectotype.]

E. trichadenia var. *gibbsiae* N.E.Br.: 524 (1911–12). Type: Zimbabwe, near Isotye, Matopos, 5 000', Feb. 1905, *Gibbs 234* (BM, holo.; K, iso.). [Brown (1911–12) cited: *Gibbs 234* (BM, K) and Victoria, Munro. Two collections of Munro have been located, namely *Munro 141* (BM) and *Munro 1467* (BM). However, he wrote 'Type' on *Gibbs 234* (BM) and 'From the type' on *Gibbs 234* (K) and nothing of this kind on the Munro collections so it is clear that the Gibbs specimens are holotype and isotype respectively.]

E. tridentata *Lam.*, Encyclopédie méthodique 2(2): 416 (1788). *Medusea tridentata* (Lam.) Klotzsch & Garcke: 251 (1859). Type: South Africa, *collector unknown* (P-LAM P00381880, holo.; K, iso.).

E. anacantha Aiton: 136 (1789). *Dactylanthes anacantha* (Aiton) Haw., Syn. Pl. Succ.: 132 (1812). Type: Illustration in J. Burm., Rar. Afric. Pl.: t. 7, fig. 2 (1738) (lecto., designated here). [This figure and one by D'Isnard (1720) were cited by Aiton (1789), The one selected as lectotype here corresponds more closely to the concept of *E. tridentata* adopted here, while that of D'Isnard is somewhat more suggestive of *E. patula*.]

E. tuberosa *L.*, Species Plantarum 1: 456 (1753). *Tithymalus tuberosus* (L.) J. Hill, Hort. Kew.: 172/3 (1768). Type: Illustration in J. Burm., Rar. Afric. Pl.: t. 4 (1738) (lecto., designated here). [This figure was cited by Linneaus (1753). It was also cited by Carter (2002) as 'T: icono' but this does not constitute valid lectotypification.]

E. crispa (Haw.) Sweet: 356 (1826). *Tithymalus crispus* Haw.: 61 (1821). Type: none located.

E. tugelensis *N.E.Br.*, Flora capensis 5(2): 335 (1915). Type: South Africa, Natal, near Tugela River, received July 1865, *Gerrard 1626* (K, holo.; W, iso.).

E. wilmaniae *Marloth*, South African Gardening & Country Life 21: 133 (1931). Type: South Africa, Cape, Boetsap, *Pagan sub Marloth 6125a* (PRE, lecto., designated here). [Marloth (1931) cited two specimens: Boetsap, *MacGregor Museum 2337* (missing); Lekkersing, *Marloth 12441* (PRE). The latter belongs to *E. celata* (Leach 1984a, Bruyns 1992). A specimen annotated exactly as the first has not been found but Wilman (1946) cited this collection as 'Boetsap, *2337 Pagan*' and so the specimen 'Boetsap, *Pagan sub Marloth 6125a* (PRE)' is strongly suspected to be this collection and is thus taken as the lectotype.]

E. planiceps Marloth ex A.C.White *et al.*: 963 (1941). Type: South Africa, Cape, farm near Griquatown, Sept. 1939, *Venter* (BOL, lecto., designated here). [White *et al.* (1941) designated the collection by Venter as 'type' and that by Mrs Cooke (missing) as 'type of inflorescence' so a lectotype is designated. The name was first used by Marloth (Wilman 1946).]

Excluded Names

E. aggregata A.Berger, Sukk. Euph.: 92 (1906a). Type: South Africa, Cape (missing). [No preserved material has been found of this species and it is difficult to be sure whether it falls under *E. ferox* or *E. pulvinata* or refers to the intermediates between them that occur widely over the eastern Karoo.]

E. curvirama R.A.Dyer, Rec. Albany Mus, 4: 104 (1931). Type: South Africa, Cape, 28–30 miles from Grahamstown towards Peddie, Apr. 1928, *Dyer 1403* (PRE, holo.; GRA, K, iso.). [This is considered to be a hybrid, possibly between *E. caerulescens* and *E. triangularis*.]

E. inconstantia R.A.Dyer, Rec. Albany Mus. 4: 93 (1931). Syntypes: Hellspoort, Oct. 1928, *Dyer 1076* (GRA); Grahamstown, Aug. 1927, *Dyer 1076* (GRA); 10 miles from Grahamstown on Queen's road, Nov. 1926, *Dyer 669* (GRA); Oct. 1927, *Dyer 1077* (GRA); Nov. 1926, *Dyer 669a* (GRA). [This is considered to be a hybid, possibly between *E. heptagona* and *E. polygona*.]

E. mamillosa Lem., Illustr. Hort. II, misc.: 69 (1855). Type: unknown. [Lemaire (1855) listed '*mamillosa* Nob.', of unknown origin, among 18 names in Sect. *Aculeatea* and provided a Latin diagnosis for it. White *et al.* (1941) listed the name as a synonym of *E. squarrosa* (= *E. stellata*), but it is hard to justify this from the details that Lemaire gave. White *et al.* (1941) also listed the name '*Anthacantha mamillosa* Lem.' and gave the same location as its place of publication, but this name does not exist.]

E. multifida N.E.Br., Fl. cap. 5(2): 253 (1915). Type: South Africa, Natal?, 1905, *Anon sub 10483* (NH, holo.). [The type of *E. multifida* consists of several inflorescences only, is of unknown origin (though suspected of coming from 'Natal') and the collector is unknown. It is not, at present, identifiable with certainty with any known species and so is placed among the excluded names.]

E. parvimamma Boiss. in DC., Prodr. 15(2): 86 (1862). [Boissier (1862) cited no material other than a sterile plant apparently in cultivation under the name *E. caput-medusae,* which may have originated at the Cape of Good Hope. The description is meagre and identification remains uncertain.]

E. scolopendria Donn, Hort. Cantab., ed. 3: 88 (1804). [Donn mentioned only that this was 'flat-leaved' and flowered Jun.–Aug. No region of origin was given.]

E. viminalis N.L.Burm., Prodr. Fl. Cap.: 14 (1768). [Both White *et al.* (1941) and Boissier (1862) cite this name. Actually no description or diagnosis was given by Burman and he merely listed *Euphorbia viminalis* of Linneaus, which is the basionym of *Sarcostemma viminale* (L.) R.Br. (Apocynaceae). Here Burman (1768) cited 'Alp. aegypt. t. 190. Dill elth. t. 368' and he appears to have copied these references directly from among the five given by Linneaus (1753) for *E. viminalis* L. (= *Sarcostemma viminale*). In fact these references are wrong. In Alpini (1735) there is no t. 190, but the figure referred to is t. 53 on page 190. This figure is the lectotype of *S. viminale*, selected by Liede & Meve (1993), though it is wrongly cited there too. Dillen's *Hortus Elthamensis* (Dillen 1732) had only 324 plates in it and here page 386 was meant, where there is no plate. This was again cited incorrectly in Liede & Meve (1993).]

E. viperina A.Berger, Monatsschr. Kakteenk. 12: 39 (1902a). Type: South Africa, Cape of Good Hope?, *collector unknown* (missing). [White *et al.* (1941) placed *E. viperina* under *E. inermis*. However, the description of Berger does not correspond closely to what we know today as *E. inermis*. No type has been located for *E. viperina*. Berger (1902a) compared the inflorescences of *E. viperina* to those of *E. caput-medusae* and *E. parvimamma*, but in fact the inflorescences of the latter were never described and it is not clear that what he called *E. parvimamma* (Berger 1899) corresponds to Boissier's concept of it.]

ACKNOWLEDGEMENTS

The curators of the herbaria B, BM, BOL, G, GRA, K, KMG, M, NBG, NY, OXF, P, PRE, S, SBT, SAM, W, WIND, WU and Z are thanked for access to the material in their care. Christiane Anderson, University of Michigan is thanked for copies of many pieces of little-known literature and for much assistance with, and helpful discussion of, nomenclatural matters concerning several of the names in Euphorbia. Paul E. Berry, University of Michigan, with funds from the U.S. National Science Foundation PBI program (award # DEB-0616533), assisted with the costs of a trip to examine specimens in some European herbaria during which many types were located. He also provided invaluable advice on the status of many other types, among extensive comments on an earlier draft of this paper. S.P.Bester, SANBI, Pretoria, helped to clear up certain problems in the collecting books of R.A. Dyer and Marloth. Gill Challen, Royal Botanic Gardens, Kew provided assistance with locating illustrations and many relevant specimens in K. Fiona Jones, Jagger Library, University of Cape Town, helped trace certain works online. Serena Marner, Oxford University, checked on specimens attributed to Haworth. This research was also partially funded by grants from the National Research Foundation and from the University Research Committee of the University of Cape Town.

REFERENCES

AITON, W. 1789. *Hortus Kewensis*, ed. 1. George Nichol, London.

ALPINI, P. 1735. *Historiae Naturalis Aegypti, pars secunda, sive, de Plantis Aegypti*. G. Potvliet, Leiden.

BENTHAM, G. 1880. *Euphorbia zambesiana* Benth. *Hooker's Icones Plantarum* 14: t. 1305.

BERGER, A. 1899. Zwei verwechselte Euphorbien. *Monatsschrift für Kakteenkunde* 9: 88–92.

BERGER, A. 1902a. Eine neue Euphorbia. *Monatsschrift für Kakteenkunde* 12: 39, 40.

BERGER, A. 1902b. Die in Kultur befindlichen Euphorbien der Anthacantha-Gruppe. *Monatsschrift für Kakteenkunde* 12: 105–110.

BERGER, A. 1902c. Die in Kultur befindlichen Euphorbien der Anthacantha-Gruppe (Schluss). *Monatsschrift für Kakteenkunde* 12: 123–125.

BERGER, A. 1906a. *Sukkulente Euphorbien*. E. Ulmer, Stuttgart.

BERGER, A. 1906b. Euphorbia dinteri Berger n. sp. *Monatsschrift für Kakteenkunde* 16: 109, 110.

BERGER, A. 1910. Einige neue afrikanische Sukkulenten. *Botanische Jahrbücher für Systematik* 45: 223–233.

BLANC, A. 1888. *Euphorbias. Catalogue and hints on cacti*, 2nd ed.: 68. Philadelphia.

BOERHAAVE, H. 1720. *Index alter plantarum quae in horto academico Lugduno-Batavo*. 1. Petrus van der Aa, Leiden.

BOISSIER, P.-E. 1860. *Centuria Euphorbiarum*. Leipzig & Paris.

BOISSIER, P.-E. 1862. *Euphorbieae*. In A.P. de Candolle, Prodromus 15, 2: 3–188.

BROWN, N.E. 1909. The Flora of Ngamiland. *Bulletin of Miscellaneous Information* 1909: 81–146.

BROWN, N.E. 1911–12. Euphorbia, in W.T. Thiselton-Dyer, *Flora Tropical Africa* 6,1: .470–603.

BROWN, N.E. 1913. Diagnoses africanae LIII. *Bulletin of Miscellaneous Information* 1913: 118–123.

BROWN, N.E. 1915. Euphorbia, in W.T. Thiselton-Dyer, *Flora capensis* 5,2: 222–375.

BROWN, N.E. 1925. Addenda and corrigenda. In W.T. Thiselton-Dyer, *Flora capensis* 5,2: 585, 586.

BROWN, N.E. 1931. A new South African Euphorbia. *Cactus & Succulent Journal* (Los Angeles) 2: 491, 492.

BRUYNS, P.V. 1992. Notes on African plants, Euphorbiaceae. Notes on Euphorbia species from the northwestern Cape. *Bothalia* 22: 37–42

BRUYNS, P.V., MAPAYA, R.J. & HEDDERSON, T. 2006. A new subgeneric classification for Euphorbia (Euphorbiaceae) in southern Africa based on ITS and psbA-trnH sequence data. *Taxon* 55: 397–420.

BURMAN, J. 1738. *Rariorum Africanarum Plantarum*. Boussière, Amsterdam.

BURMAN, N.L. 1768. *Florae indica: cui accedit series zoophytorum indicorum, nec non prodromus florae capensis*. C. Haak, Leiden.

CARTER, S. 2002. *Euphorbia*. Pp. 102–203 in U. Eggli, Illustrated Handbook of Succulent Plants: Dicotyledons. Springer, Berlin.

COMPTON, R.H. 1935. Euphorbia juglans R.H. Compton. *The Journal of South African Botany* 1: 126.

CROIZAT, L. 1933. A list of annotated observations on the remarks of Dr K. von Poellnitz Concerning A. Berger's classification of succulent Euphorbiae. *Cactus & Succulent Journal* (Los Angeles) 4: 330–334.

CROIZAT, L. 1945. 'Euphorbia esula' in North America. *American Midland Naturalist* 33: 231–243.

D'ISNARD, A-T. D. 1720. Etablissement d'un Genre de Plante appellé Euphorbe. *Memoirs de l'Academie Royale* 1720: 384–399.

DE CANDOLLE, A.P. 1805. *Plantarum historia succulentarum*. 27. Paris.

DE CANDOLLE, A.P. 1813. *Catalogus plantarum horti botanici monspeliesis, addito observationum circa species novas aut non satis cognitas fasciculo*. J. Martel, Montpellier.

DILLEN, J.J. 1732. *Hortus elthamensis*. London.

DINTER, M.K. 1909. *Deutsch-Südwest-Afrika*. T.O. Weigel, Leipzig.

DINTER, M.K. 1914. *Neue und wenig bekannte Pflanzen Deutsch-Südwest-Afrikas*. Okahandja.

DINTER, M.K. 1921a. Index, der aus Deutsch-Südwestafrika bis zum Jahre 1917 bekannt gewordenen Pflanzenarten. VIII. *Repertorium specierum novarum regni vegetabilis* 17: 258–265.

DINTER, M.K. 1921b. Index, der aus Deutsch-Südwestafrika bis zum Jahre 1917 bekannt gewordenen Pflanzenarten. IX. *Repertorium specierum novarum regni vegetabilis* 17: 303–311.

DINTER, M.K. 1923. *Sukkulentenforschung in Südwestafrika*. Verlag des Repertoriums, Dahlem.

DINTER, M.K. 1932. Diagnosen neuer südwestafrikanischer Pflanzen. *Repertorium specierum novarum regni vegetabilis* 30: 180–205.

DYER, R.A. 1931. Notes on Euphorbia species of the Eastern Cape Province with descriptions of three new species. *Records from the Albany Museum* 4: 64–110.

DYER, R.A. 1935. Euphorbia nesemannii R.A.Dyer. *Bulletin of Miscellaneous Information* 1934: 267, 268.

DYER, R.A. 1938. Euphorbia persistens. *The Flowering Plants of South Africa* 18: t. 713.

DYER, R.A. 1974. Euphorbia triangularis. *The Flowering Plants of Africa* 43: t. 1687.

FRICK, G.A. 1934. In R.W. Poindexter, New Garden Species, IV. *Cactus & Succulent Journal* (Los Angeles) 6: 74, 75.

GOEBEL, K. 1889. *Pflanzenbiologische Schilderungen 1*. N.G. Elwert'sche Verlagsbuchhandlung, Marburg.

GUNN, M. & CODD, L.E. 1981. *Botanical exploration of southern Africa*. A.A. Balkema, Cape Town.

HAWORTH, A.H. 1803. *Miscellanea naturalia, sive dissertationes variae ad historiam naturalem spectantes*. J. Taylor, London.

HAWORTH, A.H. 1812. *Synopsis Plantarum Succulentarum, cum descriptionibus, synonymis,....* R. Taylor, London.

HAWORTH, A.H. 1819. *Supplementum Plantarum Succulentarum, sistens plantas novas vel nuper introductas...* J. Harding, London.

HAWORTH, A.H. 1821. *Revisiones Plantarum Succulentarum*. R. & A. Taylor, London.

HAWORTH, A.H. 1823. Plantae rarae succulentae: a description of some rare succulent plants. *Philosophical Magazine and Journal*. 62: 380–382.

HAWORTH, A.H. 1827. Description of new succulent plants. Decas nona plantarum novarum succulentarum. *The Philosophical Magazine, or Annals of Chemistry, Mathematics, Astronomy, Natural History and General Science, Ser. 2*, 1: 271–277.

HAWORTH, A.H. 1828. Description of new succulent plants. Decas undecima plantarum novarum succulentarum. *The Philosophical Magazine, or Annals of Chemistry, Mathematics, Astronomy, Natural History and General Science, Ser. 2*, 3: 183–188.

HERRE, H. 1950. Euphorbia superans Nel *spec. nov. Desert Plant Life Magazine* 22: 15.

HIERN, W.P. 1900. *Catalogue of the African plants collected by Dr. Friedrich Welwitsch in 1853–61*, 1. British Museum, London.

HOLMGREN, P.K., HOLMGREN, N.H. & BARNETT, L.C. 1990. *Index Herbariorum, Part 1: The Herbaria of the World*. New York Botanic Garden, New York.

JACOBSEN, H. 1955. Some name changes in succulent plants—Part II. *National Cactus & Succulent Journal* 10: 80–85.

JACQUIN, N.J. 1797. *Plantarum rariorum horti caesari schoenbrunnensis descriptiones et icones*. 2. C.F. Wappler, Vienna.

JACQUIN, N.J. 1809. *Fragmenta botanica, figuris coloratis illustrata.....*, 6. M.A. Schmidt, Vienna.

JARVIS, C. 2007. *Order out of Chaos. Linnaean Plant names and their types*. Linnaean Society & Natural History Museum, London.

KLOTZSCH, J.F. & GARCKE, C.A.F. 1859. Linné's natürliche Pflanzenklasse Tricoccae der Berliner Herbariums im Allgemeinen und die natürliche Ordnung Euphorbiaceen insbesondere. *Monatsbericht der Königlichen Preussischen Akademie der Wissenschaften zu Berlin* 1859: 236–254.

KLOTZSCH, J.F. & GARCKE, C.A.F. 1860. Linné's natürliche Pflanzenklasse Tricoccae der Berliner Herbariums im Allgemeinen und die natürliche Ordnung Euphorbiaceen insbesondere. *Abhandlungen der Königlichen Akademie der Wissenschaften in Berlin* 1859: 1–108.

KOUTNIK, D.L. 1984. Chamaesyce (Euphorbiaceae)—a newly recognized genus in southern Africa. *South African Journal of Botany* 3: 262–264.

KRAUSS, C.F.F. 1845. Pflanzen des Cap- und Natal-Landes, gesammelt und zusammengestellt. *Flora* 28: 82–93.

KRAUSS, C.F.F. 1846. *Beiträge zur Flora des Cap- und Natallandes*. Regensburg.

KUNTZE, C.E.O. 1898. *Revisio generum plantarum* 3 (2). A. Felix, Leipzig.

LAMARCK, J.P.A.P.M. 1788. *Encyclopédie méthodique. Botanique*. 2, 2. Panckoucke, Paris.

LEACH, L.C. 1964. *Euphorbia* species from the *Flora Zambesiaca* area. *The Journal of South African Botany* 30: 1–12.

LEACH, L.C. 1967. *Euphorbia* species from the *Flora Zambesiaca* area: VII. *The Journal of South African Botany* 33: 247–262.

LEACH, L.C. 1975a. *Euphorbiae succulentae Angolenses*: V. Garcia de Orta, Sér. Bot. 2: 111–116.

LEACH, L.C. 1975b. *Euphorbia gummifera, E. gregaria* and a new species from Damaraland *Bothalia* 11: 495–503.

LEACH, L.C. 1984a. A new *Euphorbia* from South Africa. *The Journal of South African Botany* 50: 341–345.

LEACH, L.C. 1984b. A new *Euphorbia* from the Richtersveld. The Journal of South African Botany 50: 563–568.

LEACH, L.C. 1986a. A new *Euphorbia* (Euphorbiaceae) from the western Cape Province. *South African Journal of Botany* 52: 10–12.

LEACH, L.C. 1986b. A new *Euphorbia* from the western Knersvlakte, Cape Province. *South African Journal of Botany* 52: 369–371.

LEACH, L.C. 1988a. The Euphorbia juttae–gentilis complex, with a new species and a new subspecies. *South African Journal of Botany* 54: 534–538.

LEACH, L.C. 1988b. A new species of *Euphorbia* (Euphorbiaceae) from the Mossel Bay area. *South African Journal of Botany* 54: 539, 540.

LEACH, L.C. & WILLIAMSON, G. 1990. The identities of two confused species of Euphorbia (Euphorbiaceae) with descriptions of two closely related new species from Namaqualand. *South African Journal of Botany* 56: 71–78.

LEMAIRE, C.A. 1855. Miscellanées. Observations diagnostico-nomenclaturales sur les Euphorbes charnues du Cap. *L'Illustration horticole* 2: 65–71.

LEMAIRE, C.A. 1858. Miscellanées. Nouvelles Euphorbes. *L'Illustration horticole* 5: 63, 64.

LIEDE, S. & MEVE, U. 1993. Towards an understanding of the *Sarcostemma viminale* (Asclepiadaceae) complex. *Botanical Journal of the Linnaean Society* 112: 1–15.

LINNEAUS, C. 1737. *Hortus Cliffortianus plantas exhibens quas in hortis tam vivis quam siccis....* Amsterdam.

LINNEAUS, C. 1753. *Species Plantarum*. L. Salvius, Stockholm.

MARLOTH, R. 1910a. Some new South African Succulents. Part II. *Transactions of the Royal Society of South Africa* 1: 403–409.

MARLOTH, R. 1910b. Some new South African Succulents. Part III. Transactions of the Royal Society of South Africa 2: 33–39.

MARLOTH, R. 1913. Some new or little known South African Succulents. Part V. *Transactions of the Royal Society of South Africa* 3: 121–128.

MARLOTH, R. 1928. The Meloformia group of Euphorbias. *South African Gardening & Country Life* 18: 45, 46.

MARLOTH, R. 1930. A Revision of the group Virosae of the genus *Euphorbia* as far as represented in South Africa. *South African Journal of Science* 27: 331–340.

MARLOTH, R. 1931. Euphorbias III. *South African Gardening & Country Life* 21: 127, 128, 133.

MARX, J.G. 1996. *Euphorbia astrophora* J.G.Marx, *sp. nov.*, a new species of Euphorbiaceae from the Eastern Cape Province, South Africa. *Cactus & Succulent Journal* (Los Angeles) 68: 311–314.

MARX, J.G. 1999a. *Euphorbia suppressa* J.G.Marx and *Euphorbia gamkensis* J.G.Marx, two hitherto-unnamed species from the Western Cape Province, South Africa. *Cactus & Succulent Journal* (Los Angeles) 71: 33–40.

MARX, J.G. 1999b. The South African melon-shaped Euphorbias: the full picture as known to date. *Euphorbiaceae Study Group Bulletin* 12: 13–34.

MEERBURGH, N. 1789. *Plantae rariores vivis coloribus depictae*. Leiden.

MILLER, P. 1768. *The gardeners dictionary*, edn. 8. J. & F. Rivington, London.

MORISON, R. 1699. *Plantarum historiae oxoniensis universalis* 3. Oxford.

NEL, G.C. 1933a. Aus dem botanischen Garten der Universität Stellenbosch. *Kakteenkunde*: 134, 135.

NEL, G.C. 1933b. Aus dem botanischen Garten der Universität Stellenbosch. *Kakteenkunde*: 192, 194.

NEL, G.C. 1935. *Jahrbuch der Deutschen Kakteen-Gesellschaft* 1: 29–32, 42–43.

OUDEMANS, C.A.J.A. 1865. *Neerland's Plantentuin 1*. J.B. Wolters, Groningen.

PAX, F. 1889. Euphorbiaceae in A. Engler: Plantae Marlothianae. *Botanische Jahrbücher für Systematik* 10: 1–50.

PAX, F. 1897. Euphorbiaceae africanae. III. *Botanische Jahrbücher für Systematik* 23: 518–536.

PAX, F. 1898. Euphorbiaceae.1 in H. Schinz: Beiträge zur Kenntnis der Afrikanischen Flora. *Bulletin de L'Herbier Boissier* 6: 732–743.

PAX, F. 1899. Euphorbiaceae africanae. V. *Botanische Jahrbücher für Systematik* 28: 18–27.

PAX, F. 1900. Euphorbiaceae in A. Engler, Berichte über die botanischen Ergebnisse der Nyassa-See- und Kinga-Gebirgs-Expedition. III. *Botanische Jahrbücher für Systematik* 28: 418–421.

PAX, F. 1904. Euphorbiaceae Africanae. VII. *Botanische Jahrbücher für Systematik* 34: 368–376.

PAX, F. 1908. Euphorbiaceae. In H. Schinz, Beiträge zur Kenntnis der Afrikanischen-Flora. *Bulletin Herbier Boissier* sér. 2, 8: 634–637

PAX, F. 1909. Euphorbiaceae Africanae. IX. *Botanische Jahrbücher für Systematik* 43: 75–90.

PLUKENET, L. 1692. *Phytographia* 3. London.

ROWLEY, G.D. 1998. *Euphorbia meloformis* and *E. obesa* with two newly-assigned subspecies. *Euphorbiaceae Study Group Bulletin* 11: 93–98.

SALISBURY, R.A. 1796. *Prodromus stirpium in horto ad Chapel Allerton vigentium...* William Hooker, London.

SCHNABEL, D.H. 2011. The mysterious 'Snowflake' *Euphorbia*—a new variety of *Euphorbia polygona* Haw. *Euphorbia World* 7: 20–26.

SCHNABEL, D.H. 2012. A new variety of *Euphorbia polygona* Haw. from South Africa—*Euphorbia polygona* var. *exilis* D.H. Schnabel. *Euphorbia World* 8: 20–24.

SCHWEICKERDT, H.G.W.J. 1935. *Euphorbia aeruginosa*. In Notes on the Flora of Southern Africa VI (various authors). *Bulletin of Miscellaneous Information* 1935: 205, 206.

SCHWEICKERDT, H.G.W.J. & LETTY, C. 1933. *Euphorbia lydenburgensis*. The Flowering Plants of South Africa 13: t. 486.

SCOPOLI, J.A. 1788. *Deliciae Florae et Faunae insubricae*. 3. S. Salvatoris, Ticini.

SOJÁK, J. 1972. Poznámky K Nomenklatuře Euphorbia L. (s.l.). Časopis národního musea. *Oddíl přirodovědny* 140: 168–178.

STEUDEL, E.G. 1840. *Nomenclator Botanicus*, edn. 2, 1. J.G. Cottae, Stuttgart.

SWANEPOEL, W. 2009. *Euphorbia ohiva* (Euphorbiaceae), a new species from Namibia and Angola. *South African Journal of Botany* 75: 249–255.

SWEET, R. 1818. *Hortus suburbanus Londinensis*. J. Ridgway, London.

SWEET, R. 1826. *Hortus Britannicus*, ed. 1, 2. J. Ridgway, London.

THUNBERG, C.P. 1800. *Prodromus plantarum capensium* 2. J.F. Edman, Uppsala.

WEISS, J.E. 1893. Empfehlenswerte Cacteen. *Dr Neubert's Deutsche Garten-Magazin* 46: 286–292.

WHITE, A.C, DYER, R.A. & SLOANE, B.L. 1941. *The Succulent Euphorbieae (Southern Africa)*. Abbey Garden Press, Pasadena, California.

WIJNANDS, D.O. 1983. *The Botany of the Commelins*. Rotterdam, A.A.Balkema.

WILLDENOW, C.W. 1814. *Enumeratio Plantarum horti regii botanici berolinensis, Supplementum*. Berlin.

WILLIAMSON, G. 1995. *Euphorbia versicolores*, a new species from the northwestern Cape, South Africa. *Cactus & Succulent Journal* (Los Angeles) 67: 284–287.

WILLIAMSON, G. 2003. A new variety of *Euphorbia filiflora* Marloth from the Northern Cape Province of South Africa. *Bradleya* 21: 49–52.

WILLIAMSON, G. 2004. *Euphorbia einensis* sp. nov. (Euphorbiaceae) from the lower Orange River Valley in the northwest Richtersveld and southwest Namibia. *Haseltonia* 10: 57–66.

WILLIAMSON, G. 2007. Notes of *Euphorbia multiramosa* Nel (Euphorbiaceae) and related species. *Euphorbia World* 3. 8–18.

WILMAN, M. 1946. *Preliminary check list of the flowering plants and ferns of Griqualand West (southern Africa)*. Deighton Bell & Co., Cambridge, Alexander McGregor Mem. Museum, Kimberley.

WILSON, M.L., TOUSSAINT VAN HOVE-EXALTO, TH. & VAN RIJSSEN, W.J.J. 2002. *Codex Witsenii*. Iziko Museums, Cape Town & Davidii Media, Amsterdam.

10

New species and subspecies of *Babiana*, *Hesperantha*, and *Ixia* (Iridaceae: Crocoideae) from southern Africa; range extensions and morphological and nomenclatural notes on *Babiana* and *Geissorhiza*

P. GOLDBLATT* and J.C. MANNING**

Keywords: *Babiana* Ker Gawl., *Geissorhiza* Ker Gawl., *Hesperantha* Ker Gawl., Iridaceae, *Ixia* L., new species, southern Africa, taxonomy

ABSTRACT

Babiana **rivulicola** from stream banks in the Kamiesberg in Namaqualand and terete-leaved *Ixia* **teretifolia** from the Roggeveld, both in Northern Cape, are new species of these two largely winter-rainfall region genera. Late-flowering populations of *Hesperantha radiata* with crowded spikes of smaller flowers are segregated from the typical form as subsp. **caricina.** We also document the first record of *B. gariepensis* from Namibia, correct the authority for *B. purpurea* Ker Gawl., discuss morphologically aberrant populations of *B. tubiflora* from Saldanha, provide an expanded description for *B. lapeirousiodes* based on the second and only precisely localized collection of this rare Namaqualand species, and expand the circumscription of *Geissorhiza demissa* to accommodate a new record from the Kamiesberg, including revised couplets to the existing key to the species.

INTRODUCTION

Crocoideae, with ± 1 080 species, are the largest of the six subfamilies of Iridaceae, which now include over 2 200 species worldwide. Largely sub-Saharan African, subfamily Crocoideae is centred in southern Africa, where some 980 species occur, almost 850 of them restricted to the southwest of the subcontinent (the Greater Cape Floristic Region). The richness of the Cape flora is well established (e.g. Goldblatt & Manning 2000) and although the region has been explored botanically for over 250 years, novelties are still regularly recorded. Some are completely new discoveries but many others have reposed in herbaria, where they are recognized only after critical revisions of particular groups are undertaken. In Iridaceae alone, 97 species have been added to the family for the Cape flora over the past decade.

Here we describe one new species each of *Babiana* Ker Gawl. (now 93 spp.) and *Ixia* L. (now ± 90 spp.). Both genera are members of tribe Ixieae Dumort. (1822) as circumscribed by Goldblatt *et al.* (2006) and, apart from *Ixia*, have been revised or thoroughly reviewed in the past 25 years (Goldblatt 2003; Goldblatt & Manning 2007a, 2010). In *Hesperantha* (82 spp.), we also describe a new subsp. of *Hesperantha radiata* (L.f.) Ker Gawl and include a significant range extension into Namibia for *B. gariepensis* Goldblatt & J.C.Manning, morphological notes for variant populations of *B. tubiflora* (L.f.) Ker Gawl. and *Geissorhiza demissa* Goldblatt & J.C.Manning, nomenclatural notes for *B. bainesii* Baker, and a correction to the author citation for *B. purpurea* Ker Gawl.

---

* B.A. Krukoff Curator of African Botany, Missouri Botanical Garden, P.O. Box 299, St. Louis, Missouri 63166, USA. E-mail: peter.goldblatt@mobot.org.
** Compton Herbarium, South African National Biodiversity Institute, Private Bag X7, Claremont 7735, South Africa / Research Centre for Plant Growth and Development, School of Life Sciences, University of KwaZulu-Natal, Pietermaritzburg, Private Bag X01, Scottsville 3209, South Africa. E-mail: J.Manning@sanbi.org.za.

MATERIALS AND METHODS

We examined all relevant collections at BOL, NBG, PRE, and SAM, the primary southern African herbaria (acronyms after Holmgren *et al.* 1990).

SYSTEMATICS

New taxa

1. **Babiana** *Ker Gawl.*: Lewis's (1959) monograph of *Babiana* recognized 61 species, including one from the Indian Ocean island of Socotra, and set the stage for later studies of the genus, not least by stabilizing its complex synonymy. *Antholyza* L. (including *Anaclanthe* N.E.Br.), with two species, was added to the genus by Goldblatt (1990). In the recent revision of *Babiana* (Goldblatt & Manning 2007a) we recognized 88 species, excluding the Socotran *B. socotrana* Hook.f., which is now the new genus *Cyanixia* Goldblatt & J.C.Manning (Goldblatt *et al.* 2004). Subsequent exploration of western South Africa yielded an additional four new species (Goldblatt *et al.* 2008; Goldblatt & Manning 2010). The discovery by Cape Town biologist Nick Helme of yet another new species, described here as *B. rivulicola*, in a relatively well collected part of the Kamiesberg in central Namaqualand in 2009 is thus particularly surprising. *Babiana* now has 93 species, but additional taxa are likely to be recognized as the results of the molecular phylogenetic study (Schnitzler *et al.* 2011) are analyzed from a systematic perspective.

Babiana rivulicola *Goldblatt & J.C.Manning*, sp. nov.

TYPE.—Northern Cape, 3018 (Kamiesberg): southern Kamiesberg, Langkloof, ± 20 km from Doringkraal turnoff from N7, Farm Nartjiesdam, in rocky bed of perennial stream, (–AD), 22 Sept. 2010, *Goldblatt & Porter 13572* (NBG, holo.; K, MO, PRE, iso.).

Plants to 300 mm high, excluding longer leaves, growing in dense clumps; stem erect, simple or

1–3-branched, angled. *Corm* wedged in crevices in rock, unknown. *Leaves* mostly 6 or 7, linear, 250–450 × 5–10 mm, exceeding stem, firm but arching toward ground, shallowly pleated to almost flat, glabrous. *Spike* flexed outward with flowers initially in two ranks, mostly 6–9-flowered; bracts green below, dry and light brown toward tips, outer mostly up to 35 mm long but lowermost much longer, attenuate, inner ± 2/3 as long as outer, divided apically, attenuate, 2-keeled below. *Flowers* zygomorphic, pale blue to blue-mauve, paler in throat, lower 3 tepals with central, white, lanceolate blotch and darker violet marking toward base, darkly lined on abaxial side of throat, lightly sweet-scented; perianth tube funnel-shaped, 42–44 mm long, lower part cylindric, upper ± 10 mm expanded into wide throat; tepals lanceolate, unequal, dorsal ± 30 × 8 mm, lower three tepals united with upper laterals for ± 3 mm, arching outward, 25 × 6 mm. *Stamens* unilateral; filaments straight, suberect, ± 15 mm long, included in floral cup, reaching almost to middle of dorsal tepal; anthers ± 5 mm long, pale mauve, pollen white. *Ovary* glabrous, style dividing opposite or slightly beyond anther tips, style branches ± 5 mm long, recurved distally. *Capsules* and *seeds* not known. *Flowering* mid- to late Oct. Figure 1.

Eponymy: from the Latin, *rivulicola*, growing in a riverine habitat.

Distribution and ecology: so far *Babiana rivulicola* is known only from the southern Kamiesberg in Northern Cape, where it is restricted to the perennial stream that runs through the Langkloof, between Farms Karas (Welkom) and Doringkraal (Figure 2). Plants grow in cracks in smooth granite along the edges of the stream above the waterline. Plants are regularly inundated after heavy rains but the corms are wedged in crevices in bedrock and secure from being washed away. It proved impossible to extract corms for examination and they remain unknown. The species was first collected in October 2009 by Cape Town biologist N.A. Helme. Despite its narrow range, we see no current threat to *B. rivulicola* and suggest a conservation status of (LC), least concern, using the definitions and terminology of Raimondo *et al.* (2009).

Diagnosis and relationships: *Babiana rivulicola* most closely resembles a second, more widespread Namaqualand endemic, *B. dregei* Baker, and both species share glabrous leaves, stem and bracts, an unusual feature in a genus where pubescence of some kind on vegetative

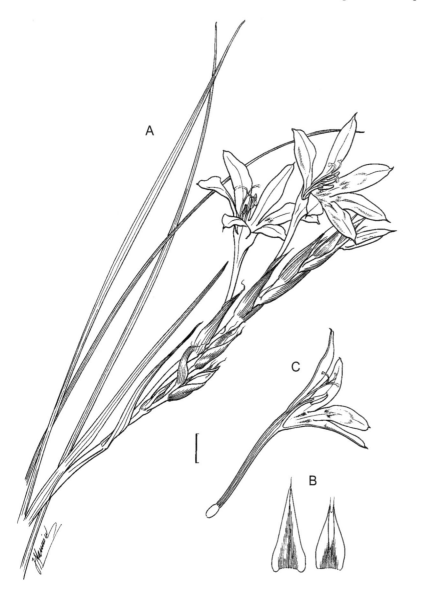

FIGURE 1.—*Babiana rivulicola, Goldblatt & Porter 13572* (NBG). A, portion of flowering plant; B, outer (left) and inner (right) bract; C, half flower. Scale bar: 10 mm. Artist: J.C. Manning.

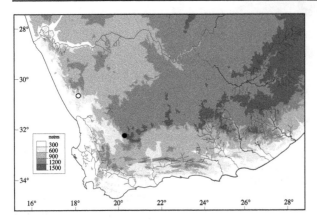

FIGURE 2.—Distribution of *Babiana rivulicola*, ○; and *Ixia tereti-folia*, ●.

organs is almost universal. The leaves of *B. dregei* are characteristically relatively broadly lanceolate and rigid with sharply pungent tips and the stem is very short and concealed among the leaves. In *B. rivulicola* the stem reaches up to 250 mm high and the linear leaves are longer and relatively softly textured, with acute but not pungent tips. The flowers in the two species are broadly similar, ± funnel-shaped with the lower two thirds of the perianth tube cylindric, but *B. dregei* has a ± cylindric perianth tube (45–)50–65(–70) mm long, curved just below the apex, unlike the shorter, funnel-shaped perianth tube of *B. rivulicola*, 42–44 mm long. Additional floral differences include the longer anthers, ± 8 mm long and shorter style branches, 3–4 mm long in *B. dregei* vs. anthers ± 5 mm long and style branches ± 5 mm long in *B. rivulicola*.

When collecting *Babiana rivulicola* in the second half of September we noted plants of *B. dregei* on adjacent slopes well past flowering and with developing capsules. Although sympatric in distribution, the two taxa are isolated by flowering time and ecology, and thus allotopic (Wiley 1981).

Additional specimen

NORTHERN CAPE.—**3018** (Kamiesberg): southern Kamiesberg, Langkloof, ± 1 km S of Farm Nartjiesdam, in rocks in stream, (–AD), 10 Oct. 2009, *Helme 6110* (NBG).

2. **Hesperantha** *Ker Gawl*.: the southern and tropical African *Hesperantha* with 82 species (Goldblatt 2003; Goldblatt & Manning 2007b), is distinguished in Crocoideae by the style dividing at the mouth of the perianth tube (rarely within the tube) into long, spreading, relatively long, more-or-less straight, laxly spreading style branches and, with a few exceptions, by hard, woody corm tunics (Goldblatt & Manning 2008). In the most widespread species of the genus, *H. radiata*, Goldblatt (2003) drew attention to collections of small-flowered plants from the Cape Peninsula and surrounding hills, as far north as Piketberg and as far east as Hermanus in the southwestern corner of the southern African winter-rainfall zone, which were then included in the more widespread and common form which extends from Namaqualand to Swaziland. These small-flowered plants have unusually crowded spikes with shortly imbricate bracts, and the uppermost leaf is always entirely sheathing and reaches almost to the base of the spike. The flowers have

tepals 7–12 mm long, a tube 6–10 mm long, anthers 4(–5) mm long, and the stems have a weakly developed neck of fine fibers around the base (Table 1). Including this morph in *H. radiata*, of which typical larger flowered populations occur in the same area, seems counter to the concept of biological species. Elsewhere, the distinction between the two forms is not always entirely clear, which argues against treating them as distinct species. As Goldblatt (2003) also pointed out, it is especially notable that the small-flowered plants bloom later than typical *H. radiata* and are sometimes sympatric but not co-blooming [allotopic (Wiley 1981)], as for example, *Oliver 4332* (NBG) collected on 21 Aug. 1973 and *4756* (NBG) collected on 17 Oct. 1973, both in the hills at Langverwacht near Stellenbosch. The August-flowering plants are typical *H. radiata* and those flowering in October are the small-flowered morph with crowded spikes. Specimens of this morph collected by C.F. Ecklon and C.L. Zeyher in the early 19th century are annotated *H. setacea* Eckl. (e.g., *Ecklon & Zeyher Irid. 233* 89.9), the name a nomen nudum and thus invalid (Ecklon 1827), while some sheets of the small-flowered plants at the Kew Herbarium are annotated *H. tenuifolia*. This is R.A. Salisbury's (1812) name at species rank for *H. radiata* var. γ *caricina* of *Curtis's Botanical Magazine* t. 790 (Ker Gawler 1804). Salisbury's epithet alludes to the characteristic narrow leaves, also found in some populations of larger-flowered plants that correspond to the type of *H. radiata*.

We propose to recognize late-blooming plants with smaller flowers as a separate subspecies, in line with our treatment of late-flowering *Moraea tripetala* subsp. *jacquiniana* (Schltr. ex G.J.Lewis) Goldblatt & J.C.Manning, which is sympatric with early flowering subsp. *tripetala* (Goldblatt & Manning 2012a).

TABLE 1.—Comparison of flowers of *Hesperantha radiata* subsp. *caricina* and subsp. *radiata*. Only plants with well pressed, fully open flowers were measured.

	Outer floral bract (mm)	Perianth tube (mm)	Outer tepal length × width (mm)	Filament length (mm)	Anther length (mm)	Style branches (mm)
Subsp. *caricina*	10–13	6–10	7–12 × 2.5–3.5	4–5	4(–5)	5–6
Subsp. *radiata*	11–15	8 × 14	13–15(–17) × 3–5	5–6	5.0–8.5	8–11

Hesperantha radiata subsp. **caricina** *(Ker Gawl.) Goldblatt & Manning* stat. nov. *H. radiata* var. γ *caricina* Ker Gawl.: t. 790 (1804). *H. tenuifolia* Salisb.: 321 (1812), nom. nov. pro *H. radiata* var. γ *caricina* Ker Gawl. (1804). *H. caricina* (Ker Gawl.) Klatt: 395 (1882), nom. illegit. superfl. pro *H. tenuifolia* Salisb. (1812). *H. angustifolia* Loudon: 91 (1841), nom. illegit. superfl. pro *H. tenuifolia* Salisb. (1812). Type: South Africa, without precise locality, illustration in Curtis's Botanical Magazine 21: t. 790 (1804).

Plants like subsp. *radiata* but corms often with a collar of fibres around base. *Leaves* 1–2 mm wide, uppermost leaf entirely sheathing, reaching close to base of spike. *Spike* crowded with outer bracts 10–13 mm long, as long as or longer than next internode and thus imbricate. *Flowers* cream to dull yellow, outer tepals reddish on reverse; perianth tube 6–10 mm long; tepals 7–12 × 2.5–3.5 mm. *Stamen* filaments 4–5 mm long; anthers 4(–5) mm long. *Style* branches mostly 5–6 mm long. *Flowering time*: mostly late Sept. and Oct., but until Dec. at higher elevations.

Diagnosis: subsp. *caricina* is usually recognized by the crowded spike of relatively small flowers, the tepals 7–12 × 2.5–3.5 mm and perianth tube 6–10 mm long, and shortly overlapping bracts 10–13 mm long. Plants are typically late flowering, in September and October, but there are collections from higher elevations that bloom as late as December. Populations of typical *H. radiata* occur throughout the range of subsp. *caricina* (from Piketberg to Caledon and Hermanus) but invariably flower earlier, from late July to mid-September, and have larger flowers, with tepals 13–15(–17) × 3–5 mm, perianth tube 8–14 mm long, and anthers mostly 4 mm long (Table 1). The leaves of subsp. *caricina* are also somewhat narrower than in subsp. *radiata*, 1–2 mm wide, and often dry at flowering time vs. 2.0–3.5 mm wide and green at flowering.

Representative specimens

WESTERN CAPE.—**3318** (Cape Town): Riebeek Kasteel, Farm Remhoogte, (–BD), 24 Oct. 1968, *Marsh 1034* (NBG); Table Mtn, top of Oudekraal Ravine, (–CD), Nov. 1972, *McKinnon s.n.* (NBG); Rosebank, Cape Town, Nov. [without year], *H. Bolus 3769* (BOL, K, NBG, PRE, Z); Langverwacht above Kuils River, (–DC), 17 Oct. 1973, *Oliver 4756* (K, NBG, PRE); Stellenbosch flats, (–DD), 30 Sept. 1958, *Perold 7* (NBG); Sept. 1964, *Holzapfel 10* (NBG); Blauwklip, (–DD), 24 Oct. 1928, *Gillett s.n.* (NBG); Uitkyk, (–DD), Oct. 1982, *Gillett s.n.* (NBG). **3418** (Simonstown): flats between Gordon's Bay and Strand on Disa Street, (–BB), 29 Sept. 2001, *Goldblatt & Nänni 11944* (MO, NBG). **3419** (Caledon): hills near Farm Die Vlei E of Bot River, (–AA), 30 Sept. 2000, *Goldblatt & Nänni 11944* (MO); western outskirts of Caledon, (–AB), 29 Sept. 2001, *Goldblatt & Nänni 11942* (MO); Fernkloof Nature Reserve, Hermanus, (–AD), 18 Oct. 1979, *Orchard 4485* (MO). Without precise locality: Piketberg, renoster hills, 9 Oct. 1922, *Marloth 11484* (NBG).

3. **Ixia** L.: restricted to South Africa and to the winter rainfall zone and adjacent areas, *Ixia* comprises an estimated 90 species in four sections. Sects. *Hyalis* (18 spp.), *Morphixia* (now 32 spp.), and *Dichone* (now 17 spp.) have recently been revised (Goldblatt & Manning 2011, 2012b) and we are currently revising the systematics and taxonomy of sect. *Ixia* (± 25 spp.). Here we describe a species in sect. *Morphixia* that we discovered on the edge of the Roggeveld Escarpment, in September 2011, and later collected in fruit in November.

Ixia teretifolia *Goldblatt & J.C.Manning*, sp. nov.

TYPE.—Northern Cape, 3220 (Sutherland): Roggeveld Escarpment, Farm Blesfontein, in dense tall grass tufts below sandstone cliffs at escarpment edge, (–AD), 25 Sept. 2011, *Goldblatt & Manning 13671* (NBG, holo., K, MO, PRE, iso.).

Plants mostly 1.2–1.5 m high. *Corm* subglobose, 12–18 mm diam., tunics of relatively fine fibres. *Stem* erect in flower, nodding in fruit; sheathed below by prominent, pale, dry-membranous, often irregularly torn cataphylls accumulating with age in a dense mass; usually with 3–5 short branches up to 4 mm long, branches subtended by pale, dry, attenuate scale-like bracts and

prophylls 2–4 mm long. *Leaves* 3, lower 2 with long terete blades ± half as long as stem, mostly 1.5–2.0 mm diam., firm-textured without evident veins when alive; uppermost leaf entirely sheathing or almost so, reaching to base of spike or to lowermost branches. *Main spike* (1)2(3)-flowered, lateral spikes 1- or 2(3)-flowered; bracts silvery membranous, translucent with narrow, dark purple veins, ± 7 mm long, outer 3 veined and 3-toothed or 3-lobed; inner 2-veined and shallowly forked at apex. *Flowers* upright to half nodding, pale blue but pale green in throat, faintly rose-scented, with bright yellow anthers prominently displayed; perianth tube broadly funnel-shaped, ± 5 mm long, narrow part ± 2 mm long; tepals spreading, fully patent distally, ovate, ± 12 × 6.0–6.5 mm, outer tepals slightly wider than inner. *Stamens* with filaments ± 6 mm long, exserted ± 4 mm from tube; anthers 4–5 mm long, ± parallel throughout anthesis, yellow; pollen yellow. *Style* dividing opposite lower third of anthers, branches ± 3 mm long, slender, extending between anthers. *Capsules* globose, ± 5 mm diam., showing outline of seeds. *Seeds* subglobose, ± 2.0 × 1.6 mm, light yellow-brown, (5)6 per locule. *Flowering time*: mostly mid-Sept.–mid-Oct., possibly later. Figure 3.

Eponymy: from the Latin, *teres*, round in section, *folia*, leaf.

Distribution and ecology: so far known from just one site, *Ixia teretifolia* grows at the base of sandstone cliffs of the Beaufort Series at the edge of the Roggeveld Escarpment west of Sutherland on the Farm Blesfontein (Figure 2). Plants grow among broken rocks, nested in dense tufts of the grass *Tenaxia* (*Merxmuellera*) *stricta* (Schrad.) N.P.Barker & H.P.Linder. Extensive searching on the escarpment itself revealed no sign of the species, but the white-flowered morph of the related *I. marginata* Salisb. ex G.J.Lewis was relatively common there, growing in mountain renosterveld vegetation together with other geophytes, including *I. trifolia* G.J.Lewis, *Geissorhiza heterostyla* L.Bolus, the deep pink-flowered morph of *Hesperantha pilosa* (L.f.) Ker Gawl. (all Iridaceae) and *Bulbinella elegans* P.L.Perry (Asphodelaceae). We estimated that the population of *I. teretifolia* consisted of about 70 mature, flowering individuals along a 200 m stretch of the escarpment edge. Although we found no plants to the south of our site on the adjacent farm Boplaas, we suspect the range of *I. teretifolia* may be more extensive further north toward Ouberg Pass, where we recommend further exploration to establish its complete range.

Diagnosis and relationships: the short, widely funnel-shaped perianth tube, ± 5 mm long, distally spreading tepals, partly exserted filaments with prominently displayed anthers, and the several short lateral branchlets place *I. teretifolia* in the informal series *Marginatae* of *Ixia* sect. *Morphixia* (Goldblatt & Manning 2011). Its tall stature and habit are shared in the section with *I. alata* Goldblatt & J.C.Manning, but the several very short branches, each with 1 or 2(3) flowers differ from that species, and recall particularly *I. linearifolia* Goldblatt & J.C.Manning. The centric leaves lacking visible veins are unique in sect. *Morphixia* and unusual in the genus, almost all species of which have plane, isobilateral leaves, apart from *I. linearifolia*, which has linear,

FIGURE 3.—*Ixia teretifolia*, *Gold-blatt & Manning 13671* (NBG). A, portion of flowering plant; B, T/S leaf blade; C, leaves and flowering stem. Scale bar: A–C, 10 mm; B, 1 mm. Artist: J.C. Manning.

fleshy leaves with the margins raised at right-angles to the blade surface. Except for their slightly smaller size, the flowers of *I. teretifolia* differ very little from other members of series *Marginatae*, all of which have a short, funnel-shaped perianth tube, distally outspread tepals and fully exserted anthers.

Internally, the leaf margins in *I. teretifolia* have paired strands of sclerenchyma separated by parenchyma tissue (Figure 3B). As in other species of *Ixia* the sclerenchyna strands are not associated with a marginal vein ((Rudall & Goldblatt 1991); Rudall's (1995) statement to the contrary for *Ixia* is an error), and there are just two opposed major vein pairs in the leaf in a lateral position, well separated internally by mesophyll. The anatomy of *I. teretifolia* does not conform to that of a truly centric (or terete) leaf as it is not radially symmetric, e.g. as seen in terete-leaved species of *Bobartia* (Rudall 1995), and the leaves are in fact internally bilaterally symmetric. Related *I. linearifolia* has fairly typical leaf anatomy for *Ixia* with a wide strand of sclerenchyma extending the width of the thickened, but slightly winged margins, themselves unusual in *Ixia*.

Additional specimen

NORTHERN CAPE.—**3220** (Sutherland): Roggeveld Escarpment, Farm Blesfontein, in dense tall grass tufts below sandstone cliffs at escarpment edge, (–AD), 4 Nov. 2011 (fr.), *Goldblatt & Porter 13723* (MO, NBG).

Morphological and nomenclatural notes and a range extension

1. **Babiana tubiflora** *(L.f.) Ker Gawl.*: a population of a *Babiana* species on the inland side of high sand dunes north of Jacobsbaai on the Saldanha Peninsula is somewhat puzzling in its unusual flowers. These plants are currently included in the widespread *Babiana tubiflora*, which extends along the coast of Western Cape from Lambert's Bay in the north to Still Bay in the east (Goldblatt & Manning 2007a), thus surrounding the Jacobsbaai population. We were unable to distinguish the Jacobsbaai plants vegetatively from specimens of *B. tubiflora* collected nearby at Langebaanweg when compared side-by-side, but the flowers are strikingly different. Largely a coastal species, *B. tubiflora* rarely extends more than 20 km inland, and has zygomorphic flowers with a white to cream-coloured perianth, the lower tepals often with a red streak on the limbs of the lower three tepals. The tepals are narrowed into an ascending claw-like lower portion with wider, spreading limbs, the dorsal tepal 18–23 × 5–7 mm and the lower tepals 15–19 × 3–4 mm. The unilateral stamens have filaments 13–16 mm long and anthers 4–5 mm long. In contrast (Table 2), the tepals of the Jacobsbaai plants are ovate

without a narrowed claw, the dorsal tepal is 12 × 6 mm, and the lower three tepals are ± 10 × 4 mm (*Goldblatt & Porter 13512* MO, NBG, 9 Sept. 2010; *Goldblatt & Porter 13572*, MO, NBG, 23 Sept. 2010). The stamens have filaments ± 6 mm long (< half the length of those of *B. tubiflora* as currently circumscribed) and anthers ± 3 mm long. The relatively short style divides opposite the base to lower third of the anthers into branches ± 2.5 mm long, whereas that in typical *B. tubiflora* divides between the middle and slightly beyond the anther tips and the style branches are 3–4 mm long. We note that the Jacobsbaai *B. tubiflora* population was sympatric and co-blooming with closely related but larger-flowered *B. tubulosa* (Burm.f.) Ker Gawl. (*Goldblatt & Porter 13566*, MO, NBG, PRE, 22 Sept. 2010).

We remain uncertain whether to regard the Jacobsbaai plants as a separate taxon or merely a trivial mutant race of no taxonomic significance. Until more is known about this distinctive population we simply report its existence and defer a decision about its status pending further study.

2. **Babiana purpurea** *Ker Gawl.*: *Ixia purpurea* Jacq. was based on a plate in his *Icones plantarum rariorum* vol. 2: t. 286 (Jacquin 1793), but with the description published earlier (Jacquin 1791). The name was later transferred to *Babiana* by Ker Gawler (1807a) and has been cited until now as *B. purpurea* (Jacq.) Ker Gawl., but it has come to our attention that *I. purpurea* Jacq. is a later homonym for *I. purpurea* Lam. (1789). Dating of Jacquin's plates is complicated because the dates on the title pages of the three volumes of the *Icones* do not represent the first publication of the individual plates, which were released in fascicles earlier than the bound volumes. The dating of the fascicles and the plates included in each was resolved by Schubert (1945). Some fascicles were predated by the publication of Jacquin's *Collectanea* and the dates on the title pages of the several volumes evidently do not reflect the actual year of publication. *Ixia purpurea* Jacq. was first published in *Collectanea* vol. 3, published in 1791 (title page 1789). Fascicle 9 of the *Icones*, which included *I. purpurea* Jacq. was, according to Schubert's analysis, published in 1793.

Jacquin's species is thus evidently predated by *I. purpurea* Lam. (1789), a valid and legitimate name typified by the plant in the Thunberg Herbarium named *I. crocata* var. c, *flore purpurea*. Lamarck was evidently providing a name to one of the variants listed by Thunberg (1783) in his account of Linnaeus's *I. crocata* (now *Tritonia crocata* (L.) Ker Gawl.), thus excluding typical *I. crocata* (which Lamarck recognized as such), and *I. purpu-*

TABLE 2.—Comparison of flowers of typical *Babiana tubiflora* and the Jacobsbaai population. Only well pressed, fully open flowers were measured.

Taxon	Perianth tube length (mm)	Tepal shape,	Dorsal tepal (mm)	Anther length (mm)	Filament length (mm)	Style branches
Typical *B. tubiflora*	45–80(–100)	Narrowly lanceolate, ± clawed below.	18–23 × 5–7	6.0–8.5	13–16	3–4 mm, dividing between middle and apex of anthers.
Jacobsbaai population	40–45	Ovate, without claws.	± 12 × 6	4–5	± 6	± 2.5 mm, dividing between base and lower third of anthers.

rea Lam. is not nomenclaturally superfluous. Thunberg's *I. crocata* var. c was later regarded by De Vos (1983) as merely a purple-flowered variant of *Tritonia crocata*.

The name *Babiana purpurea* is to be treated as a new name (McNeill *et al.* 2006: Art. 58) dating from its apparent transfer to *Babiana* by Ker Gawl., thus becoming *B. purpurea* Ker Gawl. (1807a).

Babiana purpurea *Ker Gawl.* in Curtis's Botanical Magazine 26: sub. t. 1019 (1807a), nom. nov. pro *Ixia purpurea* Jacq., Coll. 3: 268 (1791, as '1789'), hom. illegit. non Lam. (1789). *Gladiolus purpureus* (Ker Gawl.) Willd.: 198 (1797). *Gladiolus purpureus* (Jacq.)Vahl: 114 (1805), hom. illegit. non Willd. (1797). *Babiana stricta* var. *purpurea* (Ker Gawl.) Ker Gawl.: (1807b) [name misapplied to *B. villosa* (Gmelin) Ker Gawl. ex Steud.]. Type: South Africa, without precise locality or collector, illustration in Jacq., Icones Plantarum Rariorum 2: t. 286 (1793).

3. **Babiana bainesii** *Baker*: in our revision of *Babiana* (Goldblatt & Manning 2007a) we cited *B. schlechteri* Baker (1904) as a synonym of *B. bainesii* Baker (1876). Unfortunately, we overlooked the fact that this is a homonym for *B. schlechteri* Baker (1901) (= *B. nana* subsp. *maculata* (Klatt) Goldblatt & J.C.Manning). We also neglected to note that the replacement name *B. bakeri* Schinz (1906) had been provided for Baker's second *B. schlechteri*. This correction has no nomenclatural consequences at present. These names must be added to the synonymy of *B. bainesii* as follows:

Babiana bainesii *Baker* in Journal of Botany 14: 335 (1876). Type: South Africa, [Gauteng,] 'Gold Fields', Witwatersrand, near Johannesburg, 1870, *Baines s.n.* [K, lecto.!, designated by Goldblatt & Manning: 92 (2004)].

Babiana bakeri Schinz: 712 (1906), as nom. nov. pro *B. schlechteri* Baker: 1005 (1904), hom. illegit. non Baker (1901). Type: South Africa, [Mpumalanga], near Middleburg, 22 Dec. 1893, *Schlechter 4055* (Z, ?holo.).

4. **Babiana gariepensis** *Goldblatt & J.C.Manning*: currently known from three collections from the Richtersveld in Northern Cape, *B. gariepensis* is a relatively poorly known species of *Babiana* sect. *Teretifolieae* (Goldblatt & Manning 2007a: 31). It is distinguished by the unusually broad leaves mostly 70–110 × 18–25 mm, a stem produced shortly above the ground, pale, grey-green flowers, the lower tepals with white streaks edged dark red, and a relatively long perianth tube 20–24 mm long. A new collection of the species has now come to our attention from southwestern Namibia in the Namuskluft hiking trail near Rosh Pinah. Although a modest range extension, some 50 km north of the nearest station in South Africa, this represents the first record of the species in Namibia. The one of the two specimens of the collection is somewhat unusual for the species in being particularly robust and has a stem reaching ± 120 mm above the ground; including the spike the specimen is 280 mm tall.

Additional specimen

NAMIBIA.—**2716** (Witpütz): Rosh Pinah, Namuskluft, on hiking trail from campsite, (–DD), 27 July 2011, *R. & R. Saunders s.n.* (NBG).

5. **Babiana lapeirousioides** *Goldblatt & J.C.Manning*: described in 2007 and based on a specimen and painting of a plant collected by C.L. Leipoldt and grown to flowering at the National Botanical Gardens, Kirstenbosch, in 1943, *B. lapeirousioides* was then known only from the cryptic locality 'Gariep' (Goldblatt & Manning 2007a: 49). This distinctive, long-tubed, white flowered species was discovered in the wild by Helga van der Merwe and Gretel van Rooyen in October 2011 at Vaalputs, the Nuclear Waste Disposal site (between Gamoep and Platbakkies) in Northern Cape. This collection allows us to expand the description of the species, now with a specific locality (we associated the original locality, Gariep, with an area of the Richtersveld northeast of Port Nolloth and that now seems to be mistaken and we wonder if 'Gariep' was a mistranscription of Gamoep).

Babiana lapeirousioides *Goldblatt & J.C.Manning* in Strelitzia 18: 49 (2007). Type.—South Africa, [Northern Cape,] without precise locality, as 'Gariep', 30 Sept. 1943, *Leipoldt s.n.* (NBG, holo.).

Plants (5–)10–20 cm high including leaves, stem reaching shortly up to 70 mm above ground level, usually simple, rarely with 1 branch. *Leaves* sword-shaped, blade 50–150 × 5–9 mm, rigid, plicate, pungent, smooth, primary veins thickened and yellowish. *Bracts* (17–)20–25(–35) mm long, glabrous or minutely scabrid above, closely veined, dry and brownish above, inner slightly shorter than outer, forked apically. *Flowers* in a compact, 2–4-flowered spike, zygomorphic, white with red or purplish markings near base of lower tepals, presence of scent unknown; perianth tube 22–35(–40) mm long, cylindric and straight, flaring slightly toward the mouth; tepals subequal or upper three slightly larger, spreading, narrowly oblanceolate, 10–20 × 3–4 mm. *Stamens* unilateral; filaments erect, 7–8 mm long; anthers 4.5–5.5 mm, cream-coloured. *Ovary* smooth; style dividing from near base of anthers to shortly beyond anther tips, branches ± 2.5 mm long. *Flowering time*: late Sept.–early Oct.

Distribution and ecology: until now known only from the type, the species has been recollected at Vaalputs east of the Kamiesberg at the transition into Bushmanland. The type locality, 'Gariep' is unlocalized and we assumed it to refer to the Richtersveld but in the light of the recent collection we consider that it might equally refer to one or other of the inselbergs in south of the Orange River in Bushmanland. The species grows wedged in cracks among granite boulders, sometimes forming large clumps. When not in flower the rigid, pungent leaves are readily confused with *B. dregei* which has a similar habitat in granite outcrops but at higher elevations in the Kamiesberg.

Additional specimen

NORTHERN CAPE.—**3018** (Kamiesberg): Vaalputs, NECSA site, (–AB), 1 Oct. 2011, *G. van Rooyen 3048* (NBG, PRE).

6. **Geissorhiza demissa** *Goldblatt & J.C.Manning*: most closely resembling *G. aspera* in its minutely puberulous stems and radially symmetric flowers with only slightly unequal filaments, *G. demissa* has relatively small, white flowers tinged violet at the tips of the

tepals (Goldblatt & Manning 2009). A new collection, flowering after fire in the Kamiesberg, that we refer to *G. demissa*, is unusually robust with larger flowers than recorded until now for the species, the tepals ± 10 mm long, anthers ± 3 mm long and up to four flowers per spike. Floral dimensions previously recorded were tepals 7 × 3.5 mm and anthers ± 2 mm long and spikes had 2 or 3 flowers. The distinction between *G. demissa* and *G. aspera* is less clear for this population. We revise a portion of the key to subg. *Geissorhiza* (Goldblatt & Manning 2009) below.

5a Stamens unequal, one filament at least 0.5 mm shorter than other two:

6a Leaves ± plane with margins and central vein slightly to moderately thickened, but not obviously winged; margins and central vein smooth or minutely puberulous:

7a Perianth predominantly purple, pale in throat; short filament at least 4 mm shorter than long filaments; tepals 18–23 mm long . . . . 81. *G. inaequalis*

7b Perianth predominantly blue to violet or white with pale throat or purple with dark purple throat; short filament no more than 2 mm shorter than long filaments; tepals 7–20 mm long:

8a Flowers radially symmetric except for eccentric style; tepals 7–11 mm long; flowers violet to pale blue, purple or predominantly white:

9a Tepals blue to violet with white throat edged in dark violet, or uniformly white; mostly 11–14 × 4–6 mm; main spike mostly with 3–7 flowers; anthers 3–5 mm long . . . . 79. *G. aspera*

9b Tepals white, sometimes flushed dull purple outside and tepals tips light violet, 7–10 × 3.5–4.0 mm; main spike with 1–4 flowers; anthers 2–3 mm long . . . . 80. *G. demissa*

Additional specimen

NORTHERN CAPE.—**3018** (Kamiesberg): Langkloof, slopes of Rooiberg between Farms Karas and Nartjiesdam, (–AD), 22 Sept. 2010, *Goldblatt & Porter 13579* (MO, NBG)

ACKNOWLEDGEMENTS

Support for this study by grants 7103-01, 7316-02, 7799-05 and 8248-07 from the National Geographic Society is gratefully acknowledged. We extend our gratitude to Elizabeth Parker and Lendon Porter for their assistance and companionship in the field and Nick Helme for bringing his discovery of the new *Babiana rivulicola* to our attention. Collecting permits were provided by the Nature Conservation authorities of Northern Cape Province.

REFERENCES

BAKER, J.G. 1876. New Gladioleae. *Journal of Botany, London* 14: 333–339.

BAKER, J.G. 1901. Iridaceae. In H. Schinz, Beiträge zur Kenntniss der Afrikanischen Flora (Neue Folge). *Bulletin de l'Herbier Boissier*, sér. 2,1: 853–868.

BAKER, J.G. 1904. Iridaceae. In H. Schinz, Beiträge zur kenntnis der Afrikanischen-Flora. *Bulletin de l'Herbier Boissier*, ser. 2,4: 1003–1007.

DE VOS, M.P. 1983. The African genus *Tritonia* Ker-Gawler 2. *Journal of South African Botany* 49: 347–422.

DUMORTIER, B.C. 1822. *Observations Botaniques*. Tournai, Belgium.

ECKLON, C.F. 1827. *Topographisches Verzeichniss der Pflanzensammlung von C.F. Ecklon.* Reiseverein, Esslingen.

GOLDBLATT, P. 1990. Status of the southern African *Anapalina* and *Antholyza* (Iridaceae), genera based solely on characters for bird pollination, and a new species of *Tritoniopsis*. *South African Journal of Botany*. 56: 577–582.

GOLDBLATT, P. 2003. A synoptic review of the African genus *Hesperantha* (Iridaceae: Crocoideae). *Annals of the Missouri Botanical Garden* 90: 390–443.

GOLDBLATT, P. & MANNING, J.C. 2000. Cape plants: A conspectus of the Cape flora. *Strelitzia* 9. National Botanical Institute, Cape Town & Missouri Botanical Garden.

GOLDBLATT, P. & MANNING, J.C. 2004. Taxonomic notes and new species of the southern African genus *Babiana* (Iridaceae: Crocoideae). *Bothalia* 34: 87–96.

GOLDBLATT, P. & MANNING, J.C. 2007a. A revision of the southern African genus *Babiana*, Iridaceae: Crocoideae. *Strelitzia* 18. South African National Biodiversity Institute, Pretoria and Missouri Botanical Garden, St. Louis, Missouri.

GOLDBLATT, P. & MANNING, J.C. 2007b. New species and notes on *Hesperantha* (Iridaceae) in southern Africa. *Bothalia* 37: 167–182.

GOLDBLATT, P. & MANNING, J.C. 2008. *The Iris family: natural history and classification*. Timber Press, Portland, OR.

GOLDBLATT, P. & MANNING, J.C. 2009. New species of *Geissorhiza* (Iridaceae: Crocoideae) from the southern African winter-rainfall zone, nomenclatural changes, range extensions and notes on pollen morphology and floral ecology. *Bothalia* 39: 123–152.

GOLDBLATT, P. & MANNING, J.C. 2010. New taxa of *Babiana* (Iridaceae: Crocoideae) from coastal Western Cape, South Africa. *Bothalia* 40: 47–53.

GOLDBLATT, P. & MANNING, J.C. 2011. Systematics of the southern African genus *Ixia* (Iridaceae): 3. Sections *Hyalis* and *Morphixia*. *Bothalia* 41: 83–134.

GOLDBLATT, P. & MANNING, J.C. 2012a. Systematics of the hypervariable *Moraea tripetala* complex (Iridaceae: Iridoideae) of the southern African winter rainfall zone. *Bothalia* 42: 111–135.

GOLDBLATT, P. & MANNING, J.C. 2012b. Systematics of the southern African genus *Ixia* (Iridaceae: Crocoideae): 4. Revision of sect. *Dichone*. *Bothalia* 42: 87–110.

GOLDBLATT, P., DAVIES, T.J., MANNING, J.C., VAN DER BANK, M. & SAVOLAINEN, V. 2006. Phylogeny of Iridaceae subfamily Crocoideae based on combined multigene plastid DNA analysis. *Aliso* 22: 399–411.

GOLDBLATT, P., MANNING, J.C., DAVIES, T.J., SAVOLAINEN, V. & REZAI, S. 2004. *Cyanixia*, a new genus for the Socotran endemic *Babiana socotrana* (Iridaceae–Crocoideae). *Edinburgh Journal of Botany* 60: 517–532.

GOLDBLATT, P., MANNING, J.C. & GEREAU, R.E. 2008. Two new species of *Babiana* (Iridaceae: Crocoideae) from western South Africa, new names for *B. longiflora* and *B. thunbergii*, and comments on the original publication of the genus. *Bothalia* 38: 49–55.

HOLMGREN, P.K., HOLMGREN, N.H. & BARNETT, L.C. 1990. *Index Herbariorum. Part. 1: The Herbaria of the World.* New York Botanical Garden, New York.

JACQUIN, N.J. von. 1791 [as 1789]. *Collecteana botanica* vol. 3. Wappler, Vienna.

JACQUIN, N.J. von. 1793. *Icones plantantum rariorum* vol. 2: t. 286. Wappler, Vienna.

KER GAWLER, J. 1804. *Hesperantha radiata*, var. g. *caricina*. Carex-leaved hesperantha. *Curtis's Botanical Magazine* 21: t. 790.

KER GAWLER, J. 1807a. *Babiana sambucina*. Elder flower-scented babiana. *Curtis's Botanical Magazine* 26: sub. t. 1019.

KER GAWLER, J. 1807b. *Babiana stricta* var. *purpurea*. Purple-flowered babiana. *Curtis's Botanical Magazine* 26: t. 1052.

KLATT, F.W. 1882. Ergänzungen und Berichtigungen zu Baker's Systema Iridacearum. *Abhandlungen der Naturforschenden Gesellschaft zu Halle* 15: 44–404.

LAMARCK, J.B.A.P. de. 1789. *Encyclopédie méthodique* vol. 3. Pancoucke, Paris.

LEWIS, G.J. 1959. The genus *Babiana*. *Journal of South African Botany*, Suppl. 3.

MANNING, J.C. & GOLDBLATT, P. 2012. Plants of the Greater Cape Flora. 1. The Core Cape Flora. A conspectus of the vascular plants of the Cape region of South Africa. *Strelitzia: in press*.

McNEILL, J., BARRIE, F.R., BURDET, H.M., DEMOULIN, V., HAWKSWORTH, D.L., MARHOLD, K., NICOLSON, D.H., PRADO, J., SILVA, P.C., SKOG, J.E., WIERSMA, J.H. &

TURLAND, N.J. (eds). 2006. International Code of Botanical Nomenclature (Vienna Code) adopted by the Seventeenth International Botanical Congress, Vienna, Austria, July 1005. Gantner, Liechtenstein. [*Regnum Vegetabile* 146].

RAIMONDO, D., VON STADEN, L., FODEN, W., VICTOR, J.E., HELME, N.A., TURNER, R.C., KAMUNDI, D.A. & MANYAMA, P.A. (eds). 2009. *Red List of South African Plants. Strelitzia* 25. South African National Biodiversity Institute, Pretoria.

RUDALL, P.J. 1995. *Anatomy of the Monocotyledons. VIII. Iridaceae.* Clarendon Press, Oxford.

RUDALL, P.J. & GOLDBLATT, P. 1991. Leaf anatomy and phylogeny of Ixioideae (Iridaceae). *Botanical Journal of the Linnean Society* 106: 329–345.

SALISBURY, R. 1812. On the cultivation of rare plants, etc. *Transactions of the Horticultural Society, London* 1: 261–366.

SCHINZ, H. 1906. Beiträge zur Kenntnis der Afrikanischen-Flora. Liliaceae und Iridaceae. *Bulletin de l'Herbier Boissier*, ser. 2, 6: 711–712.

SCHNITZLER, J., BARRACLOUGH, T.G., BOATWRIGHT, J.S., GOLDBLATT, P., MANNING, J.C., POWELL, M.P., REBELO, T. & SAVOLAINEN, V. 2011. Plant diversification in the Cape. Causes of plant diversification in the Cape biodiversity hotspot of South Africa. *Systematic Biology* 60: 343–357.

SCHUBERT, B.G. 1945. Publication of Jacquin's *Icones plantarum rariorum. Contributions from the Gray Herbarium of Harvard University* 154: 3–23.

THUNBERG, C.P. 1783. *Dissertatio de Ixia*. Edman, Uppsala.

VAHL, M. 1805. *Enumeratio plantarum* vol. 2. Copenhagen.

WILEY, E.O. 1981. *Phylogenetics*. John Wiley & Sons, New York.

WILLDENOW, C.L. 1797. *Species Plantarum* vol. 1. Nauk., Berlin.

11

FSA contributions 20: Asteraceae: Anthemideae: *Inezia*

N. SWELANKOMO*

INTRODUCTION

The genus *Inezia* E.Phillips comprises two species endemic to eastern southern Africa, namely *I. intergrifolia* (Klatt) E.Phillips from Mpumalanga and Swaziland and *I. speciosa* Brusse from Limpopo. The type species *I. integrifolia* was originally described by Klatt (1896) in the genus *Lidbeckia* but Philips (1932) later accommodated it within a new monotypic genus *Inezia* (Table 1). A second species, *I. speciosa* was later described by Brusse (1989). The genus is placed within the subtribe Cotulinae of tribe Anthemideae (Himmelrich *et al.* 2008; Oberprieler *et al.* 2009). *Inezia* is distinguished from *Adenanthellum* and *Hilliardia* by having ovate or pinnatisect leaves and oblong, 4-angled cypselae with bicellular glands scattered on the outer surface. In the previous subtribal classification of Bremmer (1994), *Inezia* was placed within the now sub-assumed subtribe Thaminophyllinae, together with *Adenanthellum* B.Nord., *Lidbeckia* P.J.Bergius and *Thaminophyllum* Harv. Nordenstam (1976, 1987), when describing the genus *Hilliardia* B.Nord., suggested that it had affinities to *Inezia* and *Adenanthellum* on the basis of the similar ray florets with bifid or emarginate limbs and the often reduced perianth tube with a mix of eglandular hairs and bicellular glands on the margins, the distinctive branched venation and minutely papillate upper surface of the perianth limbs, and the linear, flattened, epappose cypselae with ciliate margins and scattered bicellular glands on the outer surface (Figure 1A–D). This relationship is supported by phylogenetic analyses of cpDNA ndhF sequence data (Himmelrich *et al.* 2008). Analyses of nrITS sequence data, however, suggests a close relationship between *Inezia* and *Hilliardia,* but places *Thaminophyllum* rather with *Lidbeckia.* A close relationship between *Lidbeckia* and *Thaminophyllum* was also suggested by Bremer & Humphries (1993) based on the pilose receptacles and myxogenic cells on the cypselae.

In this paper a detailed taxonomic treatment of the two species of *Inezia* is presented, including a key to the species, typification, distribution maps and illustrations.

MATERIAL AND METHODS

All the *Inezia, Hilliardia, Adenanthellum, Thaminophyllum,* and *Lidbeckia* specimens at PRE were studied. Digital images of type collections were, where necessary, accessed on the Aluka digital library (http://plants.jstor.org, accessed July 2011). *Hilliardia zuurbergensis, I. integrifolia, I. speciosa, Lidbeckia pectinata,* and *Lidbeckia quinqueloba* specimens were examined by means of the scanning electron microscope (SEM). All samples were dry and were not chemically treated before being sputter-coated with gold-palladium.

MICROMORPHOLOGY

Inezia is characterised by the presence of large sessile glands on the ovary, cypselae, perianth tube and perianth lobes of both ray and disc florets. SEM study reveals that the large sessile glands are bicellular. The epidermal cells on the upper surface of the rays of *I. integrifolia* are round and radially striate with thick walls (Figure 1A–D).

TAXONOMY

Inezia *E.Phillips* in Kew Bulletin 6: 297 (1932); Bremer & Humphries: 146 (1993); Bremer: 471 (1994); Herman *et al.*:145 (2000). Type species: *I.integrifolia* (Klatt) E.Phillips.

Subshrubs with woody rootstock, stems branching distally, terete, sparsely pilose to villous. *Leaves* alternate, entire or pinnatisect, glandular-pilose. *Capitula* solitary, terminal, radiate, pedunculate. *Involucre* campanulate; *involucral bracts* in 3 or 4 rows, imbricate, ± linear, pilose, inner bracts with membranous tip; *receptacle* conical, epaleate. *Ray florets* female-fertile or sterile, yellow or white, with bicellular glands scattered over the surface; corolla tube short or absent with mix of eglandular hairs and bicellular glands on the margins, limb obovate, 2-toothed, three to four × longer than tube. *Style* branched, truncate, with stigmatic areas lateral; *ovary* with bicellular glands. *Cypselae* linear, flattened, margins ciliate and with scattered bicellular glands on the outer surface, non-myxogenic; *pappus* absent. *Disc florets* bisexual, fertile, yellow with bicellular glands scattered over the surface, narrowly funnel-shaped, corolla tube somewhat compressed, winged, 4-lobed, two lateral lobes ± cucullate, dorsal and ventral lobes ovate. *Stamens* 4, filament collars with 6–8 rows of cells; *anthers* minutely caudate, with ovate to oblong apical appendages; *style* terete, somewhat swollen at

TABLE 1.—Differences between Lidbeckia and Inezia.

Inezia	Lidbeckia
From montane grassland in northeastern southern Africa.	From fynbos in the southwestern Western Cape.
Involucral bracts with scarious margins.	Involucral bracts without scarious margins.
Stylopodium of thick-walled cells in fruit.	Stylopodium of thin-walled cells, large and persistent.
Nectaries in disc florets smaller.	Nectaries conspicuously larger.
5–8 rows of cells in filament collars.	9–10 rows of cells in filament collars.
Cypselae non-myxogenic.	Cypselae myxogenic.

* National Herbarium, Biosystematics Research and Biodiversity Collections Division, South African National Biodiversity Institute, Private Bag X101, Pretoria, 0001. E-mail: n.swelankomo @sanbi.org.za.

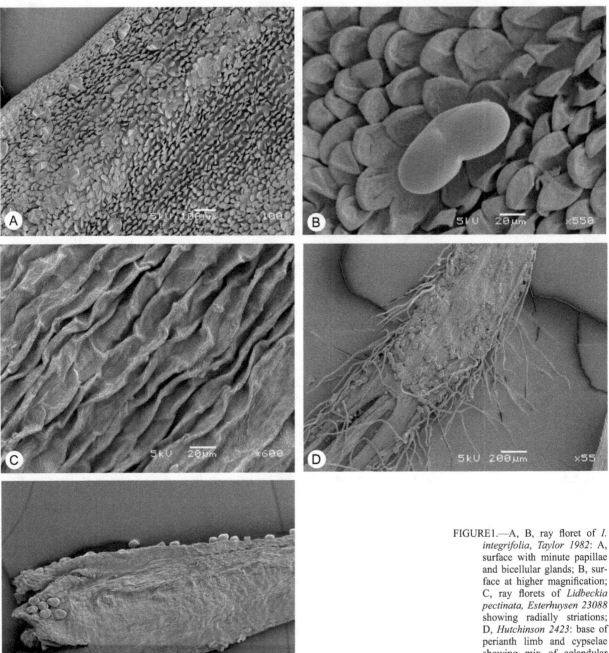

FIGURE1.—A, B, ray floret of *I. integrifolia*, *Taylor 1982*: A, surface with minute papillae and bicellular glands; B, surface at higher magnification; C, ray florets of *Lidbeckia pectinata*, *Esterhuysen 23088* showing radially striations; D, *Hutchinson 2423*: base of perianth limb and cypselae showing mix of eglandular hairs and bicellular glands; E, disc floret of *Lidbeckia pectinata*, *Esterhuysen 23088*: E, surface with bicellular glands.

base, branches oblong, truncate, laterally stigmatic; cells of stylopodium thick walled; *ovary* four ribbed, flattened, bicellular gland scattered throughout. Cypselae oblong, 4-angled, flattened when matured, with bicellular glands scattered on the outer surface, non-myxogenic; *pappus* absent.

2 spp., northeastern South Africa and Swaziland

Etymology: the genus name *Inezia* is named in honour of Ms Inez Clare Verdoorn (1896–1989), appointed to the Division of Botany and Plant Pathology as herbarium assistant in 1917 and later in charge of the National Herbarium as Senior Professional Officer (1944–1951). She is commemorated also in *Aloe verdoorniae* Rey-

nolds, *Senecio verdoorniae* R.A.Dyer, *Teclea verdoorniana* Exell & Mendonça (Glen & Germishuizen 2010).

Key to species

1. All leaves simple; peduncles up to 50 mm long; ray florets yellow; Mpumalanga and Swaziland . . . . 1. *I. integrifolia*
1.' Lower leaves pinnatisect, upper leaves simple; peduncles more than 200 mm long; ray florets white; Limpopo . . . . 2. *I. speciosa*

1. **I. integrifolia** (*Klatt*) *E.Phillips* in Kew Bulletin 6: 297 (1932). *Lidbeckia integrifolia* Klatt: 840 (1896). Type: [Mpumalanga], Saddleback Mountain, 4 000' [1 312 m], Dec.1890, *Galpin 1174*, (PRE, lecto.!, designated here; K—Aluka image!, SAM!, isolecto.). [The

collection by Galpin is chosen here from the three collections cited by Klatt in the protologue as it has more precise collection details.]

Subshrub from woody rootstock, 0.2–0.6 m high, stems branching distally, sparsely pilose to villous. *Leaves* alternate, sessile, ovate, 6.5–22.0 × 3.0–5.5 mm, entire, acuminate, dull green, moderately to densely pilose. *Capitula* solitary, pedunculate, peduncle 20–50 mm long, ribbed, glandular-pilose; *involucre* campanulate, 5–12 mm diam.; *involucral bracts* in 2 or 3 series, imbricate, sparsely to densely hairy, greyish with red-purple tips or rimmed at top with purple; outer bracts linear, 3.5–4.0 × 1.0 mm, moderately to densely pilose, inner bracts oblong with membranous tips, 4–7 × 0.5–1.2 mm; *receptacle* epaleate. *Ray florets* female-fertile, scarcely longer than involucre, yellow with bicellular glands scattered over the surface; corolla tube 1.5 mm long, base of perianth with mix of eglandular hairs and bicellular glands on the margins, limb up to 4.5 mm long, 2-lobed, adaxial epidermal cells round with thick walls and radial striations; *style* 1.0–1.1 mm long, branches 0.1 mm long. *Cypsela* 1.0–1.2 mm long, ciliate along margins, with scattered bicellular glands on the outer suface. *Disc florets* bisexual, fertile, yellow, 1.5–2.0 mm long, bicellular glands scattered over the surface, lobes 0.1 mm long; *style* 1.8–2.0 mm long, branches 0.18–0.20 mm long; *ovary* 1.0 mm long. *Cypsela* 1.2–1.5 mm long, 4-angled and laterally flattened, ciliate along margins, with bicellular glands scattered on the outer surface, non-myxogenic; *pappus* absent. *Flowering time*: Nov.–Apr. (Figures 1A–E).

Distribution and habitat: *Inezia intergrifolia* occurs in Mpumalanga, from around Piet Retief and Komatipoort, north to Pilgrim's Rest, and around Mbabane in Swaziland, from 650–2 152 m (Figure 2). The species favours montane grasslands, especially humus-rich soils.

Notes: Inezia intergrifolia is distinguished from *I. speciosa* by the consistently simple leaves along the entire length of the branch, the shorter peduncles up to 50 mm long, and the smaller involucre 5–12 mm long

with red or purple tipped bracts arranged in 2- or 3 rows, and the yellow ray florets.

Conservation status: LC (Least Concern) (Raimondo *et al.* 2009).

Additional specimens

MPUMALANGA.—**2430** (Pilgrim's Rest): Mount Sheba, (–DC), 14 Feb. 1982, *Brenan 14962* (PRE); Ohrigstad Dam Nature Reserve, (–DC), 16 Feb. 1972, *Jacobsen 2321* (PRE); Pilgrim's Rest on hillsides, (–DD), 12 Mar. 1937, *Galpin 14420* (PRE); Graskop, (–DD), 18 Jan. 1921, *Irvine & Irvine 11421* (PRE); Jan. 1920, *Rogers 23592* (PRE). **2530** (Lydenburg): Mount Anderson, summit peak, (–BA), 9 Mar. 1933, *Galpin 13757* (PRE); Mount Anderson, (–BA), 16 Jan. 1952, *Prosser 1800* (PRE); Vertroosting Nature Reserve, 12 km S of Sabie, (–BB), 25 Feb. 1972, *Muller 2469* (PRE); Nelspruit, Wonderkloof Nature Reserve, (–BC), 18 Nov. 1974, *Elan-Puttick 183* (PRE); Witklip, (–BD), 17 Jan. 1974, *Kluge 444* (PRE); Nelspruit, slopes of Amajuba Mountain (Schagen), (–BD), Dec. 1934, *Liebenberg 3320* (PRE); Kaapsehoop, Devil's Kantoor, (–DB), 11 Jan. 1924, *Pole Evans 998*, (PRE); Barberton, Nelshoogte Forest Station, (–DD), 8 Dec. 1953, *Codd 8132* (PRE); Barberton, Thorncroft Nature Reserve, (–DD), 6 Jan. 1972, *Muller 2274* (PRE); Nelsberg, (–DD), 26 Feb. 1936, *Taylor 1982* (PRE). **2531** (Komatipoort): Sheba Mine, Colombo 365 JU Farm 15 KM. S of Sheba Mine, (–CA), 1980, *Fourie 230* (PRE); mountain top between Louw's Creek and Maid of the Mist, (–CA), 5 Jan. 1929, *Hutchinson 2423* (PRE); Malelane, (–CB), Nov. 1924, *Murphy 106*, (PRE); 10 miles [16.1 km] SE Barberton on road to Havelock, (–CC), 9 Dec. 1953, *Codd 8165* (PRE); Barberton, Oosterbeek Farm, 5 km S Barberton, (–CC), 15 Nov. 1977, *De Souza 628* (PRE); on Barberton-Havelock road, 14 miles [22.4] from Barberton, 9 Feb. 1962, *Ihlenfeldt 2432* (PRE); near top of mountain behind Barberton village, (–CC), 18 Feb. 1931, *Liebenberg 2420* (PRE); Ida Doyer Nature Reserve, 38 km SE Barberton, (–CC), 7 Dec. 1971, *Muller 2050* (PRE); Barberton, (–CC), Jan. 1907, *Thorncroft TRV4973* (PRE); Barberton, Tinie Louw Nature Reserve, (–CC), 15 Jan. 1983, *Venter 9176* (PRE); W of Ngodwana, Hemlock Tulloch Mhor Nature Reserve, (–DA), 15 Feb. 2003, *McMurtry 11157* (PRE). **2730** (Piet Retief): Piet Retief, 15 km from Piet Retief on road to Amsterdam, (–BB), Mar. 1973, *Arnold 245* (PRE).

SWAZILAND.—**2631** (Mbabane): Ngwenya Hills, (–AA), 30 Jan. 1957, *Compton 26527* (PRE); Malolotja Nature Reserve, above Forbes Reef Dam, old kraal, (–AA), 2 Dec. 1986, *Heath 533* (PRE); Endingeni, Middle veld, (–AB), 18 Dec. 1970, *Barrett 551* (PRE); Hhohho, Malandzela area; road to Maphalaleni, 14 km from Nyokane (tar road) above Komati River and below Enkaba Trig beacon, (–AB), 25 Jan. 1994, *Hobson 2075* (PRE); Mbabane, (–AC), Dec. 1905, *Bolus 12012* (PRE); Black Umbuluzi Valley, 9 km from Mbabane to Balegane near mission falls, (–AC), 27 Dec. 1985, *Brusse 4344* (PRE); Mbabane, (–AC), Jan. 1905, *Burtt Davy 2868* (PRE); 17 Jan. 1951, *Compton 22463* (PRE); Mbabane, Ukutula, (–AC), 6 Jan. 1956, *Compton 25291* (PRE); Bomvu Ridge, (–AC), 5 Jan. 1962, *Compton 31193* (PRE); Mbabane Area, (–AC), 7 Dec. 1960, *Dlamini PH31853* (PRE); Usutu Forest, (–CA), Dec. 1974, *Watson 8* (PRE).

2. **I. speciosa** *Brusse* in Bothalia 19: 27–29 (1989). Type: [Limpopo], Iron Crown Mountain near Haenertsburg, 5 500' [1 804 m], *L.E.Codd 9440*, 24 Jan. 1956 (PRE, holo!; K [2 sheets]—Aluka image!, MO—Aluke image!, iso.).

Subshrub up to 0.45 m., main stem simple to 2-branched at base but well branched above, erect, hirsute, sparsely leafy, branches simple, ascending densely leafy, hirsute. *Leaves* dimorphic, lower impari-pinnatisect, upper linear-lanceolate, up to 20 mm long, moderately to densely hirsute, margins involute. *Capitula* solitary, radiate, pedunculate, peduncle 225–290 mm long, ribbed, hirsute; *involucre* campanulate, 20–25 mm long when pressed; *involucral bracts* in 3 or 4 series, imbricate, narrowly lanceolate, brownish-red, hirsute on outer surface, glabrous on inner surface, outer 5.0–6.0 ×1.0–1.5 mm, inner 6.0–9.0×1.0–1.5 mm. *Ray florets* 25–30,

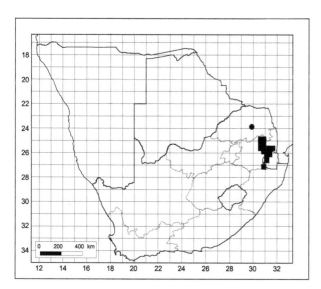

FIGURE 2.—Distribution of *Inezia integrifolia*, ■; and *I. speciosa*, ●.

female or, white with bicellular glands scattered over the surface, corolla tube 1.0 mm long, limb 3-lobed base of perianth limb with mix of eglandular hairs and bicellular glands on the margins, adaxial epidermal cells obtusely acuminate, radially slightly striate; *ovary* not well developed, with scattered bicellular glands. *Disc florets* bisexual, fertile, yellow, 2.5–3.0 mm long, with scattered bicellular glands on the outer surface, lobes 0.4–0.5 mm long; *style* 2.5 mm long, branches 0.6 mm long; *ovary* 1.0–1.5 mm long. *Cypselae* 1.5–2.5 mm long, 4-angled, laterally flattened, ciliate along margins with scattered bicellular glands on the outer surface, non-myxogenic; *pappus* absent. Flowering time: Nov.–Jan.

Distribution and ecology: Inezia speciosa is known only from the type locality near the summit of Iron Crown Mountain near Haenertsburg, 1 804 m (Figure 1), where it has been collected on two occasions. It favours quartzite rock, especially grassy slopes which are rich in forb species with scattered trees and shrubs (Mucina & Rutherford 2006).

Notes: Inezia speciosa can be distinguished from *I. integrifolia* by its dimorphic foliage, with the lower leaves pinnatisect and the upper simple, the longer peduncles more than 200 mm long, the larger involucre 20–25 mm long with bracts in 3 or 4 series, and the white ray florets.

Conservation status: **EN** B1ab (iii) + 2ab (iii) (endangered) (Raimondo *et al.* 2009).

Additional specimen examined

LIMPOPO.—**2329** (Polokwane): Wolkberg, Iron Crown, just W of summit, (–DD), 29 Nov. 1980, *McMurtry 4208* (PRE).

ACKNOWLEDGEMENTS

My colleagues from the South African National Biodiversity Institute: P. Herman for guidance and discussions and Hester Steyn for producing the distribution map. Referees are thanked for their valuable comments. The curator and staff of Harvard University Herbarium and SAM are also thanked for sending digital images of type material.

REFERENCES

BREMER, K. 1994. *Asteraceae: cladistics & classification.* Timber Press, Portland, Oregon.

BREMER, K. & HUMPHRIES, C. 1993. Generic monograph of the Asteraceae–Anthemideae. *Bulletin of the Natural History Museum, Botany Series* 23: 71–177.

BRUSSE, F. 1989. A new species of *Inezia* (Anthemideae) from the north-eastern Transvaal. *Bothalia* 19: 27–29.

GLEN, H.F. & GERMISHUIZEN, G. (compilers). 2010. Botanical exploration of southern Africa, edition 2. *Strelitzia 26.* South African National Biodiversity Institute, Pretoria.

HERMAN, P.P.J., RETIEF, E., KOEKEMOER, M. & WELMAN, W.G. 2000. Asteraceae. In O.A. Leistner, Seed plants of southern Africa: families and genera. *Strelitzia* 10: 101–170. National Botanical Institute, Pretoria.

HIMMELREICH, S., KÄLLERSJÖ, M., ELDENÄS, P. & OBERPRIELER, C. 2008. Phylogeny of southern hemisphere Compositae–Anthemideae based on nrDNA ITS and cpDNA ndhF sequence information. *Plant Systematics and Evolution* 272: 131–153.

KLATT, F.W. 1896. Tribe Anthemideae: *Lidbeckia integrifolia* Klatt. *Bulletin de l'Herbier Boisser* vol 4: 840.

MUCINA, L. & RUTHERFORD, M.C. (eds). 2006. The vegetation of South Africa, Lesotho and Swaziland. *Strelitzia 19.* South African National Biodiversity Institute, Pretoria.

NORDENSTAM, B. 1976. Re-classification of *Chrysanthemum* L. in South Africa. *Botaniska Notiser* 129: 137–165.

NORDENSTAM, B. 1987. Notes on South African Anthemideae (Compositae). *Opera Botanica* 92: 147–151.

OBERPRIELER, C., HIMMELREICH, S., KÄLLERSJÖ, M., VALLÈS, J., WATSON, L.E. & VOGT, R. 2009. Anthemideae. In V.A. Funk., A. Susanna, T.F. Stuessy. & R.J. Bayer, *Systematics, evolution and biogeography of Compositae.* International Association for Plant Taxonomy, Vienna, Austria.

PHILLIPS, E.P. 1932. *Inezia*, a new genus of Compositae from South Africa. *Kew Bulletin*: 297, 298.

RAIMONDO, D., VON STADEN, L., FODEN, W., VICTOR, J.E., HELME, N.A., TURNER, R.C., KAMUNDI, D.A. & MANYAMA, P.A. 2009. Red List of South African plants 2009. *Strelitzia 25.* South African National Biodiversity Institute, Pretoria.

12

The Cape genus *Micranthus* (Iridaceae: Crocoideae), nomenclature and taxonomy

P. GOLDBLATT[1,3], J.C. MANNING[2,3] & R.E. GEREAU[4]

Keywords: Cape flora, Iridaceae, *Micranthus*, new species, nomenclature, taxonomy

ABSTRACT

The genus *Micranthus* (Pers.) Eckl., has traditionally been treated as comprising three species, all with virtually identical, bilaterally symmetric, deep or pale blue to white flowers arranged in crowded, 2-ranked spikes and with divided style branches, but differing in their foliage. Examination of plants in the field and herbarium shows that there are four additional species. *M.* **filifolius** Goldblatt & J.C.Manning, from the Caledon District of the southwestern Western Cape, has up to six, filiform leaves, the blades of at least the lowermost terete and cross-shaped in section, and usually pale blue-mauve flowers. *M.* **simplex** Goldblatt & J.C.Manning from high elevations on Zebrakop, Piketberg, has the smallest flowers in the genus, white but tinged lilac as they age, linear leaves up to 1.5 mm wide, and undivided style branches. *M.* **cruciatus** Goldblatt & J.C.Manning, from the northern Cedarberg and Bokkeveld Mtns, has up to four leaves, the lower with linear or terete blades with heavily thickened margins and central vein and relatively large flowers, unusual in having the style dividing at the mouth of the perianth tube into particularly long branches, these deeply divided as is typical of the genus. *M.* **thereian-thoides** Goldblatt & J.C.Manning, from the Paardeberg south of Malmesbury, is unique in the genus in having flowers with an elongate perianth tube. We also document the occurrence of large populations of putative hybrids at some sites. We provide a complete revision of *Micranthus* with original observations on leaf anatomy, pollen morphology and reproductive biology and discuss its confused taxonomic and nomenclatural history and that of the three common species of the genus, known for over 150 years. In so doing, we neotypify *Gladiolus alopecuroides* L. (1756) [= *Micranthus alopecuroides* (L.) Eckl. (1827)], type of the genus, and choose lectotypes for *M. plantagineus* Eckl. var. *junceus* Baker (1892) and *Gladiolus fistulosus* Jacq. Now with seven species, *Micranthus* remains endemic to the Cape flora region, extending from its extreme northern limit in the Bokkeveld Mtns south-eastwards to Port Elizabeth. We also deal with the genera *Paulomagnusia* Kuntze and *Beilia* Kuntze with which *Micranthus* has sometimes been associated, although both are nomenclatural synonyms of *Thereianthus* G.J.Lewis, a genus close allied to *Micranthus*.

INTRODUCTION

Micranthus (Pers.) Eckl., endemic to the Cape flora region of South Africa, was first recognized as a genus when Ecklon (1827) raised *Gladiolus* subgen. *Micranthus* Pers. to generic rank. He admitted three species to the genus, in this order: *M. plantagineus* Eckl., *M. alopecuroides* (L.) Eckl. (based on *Gladiolus alopecuroides* L.), and *M. fistulosus* Eckl. The names *M. fistulosus* and *M. plantagineus* appear to be implicit references respectively to *Gladiolus fistulosus* Jacq. (1797) and *Ixia plantaginea* Aiton (1789), the latter a superfluous name for *G. alopecuroides*. An indirect reference opposite the genus name 'Watsonia Link', leads to Link (1821), in which *Watsonia plantaginea* Ker Gawl. (1803) is listed, and this work in turn cites *Ixia plantaginea*. We thus treat *M. plantagineus* Eckl. as a new name in *Micranthus* with its type that of *Ixia plantaginea*. We find no such indirect reference for *M. fistulosus* and the name must continue to be regarded as a nomen nudum and therefore invalid.

Until now, *Micranthus* has included just three species (Lewis 1950; Goldblatt & Manning 2000): *M. alopecuroides*, *M. tubulosus* (Burm.f.) N.E.Br. (1929) [with *Gladiolus fistulosus* a heterotypic synonym], and *M. junceus* (Baker) N.E.Br. (1929) [a combination based on *M. plantagineus* var. *junceus* Baker (1892) and a later name for *M. plantagineus* Eckl.]. *Micranthus* (Pers.) Eckl. (1827), itself a later homonym, is conserved against *Micranthus* J.C.Wendl. (1798), a genus of Acanthaceae, with *Gladiolus alopecuroides* L. (*M. alopecuroides* (L.) Eckl.) as its conserved type (Rickett & Stafleu 1959: 241; McNeill *et al.* 2006: 272).

These three species of *Micranthus* are mostly readily distinguished by their leaf morphology: *M. alopecuroides* has plane, ± lanceolate to falcate leaves with an evident main vein (Figure 1A); *M. tubulosus* has two or more short, inflated, terete, hollow (fistulose), falcate leaves, usually half as long as the stem (Figure 1E); and *M. plantagineus* is a taller plant with long, terete, hollow, straight foliage leaves (Figure 1F). All three species have apparently identical, small, mid to deep blue (sometimes described as violet), blue-mauve or occasionally white, bilabiate flowers arranged in congested, 2-ranked spikes subtended by dry, brittle bracts with broad membranous margins. The corms, capsules, and specialized, narrow, 3-sided, elongate seeds are also similar in all three species. Apart from their leaf differences, each species shows a modest preference for a different habitat: *M. alopecuroides* is most often found on sandy ground; *M. tubulosus* on dry, usually shale- or granite-derived soils; and *M. plantagineus* in wet habitats, often in marshes, seeps or along streams, most often in sandy or peaty soils. That said, we have seen two or even all

[1] B.A. Krukoff Curator of African Botany, Missouri Botanical Garden, P. O. Box 299, St. Louis, Missouri 63166, USA. E-mail: peter.goldblatt@mobot.org.

[2] Compton Herbarium, South African National Biodiversity Institute, Private Bag X7, 7735 Claremont, Cape Town. E-mail: j.manning@sanbi.org.za.

[3] Research Centre for Plant Growth and Development, School of Biological and Conservation Sciences, University of KwaZulu-Natal, Pietermaritzburg, Private Bag X01, Scottsville 3209.

[4] Missouri Botanical Garden, P.O. Box 299, St. Louis, Missouri 63166, USA. Email: roy.gereau@mobot.org.

three species growing together locally with only very small habitat differences, if any, so habitat preferences are far from absolute.

Several additional populations of *Micranthus* extend the range of leaf morphology in the genus. Plants at high elevations in the Piketberg (*Goldblatt & Manning 10172*, MO, NBG) have narrow, straight to falcate leaves ± 1 mm wide with one or more strongly thickened veins, small, white flowers fading pale lilac, and undivided style branches. These plants grow in an unu-

sual habitat for *Micranthus*, crevices and shallow pockets of soil on wet sandstone rocks.

A second series of populations (e.g. *Goldblatt 10438* MO, NBG) from the northern Cedarberg and Bokkeveld Mtns has long, slender, linear leaves with a heavily thickened main vein and equally thick margins, thus often cross-shaped in transverse section (Figure 1D). These slender, often tall, plants also stand out in having the style dividing at the mouth of the perianth tube, with unusually long style branches divided for less than a

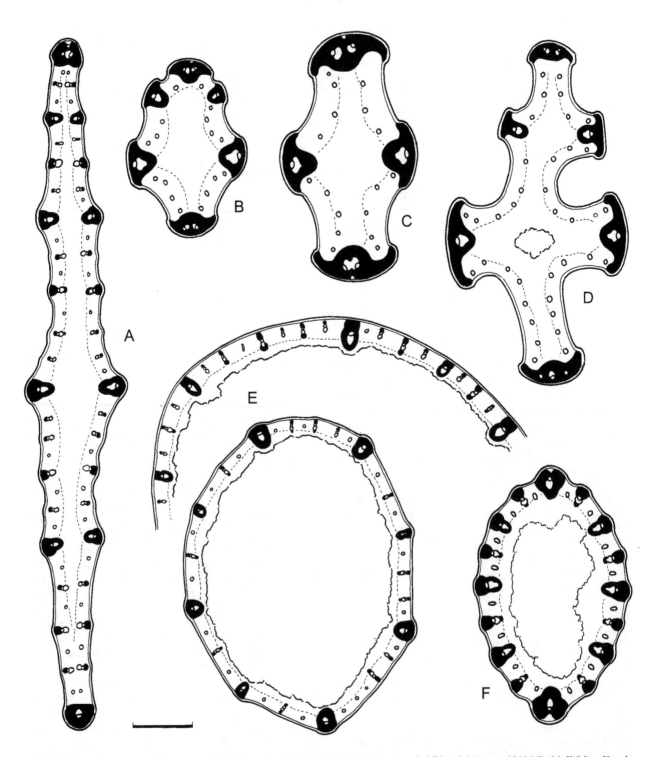

FIGURE 1.—Leaf anatomy in *Micranthus*. A, *M. alopecuroides*, Elandsberg Nature Reserve, *Goldblatt & Manning 13616*; B, *M. filifolius*, Kogelberg, no voucher; C, *M. filifolius*, Drayton, *Goldblatt & Manning 13623*; D, *M. cruciatus*, Pakhuis, *Goldblatt & Porter 13766*; E, *M. tubulosus*, Rondebosch Common, *Goldblatt & Manning 13620*; F, *M. plantagineus*, Drayton, *Goldblatt & Manning 13632*. Scale bar: 500 µm.

quarter of their length. They grow in seasonally marshy sites in peaty soil, often in moss on sandstone pavement, and appear to flower particularly well after fire.

A third series of populations from the Caledon District of Western Cape has up to 6 leaves with linear to filiform blades, often cross-shaped in section (Figure 1B, C), with the bases persisting in a well-developed fibrous neck. These plants grow on stony, loamy clay or sandy soils, usually in well drained sites that are dry at flowering time.

Lastly, a population from the Paardeberg south of Malmesbury, only discovered in 2012, has hollow leaves reminiscent of those of *Micranthus tubulosus* but is unique in the genus in having dark blue flowers with an elongate perianth tube, 20–22 mm long, thus more than twice as long as in any other species of *Micranthus*.

Consistent treatment of the genus suggests that these divergent populations should logically be recognized as separate species. The circumscriptions of the existing species cannot be expanded to accommodate these plants. We describe these new species as *Micranthus cruciatus* Goldblatt & J.C.Manning, *M. filifolius* Goldblatt & J.C.Manning, *M. simplex* Goldblatt & J.C.Manning and *M. thereianthoides* Goldblatt & J.C.Manning. Other variants, which we believe are interspecific hybrids, occur locally and we discuss these below. One of them, evidently *Micranthus plantagineus* × *M. tubulosus*, is particularly common at the foot of the Elandskloof Mtns. Plants have a flexuose stem, terete, hollow leaves and short spikes of up to 10 flowers, and appeared at first to be a separate species, so different were they from their putative parents.

We review the nomenclature of *Micranthus*, choose types for the two species currently lacking designated types, and present a systematic revision, thus dealing with collections that do not accord with the current circumscriptions of the three species included in the genus. We also deal with *Paulomagnusia* Kuntze (1891). When described, *Paulomagnusia* included two species, one a *Micranthus* and the other, *P. spicatus* (L.) Kuntze, now the type species of *Thereianthus* G.J.Lewis (1941). Our revision includes new observations on leaf anatomy, pollen morphology, and reproductive biology and pollination; these presented following the generic description and nomenclature.

TAXONOMIC HISTORY AND RELATIONSHIPS

Micranthus is most closely allied to *Thereianthus*, also endemic to the Cape flora region. Lewis (1950) first pointed out an unusual, specialized feature shared by the two genera, namely that the lowermost foliage leaf is inserted on the flowering stem as it is in *Lapeirousia* Pourret, also in tribe Watsonieae Klatt, rather than on the corm. This means that the corm tunics are formed solely from the cataphylls, without any contribution from the leaf bases as is found in *Watsonia* Mill. and some other members of the tribe. Molecular systematic studies of plastid DNA sequences confirm the immediate relationship of the two genera, which together are sister to *Watsonia* plus *Pillansia* L.Bolus (Reeves *et al.*

2002; Goldblatt *et al.* 2008), with *Lapeirousia* (*sens lat.*) retrieved as member of a second clade of the tribe, which includes *Cyanixia* Goldblatt & J.C.Manning and *Savannosiphon* Goldblatt & Marais. The close relationship of *Micranthus* and *Thereianthus* is reflected in their largely shared taxonomic and nomenclatural history.

Although *Micranthus* was maintained by most authors dealing with the genus since its recognition at generic rank by Ecklon (1827), species now recognized as *Thereianthus* have had a more chequered history, beginning with Ecklon (1827), who placed two species of that genus in '*Beilia*', then lacking a validating description. Although species of *Micranthus* and *Thereianthus* had first been referred respectively to *Ixia* L. or to *Gladiolus* L., they were included by Ker Gawler (1804) in *Watsonia*, largely because they share divided style branches with that genus. Their nomenclature subsequently became intertwined. Heynhold (1847) included the two species of *Thereianthus* known at that time in *Micranthus* as *M. spicatus* (L.) Heyn. and *M. triticeus* (Thunb.) Heyn. The British botanist and specialist in the taxonomy of Iridaceae, Baker (1877), recognized *Micranthus* in its current sense and included one species of *Thereianthus* in *Watsonia* unranked *Beilia* Eckl. ex Baker, as *W. punctata* (Andrews) Ker Gawl. (now *Thereianthus bracteolatus* (Lam.) G.J.Lewis). Baker (1892) later formalized *Watsonia* unranked *Beilia*, then with several species, as *Watsonia* subgen. *Beilia* (Eckl. ex Baker) Baker. His German contemporary, Klatt (1882), completely misunderstood the situation, and in his worldwide account of the Iridaceae, he included two species of *Thereianthus* in *Micranthus,* as *M. spicatus* (L.) Klatt (evidently referring to what is now *T. spicatus* (L.) G.J.Lewis) and *M. triticeus* (Burm.f.[sic]) Klatt [he was evidently unaware of combinations in *Micranthus* for these species by Heynhold in 1847; we also assume that the basionym attributed to Burman fil. was an error for Thunberg, as *Ixia triticea* Burm.f. is a very different species, currently *Tritoniopsis triticea* (Burm.f.) Goldblatt]. Klatt (1882) also included one species of *Thereianthus, T. juncifolius* (Baker) G.J.Lewis, in *Anomatheca* Ker Gawl. as *A. calamifolia* Klatt, and several more in *Watsonia* unranked *Beila*. Usually astute, Klatt apparently made no reference to any species we now regard as belonging to *Micranthus* but partly corrected the error when he recognized *M. plantagineus* with one variety, var. *junceus* (Klatt 1894). *Thereianthus spicatus*, however, remained in *Micranthus*.

Kuntze (1891) included both what are now *Micranthus alopecuroides* and *Thereianthus spicatus* in his new genus *Paulomagnusia*, evidently intended as a *nomen novum* for the later homonym, *Micranthus* (Pers.) Eckl. 1827 (non *Micranthus* J.C.Wendl. 1798). This genus has now been conserved, although not against *Paulomagnusia*. Kuntze (1898) validated Ecklon's '*Beilia*' at generic rank as *Beilia* Kuntze, and included in the genus only *B. spicata* (L.) Eckl. ex Kuntze, which is thus its type. Unfortunately *Beilia* is superfluous because Kuntze listed the valid *Paulomagnusia* in synonymy. It remained for Lewis (1941), over a century after Ecklon (1827) used the invalid '*Beilia*', to erect the valid genus *Thereianthus* in which she placed the two species of Ecklon's '*Beilia*' and several more then included in

Watsonia. Thereianthus now has 11 species (Manning & Goldblatt 2011).

SYSTEMATICS

Micranthus *(Pers.) Eckl.*, Topographisches Verzeichniss der Pflanzensammlung von C.F. Ecklon: 43 (1827), name conserved, non J.C.Wendl. (1898, Acanthaceae). *Gladiolus* subg. *Micranthus* Pers.: 46 (1805). *Hebea* subg./unranked *Micranthus* (Pers.) R.Hedw.: 24 (1806). Type (conserved): *Gladiolus alopecuroides* L. (= *M. alopecuroides* (L.) Eckl.).

Paulomagnusia Kuntze: 702 (1891). Type: *P. alopecuroides* (L.) Kuntze [= *Micranthus alopecuroides* (L.) Eckl.)], lectotype designated by Goldblatt & Manning: 133 (2008).

Note: although Persoon's (1805) infrageneric taxa appear at first to be unranked, the preface (ix) to his *Synopsis* has the following statement: *Melius autem judicavi, eas species (nonnullis forte tamen excipiendis) ab aliis leviter in charactere aberrantes, imprimis si genus minus amplum sit, sub divisione peculiari aut SUBGENERE, quo etiam nonnula Botanicorum recentium genera relata sunt, comprehendere, ne ultra necessitatem genera multiplicentur.* This is a clear statement that his infrageneric taxa are subgenera. [We have judged it better to include those species (with some exceptions) that in their character(s) are only slightly different from others, especially if the genus is not very large, under the 'particular division' or subgenus (which are some of genera of recent botanists), so that genera are not multiplied beyond necessity.]

Deciduous geophytes. *Corm* axillary in origin, subglobose, rooting from below; tunics coarsely fibrous. *Leaves* few, the lower 2 or 3 cataphylls, lowermost foliage leaf longest, inserted on stem above corm, blades either plane with a definite main vein and falcate or lanceolate with margins moderately to heavily thickened, or ± tubular and hollow, or terete and ± solid with heavily thickened central vein and margins separated by narrow longitudinal grooves. *Stem* erect and straight or ± flexuose, simple or few- to several-branched. *Inflorescence* a congested, 2-ranked spike, usually weakly rotated; *bracts* short, overlapping, with leathery or dry central portion and broad membranous margins, inner forked apically and shorter than to ± as long as outer. *Flowers* zygomorphic, lasting several days, blue to violet, mauve, white or flushed lilac, scentless or pleasantly scented, with nectar from septal nectaries; *perianth tube* short, curving outward, ± cylindric below, flaring in upper half; *tepals* ± equal, dorsal slightly larger and arching over stamens, lower tepals extended ± horizontally. *Stamens* unilateral and arcuate; filaments slender, free; anthers oblong, held under the dorsal tepal, splitting longitudinally. *Ovary* ovoid, sessile; style branches slender, usually deeply divided and recurved, or barely notched at apex. *Capsules* woody, small, narrowly ovoid-ellipsoid or urn-shaped with ovules in lower fourth. *Seeds* 2–4(5) per locule, 3(4)-sided below, elongate, widest at micropylar end with micropylar crest and micropyle above base, tapering and pointed at chalazal end, surface slightly wrinkled. *Basic chromosome number x* = 10.

The diagnostic features of *Micranthus* are the crowded, 2-ranked spike; small, bilaterally symmetric, tubular flowers; distinctive dry outer floral bracts with broad membranous margins; basal leaf inserted on the stem above the level of the corm (shared with *Lapeirousia* and *Thereianthus*); and small, narrow capsules, each locule containing up to four slender seeds almost as long as the locules and with a micropylar crest at the proximal end. *Micranthus* is unique among subfamily Crocoideae in having zonasulcate pollen grains (Figure 2). The sulci are distal as seen at the tetrad stage (S. Nilsson, pers. comm. Oct. 1996) and the zonasulcate condition in *Micranthus* is thus derived from the basic monosulcate grain by extension of the sulcus until it encircles the grain. Exine sculpturing is reticulate, grading to tectate-perforate close to the aperture margin. Among Crocoideae, only a few species of *Thereianthus* also have reticulate exine sculpturing (Manning & Goldblatt 2011). Most genera of Crocoideae, including *Thereianthus*, have sulcate pollen grains, with a pair (sometimes solitary) of narrow bands of exine (elongated opercula) lying parallel to one another along the long axis of the aperture. Other more complex apertures are known in *Geissorhiza* Ker Gawl. (Goldblatt & Manning 2009). *Cyanixia* and *Zygotritonia* have trisulculate grains. All these pollen types have tectate-perforate exine with small supratectal spinules.

Leaf marginal anatomy in species with plane leaves conforms to the norm for Watsonieae in combining unspecialized marginal epidermal cells and a marginal vein with a sclerenchyma cap below the epidermis. This condition prevails in *Watsonia* and the *Lapeirousia* clade (excluding *L. corymbosa* (L.) Ker Gawl. and its immediate allies), but notably not in *Thereianthus* or *Pillansia*, both of which lack a marginal vein or sclerenchyma strand below the unspecialized marginal epidermis (Goldblatt & Manning 1990; Rudall & Goldblatt 1991; Goldblatt *et al.* 2004; Manning & Goldblatt 2011).

Chromosome cytology: the basic chromosome number for *Micranthus* is $x = 10$. One population each of the four species counted, namely *M. alopecuroides*, the new *M. filifolius* (reported as *M. junceus*), *M. plantagineus* (as *M. junceus*) and *M. tubulosus*, are diploid, $2n = 20$ (Goldblatt 1971; Goldblatt & Takei 1997). The base number and karyotype, consisting of one long and nine short chromosome pairs, are matched exactly in *Thereianthus*. The related genus *Watsonia* has $x = 9$ and a derived karyotype with two long chromosome pairs. *Pillansia*, the fourth and last genus of this lineage of Watsonieae, also has $x = 10$, with its single species tetraploid, $2n = 40$ (Goldblatt 1977; not 44 as originally published by Goldblatt 1971).

Reproductive system, compatibility and pollination: virtually nothing has been reported about the reproductive system in *Micranthus*, but we infer that self-incompatibility and compatibility are important in the evolution and distribution of the genus. It is notable that three species, *M. alopecuroides*, *M. plantagineus* and *M. tubulosus* typically have all flowers producing a full complement of capsules and we infer self-compatibility and facultative autogamy for these species. In contrast, *M. filifolius* and *M. thereianthoides* exhibit lower capsule production and we infer self-incompatibility for these species. We are unable to infer compatibility relations

FIGURE 2.—Pollen morphology of *Micranthus plantagineus*, Piketberg, *Manning 2093*. Scale bar: 20 μm.

for *M. cruciatus* and *M. simplex* as good fruiting material is not available. Significantly, the putatively self-compatible species *M. alopecuroides*, *M. plantagineus* and *M. tubulosus* have the widest ranges in the genus, with *M. plantagineus* occurring over the entire range of the genus. *M. filifolius* has a modest range, entirely within the Caledon District of Western Cape but *M. cruciatus*, *M. simplex* and *M. thereianthoides* are local endemics, the latter two currently known from only one or few populations. Self-incompatibility is believed to be ancestral for Iridaceae (Goldblatt & Manning 2008). In genera that we have studied, we have found relatively few species to be facultatively autogamous, and we assume such species are specialized. Reversals from self-compatibility to incompatibility are believed to be unlikely. Thus in *Micranthus* we infer that the self-compatible *M. alopecuroides*, *M. plantagineus* and *M. tubulosus* are derived for this character. The latter two species are also derived in their hollow leaves as outgroup comparison indicates that plane, isobilateral leaves are the plesiomorphic condition (present in most members of the family and universal in sister genus *Thereianthus*). Reproduction through aerial cormlets that replace flowering on the spike axis is also known only in these three species. The arrangement of species in our account reflects our belief that the self-incompatible species are closer to the ancestral stock of *Micranthus*.

The small flowers of all species except *Micranthus thereianthoides* are so similar in size, shape and colour (perianth tube 3–5 mm long) that they almost certainly share the same generalist pollination ecology, though we have only recorded insect visitors for *M. alopecuroides*,

M. plantagineus and *M. tubulosus*. Goldblatt & Manning (2006) regarded these three species as having a generalist pollination strategy and insect visitors to these species include large-bodied bees (Apidae), bee-flies (Bombyliidae), hopliine beetles (Scarabaeidae: Hopliini) and butterflies. Among the latter are *Pieris helice* (Pieridae) (*M. plantagineus*) and *Cynthia cardui* and *Colias electo* (Pieridae) (*M. tubulosus*). New observations confirm to the generalist pattern with anthophorine bees and wasps including *Delta* cf. *caffra* (Eumenidae), a species of Sphecidae visiting *M. plantagineus*. The longer perianth tube of *M. thereianthoides*, 22–25 mm long, suggests a specialized pollination system using a long-proboscid pollinator, possibly a long-proboscid fly species or large butterfly. The nectar reward, retained in the lower part of the perianth tube, is only accessible to pollinators with a proboscis at least 18 mm long, thus excluding access to smaller butterflies, bees, bee-flies and wasps that visit flowers of other *Micranthus* species.

Key to the species

Note: plants with the lower part of the spike bearing smaller, paler floral bracts subtending one or more small cormlets may be *Micranthus junceus* or *M. tubulosus* or may be hybrids involving these two species or with *M. alopecuroides* and are not accommodated in the key.

1a Leaf blades either plane with evident central vein or one or two prominently thickened veins, or terete and 4-grooved with thickened margins, not hollow; style variously dividing between mouth of perianth tube to opposite middle of anthers:

2a Lowermost leaves plane and linear to lanceolate or falcate, (2–)4–12 mm wide; style branches divided ± halfway . . . . 1. *M. alopecuroides*

2b Lowermost leaves either plane and linear or terete and cross-shaped in section, up to 2 mm wide; style branches divided up to halfway or undivided:

3a Flowers white, outer tepals tipped palest lilac, fading slightly darker lilac; style branches undivided or barely notched at apex . . . . 2. *M. simplex*

3b Flowers pale blue, blue-mauve or deep blue or white; style branches divided up to halfway:

4a Style dividing at mouth of perianth tube opposite middle of filaments; style branches ± 2.5 mm long, divided less than one third; leaves at least 1.5 mm wide . . . . 6. *M. cruciatus*

4b Style dividing between base and middle of anthers; style branches usually ± 1.0 mm (up to 1.5 mm) long, divided up to halfway; leaves < 1 mm wide . . . . 5. *M. filifolius*

1b Leaf blades tubular and hollow, round or ± compressed in section; style dividing opposite base to middle of anthers (rarely just below anther bases):

5a Perianth tube elongate and ± twice as long as bracts, 22–25 mm long; bracts 8–15 mm long . . . . 3. *M. thereianthoides*

5b Perianth tube shorter than bracts, < 10 mm long; bracts 5–7 mm long:

6a Flowering stem ± flexed at each aerial node; stem with prominent, coarsely fibrous collar around base . . . . *M. plantagineus* × *M. tubulosus*

6b Flowering stem straight, stiffly erect; stem with or without collar of fibres around base:

7a Blade of lowermost leaf terete or oval in section, ±
   straight, green at flowering and smooth when fresh
   (with evident thickened veins when dry); stem without collar of fibres around base; capsules narrowly
   ovoid to ± urn-shaped . . . . 7. *M. plantagineus*

7b Blade of lowermost leaf tubular and inflated, falcate
   with prominent apical mucro, often dry at flowering, without prominent veins when alive or dry; stem
   with collar of fibres around base; capsules narrowly
   ovoid . . . . 4. *M. tubulosus*

1. **Micranthus alopecuroides** *(L.) Eckl.*, Topographisches Verzeichniss der Pflanzensammlung von
C.F. Ecklon: 43 (1827). *Gladiolus alopecuroides* L.:
5 (1756). *Ixia alopecuroides* (L.) L.f.: 92 (1782). *Ixia
plantaginea* Aiton: 59 (1789), nom. illeg. superfl. pro
Gladiolus alopecuroides L. [see note 1]. *Paulomagnusia alopecuroides* (L.) Kuntze: 702 (1891).Type: South
Africa, [Western Cape], Somerset West, 19 Nov. 1944,
Barker 3384 (NBG, neo., here designated; PRE isoneo.)
[see note 2].

Watsonia compacta Lodd.: t. 1577 (1830),
nom. nud.

Plants mostly 200–450 mm high, base usually
sheathed with collar of short fibres. *Corm* mostly 10–12
mm diam., tunics of dark brown, relatively coarse, reticulate fibres, drawn into short bristles above. *Stem* usually simple or 1- or 2(3)-branched, when unbranched
often with one or more scales below base of spike, usually bearing 1 or more cormlets in axil of lowermost
foliage leaf. *Leaves* 2–4(5), lowermost 1 or 2 plane,
broadly to narrowly falcate (occasionally ± lanceolate)
or linear, (2–)5–10(–15) mm wide, with moderately
prominent main vein; margins slightly or occasionally
heavily thickened (*De Vos 2288*), hyaline when dry;
upper 1 or 2(3) leaves largely sheathing. *Spike* mostly
40–80-flowered, often much congested, with internodes 1.5–3.0 mm long; bracts 5–7 mm long, outer with
broad to narrow brown centre and translucent membranous margins, inner ± as long as outer, notched apically, translucent with 2 dark veins slightly broader
toward base; lower or all nodes sometimes vegetative
and then bracts paler in colour and subtending one (or
more) cormlets in each axil. *Flowers* usually dark blue,
sometimes pale blue, often lower third to fourth of tepals
paler blue or white, distally edged with a thin darker
blue line, unscented; perianth tube ± 5 mm long; tepals
subequal, elliptic, 7–8 × ± 3 mm, with short narrow,
claw-like base. *Stamens* with filaments ± 5 mm long,
diverging in upper half; anthers oblong, 3–4 mm long,
pale mauve; pollen white to pale blue. *Style* ± 7 mm
long, mostly dividing between upper third of filaments
and lower third of anthers; branches ± 1.2–1.6 mm long,
divided for ± half their length. *Capsules* oblong to narrowly ovoid, ± 5 mm long but ± 4 mm long when dry.
Seeds angular-elongate, 3.5–4.0 mm long, 3 or 4 per locule. *Flowering time*: October in the north, November to
December in the south.

Distribution: centred in the southwestern Western
Cape, *Micranthus alopecuroides* has a relatively narrow
range, extending from the Cape Peninsula north into the
Olifants River Valley and east locally to Hermanus and
Swellendam (Figure 3). Plants typically grow on well-
drained clay or loamy, seasonally wet, slopes and flats

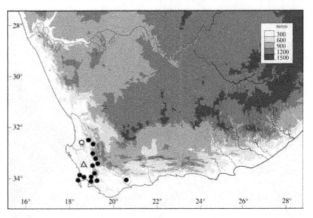

FIGURE 3.—Distribution of *Micranthus alopecuroides*,
•; *M. simplex*, ○; *M. thereianthoides*, Δ.

but have also been recorded on sandy ground. The Olifants River Valley populations grow in thin clay or
sandy gravel, often over rocky pavement that is totally
dry even before flowering commences.

Diagnosis: *Micranthus alopecuroides* is distinctive
in its plane (Figure 1A), sometimes very broad basal
leaves; the blades lanceolate to linear or falcate, sometimes up to 12 mm wide or exceptionally to 15 mm in
plants from the Roman's River area of the upper Breede
River Valley. Exceptions are numerous and there are collections with ± linear leaves 3–5 mm wide and only up
to 25 mm long (notably *Purcell s.n.*, NBG). The flowers
are typical of the genus, usually dark blue, with a perianth tube ± 5 mm long. The spikes of 40 or more flowers
are often unusually congested with the internodes ± 1.5
mm long.

Populations from the Olifants River Valley and flowering in October, at least three weeks earlier than elsewhere, stand out in their relatively lax spikes with internodes 2.5–3.0 mm long (vs. ± 1.5–2.0 mm elsewhere)
and relatively short, straight leaves, 5–8 mm wide with
particularly prominent mucronate tips. The outer bracts
of these plants also differ from those in populations to
the south in their broader translucent margins, thus with
a significantly narrower central band of green tissue
(brown when dry). These plants are typically restricted
to thin clay or light sandy soils over rocky pavement and
represent a distinctive race of the species.

Hybrids: certain collections from sandy flats south
of Malmesbury constitute a puzzle. They consist of
plants with abnormally elongated spikes up to 120 mm
long, more than 3/4 of their length bearing small, pale
bracts each enclosing not a flower but a small cormlet.
Only the top fourth of the spikes have properly formed,
dark brown bracts subtending either pale blue flowers (e.g. *Goldblatt & Manning 10431*, MO, NBG, with
tubular leaves) or deep blue flowers (*Goldblatt & Manning 10432*, MO, NBG, with plane leaves). The leaves,
either tubular or plane with a central vein, correspond to
Micranthus tubulosus or *M. alopecuroides* respectively.
We conclude that these plants constitute hybrids or a
hybrid swarm with *M. alopecuroides* as one parent and
M. tubulosus or possibly *M. plantagineus* as the other.
Microscopic examination of the pollen shows some
apparently normal grains and others smaller than normal

and evidently sterile. The available collections of this putative hybrid were made too early in the life cycle to have capsules, which if developed, would have appeared later in the season.

Similar specimens from other sites show the same striking feature (e.g. *Goldblatt 8711* MO, from Greyton; *Leighton 722* BOL from Camp Ground, Rondebosch; and *Williams 1195* MO, NBG from near Vogelgat, Hermanus). This last collection consists of plants with the sterile part of the spikes 180–250 mm long and the fertile part 20–30 mm long. Populations at Elandsberg near Bo-Hermon, with plane, narrowly lanceolate leaves (e.g. *Goldblatt & Manning 13616*) have the inflorescence sterile throughout and bearing a cormlet in all bract axils as do many individuals of the species from Rondebosch Common, Cape Town (*Goldblatt & Manning 13619*). The status of these plants is uncertain but we suspect them to have a hybrid origin.

[*Note 1.* Daniel Solander, the unacknowledged author of *Hortus kewensis* published under William Aiton's name (1789), described the new species *Ixia plantaginea* 'foliis linearibus strictis, spica disticha imbricata', based on a collection of Francis Masson and, for reasons that are obscure, at the same time cited Linnaeus's *Gladiolus alopecuroides* in synonymy. The epithet *plantaginea*, alluding to the similarity of the inflorescence to that of *Plantago* L., is no more apt than Linnaeus's recalling the resemblance to the grass, *Alopecurus* L., and constitutes an illegitimate superfluous name. Nevertheless, the broad-leaved species remained known by the later epithet *plantagineus*, until well into the 20th century (e.g. Lewis 1950) despite the leaves being described as narrow and linear in the protologue. The epithet *plantagineus* was applied to *M. alopecuroides* by, among others, Ker Gawler (1803), who evidently did not realize it applied to two different species. Baker (1892, 1896), who also used the name *M. plantagineus* for *M. alopecuroides*, compounded this error and recognized *M. plantagineus* var. *junceus* not realizing that the type of the species was in fact identical with his new variety.]

[*Note 2.* Described in 1756 by Linnaeus as *Gladiolus alopecuroides*, with the brief diagnosis 'foliis linearibus, spica disticha imbricata,' the species was transferred to *Micranthus* by Ecklon (1827), when he raised Persoon's *Gladiolus* subg. *Micranthus* to generic rank. Of the three sheets identified as *G. alopecuroides* in the Linnaean herbarium, one [LINN 59.13] is a Sparrman collection post-dating the protologue, and the other two cannot be unambiguously related to the name. One [LINN 59.15] is *M. tubulosus* and the other [LINN 59.14] may be *M. alopecuroides* but is atypical in its large size, numerous branches and particularly broad leaves that hardly accord with the protologue [leaves linear]. We prefer to choose a neotype: *Barker 3384*, which has relatively narrow leaves and conforms exactly to the protologue. This action unambiguously preserves the current application of the name to the plane-leaved species (Lewis 1950; Goldblatt & Manning 2000).]

[*Note 3. Gladiolus minutiflorus* Schrank (1822) has been associated with *Micranthus alopecuroides*, which Schrank also recognized (as *Gladiolus*), but the description is vague (flowers small, secund, tepals subequal)

and we are unable to determine the plant to genus with confidence, let alone to species. Schrank did, however, explicitly describe the leaves as short, striate and narrow, the lower ± 5 inches (125 mm) long. No authentic material has been located either at the Munich (M) or Brussels (BR) Herbarium, the institutions where the types of Schrank's species, where they exist, are believed to be located.]

Representative specimens

WESTERN CAPE.—**3218** (Clanwilliam): clay hillside S of Algeria turnoff on Clanwilliam–Citrusdal road (N7), (–BD), 13 Oct. 1974, *Goldblatt 3030* (MO). **3220** (Wuppertal): near Citrusdal on old Clanwilliam road, (–CA), 11 Oct. 1984, *Bean & Viviers 1504* (BOL); N of Citrusdal, (–CA), 16 Oct. 1935, *Taylor 1224* (BOL); clay slope near Farm Klawervlei on road to Algeria, (–CA), 11 Oct. 2011, *Goldblatt & Porter 13864* (MO, NBG, PRE). **3318** (Cape Town): Cape Peninsula, Wynberg Hill, (–CD), Nov. 1950, *Pillans 10208* (BOL, MO); Wynberg Hill, Edinburgh Drive, (–CD), 25 Jan. 2011 (fr.), *Goldblatt & Manning 13631* (MO, NBG); fields near Cape Town, (–CD), Aug.–Nov., *H. Bolus 2829* (BOL); Devil's Peak above Vredehoek, clay slopes, (–CD), 30 Oct. 1982, *Goldblatt 6637* (MO); slopes of Lions Head, (–CD), 20 Nov. 1938, *Penfold s.n. SAM53159* (SAM); Jonkershoek, Bosboukloof, (–DD), 27 Nov. 1973, *Smith 140* (NBG); Jonkershoek Valley, (–DD), 27 Nov. 1975, *Kruger 84* (NBG). **3319** (Worcester): Grootwinterhoek, (–AA), without date, *Pappe s.n. SAM21101* (SAM); Mostertshoek, (–AC), 8 Dec. 1973, *De Vos 2288* (NBG); slopes at Wabooms R., foot of Waaihoek Peak, (–AD), 11 Dec. 1948, *Esterhuysen 14822* (BOL); Bo-Hermon, Elandsberg Nature Reserve, entrance to Bosplaas, (–AC), 22 Jan. 2011 (fr. and sterile); *Goldblatt & Manning 13616* (MO, NBG); Wemmershoek, (–CC), 2 Nov. 1947, *Barker 4903* (BOL, NBG). **3418** (Simonstown): Cape Peninsula, Bergvliet Farm, E of sandpit, (–AB), 22 Nov. 1918, 5 Dec. 1818, *Purcell s.n.* (NBG); Helderberg, Somerset West, (–BB), 2 Dec. 1944, *Parker 3959* (BOL, NBG). **3419** (Caledon): Elgin Basin, Arieskraal, well drained clay ground, (–AA), 5 Dec. 1994, *Rode & Boucher 0207* (NBG); Elgin, (–AA), 19 Nov. 1944, *Barker 3369* (NBG). **3420** (Bredasdorp): Swellendam, Bontebok Park, (–AB), 2 Nov. 1965, *Grobler 552* (NBG).

2. **Micranthus simplex** *Goldblatt & J.C.Manning*, sp. nov.

TYPE.—Western Cape, 3218 (Clanwilliam): Piketberg, southwestern slopes of Zebrakop, (–DB), shallow soil on sandstone pavement, 4 Jan. 1995, *Goldblatt & Manning 10172* (NBG, holo; K, MO, PRE, iso.).

Plants (100–)140–200 mm high, base sheathed by short collar of brittle fibres. *Corm* tunics of dark brown, reticulate fibres. *Stem* usually simple, rarely 1-branched. *Leaves* (2)3, plane, linear or falcate, ± 1 mm wide, usually with 1 or 2 prominent veins, margins thickened, hyaline when dry. *Spike* 16- to 40-flowered; bracts purple-brown with broad translucent, brown-flecked membranous margins, ± 5 mm long, inner bracts ± as long as outer, membranous with 2 dark keels, notched at apex. *Flowers* white fading to lilac, outer tepals tipped pale lilac, with subapical brown ridge on reverse; perianth tube ± 3 mm long; tepals oblong, ± 4 × 1.2 mm. *Stamens* with filaments ± 2.5 mm long; anthers oblong, ± 2.5 mm long. *Style* ± 7 mm long, dividing opposite middle of anthers; branches ± 1 mm long, barely notched at apex. *Capsules* oblong, slightly warty in distal half, ± 4 mm long. *Seeds* elongate-angular, ± 3 mm long. *Flowering time*: December to at least mid-January.

Distribution: known only from the slopes of Zebrakop, highest peak in the Piketberg, *Micranthus simplex*, like *M. cruciatus*, grows in shallow soils in moss or in rock crevices on wet sandstone rocks (Figure 3). The habitat remains moist as late as January when the species blooms.

Diagnosis: unusually small for the genus, stems of *Micranthus simplex* rarely exceed 180 mm and the white flowers with lilac-tipped outer tepals are distinctive, other species having flowers in shades of deep to pale blue or blue-mauve, or occasionally white. The inflorescence has the appearance of being relatively lax, the bracts of the lower flowers of the spike not overlapping those above them, but the upper bracts are as closely set as in other species. It is one of two species of *Micranthus* with consistently plane leaves; the other, *M. alopecuroides*, is a taller plant with congested spikes of 40 to 80 flowers and broader leaves mostly 5–12 mm wide. The flowers of *M. simplex* are the smallest in the genus, the perianth tube just 3 mm long and the short anthers ± 2.5 mm long. The short, undivided style branches, ± 1 mm long, are likewise unusual for *Micranthus*, other species of which normally have the style branches somewhat to considerably longer and divided for at least one third their length.

Additional specimens

WESTERN CAPE.—**3218** (Clanwilliam): Piketberg, Zebrakop, (–DB), in moist sand, 800 m, 3 Jan. 1973, *Linder 193* (BOL).

3. Micranthus thereianthoides *Goldblatt* & *J.C.Manning*, sp. nov.

TYPE.—Western Cape, 3318 (Cape Town): Paardeberg, Vondeling, (–DB), rock cracks and sands along stream, 9 Jan. 2013, *Nicolson 995* (NBG, holo.; K, MO, iso.).

Plants 300–800(–1200) mm high, base weakly sheathed by fine fibres. *Corm* 10–15 mm diam., tunics of fine to moderately coarse, dark brown, reticulate fibres. *Stem* simple or rarely branched, with solitary cormlet in axil of second leaf and sometimes also third leaf. *Leaves* (4)5 or 6, green or drying at flowering, lowermost 2 or 3 longest, blades 100–300(–800) mm long, 2.5–5.0 (–15) mm diam., tubular and hollow, sometimes inflated, acute-mucronate, upper leaves progressively shorter and narrower, uppermost bract-like and entirely sheathing. *Spike* 10- to 40(–70)-flowered, bracts brown with broad translucent membranous margins, 8–11(–15) mm long, as long as 1.5–2.0 spike internodes, inner bracts slightly shorter than outer, forked apically, membranous with 2 dark keels broadened toward base. *Flowers* suberect, dark violet or purple, unscented; perianth tube ± cylindric, 22–25 mm long, tepals oblong, 5–6 × 1.5–2.5 mm, reverse of outer tepals with prominent subapical ridge. *Stamens* with filaments 6–8 mm long, exserted ± 3 mm; anthers oblong, 3.5–4.0 mm long. *Style* 24–27 mm long, dividing between middle and slightly beyond anthers, branches ± 1.5(–2.0) mm long, divided for ± half their length. *Capsules* ovoid, smooth, 5–6 mm long, with ± 4 seeds per locule. *Seeds* elongate-angular, tapering to points at both ends, ± 3.5 mm long. *Flowering time*: January. Figure 4.

Distribution: a highly local endemic, *Micranthus thereianthoides* is restricted to the Paardeberg near Malmesbury (Figure 3), where it grows at mid to upper altitudes along the banks of seasonal streams, the corms usually wedged among granite rocks, sometimes in humic loam, where the plants are more robust. Plants are locally plentiful along several streams on the range. The long-tubed, violet flowers are evidently adapted to pollination by long-proboscid flies. The incomplete fruit set in wild plants suggests that *M. thereianthoides* is an obligate outcrosser.

Diagnosis: *Micranthus thereianthoides* closely resembles *M. plantagineus* and some forms of *M. tubulosus* in its cylindrical leaves but is unique in the genus in the relatively large floral bracts, 8–11 mm long, and most strikingly in its dark violet flowers with elongate, cylindrical perianth tube 22–25 mm long, thus ± twice as long as the bracts (Figure 4). The species appears never to develop cormlets in the floral bract axils.

The long-tubed flowers suggest the genus *Thereianthus*, but the hollow leaves, the small, obtuse tepals, and the bracts with broad, membranous margins are characteristic for *Micranthus*. The zonasulcate pollen grains with reticulate exine conform exactly to those of other species of *Micranthus*, leaving no doubt as to its generic placement.

This extraordinary species was discovered in January 2012 by local plant enthusiasts Greg Nicholson and Dewan Roets during a botanical survey of the Paardeberg.

Additional specimens

WESTERN CAPE.—**3318** (Cape Town): Paardeberg, between Wellington and Malmesbury, Paardeberg Nature Reserve next to Malmesbury Dam, (–DB), rocky crevices near water, 10 Jan. 2012, *Nicolson & Roets 788* (NBG); Vondeling, (–DB), Feb. 2012 (fruiting), *Nicolson 994* (MO, NBG).

4. Micranthus tubulosus *(Burm.f.) N.E.Br.* in Bulletin of Miscellaneous Information, Royal Botanic Gardens, Kew 1929: 133 (1929). *Gladiolus tubulosus* Burm.f.: 2 (1768). *Ixia cepacea* Basseporte ex DC. in Redouté: t. 96 (1804), nom. nov. in *Ixia*, non *I. tubulosa* Burm. f. (= *Babiana tubulosa* (Burm.f.) Ker Gawl.). Type: South Africa, without precise locality or collector (G: Herb. Burman, holo., image seen).

Gladiolus fistulosus Jacq.: 8 (1797). *Ixia fistulosa* (Jacq.) Sims: t. 523 (1801), hom. illegit. non Andrews (1799) [= *Hesperantha radiata* (L.f.) Ker Gawl.]. *Micranthus fistulosus* (Jacq.) Eckl. ex Baker: 179 (1892), nom. superfl. pro *G. tubulosus* Burm.f. Type: South Africa, without precise locality or collector, illustration in Jacq.: t. 16 (1797), left hand plant, lectotype designated here.

Watsonia spicata Sol. ex Ker Gawl.: sub t. 553 (1803). nom. superfl. pro *G. spicatus* L. (1753), *G. tubulosus* Burm.f. (1768) et *G. fistulosus* Jacq. (1797). [The citation *Watsonia spicata* (L.) Ker Gawl. in Annals of Botany (König & Sims) 1: 229 (1804) is an error]. Type: South Africa, without precise locality, illustration in Curtis's Botanical Magazine 15: t. 523 '*Ixia fistulosa*' (1801).

Ixia teretifolia Herb. Banks ex Sims: t. 523 (1801), nom. nud. pro syn.

FIGURE 4.—*Micranthus thereianthoides*, Paardeberg, *Nicolson 995* (NBG). A, flowering plant; B, flower; C, outer (left) and inner (right) bract; D, capsule; E, seed. Scale bar: A, 10 mm; B–D, 2.5 mm; E, 1.25 mm. Artist: John Manning.

Micranthus fistulosus Eckl.: 44 (1827), nom. nud. [Probably intended as a combination but basionym not cited.]

Plants (70–)150–350(–600) mm high, base sheathed by collar of short, stiff, bristly fibres. *Corm* 14–18 mm diam., tunics of coarse, dark brown, reticulate fibres. *Stem* simple or branched, occasionally with cormlets in axil of lowermost leaf and of uppermost cataphyll. *Leaves* (2)3–5, usually dry at flowering, lowermost 1 or 2 longest, blades 50–200 mm long (to 300 mm in sterile plants), 4–7 mm diam., inflated, tubular and hollow, apex obtuse-mucronate, upper leaves progressively shorter and narrower. *Spike* 16- to 40-flowered; bracts mid to dark brown, 5–6 mm long, outer with broad translucent membranous margins, inner bracts ± as long as outer, forked apically, membranous with 2 dark keels broadened toward base; lower nodes sometimes vegetative with one or more cormlets and bracts then pale. *Flowers* pale or dark blue or white, sweetly scented; perianth tube 5–6 mm long; tepals oblong, 5–6(–10) × 1.2–2.5 mm, reverse of outer tepals with prominent subapical ridge. *Stamens* with filaments 6–8 mm long; anthers oblong, 3.5–4.0 mm long. *Style* 7–8 mm long, dividing opposite or slightly below base of anthers, branches ± 1.5(–2.0) mm long, divided for ± half their length. *Capsules* ovoid, smooth, 4–5 mm long, with 3 or 4 seeds per locule. *Seeds* elongate-angular, tapering to points at both ends, ± 3 mm long. *Flowering time*: November to December.

Distribution: typically a species of lower slopes usually on clay and granite-derived soils but also on sandstone, *Micranthus tubulosus* is restricted to the western half of Western Cape. It extends north of the Cape Peninsula as far as the northern Cedarberg, where an early (1923) collection documents its occurrence at Heuningvlei, and no further east of the Peninsula than Suurbraak near Swellendam and the Agulhas Peninsula (Figure 5). Like other species of the genus, it blooms late in the season when the hollow, inflated leaves are often dry and brown. Plants from the Pakhuis Mtns growing in moist, sandy ground are exceptional in their small size (leaves up to 100 mm long) and require additional study. A much dwarfed fragment of *Micranthus tubulosus*, said to be from Garies (*Caporn s.n.*, ex hort. Kirstenbosch (as Nat Bot Gard. 915/15) in BOL) is unlikely to be from there as no other records of the genus from Namaqualand exist.

Diagnosis: the inflated, tubular, falcate leaves (Figure 1E) are diagnostic for the species, the spikes and flowers of which differ hardly at all from those of *Micranthus alopecuroides*. A particularly distinctive feature of the leaves is the prominent brown mucro at the obtuse to ± truncate apices. The leaves are often ± dry at flowering time—Marloth's (1915: plate 41) has a particularly apt illustration of the species. As in *M. plantagineus*, one or more cormlets may be produced in the lower axils of the spike, a phenomenon first noted by Ker Gawler (in Sims 1801) and later confirmed by Lewis (1950). The condition is more frequent, although not consistent, in *M. plantagineus*. Despite its apparently preferred habitat on relatively dry slopes, *M. tubulosus* can occasionally be found on moist sandy flats, sometimes co-occurring with *M. alopecuroides* and *M. plantagineus* (*Wurts 519*

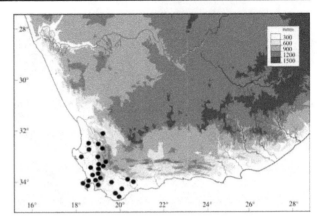

FIGURE 5.—Distribution of *Micranthus tubulosus*.

consists of just such a mixture, *M. tubulosus* and *M. plantagineus* evidently found growing in close proximity). Hybrids between these two species at shared sites blur their usually clear foliar differences. We discuss putative hybrids between *Micranthus tubulosus* and *M. alopecuroides* or *M. plantagineus* in more detail below.

Plants collected near Saron (e.g. *Schlechter 10618*) are unusually small, mostly 100–150 mm but some just 70 mm tall, and have shorter, fewer-flowered spikes than usual. They appear linked to taller, more robust specimens by a range of intermediates. In contrast, plants from Gouda (*Barker 9861*), nearby, are exceptionally robust, up to 600 mm tall, with leaves almost as long, and the white flowers have tepals 10 mm long, the outer 2.5 mm wide.

A curious feature of *Micranthus tubulosus* is that populations may consist of a mixture of some plants with entirely fertile spikes and others with the lower part of the spike sterile (e.g. *Ecklon & Zeyher Irid 192* and *190*). All three specimens of *Goldblatt 8711* and several of *Purcell 43* have spikes sterile in the lower half. Particularly short leaves in *Goldblatt 8711* are also puzzling but not unique.

History: Long known as *Micranthus fistulosus* (Jacq.) Eckl. (e.g. Baker 1896), based on *Gladiolus fistulosus* Jacq. (1797), that combination was in fact not valid, though it was used as *M. fistulosus* Eckl. (a nomen nudum assumed to be a valid combination) by Baker (1892, 1896). By citing the basionym, Baker's use of the name *M. fistulosus* becomes a valid (albeit unintended) combination, also superfluous through his citing of valid earlier synonyms, including *Gladiolus tubulosus* Burm.f. (1768). Jacquin's illustration of *G. fistulosus* has two plants: we designate as lectotype the left hand one, which has dark blue flowers and leaves typical of *M. tubulosus*. The right hand plant, which has pale blue flowers and the upper leaf more typical of *M. plantagineus*, may be a hybrid with that species. The sterile lower nodes of the spike, bearing silvery bracts, are more typical of *M. plantagineus* and represent at least a different genotype from the right hand plant.

Brown (1929) identified the type of *Gladiolus tubulosus* among specimens in Burman's herbarium, and realizing that it was an earlier name for *M. fistulosus*, provided the combination *M. tubulosus*. A fine illustration of the species in Redouté's *Les Liliacées* (1804), as *Ixia cepacea*, a name coined by the artist, Madeleine Françoise Basseporte, shows that *M. tubulosus* was cultivated in France in the late 18th and early 19th centuries. The Basseporte painting, which indeed represents *M. tubulosus*, is part of an unpublished collection of vélins (paintings on parchment), which document plants and animals in the Jardin Royal in Paris and the Ménagerie Royale in Versailles, now in the Bibliothèque Centrale of the Muséum National d' Histoire Naturelle in Paris. We treat the name as having been validated by De Candolle in 1804 in the Redouté volume. De Candolle's citation of the earlier *Gladiolus tubulosus* Burm.f. appears to render his epithet superfluous, but the name *Ixia tubulosa* Burm.f. (now *Babiana tubulosa* (Burm.f.) Ker Gawl.) prevents transfer of *Gladiolus tubulosus* to *Ixia*. *Ixia cepacea* must be regarded as legitimate and a new name in *Ixia* for *G. tubulosus*.

Hybrids: the following interspecific hybrids are known involving *Micranthus tubulosus*: *M. tubulosus* × *M. alopecuroides* and *M. tubulosus* × *M. plantagineus*. Those with *M. plantagineus* can form large populations locally, probably backcrossing with one other parental species. We discuss these separately at the end of the species account.

Additional specimens

WESTERN CAPE.—**3218** (Clanwilliam): Goedverwacht, Piketberg, (–DC), 23 Nov. 1982, *Koutnik 1037* (BOL, MO); Piketberg, top of Versfeld Pass, (–DC), 2 Nov. 2011 (in bud), *Goldblatt & Porter 13707* (MO, NBG). **3219** (Wuppertal): Cedarberg, Heuningvlei, sandy vlakte, (–AA), 23 Oct. 1923, *Pocock 586* (NBG); Olifants River near Villa Brakfontein [Citrusdal], (–?CA), Nov., *Ecklon & Zeyher Irid 190* (MO, SAM); Elandskloof, bridge ± 10 miles [± 16 km] SE of Citrusdal, (–CA), 21 Sept. 1952 (sterile), *Maguire 1832* (NBG). **3318** (Cape Town): near Hopefield, (–AB), 19 Oct. 1932, *Lavis s.n.* (BOL); Malmesbury, clay slope, (–BC), 3 Nov. 1986, *Goldblatt 8048* (MO); Cape Town, Camps Bay, (–CD), *Barker 7191* (NBG); Signal Hill, (–CD), Nov. 1939, *Lewis 707* (SAM); Rondebosch Common, dry, hard ground, (–CD), 24 Jan. 2011, *Goldblatt & Manning 13620* (MO, NBG); Wynberg Hill, (–CD), Nov. 1922, *L. Bolus s.n. (BOL17188)*; Langverwacht, Kuils River, main kloof, (–DC), 14 Dec. 1973, *Oliver 4820* (NBG); Stellenbosch Mtn, S of Paradys Kloof, (–DD), 3 Dec. 1989, *Buys 132* (NBG); Muldersvlei, (–DD). Nov. 1916, *Duthie 352* (BOL). **3319** (Worcester): near Saron, 800' [244 m], (–AC), Oct. 1896, *Schlechter 10618* (MO); Elandsberg Estate, Vangkraal road near Mountain road, (–AC), 22 Jan. 2011 (fr.), *Goldblatt & Manning 13610* (MO); Gouda, (–AC), 6 Dec. 1962 (white flowers, very robust), *Barker 9861* (NBG); Worcester [District], Waterfall, (–CC), Nov., *Ecklon & Zeyher Irid 192* (MO); Ceres, Schurfdeberg, lower slopes, (–AD), Dec. 1944, *Lewis 863* (SAM). **3320** (Montagu): Swellendam, hill below Eleven O'Clock Mtn, (–CD), 25 Nov. 1952, *Wurts 519* (mixed with *M. plantagineus*) (NBG). **3418** (Simonstown): Bergvliet Farm, flats near sand pit, (–AB), Nov. 1915, *Purcell 43* (SAM); Helderberg Nature Reserve, (pale blue or white), (–BB), 23 Dec. 1993, *Runnals 647* (NBG). **3419** (Caledon): Greyton Nature Reserve, dry flats, 1000' [305 m], 5 Dec. 1987, *Goldblatt 8711* (MO). **3420** (Bredasdorp): Suurbraak, Middelplaas, (–BA), 5 Dec. 1982 (sterile), *Viviers 274* (NBG); Struisbaai to Elim near Springfontein turnoff, hard sandy gravel, (–DB), 9 Nov. 2011 (in bud), *Goldblatt & Porter 13738* (growing with *M. plantagineus*) (MO, NBG); 4 km W of Elim (growing with *M. filifolius*), (–DA), 9 Nov. 2011, *Goldblatt & Porter 13738* (MO, NBG). Unknown locality: Leeufontein, burned veld, 28 Nov. 1908, *Pearson 3185* (BOL).

5. **Micranthus filifolius** Goldblatt & J.C.Manning, sp. nov.

TYPE.—Western Cape, 3419 (Caledon): Akkedisberg Pass, sandy hillside, (–AC), 18 Nov. 2011, *Goldblatt & Porter 13370* (NBG, holo.; MO, PRE, iso.).

Plants 180–300 cm high, base sheathed with sparse to well-developed collar of fibres. *Corm* 12–16 mm diam., tunics of relatively soft, fine or thicker fibres. *Stem* unbranched or rarely with single short branch, without cormlets in leaf axils. *Leaves* 4–6, green or beginning to dry from tips at flowering time, lowermost longest, reaching to middle of spike to shortly exceeding it, blade either ± terete and ± 1 mm diam. or ± plane and ± 2 mm wide, with heavily thickened central vein and margins, separated when dry by narrow longitudinal grooves, upper leaves shorter, with sheaths overlapping, uppermost 1 or 2 leaves sheathing for most of their length, with short free tips. *Spike* mostly 18–50-flowered, closely congested, lower bracts always subtending flowers; bracts mid to dark brown, ± 5 mm long, outer with broad translucent membranous margins, inner slightly shorter than outer, with 2 dark keels broadened toward base, notched apically. *Flowers* pale mauve or mid-blue, unscented; perianth tube ± 5 mm long; tepals oblong, ± 5 × 2.2–2.8 mm. *Stamens* with filaments ± 5 mm long, exserted ± 2.5 mm; anthers oblong, ± 3 mm long. *Style* ± 7 mm long, dividing between base and middle of anthers; branches 1.0–1.6 mm long, divided for up to half their length, rarely only notched at apex. *Capsules* narrowly ovoid, smooth, 4–6 × ± 2 mm, with up to 4 seeds per locule. *Seeds* elongate-angular, mostly 3-sided, tapering to points at both ends, 3–5 mm long. *Flowering time*: mid-November to late February. Figure 6.

Distribution: centred in the Caledon District of Western Cape, *Micranthus filifolius* is largely coastal with populations recorded from Steenbras and Cape Hangklip eastward to Hermanus and inland to Shaw's Mtns, the lower slopes of Caledon Swartberg and east to Akkedisberg Pass and Elim (Figure 7). Collections are mostly from clay and clay-loam soils, occasionally from sandy sites, but even collections from the Klein River Mtns above Hermanus at elevations of up to 400 m are from a shale band. The species is particularly abundant after fire (e.g. *Drewe 495, 1101*) but will flower in unburned veld unless shaded out by taller vegetation. Plants bloom unusually late in the season, with most flowering collections made after mid-January, and two (*Gillett 520*; *Levyns 11269*) were in mid- to late February. We have confirmed late flowering at near-coastal sites ourselves but inland populations, as from Drayton Siding, east of Caledon, and Akkedisberg Pass, flower from mid-November to early January and are in fruit before any coastal populations come into flower. We suggest that this early flowering is due to warmer and drier conditions well inland of the coast. The coarser corm tunic fibres and collar of fibres around the base of the stems in these populations are perhaps adaptations to the drier habitat. Plants sometimes co-occur with or grow close to *M. plantagineus*, which is in fruit when *M. filifolius* begins to bloom, two or three weeks after the last flowers of *M. plantagineus* have faded, both at the coast and at inland sites. We have also found *M. filifolius* growing together with *M. tubulosus*.

Diagnosis: with its narrow leaves, the lowermost of which is linear or terete (Figure 1B & C), *Micranthus filifolius* is most like *M. plantagineus* in general appearance. It differs, however, from that species in several respects, particularly in the solid leaf blades of the

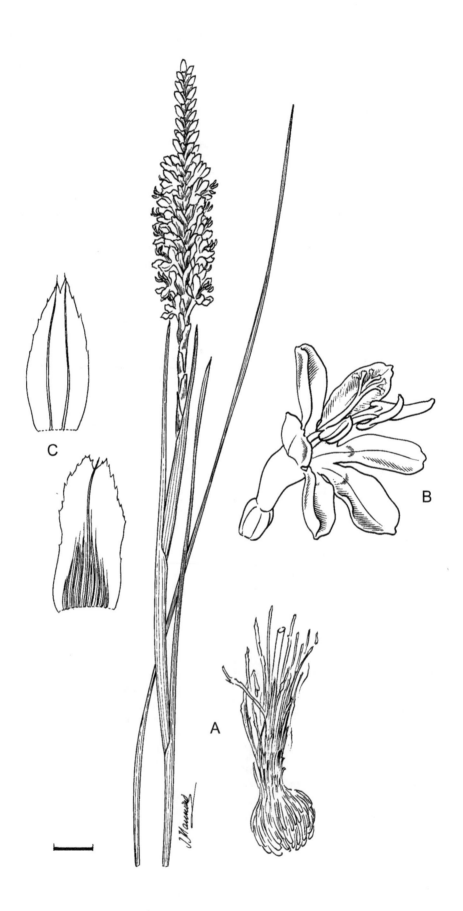

FIGURE 6.—*Micranthus filifolius*, Akkedisberg Pass, *Goldblatt & Porter 13370* (NBG). A, flowering plant; B, flower; C, outer (lower) and inner (upper) bract. Scale bar: A, 10 mm; B, C, 1.25 mm. Artist: John Manning.

lowermost and sometimes the upper leaves. The leaves closely overlap one another and sheath the stem up to the base of the spike. Plants typically have four or five leaves but several specimens have six leaves, the upper one or two largely to entirely sheathing. Unlike *M. plantagineus*, plants do not produce cormlets in the leaf axils and the corm tunics are usually soft-textured with the base enclosed by a collar of fibres, whereas *M. plantagineus* has coarser corm tunics, lacks a basal collar of fibres, and cormlet production is conspicuous in the lowermost and sometimes other leaf axils. In addition, the flowers, although typical of *Micranthus*, are somewhat smaller than in *M. plantagineus,* having a perianth tube mostly 4–5 mm long and tepals ± 4 mm long, and are more often pale mauve (drying white), although the western populations are dark blue (drying blue), the flower colour in most populations of *M. plantagineus*. Typical *M. plantagineus* has been recorded close to most localities of *M. filifolius*, flowering in November and December, supporting our conclusion that *M. filifolius* is not a local variant but a different species, flowering later, sometimes in the same habitats as *M. plantagineus*, or in drier sites.

Additional specimens

WESTERN CAPE.—**3418** (Simonstown): Steenbras, (–BB), Sept. 1944 (late fr.), *Stokoe s.n. SAM68012* (SAM); Buffels River dam area [near Rooiels], (–BD), 19 Feb. 1972, *Boucher 1822* (NBG, PRE); Cape Hangklip, peaty marsh, (–BD), *Levyns 10220* (BOL); Kogelberg Nature Reserve, (–BD), 19 Mar. 1983, *Kroon 10200* (PRE); Betty's Bay, sandy slopes, (–BD), 13 Feb. 1962, *Levyns 11269* (BOL). **3419** (Caledon): Drayton Siding, pale blue (–AB), 16 Dec. 1968, *Goldblatt 395* (BOL); field E of Drayton siding, (–BA), 25 Jan. 2011, *Goldblatt & Manning 13623* (MO, NBG); Kleinmond, near Palmiet River mouth, (–AC), 31 Jan. 1933, *Gillett 615* (NBG); Kleinmond, road to reservoir, (–AC), 27 Jan. 1947, *De Vos 485* (NBG); Hermanus, (–AC), Jan. 1920, *Burtt Davy 18711* (BOL); top of Shaw's Pass, (–AD), Jan. 1957, *Lewis 2904* (SAM); Shaw's Pass, east side, (–AD), 29 Dec. 1955, *Lewis 4454* (SAM); Hemel-en-Aarde, mountain side, (–AD), 15 Jan. 1933, *Gillett 520* (NBG); Vogelgat, Hermanus, Vogelpool to Fernkloof, S slopes on shale band, (–AD), 2 Jan. 1979, *Williams 2710* (NBG); Fernkloof, Hermanus, 350 m, clay area, 1 year after fire, (–AD), 11 Jan. 1987, *Drewe 495* (MO); 400 m, shale band, after fire, 25 Jan. 1996, *Drewe 1101* (MO); 4 km W of Elim, stony clay in renosterveld, (growing with *M. tubulosus*), (–DA), 9 Nov. 2011 (in bud), *Goldblatt & Porter 13745* (MO, NBG).

6. **Micranthus cruciatus** *Goldblatt & J.C.Manning*, sp. nov.

TYPE.—Western Cape, 3219 (Wuppertal): Pakhuis Mts, trail to Heuningvlei, (–AA), local in wet seep on rocky sandstone slope, 19 Dec. 1995, *Goldblatt 10438* (NBG, holo.; MO, iso.).

Plants 300–450 mm high. *Corm* globose, 8–10 mm diam., tunics of brown, soft membranous layers not accumulating. *Stem* simple or 1-branched, usually with 1 or 2 small cormlets in lowermost leaf axil. *Leaves* (3)4 or 5, lower 3 linear to subterete, ± 1.5 mm wide, margins and midrib heavily thickened with narrow longitudinal grooves between (often cross-shaped in section with 4 narrow longitudinal grooves), reaching to base or middle of spike, uppermost leaf sheathing stem almost to base of spike, with short free portion. *Spike* up to 70-flowered; bracts brown or straw-coloured with broad translucent membranous margins, ± 4 mm long, inner ± as long as outer, membranous with 2 dark keels, apically notched. *Flowers* pale blue-mauve (drying ±

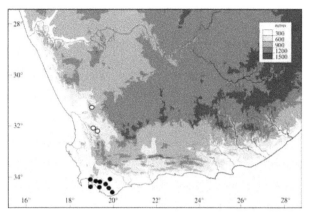

FIGURE 7.—Distribution of *Micranthus filifolius*, ●; *M. cruciatus*, ○.

white) or deep blue, perianth tube ± 3 mm long, tepals oblong, ± 4 × 1.5 mm. *Stamens* with filaments ± 5 mm long; anthers oblong-linear, ± 3 mm long. *Style* ± 4 mm long, dividing ± at mouth of tube opposite middle of filaments, branches ± 2.5 mm long, divided for ± one third their length. *Capsules* oblong, smooth, 5.0–5.5 mm long. *Seeds* elongate-angular, mostly 3-sided, tapering to points at both ends, ± 3 mm long. *Flowering time*: mid-November to late December, possibly lasting into January.

Distribution: restricted to the northern Cape flora region, *Micranthus cruciatus* is known from the northern Cedarberg immediately south of Pakhuis Pass and in the Bokkeveld Mtns southwest of Nieuwoudtville (Figure 7). Plants grow on rocky slopes, in seeps on thin sandy soil over sandstone pavement, flowering in December as the habitat dries out in the hot weather. No doubt the species is rare but the very few collections are probably due to its midsummer flowering when little plant collecting is undertaken. We suspect that *M. cruciatus* occurs in suitable sites between its few stations, thus in the southern Bokkeveld Mtns and the Gifberg/Matsikamma Mtn complex and perhaps elsewhere in the Cedarberg. First collected by the late Elsie Esterhuysen in 1941 according to available records, *M. cruciatus* has elicited no attention until now and was assigned to the broadly similar *M. plantagineus* (as *M. junceus*) in herbaria.

Diagnosis: linear- to terete-leaved *Micranthus cruciatus* is immediately distinguished by its solid, narrow leaf blades, ± 1.5 mm wide, with heavily thickened veins and margins separated by narrow longitudinal grooves (Figure 1D). Leaves are either linear or terete becoming cross-shaped in section distally with only the margins and central vein thickened. Plants broadly resemble *M. plantagineus* although they are more slender than is usual in that species, which has hollow leaves 2–3 mm diam. and is conspicuous in the production of cormlets in the lowermost and sometimes upper leaf axils. In contrast, *M. cruciatus* has no more than one or two small cormlets, these borne in the axil of the lowermost leaf. The pale mauve-blue or sometimes dark blue flowers are typical of the genus in shape but notable in the short perianth tube, ± 3 mm long, in the style dividing at the mouth of the perianth tube and in the unusually long style branches up to 2.5 mm long, divided for up to one third their length. Most species of *Micranthus* have the style dividing opposite the base to middle of the anthers and style branches typically less than 1.6 mm long. The narrow, heavily thickened leaf in *M. cruciatus* is convergent with that in *M. filifolius*, but in other critical details the two are very different, the latter with coarsely fibrous corm tunics and a collar of fibres around the base of the stem.

It is noteworthy that typical, hollow-leaved *M. plantagineus* with dark blue flowers also occurs in the Pakhuis and Bokkeveld Mtns (e.g. *Leipoldt 3596* BOL, NBG, PRE) as well as in the Cedarberg, but it has not been recorded growing near *M. cruciatus* and they evidently have somewhat different habitat preferences.

Additional specimens

NORTHERN CAPE.—**3119** (Calvinia): sandstone slope between Nieuwoudtville and Vanrhyns Pass on road to Keyserfontein, (–AD), 27 Nov. 1985, *Goldblatt 7399* (MO, PRE).

WESTERN CAPE.—**3218** (Clanwilliam): Zandfontein, Farm Verkeerde Vley (Klip-op-mekaar), 12 km N of Pakhuis Pass, (–BB), July 2013 (fr.), *Helme 7778* (NBG). **3219** (Wuppertal): Cedarberg, Pakhuis to Heuning Vlei (–AA), 28 Dec. 1941, *Esterhuysen 7436* (BOL); Pakhuis Pass, Kliphuis campsite, wet seep on sandstone pavement, (–AA), 15 Nov. 2011 (in bud), *Goldblatt & Porter 13766* (MO, NBG).

7. **Micranthus plantagineus** Eckl., Topographisches Verzeichniss der Pflanzensammlung von C.F. Ecklon: 43 (1827), nom. nov. pro *Ixia plantaginea* Aiton: 59 (1789), nom. illeg. superfl. pro *Gladiolus alopecuroides* L. *Watsonia plantaginea* Ker Gawl.: t. 553 (1803), nom. nov. pro *Ixia plantaginea* Aiton et nom. illeg. superfl. pro *G. alopecuroides* L. *Gladiolus plantagineus* Pers.: 46 (1805), nom. nov. pro *Ixia plantaginea* Aiton et nom. illeg. superfl. pro *G. alopecuroides* L. Type: South Africa, without precise locality, *Masson s.n. BM922008* (BM, holo.!— narrow-leaved specimens mounted with *Nelson 1777* with broad, flat leaves [= *M. alopecuroides*]).

Phalangium spicatum Burm.f.: 3 (1768), nom. nud. [cited illustration, Plukenet: t. 310, f. 1 (1694) lacks text or figure analysis; it probably represents *Ixia scillaris* L.; specimen in G: Herb. Burman is *Micranthus* and designated the 'type'.]. *Micranthus spicatus* (Burm.f.) N.E.Br.: 138 (1929), nom. inval.

Phalangium spicatum Houtt.: 115 (1780). Type: South Africa, without precise locality or collector, illustration in Houtt., Nat. Hist. ed. 2, 12: t. 80 f. 2 (1780).

Micranthus plantagineus var. *junceus* Baker: 179 (1892). *Micranthus junceus* (Baker) N.E.Br.: 138 (1929). Type: South Africa, [Western Cape], Groenekloof and vicinity, *Zeyher 1611* (K, lecto!, designated here, K000320508; PRE!, isolecto.; other collections numbered *Zeyher 1611* in PRE and SAM are from Klipfontein or Tulbagh, thus not type material).

Plants 200–400(–650) mm high, base without collar of fibres. *Corm* globose, 12–15 mm diam., tunics of dark brown, medium-textured, reticulate fibres drawn into fine points above. *Stem* erect, simple or with up to 9 short branches, with cormlets in axil of lowermost leaf and sometimes of upper cataphyll and rarely other leaf axils. *Leaves* (2)3(4), green at flowering, lower-most leaf longest, blades terete or oval in section, hollow, 2.0–3.5 mm diam., smooth when fresh with translucent veins, when dry, veins appearing thickened with homologue of marginal vein pair more prominent, usually reaching to middle of spike to shortly exceeding spike, uppermost 1 or 2 leaves sheathing for most of length, with free part often slightly longer than sheath. *Spike* (16–)40–100-flowered, lower bracts sometimes subtending cormlets; bracts mid- to dark brown, ± 6 mm long but slightly smaller if subtending cormlets, outer with broad translucent margins, apices sharply acute and ultimately curved outward, inner bracts ± as long as outer, with 2 dark keels broadened toward base. *Flowers* usually dark blue, occasionally pale blue or white, evidently sometimes slightly sweetly scented, perianth tube 6–7 mm long, tepals oblong, with thickened subapical ridge on reverse, (4–)6–7 × ± 1.2 mm. *Stamen* filaments ± 5 mm long; anthers oblong, 3–4 mm long. *Style* ± 7 mm long, dividing between lower one third and middle of anthers (rarely ± 1 mm below anther bases), branches 1.0–1.8 mm long, divided for one third to half their length. *Capsules* smooth, ± urn-shaped or narrowly ovoid, (3)4–5 × 2–3 mm, with (2)3 or 4 seeds per locule, 5 mm long. *Seeds* elongate-angular, 3(4)-sided, tapering to points at both ends. *Flowering time*: October to December (rarely in May).

Distribution: *Micranthus plantagineus* has a wide range across the Cape flora region, extending from the Bokkeveld Plateau near Nieuwoudtville south to the Cape Peninsula and east to Port Elizabeth (Figure 8). An isolated population from the Anysberg Nature Reserve in the Little Karoo (*Vlok 2545*) appears typical of the species except for the shorter perianth tube, ± 4 mm long. Plants typically grow in seasonally wet habitats, often in marshy sites, along streams, or at least in places that are waterlogged in the winter months.

Diagnosis: the elongate inflorescence with up to 100 flowers and a perianth that is often deep blue, but sometimes pale blue or white, are unexceptional for the genus and identification of *Micranthus plantagineus* depends on leaf morphology. The two to four leaves are straight, stiffly erect, hollow and terete to oval in section, ± 2–3 mm diam., and reach or shortly exceed the spike (Figure 1F). When alive the leaves are smooth with the veins evident as paler, translucent lines. On drying, the veins appear hyaline and the veins at the adaxial and abaxial poles are somewhat more prominent. In addition, the stem is often branched, and as many as four (exceptionally nine) short branches may be produced shortly below the base of the main spike, these seldom exceeding half the length of the main spike. Lewis (1950) noted that the lower flowers of the spike are often replaced by cormlets [as many as five may be present in an axil]. That feature is not universal and many otherwise typical plants may have normal flowers and capsules from base to apex of the spike. A second characteristic feature of *M. plantagineus* is the presence of one or more cormlets in the lowermost leaf axil (not invariably present in other species) and occasionally in the axils of the upper cataphyll and one or more of the upper leaves. Plants lack a collar of fibres around the base (in contrast to superficially similar *M. filifolius* and *M. tubulosus*).

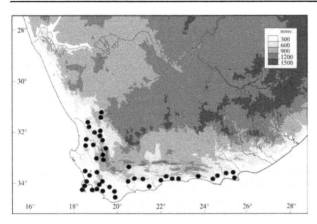

FIGURE 8.—Distribution of *Micranthus plantagineus*.

There are two, somewhat poorly defined, morphs of the species. One has narrow leaves, ± 2 mm diam. when alive, and slender capsules, ± 4.3 × 1.5 mm. Plants with mature capsules have 2(1) fusiform-angular seeds per capsule, ± 3.7–4.0 mm long. The second morph has broader leaves, 2–3 mm diam. when alive, and urn-shaped capsules ± 5 × 2–3 mm. Capsules have 4 seeds per locule, these flattened-angular, 3.0–3.5 mm long.

Collections of *Micranthus plantagineus* from Grootwinterhoek Forestry Station exemplify this situation: *Goldblatt 10451* with dark blue flowers represents the robust morph of the species, but slender-leaved plants (*Goldblatt 10452*), growing adjacent to stands of the robust morph, have white flowers. Other slender-leaved *M. plantagineus* (*Goldblatt10453*) growing nearby have blue flowers. These last two collections consist of shorter, less robust plants, 200–250 mm tall, and have more slender, but still hollow leaves ± 1.5 mm diam. No other *Micranthus* species were found in the area making the possibility that hybridization has played a role in this pattern of variation unlikely.

We must also mention a collection made by T.M. Salter in May 1935 from Viljoen's Pass. The morphology conforms closely to *Micranthus plantagineus* in the three terete and hollow foliage leaves, stems without a collar of fibres at the base, cormlets in axils of all leaves and in floral dimensions. Thus the only difference we see is the flowering time, noted as anomalous on the specimen label by Salter. The marshy habitat likewise conforms to the species. A search for the population in May 2013 failed to find any *Micranthus* species in bloom at this time of year, but the site may simply be lost to farming activity or dam construction.

Putative hybrids between *Micranthus plantagineus* and *M. alopecuroides* are discussed under the latter species.

Related species: until now, several collections of plants with filiform-linear and plane or terete, but not hollow, leaves with a thickened central vein and margins (cross-shaped in section) have been included in *Micranthus plantagineus* (as *M. junceus*) in herbaria. We believe these are separate species. The several southern Western Cape populations with this leaf type, here referred to *M. filifolius*, always have four or five, rarely

six leaves, the blades ± 1 mm wide vs, mostly three (rarely two or four) leaves in *M. plantagineus*. These plants have flowers with a consistently shorter perianth tube ± 4 mm long and shorter tepals, also ± 4 mm long vs, both perianth tube and tepals mostly ± 6 mm long in *M. plantagineus* and often a pale blue mauve to almost white (less often deep blue) perianth. *M. filifolius* is rarely branched and we have seen no specimens with the lower flowers aborted and replaced by cormlets, both common but not universal traits of *M. plantagineus*.

Other collections with this derived leaf type are known from the northern Cedarberg and Bokkeveld Mtns and are here treated as the new *Micranthus cruciatus* Goldblatt & J.C.Manning. These plants have only four, rarely five leaves, the two basal with linear or terete, four-grooved blades, ± 1.5 mm wide, similar to but broader than those of *M. filifolius*. An important associated character is the style, which divides at the mouth of the perianth tube into unusually long branches 2.0–2.5 mm long, divided for ± one third their length. Other species of *Micranthus* have the style dividing between the base and middle of the anthers and the style branches never exceed 1.5 mm.

History: long known as *Micranthus junceus* (Lewis 1950; Goldblatt & Manning 2000), *M. plantagineus* was evidently first recognized as a distinct species, called *Phalangium spicatum* by Burman (1768), at least as to the specimen in his collection (now at the Delessert Herbarium, Geneva). Burman provided no validating description, instead merely citing Plukenet's (1694) illustration in part 3 of the *Phytogeographia*. There is no accompanying text or even polynomial identifying the illustration, plate 310, f. 1., nor does the figure constitute a validating illustration with analysis. *Phalangium spicatum* Burm.f. is thus a nomen nudum and invalid. Plukenet's illustration is of a broad-leaved plant and does not, in our opinion, represent any species of *Micranthus* but is probably *Ixia scillaris* L. Even if any text associated with this illustration is found and if the name is lectotypified on the specimen rather than the Plukenet illustration, the combination *M. spicatus* (L.) Heyn. (1847) (= *Thereianthus spicatus*) prevents the use of Burman's epithet at species rank in *Micranthus*.

Curiously, *Phalangium spicatum* Houtt. (1780), typified by a good illustration, marks this as the first valid naming of *M. plantagineus*. Although seeming to refer to Burman's *P. spicatum*, Houttuyn makes it clear this is his species (*Phalangium scapis spicatis mihi*, i.e. Houttuyn) and that the Plukenet figure cited by Burman is an entirely different plant. As noted above, Heynhold's combination *M. spicatus* (L.) Heyn. bars transfer of Houttuyn's epithet to *Micranthus*.

In Aiton's (1789) *Hortus Kewensis*, Daniel Solander, the unacknowledged author of the species in this work, described *Ixia plantaginea* based on a collection of Francis Masson. The sheet at BM includes two plants with narrow, stiffly erect, centric leaves (the Masson collection) and three specimens with shorter, plane leaves (*Nelson 1777*) that are *M. alopecuroides*. The Masson specimens conform to the diagnosis, '*foliis linearibus strictis, spica disticha imbricata* [leaves linear, straight and upright] and constitute the holotype. The name is unfortunately superfluous as *Gladiolus alopecuroides* was cited in synonymy and, likewise, transfers of *Ixia plantaginea* to *Watsonia* (Ker Gawler 1803), and *Gladiolus* Pers. (1805), are superfluous as both authors cited *Gladiolus alopecuroides* as synonyms. Ecklon (1827), however, intended to transfer the species to *Micranthus*, where it becomes valid and is treated as a new name from that date rather than a new combination based on *I. plantaginea* Aiton. Ecklon (1827) recognized *M. alopecuroides* as a separate species, the first author to differentiate it from *M. plantagineus*, but whether deliberately or by accident is uncertain.

Baker (1892) described *Micranthus plantagineus* var. *junceus*, citing no specimens, but later listed several exsiccatae (Baker 1896), all of which were available to him in 1892. We choose a lectotype from among these, *Zeyher 1611*, a specimen in good condition and representative of the species. The taxon was raised to species rank by Brown (1929), who, at the time also identified *Phalangium spicatum* Burm.f. as the same species, at least as to the specimen in Burman's herbarium. *M. junceus*, a name used until now for this plant, becomes a synonym of *M. plantagineus*.

Additional specimens

NORTHERN CAPE.—**3119** (Calvinia): Nieuwoudtville waterfall, damp washes along stream, on sandstone, (–AC), 5 Dec. 1996, *Manning 2129* (NBG); Nieuwoudtville Escarpment, small vlei in arid fynbos, (–AC), 28 Nov. 1993, *MacGregor s.n.* (NBG153534); Oorlogskloof Nature Reserve, (–AC), 14 January 2000, *Pretorius 664* (NBG).

WESTERN CAPE.—**3118** (Vanrhynsdorp): Matsikammaberge, among sandstone rocks, (–DB), 11 Nov. 1985, *Van Jaarsveld & Bodenstein 8283* (NBG); top of Gifberg Pass, Farm Van Taakskom, (–DD), 11 Nov. 1985, *Snijman 946* (NBG). 3218 (Clanwilliam): Piketberg, road to Sun Mtn, (–DA), 16 Nov. 1993, *Manning 2093* (NBG). 3219 (Wuppertal): Pakhuis Mtns above 3500 ft [1 065 m], (–AA), 30 Dec. 1940, *Leipoldt 3596* (BOL, PRE), Nov. 1929, *Thode A2141* (PRE); Biedouw Valley, (–AA), 25 Nov. 1955, *Middlemost 1897* (MO, NBG); Wuppertal, (–AA), Oct. 1929, *Thode A2083* (NBG); Driehoek Vlei, Cedarberg, (–AC), 3 Dec. 1934, *Compton 4798* (NBG); banks of the Olifants River at Citrusdal, sandy ground, (–CA), 5 Nov. 1982, *Goldblatt 6707* (MO), Feb. 1982, *Goldblatt 6556* (fr.) (MO); Citrusdal, Farm Kleinplaas, moist hillocks in loamy clay among restios, (–CA), 11 Dec. 1997, *Hanekom 2972* (MO, NBG, PRE); Gonnafontein, seasonally damp sand, (–CB), 3 Dec. 2000, *Pond 254* (NBG); Leeu River, Ceres, (–CD), 18 Dec. 1944, *Compton 16741* (BOL, NBG). 3318 (Cape Town): Darling Flora Reserve, (–AD), 17 Nov. 1964, *Thompson 76* (NBG); 13 Nov. 1956, (–AD), *Winkler 166* (BOL); Kenilworth Racecourse, low lying areas wet in winter, (–CD), 5 Jan. 1970 (fr.), *Esterhuysen s.n.* (MO); Devil's Peak, 300 ft [± 90 m], Dec., *Pappe s.n.* (SAM). **3319** (Worcester): Groot Winterhoek Forest Station, rocky sandstone flats, (–AA), 27 Dec. 1995. *Goldblatt 10451* (MO, NBG); Keerom hills at foot of Twenty Four Rivers Mtns, (–AA), 3 Dec. 1950, *Esterhuysen 17869* (BOL, PRE); wet flats 9.4 miles [± 14 km] NE of Hermon Station, (–CC), 18 Oct. 1959, *Acocks 20744* (MO, PRE). 3320 (Montagu): Anysberg Nature Reserve, deep loamy sand, edge of seep, 12 Oct. 1991, (BC), *Vlok 2545* (MO); Swellendam, hill below Eleven o'Clock Mtn, (–DC), 25 Nov. 1952, *Wurts 519* (mixed with *M. tubulosus*) (NBG); Langeberg between Lemoenshoek and Naauwkranz, Farm Strawberry Hill, (–DD), 11 Jan. 1957, *Stokoe s.n.* (NBG). 3321 (Ladismith): Garcias Pass, 1300 ft, (–CC), Dec. 1904, *Luyt s.n.* (BOL). 3323 (Oudtshoorn): Saasveld, George, (–DC), 1 Dec. 1985, *Vlok 1299* (MO, NBG). 3418 (Simonstown): Bergvliet Farm, E of sand pit, (–AB), 5 Dec. 1918, *Purcell s.n.* (SAM90106); Cirkels Vlei, Cape Peninsula, (–AB), 15 Jan. 1946, *Barker 3954* (NBG), *Lewis 1495* (SAM); Betty's Bay, (–BB), 5 Jan. 1962, *Tijmans 25B1962* (NBG); Cape Hangklip, marsh (with *M. filifolius*), (–BB), 25 Jan. 2011, *Goldblatt & Manning 13625* (MO, NBG). 3419 (Caledon): Riviersonderend Bridge, foot of Franchhoek Pass, (–AA), 1 Jan. 1936, *Barker s.n.* (BOL45075); Nuweberg Forest Reserve, below Forestry offices, (–AA), 31 Dec. 1989, *Goldblatt 9035* (MO); Viljoen's Pass, in marsh, (–AA), 4 May 1935, *Salter 5255* (BOL, K). Drayton siding, Caledon, near stream, (–BA),

25 Jan. 2011, *Goldblatt & Manning 13622* (MO, NBG); Fernkloof Nature Reserve, Hermanus, deep sand, (–AD), 5 Dec. 1975, *Orchard 349* (MO, NBG); Fairfield Farm, W of Napier, clay ground, (–BD), 9 Dec. 1994, *Kemper IPC750* (NBG). **3421** (Riversdale): Stilbaai, Farm Klipfontein, shale ground near water, (–AD), 26 Nov. 1990, *Bohnen 9152* (NBG). **3422** (Mossel Bay): Mossel Bay, grassy plains, (–AA), Jan. 1926, *Taylor 316* (BOL); inland of Oubaai, George, (–AB), 3 Jan. 1994, *Victor 558* (BOL); Belvedere, churchyard, (–BB), 30 Dec. 1928, *Duthie s.n. STE29795* (NBG). Without precise locality, as Stellenbosch, Somerset [West] and Hottentots Holland, without date, *Ecklon & Zeyher Irid 193* (83) (SAM).

EASTERN CAPE.—**3324** (Steytlerville): Honeyville Farm, 10 km along Humansdorp-Hankey road, (–DC), 9 Feb. 2009, Van Wyk FBG293/CR3761/ (NBG); 'Galgebosch, Uitenhage' [near Hankey], (–DD), 1935, *MacOwan s.n.* (SAM). 3325 (Port Elizabeth): Loerie Forest Reserve, (–CC), 21 Dec. 1933, *Long 1* (NBG); Uitenhage Division, between Vanstadensberg and Bethelsdorp, (–CD), 1840, *Drège 8445* (K); between Port Elizabeth and Thornhill, (–CD), 31 Dec. 1939, *Barker 604* (NBG). 3424 (Humansdorp): Witte Els Bosch, flats, (–AA), Dec. 1920, *Fourcade 1025* (BOL, NBG, SAM); Humansdorp, (–BB), Jan. 1932, *Wagner s.n. STE17114* (NBG). Without precise locality, as 'Uitenhage,' Dec., *Ecklon & Zeyher Irid 194* (MO, SAM).

Hybrids

Interspecific hybrids are not uncommon in *Micranthus* and are likely to occur when two or more species co-occur. Most striking of the hybrids is that between *M. plantagineus* and *M. tubulosus*. The two species flower together at the foot of the Elandskloof Mtns in Elandsberg Nature Reserve and present a remarkable sight. The hybrids are locally very common growing with typical *M. tubulosus* and are always slightly shorter than the parent, 100–150 cm high, and like it have a well-developed collar of fairly coarse fibres around base. The other parent is less common, but present in small clumps, recognized by its erect habit, straight leaves and pale blue flowers. The hybrid is evidently fertile (plants in fruit have well developed capsules with apparently normal seeds) and stand out in having a slightly flexuose stem and narrower leaves than either parent. Unlike *M. tubulosus*, which they otherwise most closely resemble, hybrid individuals bear small cormlets at aerial nodes and sometimes at the base of the spike. We have seen similar hybrid plants near Elim where *M. tubulosus* and *M. plantagineus* also grew side-by-side.

WESTERN CAPE.—**3319** (Worcester): Elandsberg Estate, foot of the Elandskloof Mtns, (–AC), 2 Mar. 2000 (sterile), *Goldblatt & Manning 11281* (MO, NBG), *13617*(fr.) (MO, NBG); Jan. 2011 (fr.; growing with *M. plantagineus* and *M. tubulosus*), *Goldblatt & Manning 13605* (MO, NBG), 22 Jan. 2011 (sterile), *13609* (MO, NBG, 11 Nov. 2011, *Goldblatt & Manning 13751* (MO, NBG, PRE).

Less common are hybrids between *Micranthus tubulosus* and *M. alopecuroides*, but at Elandsberg Nature Reserve we noted both species growing close to one another with apparent hybrids among them. The putative hybrids have short, plane leaves, in outline like those of *M. tubulosus* but not round in section, although the leaves have an airspace between the two surfaces and lack the visible main veins of *M. alopecuroides*.

WESTERN CAPE.—**3319** (Worcester): Elandsberg Estate, foot of the Elandskloof Mtns, Vangkraal road, (–AC), 22 Jan. 2010 (sterile), *Goldblatt & Manning 13611* (MO, NBG).

ACKNOWLEDGEMENTS

We thank Elizabeth Parker and Lendon Porter for their assistance and companionship in the field; Mary Stiffler, Research Librarian, Missouri Botanical Garden, for providing copies of needed literature; and Clare Archer, SANBI (Pretoria) for help with several questions relating to collections at PRE. We also thank Nick Helme for helping with field observations of *M. plantagineus*. Collecting permits were obtained from the Nature Conservation authorities of Western Cape, South Africa. We thank the curators of the following herbaria for access to their collections or loans of types and other specimens: BOL, MO, NBG, PRE, SAM.

REFERENCES

AITON, W. 1789. *Hortus Kewensis*, vol. 1. George Nicol, London.

ANDREWS, H. 1799. *Ixia fistulosa. The botanists repository* 1,3: t. 59.

BAKER, J.G. 1877 [as 1878]. Systema iridearum. *Journal of the Linnean Society, Botany* 16: 61–180.

BAKER, J.G. 1892. *Handbook of the Irideae*. George Bell and Co., London.

BAKER, J.G. 1896. Iridaceae. In W.T. Thiselton-Dyer, *Flora Capensis* 6: 7–71. Reeve, Ashford.

BROWN, N.E. 1929. The Iridaceae of Burman's *Florae capensis prodromus. Bulletin of Miscellaneous Information, Royal Botanic Gardens, Kew* 1929: 129–139.

BURMAN, N.L. 1768. *Prodromus florae capensis*. Cornelius Haak, Leiden.

ECKLON, C.F. 1827. *Topographisches Verzeichniss der Pflanzensammlung von C.F. Ecklon*. Reiseverein, Esslingen.

GOLDBLATT, P. 1971. Cytological and morphological studies in the southern African Iridaceae. *Journal of South African Botany* 37: 317–460.

GOLDBLATT, P. 1977. Chromosome number in *Pillansia* (Iridaceae). *Annals of the Missouri Botanical Garden* 64: 36, 37.

GOLDBLATT, P. & MANNING, J.C. 1990. Leaf and corm tunic structure *Lapeirousia* (Iridaceae–Ixioideae) in relation to phylogeny and infrageneric classification. *Annals of the Missouri Botanical Garden* 77: 365–374.

GOLDBLATT, P. & MANNING, J.C. 2000. Cape plants. A conspectus of the Cape flora of South Africa *Strelitzia* 9. National Botanical Institute of South Africa, Cape Town and Missouri Botanical Garden, St. Louis, Missouri.

GOLDBLATT, P. & MANNING, J.C. 2006. Radiation of pollinations systems in the Iridaceae of sub-Saharan Africa. *Annals of Botany (London)* 97: 317–344.

GOLDBLATT, P. & MANNING, J.C. 2008. *The Iris Family: natural history and classification*. Timber Press, Oregon.

GOLDBLATT, P. & MANNING, J.C. 2009. New species of *Geissorhiza* (Iridaceae: Crocoideae) from the southern African winter-rainfall zone, nomenclatural changes, range extensions and notes on pollen morphology and floral ecology. *Bothalia* 39: 123–152.

GOLDBLATT, P. & TAKEI, M. 1997. Chromosome cytology of Iridaceae, base numbers, patterns of variation and modes of karyotype change. *Annals of the Missouri Botanical Garden* 84: 285–304.

GOLDBLATT, P., MANNING, J.C., DAVIES, T.J., SAVOLAINEN, V. & REZAI, S. 2004. *Cyanixia*, a new genus for the Socotran endemic *Babiana socotrana* (Iridaceae–Crocoideae). *Edinburgh Journal of Botany* 60: 517–532.

GOLDBLATT, P., RODRIGUEZ, A., POWELL, M.P., DAVIES, T.J., MANNING, J.C., VAN DER BANK, M. & SAVOLAINEN, V. 2008. Iridaceae 'out of Australasia'? Phylogeny, biogeography, and divergence time based on plastid DNA sequences. *Systematic Botany* 33: 495–508.

HEDWIG, R.A. 1806. *Genera Plantarum*. Reclam, Leipzig.

HEYNHOLD, G. 1847 [as 1846]. *Alphabetische und synonymische Aufzählung der Gewächse*. Dresden & Leipzig.

HOUTTUYN, M. 1780. Natuurlyke Historie, ed. 2, 12. F. Houttuyn, Amsterdam.

JACQUIN, N.J. 1797. *Plantarum rariorum horti caesarei schoenbrunnensis* 1. C.F. Wappler, Vienna.

KER GAWLER, J. 1801. *Ixia fistulosa*. Hollow-leaved Watsonia. *Curtis's Botanical Magazine* 15: t. 523.

KER GAWLER, J. 1803. *Ixia plantaginea*. Small-flowered Watsonia. *Curtis's Botanical Magazine* 16: t. 553.

KER GAWLER, J. 1804 [as 1805]. Ensatorum ordo. *Koenig & Sims Annals of Botany* 1: 219–247.

KLATT, F.W. 1882. Ergänzungen und Berichtigungen zu Baker's Systema Iridacearum. *Abhandlungen der Naturforschenden Gesellschaft zu Halle* 15: 44–404.

KLATT, F.W. 1894. Irideae. In T. Durand, T. & H. Schinz, *Conspectus florae africae* 5: 143–230. Charles Van de Weghe, Brussels.

KUNTZE, O. 1891. *Revisio Generum Plantarum*, vol. 2. Felix, Leipzig.

KUNTZE, O. 1898. *Revisio Generum Plantarum*, vol. 3, part 3. Felix, Leipzig.

LEWIS, G.J. 1941. Iridaceae. New genera and species and miscellaneous notes. *Journal of South African Botany* 7: 19–59.

LEWIS, G.J. 1950. Iridaceae. In R.A. Adamson & T.M. Salter (eds), *Flora of the Cape Peninsula*, pp. 271–265. Juta, Cape Town.

LINK, J.H.F. 1821. *Enumeratio plantarum*, etc. Reimer, Berlin.

LINNAEUS, C. 1753. *Species Plantarum*. Salvius, Stockholm.

LINNAEUS, C. 1756. *Centuria Plantarum II*. Höjer, Uppsala.

LINNAEUS, C. fil. 1782 [as 1781]. *Supplementum plantarum*. Orphanotropheus, Braunschweig.

LODDIGES, G. 1830. *Watsonia compacta. The Botanical Cabinet* 16.

MANNING, J.C. & GOLDBLATT, P. 2011. Taxonomic revision of the genus *Thereianthus* G.J.Lewis (Iridaceae: Crocoideae). *Bothalia* 41: 239–267.

MARLOTH, R. 1915. *Flora of South Africa. 4. Monocotyledons*. Juta, Cape Town.

MCNEILL, J., BARRIE, F.R., BURDET, H.M., DEMOULIN, V., HAWKSWORTH, D.L., MARHOLD, K., NICOLSON, D.H., PRADO, J., SILVA, P.C., SKOG, J.E., WIERSEMA, J.H. & TURLAND, N.J. 2006. International Code of Botanical Nomenclature (Vienna Code) adapted by the 17th International Botanical Congress, Vienna, Austria, July 2005. *Regnum Vegetabile* 146: i–xviii, 1–568. Koeltz, Königstein.

PERSOON, C.H. 1805. *Synopsis plantarum* 1. Cramer, Paris.

PLUKENET, L. 1694. *Phytogeographia*. Part 4. London.

REDOUTE, P.J. 1804. *Les Liliacées* 3. Didot Jeune, Paris.

REEVES, G., CHASE, M.W., GOLDBLATT, P., RUDALL, P., FAY, M.F., COX, A.V., LEJEUNE, B. & SOUZA-CHIES, T. 2002. Molecular systematics of Iridaceae: evidence from four plastid DNA regions. *American Journal of Botany* 88: 2074–2087.

RICKETT, H.W. & STAFLEU, F.A. 1959. Nomina generica conservanda et rejicienda spermatophytorum. *Taxon* 8: 213–243.

RUDALL & GOLDBLATT, P. 1991. Leaf anatomy and phylogeny of Ixioideae (Iridaceae). *Botanical Journal of the Linnean Society* 106: 329–345.

SCHRANK, F. DE P. DE. 1822. Commentarius in Irideas capenses. *Denkschriften der Königlich-Baierischen Botanischen Gesellschaft in Regensburg* 2: 165–224.

SIMS, G. 1801. *Ixia fistulosa*. Hollow-leaved ixia. *Curtis's Botanical Magazine* 15: t. 523.

WENDLAND, J.C. 1898. *Botanische Beobachtungen*. Gebrüdern Hahn, Hannover.

Permissions

List of Contributors

A. N. Moteetee and B.-E. Van wyk
Department of Botany and Plant Biotechnology, University of Johannesburg, P.O. Box 524, Auckland Park, 2006 Johannesburg

R. R. Klopper
Biosystematics Research and Biodiversity Collections Division, South African National Biodiversity Institute, Private Bag X101, 0001 Pretoria / Department of Plant Science, University of Pretoria, 0002 Pretoria

N.R. crouch
Ethnobotany Unit, South African National Biodiversity Institute, P.O. Box 52099, Berea Road, 4007 Durban / School of Chemistry, University of KwaZulu-Natal, 4041 Durban

C. Carbutt
Scientific Services, Ezemvelo KZN Wildlife, P.O. Box 13053, Cascades 3202, South Africa

Fotouo makouate and M.W. Van rooyen
Department of Plant Science, University of Pretoria, 0002 Pretoria

C.F. Van der merwe
Laboratory for Microscopy and Microanalysis, University of Pretoria, 0002 Pretoria

J.C. Manning
Compton Herbarium, South African National Biodiversity Institute, Private Bag X7, Claremont 7735, South Africa / Research Centre for Plant Growth and Development, School of life Sciences, University of KwaZulu-Natal, Pietermaritzburg, Private Bag X01, Scottsville 3209, South Africa

P. Goldblatt
B.A. Krukoff Curator of African Botany, Missouri Botanical Garden, P.O. Box 299, St. Louis, Missouri 63166, USA

J.C. Manning
Compton Herbarium, South African National Biodiversity Institute, Private Bag X7, Claremont 7735, South Africa / Research Centre for Plant Growth and Development, School of life Sciences, University of KwaZulu-Natal, Pietermaritzburg, Private Bag X01, Scottsville 3209, South Africa

P.V. bruyns
Bolus Herbarium, University of Cape Town, 7701

N. Swelankomo
National Herbarium, Biosystematics Research and Biodiversity Collections Division, South African National Biodiversity Institute, Private Bag X101, Pretoria, 0001

R.E. Gereau
Missouri Botanical Garden, P.O. Box 299, St. Louis, Missouri 63166, USA